Benchmark Papers in Geology

Series Editor: Rhodes W. Fairbridge
Columbia University

RIVER MORPHOLOGY / *Stanley A. Schumm*
SLOPE MORPHOLOGY / *Stanley A. Schumm and M. Paul Mosley*
SPITS AND BARS / *Maurice L. Schwartz*
BARRIER ISLANDS / *Maurice L. Schwartz*
ENVIRONMENTAL GEOMORPHOLOGY AND LANDSCAPE CONSERVATION, VOLUME I: Prior to 1900 / VOLUME II: Urban Areas / VOLUME III: Non-Urban Regions / *Donald R. Coates*
TEKTITES / *Virgil E. Barnes and Mildred A. Barnes*
GEOCHRONOLOGY: Radiometric Dating of Rocks and Minerals / *C. T. Harper*
MARINE EVAPORITES: Origins, Diagenesis, and Geochemistry / *Douglas W. Kirkland and Robert Evans*
GLACIAL ISOSTASY / *John T. Andrews*
GLACIAL DEPOSITS / *Richard P. Goldthwait*
PHILOSOPHY OF GEOHISTORY: 1785–1970 / *Claude C. Albritton, Jr.*
GEOCHEMISTRY OF GERMANIUM / *Jon N. Weber*
GEOCHEMISTRY AND THE ORIGIN OF LIFE / *Keith A. Kvenvolden*
GEOCHEMISTRY OF WATER / *Yasushi Kitano*
GEOCHEMISTRY OF IRON / *Henry Lepp*
GEOCHEMISTRY OF BORON / *C. T. Walker*
SEDIMENTARY ROCKS: Concepts and History / *Albert V. Carozzi*
METAMORPHISM AND PLATE TECTONIC REGIMES / *W. G. Ernst*
SUBDUCTION ZONE METAMORPHISM / *W. G. Ernst*
PLAYAS AND DRIED LAKES: Occurrence and Development / *James T. Neal*
PLANATION SURFACES: Peneplains, Pediplains, and Etchplains / *George Adams*
SUBMARINE CANYONS AND DEEP-SEA FANS: Modern and Ancient / *J. H. McD. Whitaker*
ENVIRONMENTAL GEOLOGY / *Frederick Betz, Jr.*
LOESS: Lithology and Genesis / *Ian J. Smalley*
PERIGLACIAL DEPOSITS / *Cuchlaine A. M. King*
LANDFORMS AND GEOMORPHOLOGY: Concepts and History / *Cuchlaine A. M. King*
METALLOGENY AND GLOBAL TECTONICS / *Wilfred Walker*
HOLOCENE TIDAL SEDIMENTATION / *George deVries Klein*
PALEOBIOGEOGRAPHY / *Charles A. Ross*
MECHANICS OF THRUST FAULTS AND DÉCOLLEMENT / *Barry Voight*
WEST INDIES ISLAND ARCS / *Peter H. Mattson*
CRYSTAL FORM AND STRUCTURE / *Cecil J. Schneer*
METEORITE CRATERS / *G. J. H. McCall*
AIR PHOTOGRAPHY AND COASTAL PROBLEMS / *Mohamed T. El-Ashry*
DIAGENESIS OF DEEP-SEA BIOGENIC SEDIMENTS / *Gerrit J. van der Lingen*
DRAINAGE BASIN MORPHOLOGY / *Stanley A. Schumm*
SEA WATER: Cycles of the Major Elements / *J. I. Drever*
ANCIENT CONTINENTAL DEPOSITS / *Franklyn B. Van Houten*
BEACH PROCESSES AND COASTAL HYDRODYNAMICS / *John S. Fisher and Robert Dolan*

Benchmark Papers in Geology / 39

A BENCHMARK® Books Series

BEACH PROCESSES AND COASTAL HYDRODYNAMICS

Edited by

JOHN S. FISHER

and

ROBERT DOLAN

University of Virginia

Dowden, Hutchinson & Ross, Inc.

Stroudsburg, Pennsylvania

Benchmark Papers in Geology, Volume 39
Library of Congress Catalog Card Number: 77–9329
ISBN: 0–87933–137–2

79 78 77 1 2 3 4 5
Manufactured in the United States of America.

LIBRARY OF CONGRESS CATALOGING IN PUBLICATION DATA

Main entry under title:
Beach processes and coastal hydrodynamics.
(Benchmark papers in geology; v. 39)
Includes bibliographical references and indexes.
1. Beaches—Addresses, essays, lectures. 2. Coast changes—Addresses, essays, lectures. 3. Hydrodynamics—Addresses, essays, lectures. I. Fisher, John Stanley, 1945- II. Dolan, Robert.
GB451.2.B42 551.4′5 77–9329
ISBN 0–87933–137–2

Exclusive Distributor: **Halsted Press**
A Division of John Wiley & Sons, Inc.
ISBN: 0–470–99304–9

SERIES EDITOR'S FOREWORD

The philosophy behind the "Benchmark Papers in Geology" is one of collection, sifting, and rediffusion. Scientific literature today is so vast, so dispersed, and, in the case of old papers, so inaccessible for readers not in the immediate neighborhood of major libraries that much valuable information has been ignored by default. It has become just so difficult, or so time consuming, to search out the key papers in any basic area of research that one can hardly blame a busy man for skimping on some of his "homework."

This series of volumes has been devised, therefore, to make a practical contribution to this critical problem. The geologist, perhaps even more than any other scientist, often suffers from twin difficulties—isolation from central library resources and immensely diffused sources of material. New colleges and industrial libraries simply cannot afford to purchase complete runs of all of the world's earth science literature. Specialists simply cannot locate reprints or copies of all their principal reference materials. So it is that we are are now making a concerted effort to gather into single volumes the critical material needed to reconstruct the background of any and every major topic of our discipline.

We are interpreting "geology" in its broadest sense: the fundamental science of the planet Earth, its materials, its history, and its dynamics. Because of training and experience in "earthy" materials, we also take in astrogeology, the corresponding aspect of the planetary sciences. Besides the classical core disciplines such as mineralogy, petrology, structure, geomorphology, paleontology, and stratigraphy, we embrace the newer fields of geophysics and geochemistry, applied also to oceanography, geochronology, and paleoecology. We recognize the work of the mining geologists, the petroleum geologists, the hydrologists, the engineering and environmental geologists. Each specialist needs his working library. We are endeavoring to make his task a little easier.

Each volume in the series contains an introduction prepared by a specialist (the volume editor)—a "state of the art" opening or a summary of the object and content of the volume. The articles, usually some twenty to fifty reproduced either in their entirety or in significant extracts, are selected in an attempt to cover the field, from the key papers of the last century to fairly recent work. Where the original works are in foreign languages, we have endeavored to locate or commission translations. Geologists, because of their global subject, are often acutely aware of the oneness of our world. The selections cannot, therefore, be restricted to any one country, and whenever possible an attempt is made to scan the world literature.

To each article, or group of kindred articles, some sort of "highlight commentary" is usually supplied by the volume editor. This commentary should serve to bring that article into historical perspective and to emphasize its particular role in the growth of the field. References, or citations, wherever possible, will be reproduced in their entirety—for by this means the observant reader can assess the background material available to that particular author, or, if he wishes, he, too, can double check the earlier sources.

A "benchmark," in surveyor's terminology, is an established point on the ground, recorded on our maps. It is usually anything that is a vantage point, from a modest hill to a mountain peak. From the historical viewpoint, these benchmarks are the bricks of our scientific edifice.

RHODES W. FAIRBRIDGE

PREFACE

This volume is another of the growing list of titles in the "Benchmark Papers in Geology" series that are concerned with the coastal land–sea interface. *Barrier Islands* and *Spits and Bars*, both edited by Maurice Schwartz, address subjects virtually inseparable from our contribution, and several additional titles are in preparation, including a volume on coastal sedimentation by Harold Palmer and Donald Swift. It is easy to identify the common ground for these collections, but perhaps more difficult to separate them and identify their differences. By identifying the beach and its nearshore boundary as a focal point for this volume, we are suggesting that there is an important body of literature that merits recognition and needs to remain available to students and the research community. Even within the limited context of this subject, however, we encountered a wide range of interests and suggested articles to be included in this volume. Our discussions with other shoreline investigators rather dramatically restated the concept that neither the beach nor its immediate wave environment can be treated to the exclusion of the other. The titles selected here are a modest attempt to bridge the trough between these elements.

Obviously, our subject is a part of coastal sedimentation, often on barrier islands, perhaps dealing with nearshore bars. So we can not claim to have avoided the territory of these other volumes entirely. We leave it to the reader to judge the propriety of our boundaries.

Wilma Le Van and Libby Scopino assisted in the preparation of this volume. In addition, we offer thanks to our colleagues and students who have helped us in the selection of articles.

JOHN S. FISHER
ROBERT DOLAN

CONTENTS

PART III: BEACH PROCESSES

CONTENTS BY AUTHOR

BEACH PROCESSES AND COASTAL HYDRODYNAMICS

INTRODUCTION

In its most general terms, a beach marks the interface of land and sea, often characterized by a deposit of unconsolidated sediment. These deposits assume many forms and comprise a wide variety of materials. The physical processes of beach mechanics are generally the same, however, regardless of morphology or composition. This volume presents a small selection of the numerous contributions that have developed and refined the present ideas and models of these processes. In these days of heightened concern for the responsible use and management of natural resources, the need for the further development and refinement of our understanding of a beach's physical nature is obvious.

Although beaches are relatively narrow geological deposits that may appear uncomplicated, they have proved difficult to study. The entrainment and transport of coastal sediments by waves, tides, currents, and wind is an exceedingly complex process, and field investigations of beach processes are frustrated by the difficulties of making observations and measurements in breaking waves. As we review these papers by the early coastal investigators, we are reminded of their inventiveness and tenacity in field studies. Even today, with the aid of sophisticated electronic devices, the measurement of basic beach processes remains cumbersome and often impossible. To make this problem even more complex we are often interested in studying basic mechanics during high energy conditions. A beach will often change more within a few hours during a storm than within days or even weeks of more routine weather. The problems of arranging to be at a research site at the correct time are exceeded only by the difficulty of collecting data during storms. Many investigators' instruments have been swept to sea, sunk on the sea floor, or rendered useless by exposure. While these events add drama and excitement to coastal research, they also demonstrate the strength of physical forces.

Because of the nature of the beach environment and its dynamic and rapidly changing elements, a large portion of the early literature is descriptive. These initial contributions provide us with a vocabulary and physical context for the more quantitative studies of recent years. We should recognize, however, that although these early authors approached coastal problems lacking the analytical tools of today, their fundamental insights often have proved correct and still remain parts of the literature. Indeed, the first paper presented in this volume, Paolo Cornaglia's "On Beaches," contains the basic framework for most of our present understanding of the relationships between waves and beach profiles.

Historically, there has been a degree of separation between the geologist, who studies the beach as a sedimentary deposit, and those concerned with fluid forces acting on beach sediment. Although this division has become less notable in recent years, these studies are still commonly called beach processes *and* coastal hydrodynamics. The sources of the papers presented here are both geology and geophysical journals, and journals and proceedings of engineering societies. In fact, one of the most established and widely recognized forums for international discussion of this subject is the biennial Coastal Engineering Conference, which is attended by a wide range of scientists and engineers.

Obviously, the literature in either beach processes or coastal hydrodynamics is too broad to be covered in a single Benchmark collection. The guideline for this volume has been to present papers that primarily treat sediment transport processes and the resulting morphology of the active beach deposit. Definitions vary, but in general this zone extends from the area of breaking waves to the upper limit of swash. Although some of the papers included in this volume might appear to cover areas landward or seaward of this zone, we have tried to focus on this narrow band within the shoreline.

As editors, we admit to a certain bias in the selection of these papers, which emphasize the nature of the processes rather than the morphology of the deposits. Two other Benchmark volumes, *Spits and Bars* and *Barrier Islands,* complement the topics considered here.

As with the other volumes in this Benchmark series, some of the papers selected have become difficult to obtain, although they continue to be assets to reference libraries. In addition, we have included a new translation of Cornaglia's above-mentioned treatise, in the hope that it will gain further exposure and recognition. Of course, some of the papers presented here are neither old nor difficult to obtain; they are included as further evidence of their established merit, as well as to provide both a unified and a convenient volume on this subject. By balancing these newest contributions with the more classical works, we are once again reminded of the value and significance of the latter, and of the foundations of the former. It is always rewarding to return to these earlier discussions as new ideas and analyses push the state of the art forward.

We have divided this collection into three parts: "Historical Milestones," "Coastal Hydrodynamics," and "Beach Processes." The last two sections admit to our own failure at integrating these disciplines; nevertheless, our intention has been to explore the fluid motion first, and then to consider the resulting beach processes. For Part I ("Historical Milestones") we have chosen four classic papers that illustrate the insight and resourcefulness of some of the first coastal scientists. The early observations of both Cornaglia and the Thompsons have remained fundamental to the present body of knowledge. The Thompsons' work, while a treatise on marine biology, is a thoroughly solid and fundamental discussion of swash zone processes. M. P. O'Brien's summary of wave theory and W. C. Krumbein's analysis of beach processes provide the new reader with the background and terminology to understand the level of the present state of the art.

It is interesting to note World War II's role, and the reports of the Beach Erosion Board in particular, in providing the foundation for this relatively modern discipline. That some of these reports were once classified as military secrets illustrates both the importance and the unusual circumstances that prompted these studies. Many of the investigators who contributed to the war effort remained interested in coastal studies, as did the laboratories that developed with them. The group at the University of California, under the leadership of M. P. O'Brien, is particularly notable in this regard. We are fortunate to enjoy the contribution of both O'Brien and Krumbein today, and we can expect their recent studies to become the Benchmarks of the future.

The second part ("Coastal Hydrodynamics") begins with J. W. Johnson's review of the dynamics of the nearshore zone. This paper also serves as an introduction to the final section ("*Beach Processes*"). The other selections trace the development of our present understanding of the nature of waves, currents, and tides as they relate to the nearshore environment. Because a beach's final form is a result of the complex interactions of the nearshore circulation and swash dynamics, this part presents a rather wide range of subjects and physical scales. The contribution and significance of each individual paper is often best judged in the light of the subsequent analysis that draws from the former one. It is the total synthesis of these works and the many papers omitted that provide a physical framework for the final section.

The last part explores the response of beach sediments to the forces described in Part II. But because of the complexity of sediment transport processes, these articles do not provide a picture as unified as in Part II. One of course could debate the relative difficulties of unraveling the complexities of the motions within a breaking wave or the motions of the sand particles beneath the wave. Regardless, one must conclude from the present body of literature that we are more confident about predicting the wave dynamics than the resulting beach configuration. With the present research emphasis on this area of coastal studies, we can anticipate significant advances in the near future and a closing of the gap between our understanding of the hydrodynamics and the resulting sediment transport.

Our experience in teaching this subject at both the graduate and the undergraduate level has suggested the need for a set of readings to supplement basic textbooks. This need has been a major influence in the selection of the papers in this volume. The papers selected are well suited to a course developed from Cuchlaine A. King's *Beaches and Coasts* or Arthur T. Ippen's *Estuary and Coastline Hydrodynamics,* and Paul D. Komar's *Beach Processes and Sedimentation* as three examples.

As in any collection of reprints, many fine papers have had to be omitted and thus several important investigators excluded. Douglas Wilson Johnson's *Shore Processes and Shoreline Development* includes a comprehensive review of the literature of the late 1800s and early 1900s. V. P. Zenkovich's *Coastal Development* contains an excellent review of European literature, especially Russian. *Waves on Beaches and Resulting Sediment Transport,* edited by R. E. Meyer, is an invaluable reference, as is *Coastal Engineering* by Richard Silvester.

REFERENCES

Bascom, Willard (1964). *Waves and Beaches: The Dynamics of the Ocean Surface.* New York: Anchor Books, Doubleday.

Bird, E. C. F. (1968). *An Introduction to Systematic Geomorphology,* vol. 4. Cambridge, Mass.: M.I.T. Press.

Coates, Donald R. (1973). *Coastal Geomorphology.* Publications in Geomorphology, State University of New York, Binghamton.

Davies, John Lloyd (1973). *Geographical Variation in Coastal Development.* New York: Hafner.

Davis, Richard A., and R. C. Ethington (1976). *Beach and Nearshore Sedimentation.* Society of Economic Paleontologists and Mineralogists. Tulsa, Okla:

Hails, John, and Alan Carr (1975). *Nearshore Sediment Dynamics and Sedimentation.* London: John Wiley & Sons.

Ippen, Arthur T. (1966). *Estuary and Coastline Hydrodynamics.* New York: McGraw-Hill.

Johnson, Douglas Wilson (1965) *Shore Processes and Shoreline Development.* New York: Hafner.

King, Cuchlaine A. M. (1972). *Beaches and Coasts.* New York: St. Martins Press.

Komar, Paul D. (1976). *Beach Processes and Sedimentation.* Englewood Cliffs, N.J.: Prentice-Hall.

Le Mehaute, Bernard (1976). *An Introduction to Hydrodynamics and Water Waves.* New York: Springer-Verlag.

Meyer, R. E., ed. (1972). *Waves on Beaches and Resulting Sediment Transport.* New York: Academic Press.

Silvester, Richard (1974). *Coastal Engineering.* New York: Elsevier Scientific Publishing Company.

Steers, J. A. (1971a). *Applied Coastal Geomorphology.* London: Macmillan.

Steers, J. A. (1971b). *Introduction to Coastline Development.* London: Macmillan.

U.S. Army Coastal Engineering Center (1973). "Shore Protection Manual."

Wiegel, Robert L. (1964). *Oceanographical Engineering.* Englewood Cliffs, N.J.: Prentice-Hall.

Zenkovich, V. P. (1967). *Processes of Coastal Development,* New York: Interscience.

Part I

HISTORICAL MILESTONES

Editors' Comments on Papers 1 Through 4

1 CORNAGLIA
On Beaches

2 THOMPSON and THOMPSON
The Spawning of the Grunion

3 O'BRIEN
Excerpt from *A Summary of the Theory of Oscillatory Waves*

4 KRUMBEIN
Shore Processes and Beach Characteristics

Each of the four papers we have selected to initiate this collection represents a different interpretation of the concept of a historical milestone. Our primary goal is to recognize and applaud the achievements of a few of the early researchers as well as to establish a context for the papers that follow in the other two parts of this volume. There is no universal criterion for a milestone, and we admit to our own biases in these selections. While Cornaglia's treatise is certainly one of the earliest on this subject, we recognize that it is not *the* earliest. The Thompsons' contribution is often overlooked in the current literature; its place in this milestone section will be apparent, and we are pleased to be able to give it a prominent place. O'Brien and Krumbein are widely recognized as two modern leaders of the study of coastal processes. While they are important in their own right, these early reports are just reminders of the more than quarter-century that these researchers have contributed to the understanding of this subject. One need only refer to the recent literature to realize that we continue to be rewarded by their work and achievements.

By limiting this section to four papers, we had to omit many contributions that rightfully deserve equal recognition. The brief list of references includes a few of the ones we also considered. It is appropriate to open this Benchmark collection with Paolo Cornaglia's treatise on waves and beaches. His paper was one of the earliest comprehensive discussions of the interaction of wave forces, currents, and beach sediment distribution. Many of his concepts have withstood the test of future research and are important principles in the more recent literature. Of these, the theory of the neutral line (or null point), where there is no net transport for given wave and sediment distributions, is perhaps his most important contribution. However, an inspection of this remarkable paper will reveal a thorough documentation of the general mechanics of coastal sediment transport. It should be recalled that Cornaglia had little previous literature to draw on in developing this conceptual model of coastal processes. While it is apparent that certain specifics of Cornaglia's explanations on wave mechanics have been proved false, his basic presentation of the relationship of the shoaling wave and its ability to transport beach material is still defensible. This test of time is the best argument for its recognition as a milestone in coastal hydrodynamics.

Cornaglia's original paper is not readily available; the several English translations are also difficult to locate. This new translation was carefully prepared by Wilson N. Felder, and we express our gratitude for his effort.

Another important early contribution is that of a pair of marine biologists. When the Thompsons published their paper on the spawning of grunion, they succeeded in examining not only an important biological phenomenon, but also an impressive set of observations and conclusions on the physical nature of a beach.

Their presentation (Paper 2) is one of the earliest and certainly one of the most complete statements of the role of tidal action in systematic changes of the beach profile as cut and fill occurs by swash action. Figure 7 in their text illustrates a classic set of observations of beach change, and it is an obvious predecessor of more recent studies of this process. Their discussion of the separate roles of uprush and backwash in the overall budget of beach sediment presents a clear and solid framework of this mechanism of erosion and profile readjustment.

Although this paper was chosen for its early contribution to our understanding of physical processes, it should remind the modern investigator of the importance of recognizing interrelated disciplines. It is perhaps unfortunate that the more recent literature dealing with coastal processes has failed to maintain the spirit of the Thompsons' scientific efforts.

There is a small group of coastal investigators whose contributions span almost the entire period of contemporary developments. We believe that Paul O'Brien is one of the most noteworthy leaders of this group. O'Brien and his students at the University of California have an impressive list of contributions in coastal engineering. The paper presented here (Paper 3) is one of O'Brien's early contributions, prepared for the U.S. Army's Beach Erosion Board. The board and its functions were reorganized in 1963 to become the Coastal Engineering Research Center, an important agency for coastal research.

When O'Brien prepared this summary of oscillatory water wave theories for the Beach Erosion Board, no general coastal hydrodynamic text was available. Today there are many such references, including those cited in the introduction to this volume. O'Brien's report is certainly a benchmark paper, but his discussion is still of value and it provides an excellent introduction to this field of study.

Because of space constraints, only the first three sections of O'Brien's report have been reproduced here. The paper continues with an analysis of waves of finite amplitude, breaking criterion, energy transmission, and other topics. The first three sections provide sufficient background for most of the other papers in this volume, an interesting comment both on the breadth of O'Brien's summary and on the applications of small amplitude wave theory to the bulk of the literature concerning coastal processes.

It is difficult to discuss the design of a coastal field study or the analyses of beach sediment without referring to one of the many contributions of W. C. Krumbein. His papers and texts remain relevant sources for the analysis of geologic data. The report presented here (Paper 4) is one of Krumbein's earliest, prepared for the Beach Erosion Board. It is also one of the earliest and most complete investigations of a beach. Note the careful, systematic treatment of the topic, a characteristic of Krumbein's papers that is appreciated by both students and established researchers. The experimental design, analysis, and discussion of data are of a quality not often found in the literature.

The scope of this report provides a thorough description of the beach environment as well as some fundamental ideas on the geophysics involved. In addition, Krumbein's discussion on the statistical analysis of beach sediment is a brief but excellent introduction to this subject. His presentation of a beach slope and grain size relationship, along with the contribution of W. Bascom, also included in this volume, remains one of the significant data sets today.

We have omitted the four photographic plates from the original report. In addition, the large bathymetric chart, Figure 1, has been excluded.

REFERENCES

Cornish, V. (1898). "On Sea and Sand Banks." *Geog. Jour.* 11:628–651.

Eaton, Richard O. (1951). "Littoral Processes on Sandy Coasts." *Proceedings* of the First Conference on Coastal Engineering, pp. 140–154.

Evans, O. F. (1940). "The Low and Ball of the East Shore of Lake Michigan." *Jour. Geology* **48**:476–511.

Fenneman, N. M. (1902). "Development of the Profile of Equilibrium of the Subaqueous Shore Terrace." *Jour. Geology,* (1):1–32.

Zenkovich, V. P. (1946). "On the Study of Shore Dynamics." *Trudy. Inst. Okeanologie* 1:99–112.

1

ON BEACHES

Paolo Cornaglia

This article was translated expressly for this Benchmark volume by W. N. Felder, University of Virginia, from "Delle Spiaggie," in Accad. Naz. Lincei Atti. Cl. Sci. Fis., Mat. e Nat. Mem. **5**, *ser. 4:284-304 (1889), with the permission of the publisher, Accademia Nazionale dei Lincei.*

Translator's Note: Two terms used in the present translation deserve some comment. Cornaglia repeatedly used the word *traversia* in describing the action of waves on a coast. The word can be translated literally as "cross-sea", but clearly it has a deeper meaning. In Italian coastal engineering practice, it has been found useful to describe sectors of wave-approach angle (*settori di traversia*), which correspond to the directions of approach of the most frequent waves, highest waves, and waves of greatest effect. Cornaglia probably intended *traversia* to refer to the last two categories, which is why the word has been translated as "sector of effective wave approach." In describing the mechanics of particle motion within waves, Cornaglia speaks of "fluxes." Clearly, Cornaglia thinks of these fluxes as waves of a sort. But because of the general nature of their behavior and because of the difficulty of finding an appropriate "wave" word to describe them, I prefer to use the literal translation of the Italian word *flussi*.

1. BEACH SEDIMENT IS ESSENTIALLY MOVED BY WAVE ACTION

If an obstacle, such as a breakwater or a wharf, is placed across a marine beach such as those of Italy, it is generally observed that in the corners at either side of the obstacle a certain amount of material is accumulated, distributed so that at this point the beach assumes a broad and open curve, concave offshore, which joins the side of the obstacle to the remainder of the beach. This material is essentially accumulated and distributed by the sea, since the phenomenon is also observed—and in fact is better defined—when there are no currents produced by other causes. Where currents do exist, they are not strong enough to move the material, as is the case with a littoral current. Furthermore, it must be understood that the sea accumulates and distributes this material through the action of the waves along the bottom; it is precisely, and only, under the impulse of the moving fluid touching the bottom that the bottom material is moved.

Since this material is moved under the action of the sea, it is accumulated in greatest amount on the side from which the largest waves arrive, especially if the waves come from the direction of effective wave approach: the general trend of the beach is built out farther on that side of the obstacle than on the other. Thus all the numerous breakwaters built on the western Ligurian coast to improve the beaches are buried deeper on the western side because the effective wave approach is from the southwest, and because the waves reaching the beach form an acute angle to the west. This matter is illustrated more clearly by the fact that, in general, the headlands along sea coasts are buried more on the updrift side, while on the other side they are more or less scoured because the sea on that side, having less force, collects a lesser amount of material. If the prominence is not too pronounced, so that the adjacent beach on the downdrift side can still be struck sufficiently by the most effective waves, they will entrain the material and move it away from the prominence.

Any coastal chart will verify this, but the Calabrian coast of the Tyrrhenian Sea is a particularly clear example. Here the headlands are buried more on the side toward Reggio in the section between Cape Vaticano and the Gulf of Saint Eufemia; but are buried more on the Naples side in the section between the gulfs of Saint Eufemia and Policastro. In its general trend, the coastline in the first section strikes E 18° N and in the second N 17° W. Given the coast's diverse placement and orientation, the effective wave angle is from the west-northwest in the first section and from the west in the second. The same waves form an acute angle with the coast toward Reggio in the first section, and an acute angle with the coast toward Naples in the second section; therefore they cause material from everywhere to move toward the Gulf of Saint Eufemia, as if it were a sort of dead end. Further, the material that the sea gathers on one side of the obstacle is generally left in place by waves coming from the other side, precisely because the material is in the lee of that obstacle.

From all this it is deduced that as long as the sea has sufficient force, it moves all or some beach sediments such as sand, gravel, cobbles, etc., and that it moves them now one way and now another way according to its direction. In the end, however, the sediments move principally in the direction from which the sea has its prevalent action, which generally—as has been said—is that of the effective wave approach.

2. THE WEIGHT OF MATERIAL MODIFIES THE EFFECTS OF MOVEMENT CAUSED EXCLUSIVELY BY WAVES

It has been noticed that some beaches are prograding, while others are being eroded. In the first category are the beaches of Ferrara and Ravenna, of western Calabria, and of Pisa, among others in Italy. Among the second are the beaches at the quarries of Monte Conero near Ancona, and many of the western Ligurian coast, where groins were built in an attempt to control the erosion. Now, since the laws of nature are undoubtedly everywhere the same, it must be concluded that at the shore: (1) material is not exclusively carried on shore or dragged offshore, but is alternately moved now in one direction, now in the other, as is likewise observed of waves at the shore; and (2) in certain circumstances movement toward shore prevails. This latter statement implies another: some process must intervene to modify the effect that the waves alone would have on the ocean bottom. It is found that all prograding beaches are shallow, while all truly eroding beaches are more or less deep—that is to say, they have steep bottoms. I say "truly eroding" because after some beaches have prograded for a certain time, they recede; the beaches east of the old port of Pesaro and in Port Maurizio, for example, have done so.

This happens on shallow beaches where, given certain conditions, submarine banks are formed that intercept the waves from offshore. The beaches along the shore recede, but they continue to prograde under water, at some distance offshore as will be seen below. However, instead of contradicting the general rule that shallow beaches prograde and deep beaches erode, this case confirms it.

The Royal Navy hydrographic charts for shallow beaches, and profiles made specially or on occasion of the execution of marine works on deep beaches, show in fact that the distance from the shore to the 10-m contour line is as follows for the beaches listed:

Shallow beaches:

at Ferrara	2700–5000 m
at Ravenna	4200–6500 m
in W. Calabria	250–800 m
at Pisa	1000–2800 m

Deep beaches:

at quarries of Mt. Conera	100 m
at Chiavari	400 m
at Cornigliano	150 m
at Pegli	400 m
at Voltri	350 m
at Varazze	180 m
at Celle	275 m
at Bergeggi	125 m
at Noli	100 m
at Finalmarina	130 m

From this data it is evident that the process intervening to modify the effect of the waves on the bottom is the component of the weight of the material parallel to the bottom which always acts according to the line of maximum slope of the bottom, and which is not brought into play except when there is sufficient energy to make the material move. On the other hand, the waves act in the direction of their own travel, which is demonstrated by the fact that sediment moves now in one sense, now in the other, according to the direction of the waves. Now, all sediment movement on the bottom that is exclusively generated by waves can be broken down into two components: the first, parallel to the bottom contours; the second, normal to them. The first is what makes materials move along shore. The second component, combined with the component of the weight of the materials parallel to the bottom, constitutes the force that impels the sediment onshore or carries it out to sea. It will be supposed in the text that follows that the bottom slopes offshore; this can always be verified at a greater or lesser distance from shore according to the circumstances. If the bottom does indeed slope offshore, then the component of force due to sediment weight is always subtracted when the force of the waves is directed toward shore, and it is always added if that force is directed offshore. Over a bottom with a shallow slope, this component has a very small value, while on a steep bottom it has a large value. In the first case the component subtracted from the force of the waves toward the shore and added to the force directed offshore is not capable of rendering the magnitude of the former less than that of the latter. The opposite happens in the other case, in which the downward force, with the addition that it receives, results in a magnitude greater than that of the upward force, with the reduction that it receives. From this it also follows that the shoreward wave force on the bottom must prevail over the offshore wave force.

If the forces were of equal intensity, the offshore force would always prevail, due to the effect of the weight component, and beaches would always be eroded. The same thing would happen if the offshore force were to prevail over that shoreward force, and for greater reason.

3. SEDIMENTS BROUGHT TO THE SEA REMAIN IN THE BASIN OF THEIR ORIGIN

Another important fact is observed in nature: that the detrital material or sediments that streams bring to the sea, or that waves erode from the coast, do not travel, so to speak, beyond their basins of origin. I say "so to speak" because occasionally sediments will travel a greater or lesser distance along the shore, becoming more and more rare and at a certain point disappearing completely. The western Ligurian coast, with its great variety of geologic terrain and its oblique orientation to storm waves, offers remarkable and eloquent examples in this regard.

Looking at the western Ligurian coast in general, with the exception of certain beds of Quaternary, of Pliocene, and of Miocene here and there, especially close to the sea, the geologic makeup of the shore (going toward Genoa as is shown on the geologic map of A. Issel, L. Mazzuoli, and D. Zaccagna) is as follows: At Point Mortola, near the border of the kingdom, the Jurassic and the Cretaceous terminate. (These are compact limestones, for the most part of light color.) From there to Albenga is found a great deposit of Eocene (calcareous marl, clayey schists, and sandstone, all brown, occasionally very dark) with one or another of these three rock types prevailing at various points. From Ceriale just past Albenga up to Bergeggi appears an extensive deposit of lower Triassic (waxy limestones and marbles, quartzites and aragonites, calcareous schists, and micaceous and talcose schists) that alternates with middle Triassic (chalky grey and blue limestone and dolomitic limestone). From Bergeggi to Albissola are a series of Precambrian rocks (quartzites, talcose and chloritic schists, and appinites). Upstream in the valley of Letimbro, which flows a little west of Savona, and the valley of the Sansobbia, which flows immediately eastward of Albissola, we find more or less extensive masses of middle Triassic limestone. From Varazze to Cornigliana, there is a large formation of serpentine intermixed with talcose micaceous rocks of the lower Triassic, principally in the valleys that emerge at Voltri, with outcrops of dolomitic limestone (middle Triassic) at Cogoleto, Voltri, and Sestri-ponente. Thence to Genoa and beyond, we again find Eocene sediments as in the first section. (Point Mortola to Albenga) . The Quaternary (from Albenga to Pietraligure) is represented by deposits of gravel and cobbles. The Pliocene (in several outcrops from the first section and in the flat area between Vada and Savona) consists of conglomerates, sands, marls, and clays, for the most part of light color; and the Miocene (from Albissola almost to Varazze) is formed of sandstones and conglomerates.

Now, the beach sediments with the exception of those between Bordighera and Albenga are only of the Eocene, and in fact, in general they are either sandstone or marly limestone like the adjacent coast. The beach at Noli is composed of fragments of quartzite, talcose schist, and limestone of the Mesozoic. The beach of Vado to the south of the Fort of San Lorenzo is composed of fragments of quartzose rocks, including some with the gneissic structure of Precambrian rocks, and there are no limestone fragments. The weak effervescence of these sediments in acid is due to the presence of shells and the small amount of calcite in certain siliceous rocks.

At Pegli in 1880—that is, before the eastern breakwater was extended and before the western breakwater was constructed around 1883 with masses of natural talcose and serpentine rock—the beach was exclusively made up of gravel lacking in serpentine. This is because proceeding toward Genoa the elements of the Jurassic and Cretaceous limestone, encountered on the beaches of Ventimiglia and Bordighera, do not extend beyond the Cape of St. Ampeglio. The Eocene material is carried beyond Loana and possibly also around the Cape of Caprazoppa, but very little Eocene sediment passes this last point, and none at all is carried around Cape Noli. Fragments of Mesozoic limestone that are still abundant south of the breakwater at the Bergeggi quarry do not pass it. Likewise the talcose schist sediments that are still very abundant to the west of the so-called Pria Pula do not pass beyond onto the nearby beach of Pegli. Not only do certain headlands of the coast intercept the drift of sediment, but so do certain of its sharper embayments. We find certain elements of Mesozoic limestone carried in by the Letimbro on the beach to the south of the port of Savonna and carried to the beach of Albissola north of the port itself by the Sansobbia. However, no Mesozoic limestone is found at all on the beaches along the coast from Savona to Albissola; and if some effervescence is found in the sand of this shoreline, it is due to the same causes as that of the sand making up the beach to the south of the Fort of San Lorenzo at Vado. Now, in general, the particularities of the shoreline—or more precisely, the width of the beaches and the distance the 10-meter contour passes from the capes and headlands—are as follows:

	Development of the beach in km	*Distance to the 10-m contour line*
From Point Mortola near the border of the kingdom to beyond Bordighera	10	
Promontory at Ventimiglia	4	375
Cape Saint Ampeglio		245
From Albenga to Cape Caprazoppa	18	
With a small headland at Loano	10	470
The Cape of Caprazoppa		150
Cape Noli		95
Axis of the breakwater at Bergeggi		45
From Voltri to Pria Pula and beyond	4	
To the shore of Pria Pula		340
From the rocks awash off Pria Pula		100

The same thing is found going toward the French border; the sediment on the beaches is carried around certain headlands that are less pronounced, but it does not pass certain other headlands. A similar situation prevails along the French coast of the Mediterranean where Delesse (*Lithologie des mers,* Chapter X) concluded: "Each littoral deposit in the Mediterranean tends to be localized on the coast which furnished its sediments," and "In the Mediterranean littoral deposits are essentially local."

4. WAVE MOTION MAY BE GREAT EVEN AT CONSIDERABLE DEPTHS

The fact that sediment does not pass certain points made many believe that wave motion, diminishing rapidly beneath the ocean surface, ceased to be felt even in the strongest storms at a certain depth. It was said that the waves could no longer move sediment because on the beaches in the Mediterranean in general, regardless of their state of exposure, form, or composition, sediment is no longer carried onshore when it is at a depth of about 10 meters.

This belief is in fact erroneous, since actually wave motion can be great even at considerable depths. Several facts prove this. A ship passing over the Caccia bank, which is found at a depth of 23 meters between the islands of Corsica and Minorca, took on board several waves loaded with sand from the sea bottom. The frigate Tethys sank in water 22 meters deep off Cape Frio in South America. In the recovery of the ship's precious cargo, it was found that the frigate was torn to pieces, and that the pieces were scattered here and there over the bottom and buried under great masses of rock.

Ships that pass over the Terranova bank feel strong blows on their keel produced by the reaction of the bank, although it is 80 to 160 meters under water. In this case the upward movement that is generated by the fluid on the escarpment of the bank produces a jet in the direction of the escarpment itself, which, leaving the bank, passes through the entire overlying fluid mass and comes to the surface with enough force for ships to feel its effect. In fact, in very strong storms the jet comes to the sea surface with such force as to produce great breakers. The same thing takes place on the "Aventura" bank that faces toward Cape Bon from Cape Frinitola in Sicily.

5. FLUID MOTIONS ON THE BOTTOM GROW IN INTENSITY AS LAND IS APPROACHED

On the other hand, as great as wave motion may be, it ceases to be felt at sufficiently great depths, although it never ceases to exist (as will be seen below when the behavior of waves is discussed). Sediments recovered with sounding lines prove that the most severe Atlantic storms do not have the force to cause sufficient stress on the bottom of the so-called Telegraphic Plain to disturb the very delicate organisms that are deposited there, nor to move them, nor to mix any grains of sand with them. Therefore wave motions on the bottom diminish in intensity as they go farther from shore and into greater depths until they are no longer felt. On the shore, on the other hand, truly formidable effects are produced in certain circumstances, so much as to move, and violently at that, masses of enormous volume. This occurred, for example, on the western breakwater at Naples where the unusually severe storm of 3 December 1872 shifted a mass of masonry 128 cubic meters in volume that lay 1.4 meters above mean sea level. Such effects as this are surely greater than those that the simple translatory movement of the waves could produce. In fact, the greatest translatory velocity of the waves is 20 meters per second measured at Saint Giovanni di Luz in Gascogne. Now, to produce the vertical jets of water 50 or more meters high that are seen occasionally at the Eddystone Light, the initial velocity must be far greater than 20 meters per second, since even if losses due to friction are ignored, the velocity needed to reach that height is 31.3 meters per second. The aforementioned effects at the shore are surely much greater than those that could be produced by orbital movements developed in the molecules of the fluid mass by the passing of waves. In fact, the waves of greatest power, such as those Wilkes observed south of Cape Horn—for which a height of 9.7 meters was measured with a wave length of 116 meters and a translational velocity of 13.7 meters per second—had a maximum vertical velocity of 3.6 meters per second developed in the orbital movement at the surface of the fluid. This velocity is surely much below the required velocity of 31.3 meters per second. In addition, in the horizontal direction the maximum velocity of the orbital movement would be even less, i.e., since the displacement of the molecules in this direction is much less than in the vertical direction as observation has demonstrated. Furthermore, inside the fluid mass the velocities of the orbital movements will be even less, for whatever the law according to which the velocity varies, velocity will without doubt decrease below the surface of the fluid with increasing depth. In addition, some facts prove that the agitation of the fluid mass diminishes as one descends lower, as will be seen below.

6. THE WAY THAT WAVES ACT ON THE BOTTOM

From all this it must be concluded that wave motion at the bottom is not simply the effect of a simple propagation of the disturbance that is observed on the surface, but rather is due to a much more powerful process. On this account it is to be noted that from each wave on the surface a certain motion is propagated all the way to the bottom; because the bottom is not elastic, it impedes the movement that the molecules at the boundary would otherwise have normal to the bottom, and it allows them to move only parallel to the bottom itself. In other words, in advancing, each wave produces in the direction of its own travel along the bottom a motion that is propagated farther according to some physical law. In its successive position, the wave produces an analogous movement a little bit farther ahead on the bottom, which accelerates material which has been, according to the circumstances, moved from the first point to the second. Thus the movement produced tangential to the bottom starts at the origin of the wave offshore, and from this point it is infinitesimally small for a certain distance greater or lesser according to the circumstances. But in general it increases successively in energy as it proceeds, until it finally explodes at the shoreline with maximum intensity. And in fact, according to the experience of mariners, storms are more violent if the sea is longer and deeper. The requirement for depth is shown by the fact that however brisk the wind and however far away the disturbance comes from, waves are never very high when they travel parallel to the shore. The disturbance increases in energy only in general because special factors may intervene and cause the effect to diminish more or less rapidly over certain areas, as will be seen below. Furthermore the fact that, other conditions being equal, the severity of storms depends on the depth of the water demonstrates again that the reaction on the bottom is much greater when the bottom is steeper. In the opposite case—that is, when the bottom is less steep—the fluid column would have lesser height and the propagation of the disturbance from the surface would be felt more greatly on the bottom. In this regard we cannot admit DeTessan's idea cited by Delesse (*op. cit.,* Chapter VII), concerning the method of wave action: "The enormous energy of a great wave must therefore be concentrated instead in a smaller and smaller mass of water, and render it capable of producing great mechanical effects." To prove that this idea is not admissible it is sufficient to cite the fact that in bad weather ships always find a relative calm over the bank of Terranova, almost as though they were in port, even though the depths are much less than off the bank. The line of breakers that are found on the edge of the bank signals the line between the calm and the stormy sea. And further, it is to be noted that in line with the other conditions, on gently sloping beaches the waves are never as high relatively as on those shorelines with deep water.

7. MOTIONS TANGENTIAL TO THE BOTTOM CEASE AT A CERTAIN HEIGHT OFF THE BOTTOM

The motions generated by waves on the bottom must be propagated to the overlying mass of fluid so that they vanish at a certain height from the bottom, proportional to the energy of the movements themselves and the height of the fluid column.

This is proved by the fact that near the coast, where the water is deep enough to prevent anchoring but still relatively shallow, ships can hold their position quite well over a sea anchor lowered to a depth of about a hundred yards. The sea anchor is made from a cross on which a sail is stretched tightly, and it is suspended by a line attached to the four extremeties of the cross. This method of anchoring does not work if the anchor is lowered only a few meters. The difference in pressure on the two faces of the anchor is always the same as the weight of the water displaced, whatever the depth to which it is lowered. On the other hand, the resistances of the ship and of the floating anchor in the direction of the cable joining them are opposed to each other and in each case proportional to the square of their relative velocity with respect to the fluid. From this it must be concluded that in these circumstances—that is to say, at that depth and at a certain distance from the bottom —the anchor must be located in an ambient fluid sufficiently calm to hold. This proves that the sea anchor is at a point where on the one hand the disturbance being propagated from the surface is very slightly felt, and on the other hand the powerful motions produced by the waves on the bottom are not felt. This also serves to prove that the disturbance propagated from the surface loses much strength descending to a certain depth, and therefore it is incapable of directly producing movements at the bottom. Another proof that the movements on the bottom do not extend very far above it is that certain harbors are very secure for anchoring although they are apparently exposed to storm waves. This happens when the bottom presents a vertical wall, a ridge of rocks, or other similar obstacles of sufficient extent that shelter the harbor in that direction, having the capability to sufficiently intercept the bottom motion. This could happen only if the motion is restricted to within a short distance of the bottom. Meanwhile, it is important to note that these movements vanish at a certain height above the bottom and therefore diminish rapidly in intensity according to some law, as we have seen.

8. OFFSHORE THE MOVEMENTS ON THE BOTTOM ALTERNATE AS DO THOSE INSHORE

The motions caused by waves on the bottom offshore also alternate backward and forward, like those that are observed at the shore. This is shown by the following quotation from a report published in the *Nautical Magazine and Naval Chronicle* (London 1832), concerning the recovery of H. M. S. Tethys: "Even in favorable weather there is great difficulty and danger in descending to the bottom of the bay since the bell frequently takes a swing or sweep of four or five feet and then knocks against the rocks with such great force that it would be fatal if it hit us and which no bell of cast iron could withstand." Another proof is furnished on the bank of Terranova, where ships are carried off the bank by a movement of the water when they come close to the edge. In this regard we have seen that the upward motion that is generated on the edge of the bank produces a jet that can come to the surface of the sea in certain circumstances and in fact sometimes erupts from the sea. This jet would push ships onto the bank, and thus it must be succeeded by another force in the opposite direction. This other force is due to the downward motion generated at the same location, and it produces in the overlying fluid a suction that is inevitably felt at the sea surface because the mass of fluid overlying the bank within a certain sector is all drawn in one direction. And since the jet penetrates into the overlying fluid mass with motion different from the motion of the mass itself, while in the suction no other liquid is introduced, thus in the jet there must be losses of energy that do not occur in the suction. Therefore all other conditions being equal, the effects of the suction must be much more noticeable than the jet's effects; it is precisely this suction that carries ships off the bank. Thus movements generated on the bottom are wavelike fluxes with alternate movement, first in the sense of travel of the waves, and then in the opposite sense; therefore we shall continue to call them bottom fluxes, or simply fluxes, a term that has commonly been used to describe them.

9. THE DIRECTION OF THE MOTION OF THE FLUX IS IN PHASE WITH THE WAVES

Given that the fluxes possess an alternate motion, at each change of direction they must for an instant have zero velocity. In addition, both the forward and reverse movements have maximum velocity at a given instant. Regardless of the precise instant when the zero velocity and the forward or reverse maximum velocities take place, it is notable that in general the direct movement of the flux corresponds vertically to the crest of the wave and the reverse to the trough, such that the flux has a movement in the same sense as the direction of travel of the waves underneath the crest and in the opposite direction under the trough. The following facts are sufficient to prove this. Whenever the bottom is sandy, it assumes a rippled aspect with many small ridgelike rises nearly parallel in their general trend. Now, if fairly shallow water that is clear enough for the bottom to be seen is silty and if the sea is sufficiently rough, as the crest of the wave passes, the bottom below it becomes turbid, beveling the tops of the ridges. Each parcel of turbidity follows the wave's movement below the crest and then returns during the trough's passage. Regardless of the state and the depth of the sea, the jet that is expelled from the fluid mass at the shore erupts at the same moment and in the same direction as the wave crest. By the same token, if there is a steep rise in the bottom offshore, and if this rise is found at shallow enough depth and if the sea is sufficiently rough, when the crest of a wave passes over the edge of the bank, a breaker is formed. The breaker accompanies the wave for some distance, according to the circumstances. All this is due to the effect of the jet produced by the forward flux; emerging from the steep scarp of the offshore bank, the jet strikes the wave on its back, raises it, and causes it to break by toppling it. Under these circumstances the jet becomes detached from the bottom at a point near enough to the surface and close enough to the vertical to strike the wave near the point directly above the point it issued from.

10. FORWARD FLUXES PREVAIL OVER BACKWARD FLUXES

It has already been seen that near the shore the upward, or forward fluxes prevail over the downward, or backward, fluxes. A proof, other than the progradation of certain beaches, is that when ships come too close to land they are thrown onto the shore. With shallow depths and stormy seas, the bottom flux can reach the keel of these ships, and its upward resultant pushes them ashore. But the prevalence of forward over backward fluxes happens not only near the shore but also offshore. This is proved beyond doubt by the fact that detritus whose specific weight is only slightly greater than that of water (such as certain animal or vegetable remains) is carried onshore even from great depths. Only this type of detritus is thus thrown up from those depths, never fragments of rocks whose specific weight is very much greater than that of water. When the specific weight of the material is only slightly greater than water's the component of its absolute weight parallel to the bottom (lessened of course by the weight of water displaced by it) has a very small value, even though the slope of the bottom may be great. Hence what is taken from the force of the upward fluxes and is added to the force of the downward fluxes is so small that it is almost negligible; the result is that the upward component is still clearly greater than the downward component, although the upward component is no longer undiminished. For that matter, the prevalence of the upward fluxes over the downward fluxes is quite natural if it is considered that the wave motion propagated to the bottom must in some way carry the fluid forward through its translational movement along the upward-sloping bottom. The raising of the water surface that is observed in stormy weather must be attributed in part to the prevalence of the upward fluxes, since the transport of fluid toward shore is not observed due to wave motion. The bottom fluxes must transport a certain amount of fluid through the prevalence of the ascending fluxes, as has been noted in the discussion of the significance commonly attributed to the fluxes. The prevalence of the forward over the backward fluxes must increase in proportion to the fluxes themselves as the shore is approached. This is shown by the fact that even when the slope of the bottom is greater at the shore, the component of sediment weight parallel to the bottom must not be sufficient to overcome this prevalence, since sediment is transported onshore.

11. SEDIMENTS ARE MOVED ALONG SHORE

This relationship between the fluxes means that offshore, in the direction parallel to the shoreline, sediment must move in the direction of the waves just as it does near shore, since the component of sediment weight parallel to the bottom does not influence movement generated by waves parallel to the bottom contours. In this manner sediment travels to and fro according to the direction of the fluxes, and in the long run it is transported away from the direction of the prevailing fluxes' movement. Fluxes are said to be prevailing when they dominate in terms of energy, frequency of occurrence, duration, and angle of incidence, upon which their component parallel to the contours depends.

Sediment is transported principally, but not wholly, in the direction of the prevailing fluxes; because according to the energy and the frequency of the fluxes from each side and according to the local circumstances (slope of the bottom in the longitudinal direction—that is, parallel to the shore, exposure of the shoreline to fluxes from different directions, etc.), considerable sediment may also move in the direction opposite to the prevailing fluxes, as will be seen in Bari's case. On the western Ligurian Riviera, for example, fluxes from the southwest prevail although they are rarer than those from the southeast, which are much weaker. In general, it may be said that the sediment along a coast is transported principally in the direction of effective wave approach. In any case the degree of movement of sediment parallel to the shoreline is moderated by the fact that waves that have a direction of approach at an angle to the coast do not travel in a straight line, but rather are bent near the shore in a direction that becomes more and more normal to the shore itself (this is a very pronounced natural tendency).

12. THE NEUTRAL LINE

What is the significance of the fact that beach material is transported around certain promontories such as those as Loano and Ventimiglia, but not around others that are more pronounced, such as Cape St. Ampeglio, Cape Noli, the breakwater at Bergeggi, and Pria Pula? Simply, in the first case sediment must pass over certain depths from which it may still be carried ashore, while in the second case, where the depths over which it must pass are greater, it may be carried offshore. The same holds for indentations and even moreso for sharp embayments. If these are sufficiently deep, such as the entrance to the port of Savona, the sediment that passes them does not reappear again onshore. This effect is even greater in front of headlands, such as Cape Noli, that present to the sea an extensive steep scarp, because here the backwash also carries sediment offshore. Where there are no sufficiently pronounced headlands or embayments, beach material may appear along the shoreline for great distances, such as at Bari, where abundant volcanic sands from the Vulture River in the Ofanto Valley are found in that city's new harbor. The Vulture empties into the sea about 60 km west of Bari. By contrast, the prevailing waves cause other sediment to move in the opposite direction, that is, from Bari to Ofanto. There must therefore be a point, which will be called *neutral*, where the ascending flux, less the component of weight parallel to the bottom, is in perfect equilibrium with the descending flux added to that same component. The result will be the establishment along the shoreline of a *neutral line*, shoreward of which sediment is carried toward shore, and seaward of which it is moved offshore. Of course the position of the neutral line depends on all the factors that act to fix its position, and primarily on the force of the fluxes. This is shown by the fact that the fluxes coming from the east along the western Ligurian Riviera cannot carry onto the beach at Pegli the dolomitic limestone sediments that are very abundant on the beach at Sestri-ponente, even though the two beaches are contiguous and are not separated by true headlands, and even though the 10-m contour line is at almost 800 m from shore. Furthermore, the compact Mesozoic limestone sediments on the beach at Albenga do not surpass the small point immediately westward of the delta of the Centa, even though the point is less than 2 km away from the mouth of the stream and even though the 10-m contour line is still 460 m offshore. Therefore sediment that can be brought onshore from a certain depth by the southwest waves can no longer be brought onshore from lesser depths by the southeast waves, which tend to be smaller.

The position of the neutral line varies greatly with the slope of the bottom and with the specific weight of the sediment, as has been seen. It also varies with the size of the sediment itself. In this regard it is to be recalled that the intensity of the flux over a given point on the bottom diminishes rapidly with distance from the bottom so that the flux acts more weakly against the upper parts of the sediment, which may project above the bottom flux's layer of action if it is sufficiently large. Now for bodies of similar shape and disposition, weight grows more rapidly than the force imparted by the fluxes, since weight increases as the cube of the diameter and the force increases less than the square of the same since the flux does not have the same intensity over the whole depth. Therefore although larger sediments have greater stability, according to what has been said above other conditions being equal, they will have a shallower neutral line. To recapitulate, the neutral line deepens with increasing flux strength, as the slope of the bottom increases, and as the specific weight and volume of the sediment decrease. Therefore close to shore the same sediment that is carried offshore under the action of fluxes may on the other hand be brought onshore by other fluxes of greater intensity, because the first have a neutral line higher than the others. In the Mediterranean, during the strongest storms on the most eroded beaches the neutral line for the sediment that ordinarily constitutes such beaches probably does not extend deeper than 10 m. Remembering generally that at a certain depth wave action is no longer noticeable, in order for the statement of this maxim to be exact, we must complete it with the words "for the purposes of sedimentation." Further, as has been said, sediment remains in the basin of its origin, but only insofar as it is found above its particular neutral line, since below this neutral line, the sediment can pass beyond the most pronounced projections of the coast as it moves offshore, and in this way it can be transported far along the shoreline. Finally, the fluxes associated with the effective wave sector, being the most forceful and having therefore the deepest neutral line, carry onshore sediment from greater depths, accumulating it at times to great heights. Therefore if the effective wave flux does not always prevail over other types of flux moving sediment along the shoreline, it does prevail in accumulating sediments at the shore, with the exception of what will be said concerning beaches undergoing erosion.

13. CONTINENTAL SHELVES AND TERRACES

In other words, through their alternate movement and with the concurrence of the component of weight parallel to the bottom of the sediment put into movement by them, the bottom fluxes carry away sediment that arrives at the shore. With their resultant shoreward component the same fluxes hold back as much sediment as possible toward the shore and distribute and arrange this sediment along the shoreline. If the circumstances are such that the quantity of sediment brought to the shore is less than what the fluxes remove, then the beach will erode, because even if there is some point along the shore where sediment is being carried landward, the beach will retreat because it lacks footing. In the opposite case the sea fails to carry away a certain amount of sediment, and this accumulation causes the slope of the bottom to increase and thus helps the waves to carry the remainder of the material offshore. The largest sediment is always deposited at the shoreline because the energy of the fluxes diminishes offshore, and therefore there will be a point where the larger sediment can no longer be moved, while the smaller sediment will continue to be carried offshore to the point where the fluxes can no longer move it. Wherever sediment comes to rest, it will then be oriented parallel to the fluxes because it is less disturbed by them. Thus continental shelves that are gentle in slope and that form the submarine continuation of the beaches are found in front of coasts where sedimentary material is highly abundant. Naturally, the extent of these shelves is in direct proportion to the supply of sediment and their age, but at a certain point offshore they always fall off—that is, the slope becomes relatively steep and therefore the shelves take the form of true terraces. Now, as is seen from the table below, while breaks in slope of the continental shelf at Cape Noli in Liguria is 95 m offshore and at Carlovassi, on the island of Samos, 210 m, in front of Falmouth on the Atlantic the break does not occur until 260 km.

Continental shelf		*Depth to break (m)*	*Distance in slope (m)*
Cape Noli		10	95
Carlovassi (in front of the Customs house)		10	210
Port Maurizio		110	5,000
Mouth of the Tiber		170	15,000
Belgian and Dutch Coast		5	20,000
French Atlantic Coast	Gulf of Gascogne	200	60,000
	Islands of Yeu	200	165,000
English Coast at Falmouth		200	260,000

It has been pointed out that the continental shelf changes slope at a certain point after having maintained a gentle slope over some distance so that it presents the form of a terrace. It is sufficient to give as an example what is observed off the Island of Yeu in France where the continental shelf is without doubt average for that shoreline. Here it is necessary to go 165 km offshore to reach a 200-m depth, whereas only 55 km farther the depth is 4000 m. In the first section the average slope of the bottom is 1.2 parts per thousand and in the second, 69.1. Starting at the point where the break in slope occurs, it forms a scarp on which the greater slope of the bottom not only overrides the lessened intensity of the landward flux to carry sediment offshore, but in fact favors the discharge seaward since on that scarp a small force is sufficient to move sediment downward. The slope of the terrace and other conditions being equal, sediment can move over the scarp more easily and therefore it is discharged in greater abundance if the terrace is narrow. The width of the terrace therefore also influences the speed of progradation or erosion of the subaerial beach according to the circumstances (except, of course, in view of what will be said in Section 14 with respect to the fluxes that do not arrive freely at the shore—that is, do not arrive there with all the energy that they possess offshore). Examining the composition of the continental shelf across the island of Yeu yields the following results:

Distance offshore	*Composition*
To 30 km	Boulders, rocks, and sand
30 to 50 km	Sand and gravel
50 to 105 km	Clay
105 to 120 km	Sandy clay
120 to 175	Sand
Beyond 175 km	Clay

The break in slope is therefore at 165 km, but the increase in slope begins at 105 km. Thus starting at this point, the bottom is sandy clay, a transition between the mud that is found on the terrace and the sand that is on the scarp, because the mud precipitates at the foot of the scarp. Furthermore this shows that sand is carried a considerable distance offshore. Overlapping layers of sand are deposited on the scarp in much the same way as that practiced in the construction of highway embankments where material is deposited successively at the end of the already built embankment. While the shelf under water progrades offshore, the subaerial beach can also prograde and the bottom becomes raised by the deposition of successive strata. Progradation may not necessarily take place, however, because frequently—as will be seen in Section 14—the subaerial beach assumes a certain permanence of position, if not of sediment type, that may be replaced. As the subaerial beach progrades, its profile advances parallel to itself and parallel to the profile of the offshore shelf. This is only true up to a certain point because the form of the profile depends on the state of the sea and on the nature, size, and abundance of sediment as well as on the extent of the submarine shelf, all of which can change with time. Not only will the submarine shelf certainly prograde, but it will always do so more than the subaerial beach (given that this advances), because the zone where the downward fluxes dominate is by and large greater, even though the action may be weaker than in the zone where upward forces dominate.

14. SHOALS, BARRIER BEACHES, AND SUBAQUEOUS ANTIDUNES

Whenever the slope of the subaqueous beach is so gentle—that is, so shallow—that the fluxes cannot come freely to the shore, the fluxes rapidly lose much of their intensity at a certain point. At that point the fluxes offshore and landward of the point of rapid loss of intensity behave as two distinct fluxes, each of which according to its individual intensity has its own neutral line. In other words, the flux then has two neutral lines, of which the one corresponding to the landward flux is found at a shallower depth and closer to shore than the one that corresponds to the seaward flux. And if the point where the flux loses energy is found in shallow water, as is usually the case, the sediment from the point seaward to the lower neutral line is carried onshore because of this neutral line. The sediment between the same point and the higher neutral line, however, is carried offshore because of this other, higher neutral line. Furthermore, if the flux on the landward side has sufficient energy it levels the sediment on which it acts; but the flux on the seaward side does nothing more than prevent the sediment from moving offshore. Thus near the shore on top of the submarine shelf there is formed another similar but smaller shelf with a seaward facing scarp. This is the case in many Calabrian beaches on the Tyrrhenian Sea. This ridge that runs alongshore and has the form of a large, rounded step will be called a littoral ridge to distinguish it from other ridges of different forms and extent.

If instead the flux on the landward side is weak compared with the flux offshore, the offshore flux entrains sediment carried out by the landward flux and at their meeting point a rise forms on the bottom in the form of a ripple—that is, a subaqueous antidune. This is true on the beaches of Tortoli in Sardinia, of Alassio in Liguria, and in many beaches in the Marche. If the sea is sufficiently rough, there will be breakers at the surface of the fluid, almost vertically above the seaward edge of both the littoral ridge and the subaqueous antidune. This is the flux that produces a jet in the direction of the ridges followed by a suction—just as has already been noted concerning the bank of Terranova—a jet that, striking the wave from below at a certain angle, makes it topple and break, as has already been noted. At the point where the jet becomes detached from the ridge, the flux rapidly loses intensity. When circumstances are such that the offshore flux accumulates additional material on the antidune, this becomes elevated and grows so that it emerges above the water surface. Thus it is that so-called barrier beaches are formed with lagoons, marshes, and swamps at their backs, of which there are many examples. In every case when the flux that formerly arrived at the shore with all its energy arrives weakened due to the changed conditions of the bottom, the beach that was formerly prograding, now retreats at least for some time, as was mentioned above concerning the beaches eastward of Port Maurizio and of the old port of Pesaro. But the beach will retreat only above water because there remains to seaward an expanse of water that is still quite shallow. Offshore the submarine shelf continues to prograde and rise as in the general case. Thus Belpaire's observation is explained: "It is remarkable that almost everywhere the coast is low and sandy, the extreme edge of the country is formed by marshes. This is what is seen from Bayonne to Bordeaux, and on the southern edge of the Baltic, primarily between Danzig and Koenigsberg." [*De la plaine maritime depuis Boulogne jusqu'au Danemark*]

15. THE SAME OBSERVATIONS CAN BE APPLIED TO LAKES

Since the laws of nature work everywhere in the same way, whatever can be said for the sea can be repeated for lakes with the exception of the smaller scale of the effects.

16. CONCLUSIONS

Recapitulating, we come to the following conclusions:

1. Waves in a fluid generate an alternating motion on the bottom, first according to the direction of travel of the waves, then in the opposite direction. This alternate movement is precisely what is meant by "bottom flux" or simply "flux." The flux will act with intermittent force on bodies situated on the bottom.

2. In general, the movement of the bottom flux is forward under the crest of the waves and backward under their trough.

3. The intensity of the bottom flux is greater as the waves that generated it come from farther away and from greater depths. The energy of the flux can be great even at great depths. In every case the flux, diminishing rapidly in energy with distance from the bottom, is felt only over a fluid stratum of a small variable thickness.

4. On a bottom sloping upward toward land, the forward, or ascending flux prevails over the backward, or descending, flux. The forward flux dominates more and more as land is approached.

5. The weight component parallel to the bottom of bodies exposed to the fluxes may weaken or even nullify the dominance of the forward flux.

6. The locus of points where, along the gravitational component, the two fluxes are exactly in balance along the line of maximum bottom slope establishes on the bottom a "neutral line." Other conditions being equal, the neutral line is deeper with increasing wave disturbance, increasing bottom slope, and decreasing size and specific weight of sediment. In the Mediterranean the neutral line is rarely deeper than about 10 m.

7. Landward of the neutral line, bottom sediment is carried toward shore, while seaward of the neutral line, it is carried offshore.

8. Parallel to the shoreline, sediment moves everywhere away from the direction of wave approach. Thus sediment moves to and fro but ultimately it is transported mainly in the direction of the prevailing flux that usually corresponds to the direction of effective wave approach.

9. With their back-and-forth motion and with the help of the gravitational component, the fluxes distribute the sediment that comes to the sea by carrying it offshore. The dominance of the forward flux spreads the material along the shoreline, keeping it concentrated near land and depositing it in part against the shore according to the circumstances. In this manner the fluxes arrange material and by successive overlaid strata the sedimentary terrains of the sea and of lakes are formed.

? J. W. Johnson ?

2

Reprinted from *Fish Bull. No. 3*:2–29 (1919)

THE SPAWNING OF THE GRUNION.

(Leuresthes tenuis.)*

By Will F. Thompson, assisted by Julia Bell Thompson.

On moonlit nights during the high tides of March, April, May and June, a small smelt comes in on the long sandy beaches of California. It comes in with the sweep of the water up the beach as the waves break, and lies for a moment glittering in the faint light, then squirms and flops back into the wash of the next wave. Along the whole magnificent sweep of broad sandy shore at Long Beach, crowds of people gather to pick up these fish, by the light of the moon, and of bonfires, and of flash-lights. Some content themselves with picking up the stranded fish; others utilize wire screens, or even portions of beach seines, catching the smelt as they venture inshore. The fish they obtain are less than the length of one's hand, slender, with a broad lateral stripe, and very plainly of the smelt family. At Long Beach no name other than "grunion" is ever heard, although one gleans from scientific works such names as "silver-sides" and "little-smelt." Those who gather this "grunion" know that it comes in to spawn its eggs, but of the marvelous story that lies ready to discovery, they know not a whit. The crowds of bathers who follow in their footsteps on succeeding days, little think that four inches below their feet is unfolding one of the really remarkable stories in the annals of natural history.

Surely the grunion has a purpose in thus venturing out of its native element, to lay its eggs in the sand. In some way the act must serve the species, must aid it in its survival. It must escape its enemies, or obtain favorable conditions for development. Other smelts lay their eggs very differently, attached to the rocks or the bottom of the ocean by slender stalks or filaments. Many species migrate into brackish or even fresh water to spawn, while other genera or species are entirely confined to fresh water. Even in species apparently closely related, the spawning habits have become diverse, perhaps under the pressure of the struggle for survival. Fish, smelts and others, lay their eggs in every conceivable marine locality where suitable conditions may be obtained, but it remains for the "grunion" to utilize what is practically dry land.

*California State Fisheries Laboratory. Contribution No. 8.

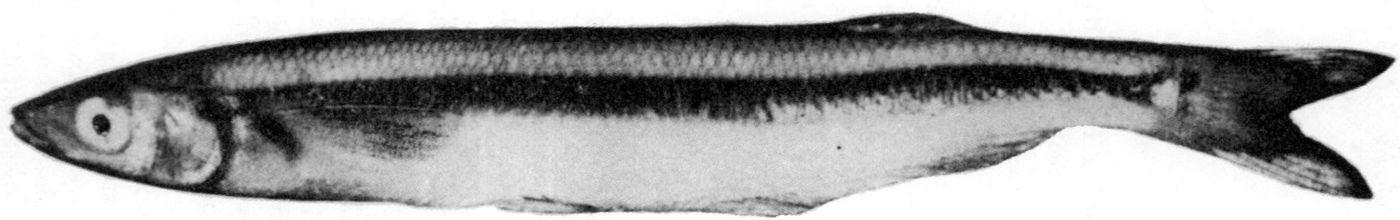

Fig. 1. The Grunion, *Leuresthes tenuis.* Male, length 5¾ inches, taken April 16, 1919, at Long Beach, California.

For that is actually what it is doing when it ventures inshore as far as the high tides will carry it. The eggs are laid in the sand as far down as the fish is able to bury them, and far above the level of the average tide. The way in which it does this, the history of the eggs in the sand, and the story of their escape are interest-compelling.

The grunion comes the second, third, and fourth nights after the full of the moon, according to popular tradition; therefore but once a month and shortly after the highest tides which accompany the full moon. The tides are then highest at about nine o'clock (ten o'clock according to summer time). Shortly before the tide is farthest in, the grunion may be taken with a beach seine placed athwart the wash of the waves. At the same time, occasional fish may be picked up on the beach as they are left exposed at the highest point reached by the waves. Very shortly after the time of the highest tide, and when the moon is well up above the horizon the real run commences, and for an hour or more fish may be found in numbers, especially where there is a light run-off or curve in the beach which produces a swirl in the wash of the waves. Occasionally thousands of fish are within sight at one time. The run diminishes gradually, and finally stops. Our first observations on this run were in April of 1919, on the fifteenth when the moon was full. The first fish were taken on the sixteenth and the last on the night of the eighteenth, thus confirming in a measure the popular belief of the crowds who gather expectantly at the proper time, as given by Mr. J. B. Joplin of Santa Ana, who says, "Three months during the year, usually March, April, and May, on the second, third, and fourth nights after the full moon, at full tide, great schools of them come out in the breakers * * *."*

The Spawning Run.

The schools of fish which come in seem to work back and forth on the edge of the beach, and they may be taken some time before they commence to run up the beach. By the use of a short piece of seine, with cork line at the top and lead line below, numbers of fish were taken before high tide time. When the waves washed up the beach the net was lifted, but when the water returned in a torrent the net was dropped and held firmly against the bottom so as to make a manner of bag into which the smelt were carried. Two men captured by this method a bucketful of fish in an hour's work. It seemed possible to do as well by picking up the fish on the beach, but so many people were doing this as to render it very difficult.

*However, we also observed a spawning run in June, the fourth during the season.

The males were nearly twice as abundant as the females in this catch by the seine, as will be evident from the following table:

TABLE 1.

Body length	Number of fish of given length	
	Male	Female
10.5—10.9 cm.	1	----------
11.0—11.4 cm.	4	----------
11.5—11.9 cm.	8	1
12.0—12.4 cm.	11	1
12.5—12.9 cm.	31	----------
13.0—13.4 cm.	28	8
13.5—13.9 cm.	19	9
14.0—14.4 cm.	8	13
14.5—14.9 cm.	2	11
15.0—15.4 cm.	1	6
15.5—15.9 cm.	----------	2
16.0—16.4 cm.	----------	2
Total number	113	53
Average size	12.95 cm.	14.25 cm.

The males were also much the smaller fish, averaging 12.95 centimeters in length without the tail fin, whereas the females averaged 14.25 centimeters. This means 15 centimeters or 5.9 inches total length for the males and 16.6 centimeters or 6.5 inches for the females, a difference somewhat more than half an inch in favor of the females. This difference is, however, much more apparent when the bulk of the fish is considered, for although the length of the average male is roughly 90 per cent of the length of the female, its weight is but 70 per cent, because the weight increases approximately according to the cube of the length. This difference in size allows one to decide readily on the sex of the fish in most cases, and this may be quickly confirmed by squeezing out some of the milt or roe.

The difference between sexes being known, further observations were made on the actions of the fish which had ventured high on the beach.

The Spawning. A station was taken near one of those slightly uneven portions of the beach where the water swirled and gathered as it ran off, and in which the fish gathered most abundantly, and careful watch was kept for incoming fish. As soon as a high wave (one in every seven, the passers-by informed us) had passed well up the beach and then receded, fish were to be found lying a few feet below the highest point reached. At first single fish, or perhaps a number of fish, were taken lying on the sand and an occasional one was captured buried tail downwards to the level of the pectoral fins. The latter proved to be females, invariably, and by acting quickly, it was possible to take one, or perhaps more, males in the vicinity of each buried female. If the male, lying flat on the sand, could be seen and captured before it moved, it was very often, indeed almost always, found with its vent lying very close to the buried female. The conclusion is irresistible that actual pairing takes place.

But a further conclusion follows, for surely this elaborate habit must be for a purpose. The females, in thus burying themselves, must be doing so in order to deposit the eggs. The males closely attend the females, and when the latter deposit their eggs the milt is extruded in the sand as the female agitates it. Close, quick observation showed the whole process of pairing and egg laying. In a typical case, two fish were observed lying side by side, heads up the beach, in a small gully near the upper part of the area covered by the waves. The fish, moving very slightly, seemed quiescent for a moment; then the female, stirring the sand with her tail and fins until it became fluid, buried the posterior part of her body, ultimately assuming a vertical position, while the male lay arched around her and in contact with her side. The female twisted slightly to and fro, her body revolving considerably on its long axis, until she had sunk to the level of the pectoral fins. About this time the male flapped away, and was promptly captured for examination. The female then began to bend from side to side slowly and as though fatigued until she emerged from the sand, obviously spent, her abdomen flaccid. The whole process took an average time of twenty to thirty seconds in five cases timed. By digging up the sand the eggs were quickly obtained, buried at a depth of two and a half to three inches at most. In two cases observed, the female actually buried herself an inch and a half or two inches completely under the sand. These completely buried ones were, however, but two among the thirty-five or forty that were watched sufficiently; hence, but a small percentage.

The affair is, however, not a real pairing, for frequently there are four or five males surrounding one female, and in one case two females were seen to mate with one male which lay between them. In fact the mating is accomplished in a casual way, the fish happening to come to rest in the same slight hollow, or in a small group as the swirl of the water left them. Nor do males and females always find each other, for females turgid with eggs may be picked up in numbers after the waves, energetically pursuing their course back into the water. There are certainly, however, the best of reasons to consider the pairing, or mating, as necessary before spawning can occur, for in observing six runs of fish not a single female was caught in the act of burying herself which did not have, or had not had, a male near her.

That this burial was the sole method of laying the eggs could hardly be questioned. A close examination of the sand with the aid of a bright flashlight showed no such number of eggs on the surface as would be expected to be laid by a female, and there were certainly pods of eggs deposited in the sand at varying depths.

To corroborate these observations search was made for the pods of eggs during the next day. It was judged that they should be sought at about the depth at which the eggs would be buried by a fish of the given size, at most the distance between the vent and the pectoral fins. Measurements indicated that this was approximately two inches. It proved possible, indeed, to find the eggs, but very strangely they were not two inches down. The eggs were found in small masses, or "pods" as they may be termed, unmixed with the sand save at the surface of the pods, and no more moist than their surroundings. But these pods were in various cases the following distances below the surface: $3\frac{1}{2}$ inches; 4; $4\frac{1}{2}$; $4\frac{1}{4}$; $4\frac{1}{4}$; $2\frac{3}{4}$; 3; $2\frac{3}{4}$; $4\frac{1}{4}$; $3\frac{3}{4}$; 2; 6; 6; $5\frac{1}{2}$. These average 4 inches down. Careful comparison of the eggs squeezed from females and the eggs in these pods proved them identical. There was no doubt whatsoever regarding the fact that the pods were those laid by *Leuresthes*, yet the attainment of a depth of $4\frac{1}{2}$ inches would certainly have buried the fish with the tip of its snout two or three inches under the surface of the sand, something which was observed only occasionally. But where was the error in observation? Certainly if the female had not buried herself that deeply, but had forced the eggs down by the action of her anal and caudal fins, they would have been mixed with sand. This circumstance demanded explanation, which was fortunately obtained and is given later.

The Position of the Eggs.

Pending this, careful examination was made of the pods and their position. A female 17.1 centimeters in total length, captured in the seine before it had spawned, was opened, and the eggs counted, giving 2,528 eggs. Six pods were carefully dug up on the twenty-third, six or seven days after they were laid, and the counts of eggs were as follows: 1,911; 1,514; 1,479; 2,540; 2,705, with an average of 2,200. It is evident enough that all the eggs in a female of large size were laid in some of the pods, but in the others of lesser number it is possible that only part of the eggs were laid, or that small females laid them. In a female 11.8 centimeters in body length (13.7 centimeters total length) but 475 were counted. Pods containing as few eggs as this were certainly to be found, but in each case were open to the suspicion that parts of them were lost. However, these numbers confirmed the evidence derived from the examination of the structure of the eggs that the pods were laid by *Leuresthes*.

These pods were, moreover, located in those precise regions in which the adults had been observed to spawn. As has been mentioned, the spawning takes place just below the line (two or three paces) reached by the very highest of the waves, and at about the average height reached by good-sized waves at the height of the tide. At that point the sand is lifted and lightened by the water of the high waves until it is very easy to penetrate, just as any soil will become softened by moisture—in this case excessive. On the most even of beaches there are of course inequalities, and in the gullies the water will collect during the run-offs and remain a trifle longer in a swirl, thus allowing the fish a better chance to burrow before the sand settles. In these localities the pods were most abundant. In the accompanying plate (Figure 2) the

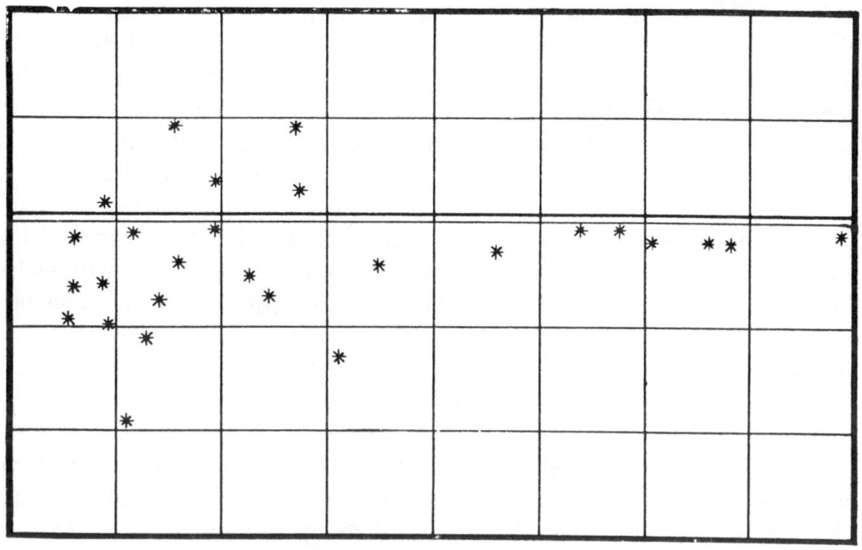

Fig. 2. Locations of pods of *Leuresthes tenuis*, collected April 21, 22, and 23, 1919 at Long Beach. Average highest points of waves during high tides (5.3 feet) of April 16, 17 and 18 indicated by the double line. Squares three feet on each side.

area dug over during three days of collecting samples is shown, the level reached by the average good sized wave of a 5.4-foot tide being indicated by a double line. The area in the left-hand side of the chart covered a slight depression and in it more pods were found than on a more even beach. No pods were to be found above or below those shown on the chart, although diligent search was made. Although these pods were laid in sand which was moistened by the larger waves of the high tide, yet they were practically in an environment far from marine in character. The conditions under which the eggs existed were beyond doubt unendurable for eggs such as are taken in the plankton, floating freely in the surface waters of the ocean. Temperature, salinity, and enemies would naturally be expected to vary widely from marine conditions.

In searching for pods, note was made of possible enemies, and it was found that numerous species of animals inhabited the sand. Certain of these were marine, others terrestial, the beach at high tide line representing an area apparently invaded by hardy members of each group. Annelids (fish worms) and amphipods (sand-fleas) were abundant, but in no case were they connected with the destruction of fish eggs. An isopod,[1] however, was found in the very center of a pod of eggs. Whether it was feeding on the eggs or simply happened to stray in is unknown.

The Enemies of the Eggs.

The case is different with a beetle, *Saprinus sulcifrons* Lec., a member of the family *Histeridæ*. This species was found very frequently in the immediate vicinity of the pods, and very often in the midst of the eggs. Occasionally warning of the presence of eggs near by was given by the discovery of the beetle. Its powerful legs render it a most efficient borer in the sand and it apparently ranged widely in search of food. When placed in a vial of sand with a numbered lot of eggs it soon consumed them, thus proving without doubt the cause of its presence in the pods. It was not to be found during the days immediately following the depositing of the eggs, but appeared later, in so far as the limited observations indicated. Blatchley,[2] a specialist on beetles, says of this group, "Members of this genus live mostly in carcasses, especially those of dead fish along shores and lakes." Also, "These beetles were formerly thought to be scavengers, but it is now believed they are predaceous in all stages, devouring larvæ of diptera which are feeding upon decaying matter." This beetle is apparently the only serious enemy of the eggs, and its attacks on the eggs may be the result of straying from the normal habitat. Surely the escape of the eggs must involve the destruction of their enemy, either adults or young, for the presence of water in volume sufficient to sweep eggs and all out to sea must be a necessity. However this may be, either the beetle or the fish would seem to have made a mistake in its wanderings.

Two pods of eggs were taken with the larvæ of flies within them, and these larvæ were allowed to develop in the laboratory. In one case the pod had been dug up and reburied so that it is possible the fly concerned

[1]This isopod was, through the courtesy of Waldo F. Schmitt, submitted to the United States National Museum for identification, and the following information obtained:

"The isopod crustacean submitted is *Tylos punctatus* Holmes and Gay, described in 1909, type specimen from San Diego, California, with note 'burrows in sand.' This species is the only one of the Tylidæ so far recorded from the west coast of America, and there are but two specimens in the collections of the National Museum."

[2]We are under obligations to Dr. M. I. McCracken of Stanford University for quotations from the literature and reference of the specimens to Dr. E. C. Van Dyke for identification. Dr. E. C. Van Dyke states: "The beetles are specimens of *Saprinus sulcifrons* Lec., one of the *Histeridæ*, a family of steel-clad predaceous beetles which are generally to be found about dead animal and vegetable substances. * * * Most of the *Histeridæ* are more or less restricted in their habitat and general life history. Such is the case with the species sent by you. This is a seacoast species entirely, being widely distributed along the seashore of both Lower and Upper California. It is generally to be found under seaweed or old dead animals that are lying on the sand a little above the normal tide line. They are never to be found on the sand dunes back from the seacoast."

had laid its eggs while the pod was exposed. In the other case this could not have been true, the pod being exposed for the first time when the maggots were discovered. The two pods contained different species of flies, which have not been identified.

But these natural enemies are obviously very few in number, and the greatest mortality would seem to be inevitably from physical causes.

The Surrounding Conditions.

This would be dependent on the capacity of the eggs to withstand the variations in the environment. That they are able to withstand great changes in the salinity of their surroundings was easily ascertained by dropping the eggs in fresh water after the beating of the heart had become apparent. Eggs placed in tap water on the twenty-seventh of April were still alive on the fifth of May, although the heart beat was spasmodic. This was a period of eight days, and fully a week later than the normal hatching time of the eggs. The delay in hatching is not the result of the action of the fresh water, as will be seen later. Upon being placed in salt water one of these eggs hatched, although the larva proved incapable of movement. The larvæ from normally raised eggs (24 in number) placed in fresh water, died in the space of an hour (from 1.58 p.m. to 2.55 p.m. on the twenty-ninth of April), whereas their brothers in sea water lived for days. Evidently the egg capsules were efficacious in protecting the larvæ against such accidents as heavy rainfalls or freshets from the land.

However, the egg capsules seemed unable to protect the larvæ against desiccation, such as might ensue during days of great heat and consequent rapid evaporation. During the first days succeeding the tides the sand was fairly moist and firm—somewhat as it would be immediately after immersion in water and subsequent drainage, but ten or eleven days after the eggs were deposited the sand was exceedingly friable and on the surface completely dried out. The eggs were, however, in every case moist enough for life, and samples continued to exist in sand allowed to become much drier. To test the capacity to withstand such desiccation as might result from complete dryness, two dozen eggs were placed on a blotter and exposed to the air. These eggs were twelve days old, and well eyed. Placed upon the blotter at 3.38 p.m., the heart was still beating at 8.20 p.m., although the capsule was invaginated greatly, forming a cup. Shortly thereafter the beating of the heart ceased, the eggs appearing completely desiccated. Other eggs placed in a covered dish seemed to show no effects, moist atmosphere seemingly being sufficient. Still others, allowed to lie on the surface of the sand in closed bottles, existed for several days in the laboratory, and could not at the end of that time be distinguished from those which were buried. They are certainly not proof against evaporation, but are at the same time able to withstand as much as they would normally be exposed to on a beach. Even above the level of the highest tides the sand at a depth of four inches is always moist enough to enable the eggs to live. The danger from excessive evaporation does not seem very great.

But, perhaps, they might be compelled to undergo unusual temperature conditions. To ascertain what these were, a series of observations

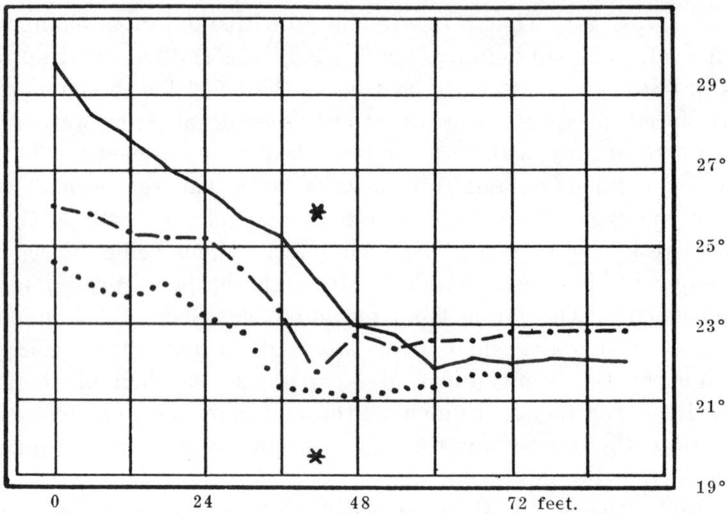

Fig. 3. Temperatures of beach across high tide line, the left of the figure the upper part of the beach, 6 p.m. April 20, 1919. Location of double line on Figure 2.✶

were made at stations three feet apart across the beach on the twentieth of April, four days after the first eggs were laid. A thermometer, the usual chemical type with a bare bulb and stem, was thrust down into the sand, first just below the surface, then three inches down, and lastly six inches. The accompanying figure (Figure 3) shows the results of this, the left-hand side of the chart representing a distance of fifty-two feet above high tide line and the right-hand forty-eight below. The high tide line is the same as that shown in Figure 2. It will be noticed that the area occupied by the eggs was not markedly higher in temperature than that of the beach nearer the water, that at six inches down being actually less. Unfortunately the only thermometer then available broke shortly before the completion of the observations at a depth of six inches. The day was fair and warm, as is usual in southern California at the period of the year.

During the high tides beginning about May 13 two series of similar observations were made across the beach at stations three feet apart.

One of these series was made two hours after the high tide, which occurred at 9.06 p.m., and the other was made at 4.30 p.m. the next day. These should provide a contrast between the extremes of temperature which the eggs might meet with on such a day. In Figure 4 the two series are plotted. The space between the marks denoting

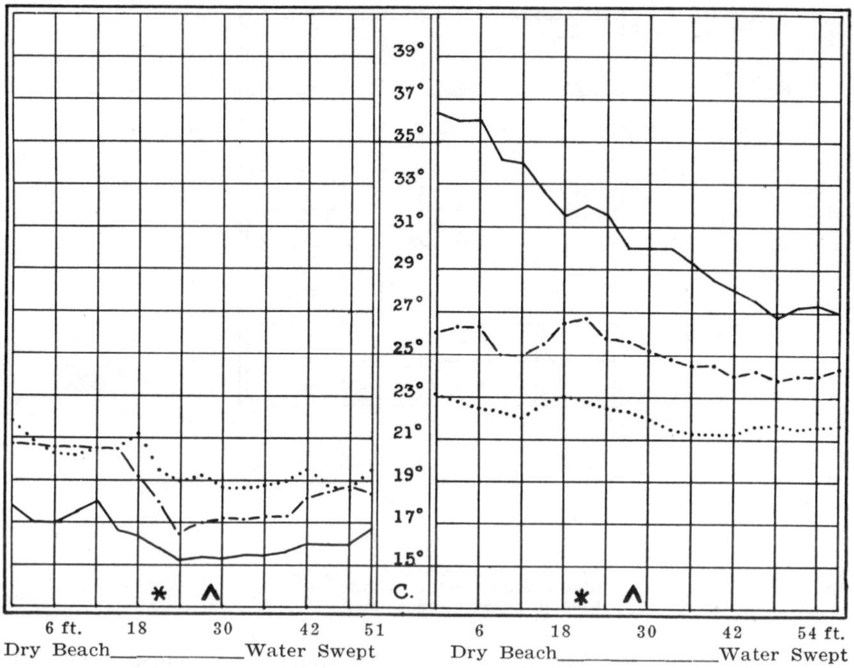

Fig. 4. Temperatures of beach across high tide line. Left hand of figure: 11 p.m. May 13, 1919. Air temperature 15.2° C. Water 17.9° C. Two hours after high tide of 5.6 feet. Right hand of figure: 4.30 p.m. May 14, 1919. Air temperature 18.5° C. Water 24.0°. Seven hours after high tide of 3.6 feet, seventeen hours after tide of 5.6 feet.

At surface: ————
At 3 inches down: .——.——.
At 6 inches down: ....
Line of dry surface sand: ✶
Line of loose sand: Λ

the dry and the loose sand is the area within which the eggs, laid during the slightly higher subsequent tides, were expected to be. In Figure 5 the differences between the readings at each post are shown, revealing the rather surprising fact that at the three- and six-inch levels the fluctuation was greater near the line of loose sand than at any other part of the beach. It would seem from this observation that the eggs are laid in an area unfavorable in so far as evenness of temperature is desirable, although the depth in the sand avoids the great range of temperature found at the surface. In the spawning area there were, therefore, considerable variations in such conditions as temperature and amount of moisture present, to which the eggs were adapted. It would seem that a lower part of the beach would present

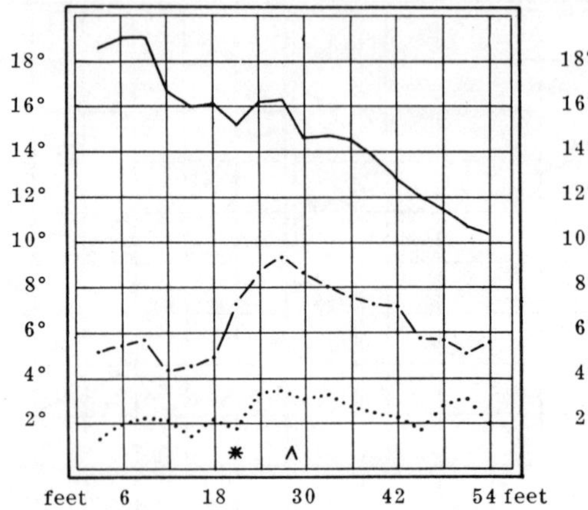

Fig. 5. Range in terperature between measurements of Figure 4, showing amount of fluctuation during one day across beach.

At surface: ————————
At 3 inches down: .——.——.
At 6 inches down: . . . .
Line of dry surface sand: ✶
Line of loose sand: ⋀

more even conditions, but there is, as will shortly be seen, a very convincing reason in the mechanism of liberation against the utilization of such parts.

When the pods of eggs were first found, there arose immediately the question of the manner of their escape. In order that direct evidence

The Escape of the Larvæ.

might be obtained as to this, five pods were dug up, placed in bags made of bobinette, with sufficient sand to weight the bags well, and reburied in the localities and depths from which they were taken. Another was reburied just before the following high tides. It was confidently expected that the eggs would work themselves free from sand, in the same manner in which they were separated from the sand in the laboratory. When eggs mixed with sand were placed in sea water and the whole agitated, the eggs, because of their specific gravity being less than that of the sand and more than that of the water, collected on top of the sand and could be picked off. Hence, the reburied eggs were expected to collect at the upper side of the bags. However, when the very high tides of the last of April had passed (see Figure 6), it was impossible to find all

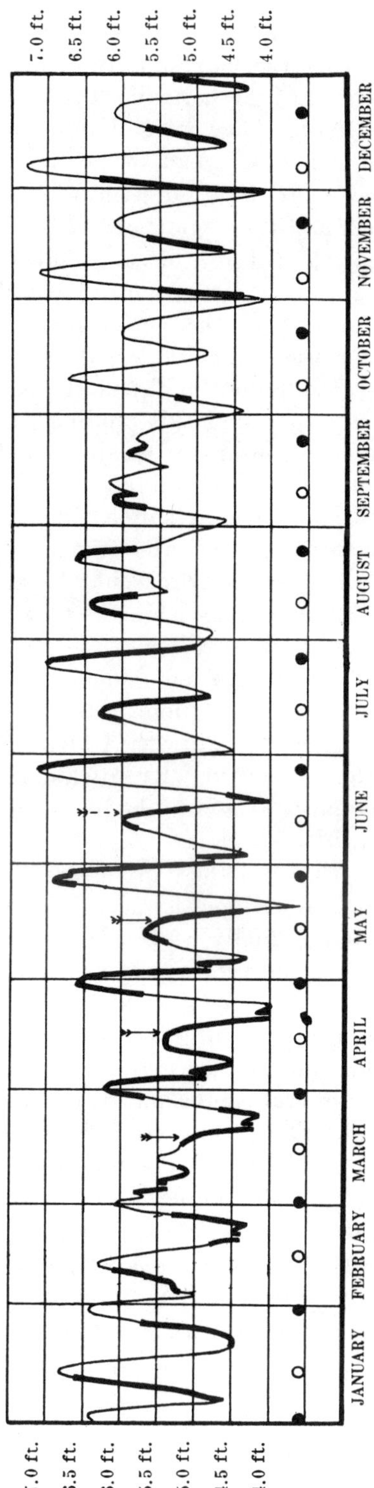

Fig. 6. Highest levels reached by tides each day at San Diego. Those occurring at night shown by heavy lines, those during the day by light lines. The phases of the moon indicated. The observed spawning times of *Leuresthes tenuis* indicated by arrows. Year 1919.

the buried bags. Four of them had been swept out to sea, and in the other two, picked up during the height of the tide before they were swept out, the eggs were still at the bottom. To confirm these observations eggs were dug each day. They were obtained at the usual depth until the day after the first high tide had swept over them. Then they were found, still in their pods unmixed with sand, at a depth of an inch and a half or an inch and three-quarters. Plainly the presence of water more than sufficient to saturate the sand and seriously disturb it had proved unequal to the task of liberation. The following day, after another tide had swept over them, there were no eggs whatever to be obtained. The conclusion seems inevitable that the eggs, when they were liberated, were literally dug out of the sand.

This, of course, did not prove that this eroding action of the waves was invariable, and a sufficient means of escape. A direct observation of the action of the waves and tides was all that could settle such a question. To observe the action of a single tide, a double series of twelve posts were driven into the sand three feet apart across the beach from the highest point which the waves would be expected to reach.[1] These were driven in until their tops were thirteen centimeters above the sand. The position of the sand below these posts was read every fifteen minutes from 6.15 p.m. (7.15 p.m. summer time) to 10.15 p.m. (11.15 p.m. summer time), the time of high tide being about 8.10 p.m., and its height 5.6 feet, according to the United States Coast and Geodetic Survey tide tables for 1919. By calculating on the basis of the original height of the posts as a straight line, a cross section of the beach was made for each fifteen-minute interval. Tables 6 and 7 give the data obtained, and in Figure 7 these are shown graphically for one of the series of posts. The erosion is shown in stipple, while the deposited sand, as well as the finally replaced sand, is shown in black. In order to give a record of the progress of the tide, the average post reached by six successive waves during each time interval is shown by the circles. It will be seen that there was a well-marked area in which the waves deposited sand, and another in which erosion predominated. There was, as a result, a gradual encroachment of the eroded area until the height of the tide had passed, while sand was deposited at a higher level. As the tide receded the deposited sand was left and the eroded area was partially refilled. There is, therefore, an area of erosion, and a depositing lip, as one may term it.

The area of erosion is characterized by a steep slope, and by gullies, sometimes six or eight inches deep. It may therefore be easily distin-

[1]These observations were made in conjunction with Professor F. W. Weymouth of Stanford University, who was interested because of the possible bearing of the phenomenon on the clams and crabs he was engaged in studying for the Fish and Game Commission.

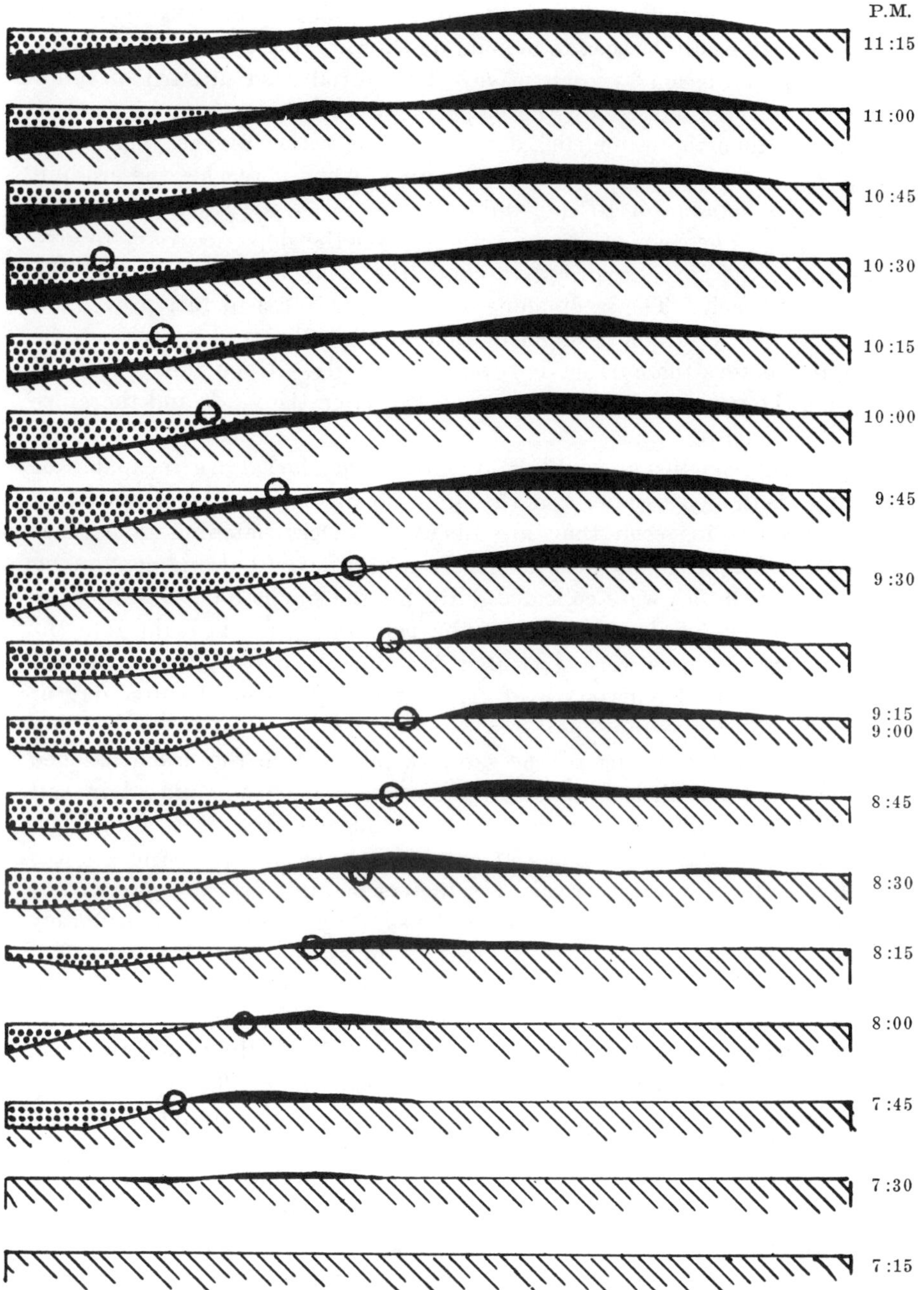

Fig. 7. A series of cross-sections of the beach, the ocean on the left-hand side, to show erosion and replacement by the tides. Taken at fifteen-minute intervals during a high tide of 5.6 feet on May 13, 1919, at Long Beach. Erosion indicated by dotted area, deposited or refilled sand by solid black, and average extent of wave wash by circles. Basis of measurement, the profile of the beach before the tide, shown as a straight line at 7:15 (summer time).

guished, and its invariable presence demonstrated by observation. The process may easily be seen to depend on the carrying power of the water swept up the beach by each wave. The initial rush upward of sand-filled water is stopped by the slope of the beach so that the sand is dropped; then the water runs down the slope gathering carrying power as it goes and increasing in quantity, until, when it reaches the eroding area, it is capable of lifting and carrying away the sand it brought, and more. The coarser the sand, the steeper the slope necessary to give the water the speed required to carry the sand and the narrower the resultant beach. The steep slope of the eroded area in fact prevents a pause of the water for a sufficient time to allow the sand it carries to be dropped before the current is reversed strongly. Upon this process of erosion the grunion depends for its escape from the sands and therefore for its existence.

But this conclusion could hold only if the larvæ are incapable of aiding their own escape, and the question arises as to what condition the eggs are in when they are liberated. Does hatching take place before or after? This question received an unexpected and convincing answer. Samples were collected daily for the purpose of obtaining the various stages of development of the fish, and when brought into the laboratory the eggs were separated from the sand by agitating in salt water. On the twenty-sixth of April, ten days after the first day of the preceding spawning, and one day before the first of the succeeding high tides, a large part of the samples taken from the beach hatched while being handled, four or five minutes after being placed in salt water. The samples, five in number, collected the succeeding day, hatched to the last fertile egg after a similar period of exposure to salt water. That they were ready for their scheduled escape could not be gainsaid. The eggs, washed out of the sand by its erosion, immediately hatch, and the vigorous larvæ are free to escape into the surf. The period of development is adapted to the tides.

However, if the larvæ are thus ready for their escape, it is possible that when the sand around them is moistened and lightened, they can hatch and make their own way into the open water. The truth of this was easily determined, by placing a pod with its surrounding sand in a dish and flooding it with water. There were a number of eggs—perhaps 30 per cent of the total—which were left on the surface of the sand. These were picked off and placed in a separate dish, where they quickly hatched. Of the others, one lot, buried in the sand, was left over night, then separated from the sand. None hatched before being freed. Another lot, left an hour buried in wet sand, hatched only when released from the sand. In still another lot, small numbers of eggs were freed at intervals, when they hatched; but those buried awaited their liberation. The eggs in all these cases were frequently visible among the grains of sand, being but partially buried. The experiments were repeated many times in the process of preserving samples of eggs, with the same results each time. Complete freeing was necessary, the larvæ not aiding their own escape in any way whatsoever. In other words, they were completely dependent on the action of the waves for their existence.

Knowing these conditions under which the eggs live and are liberated, it is possible to return to a consideration of the conditions under which they are laid, and the adaptations made to meet the circumstances. A consideration of the runs in relation to the waves; to the individual tide; to those of succeeding days; to the series of tides accompanying the phases of the moon; and to the season, will be taken up in order and will bring out the details.

The Adaptation of the Spawning to Physical Phenomena.

The habit of the fish of running in with the highest waves is plainly adapted to allow them to reach the area in which the sand is being deposited. The females caught were invariably laying their eggs at the upper edge of the eroded area, where the sand was soft and easily penetrated. To reach this during either the full or the ebbing tide, the heavier waves were necessarily utilized. If the eggs had been laid at a lower level the same or succeeding tides would have swept them away. Laid where they were, the sand was actually deposited over them, and in this circumstance is found the explanation of the discrepancy between the depth at which the pods were found and the size of the female. But the certainty of being swept out to sea also existed if the earlier parts of the tides were used. The eggs, buried in the upper edge of the area of erosion, would be unearthed by its advance. However, it has been repeatedly observed by the writer and others—indeed it is common knowledge—that the run does not begin until the tide has reached its height, and continues during the first stages of its ebb. Observations thus far made by the writer show that the run lasts about an hour, beginning near the time of high tide. This is given in the following table, showing the approximate duration and times of the runs of the grunion.

TABLE 2.

Tide beginning on	Run beginning	Run ending	Time of high tide
April 18	10.30 p.m.*	11.40 p.m.	10.15 p.m.
May 16	9.15 p.m.	10.00 p.m.	9.25 p.m.
May 17	10.00 p.m.	11.00 p.m.	9.54 p.m.
May 18	10.20 p.m.	12.01 a.m.	10.25 p.m.

*Add one hour for summer time.

The data given in this table are, however, somewhat unreliable. The runs were not at all continuous, and were found to be made up of various small runs, some of which came in on but parts of the beach, making it extremely difficult to be sure that all were observed. Furthermore, isolated pairs were observed to come in long after the larger schools seemed to have departed. That the times given are approximately correct was judged by the actions of the crowds of "fishermen," who were thrown into great activity by the start of the run and departed upon its conclusion. In general, it may be said that the runs began about the turn of the tide and continued for about an hour. This allowed the depositing of the eggs when the eroding area had reached its greatest extent (see Figure 7) and in a region which would not be reached by it in succeeding tides. Eggs laid an hour too early would have been laid at too low a level and would have been washed out; while eggs laid two or three hours after the tide had turned might also have been laid too low and have been washed out by the succeeding tide. That the period of each run is thus approximately timed admits of no doubt, however difficult it may be to explain the exquisite nicety of the adaptation.

But adaptation must also have a relation to the varying height of the tides from day to day, as well as to the height of the individual tide. It would avail little to have the eggs laid where succeeding tides would wash them away during the extension of the eroded area. Just as the spawning must occur at that stage of the individual tide at which the ebb commences, so must the spawning runs occur at those times when the succeeding tides become lower. The highest tides occur at about the full, or the dark of the moon, hence these lower tides *follow* the full of the moon, or the dark.

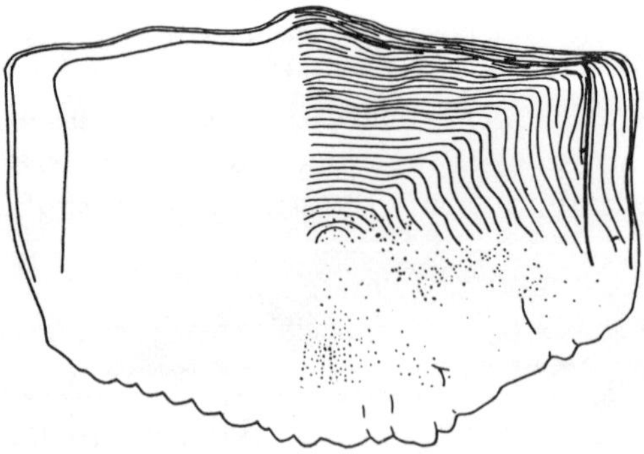

Fig. 8. Scale of *Leuresthes tenuis*, the Grunion. Female 17.1 cm. long taken April 17, 1919, at Long Beach, California. Showing what seems to be two winter marks, one of which is at the very edge.

And in the adaptation to this sequence of the tides, is seen the explanation of the fact that the runs occur after the full moon. Mr. Joplin said: "On the second, third, and fourth nights after the full moon." However, in so far as the present limited observations go, this is only approximately correct. Thus the moon was full April 15, at 8.25 a.m., Greenwich mean civil time, or about 12.30 a.m. our local time. The fish began to run April 16 at 9.28 p.m. (10.28 p.m. summer time), therefore the second night, and the second tide of the series after the full moon. Again, the moon was full May 15, at 1.01 a.m. Greenwich mean civil time, or May 14, at about 5.00 p.m. local time, whereas the grunion ran at 9.25 p.m. (10.25 p.m. summer time) on the sixteenth. This was the third night, and the third high tide of the series after the full of the moon. In both cases the grunion ran on the sixteenth, seventeenth, and eighteenth, three days. Therefore, probably the only exact statement which can be made is that the spawning runs come shortly *after* the full of the moon, in other words *after the highest tide* of the series, which is the really significant fact. The popular saying is merely an attempt to state exactly a very remarkable fact.

To illustrate the varying effect of the advancing and receding tides of a series, measurements of the depth of the sand were taken along a line of posts transverse to the beach, and also along a line parallel to the beach. The former were at three-foot intervals, post 1 being above high tide. The results are shown in Table 3. Measurements were taken at

TABLE 3.

Showing Erosion and Replacement of Sand by Successive Tides. Also Contrasting Erosion Shown at Full Tide With That Left After Its Ebbing. Line of Posts Across Beach for Measurements.

Height of last tide covering posts, ft.	6.2	6.5	6.5	6.6	6.6	6.5	6.2	5.6
Date	4/29	4/29	4/30	4/30	5/1	5/2	5/3	5/4
State of tide	Low	Full	Low	Full	Low	Low	Low	Low
	cm.	cm.	cm.	cm.	cm.	cm.	cm.	cm.
Post 1	0	—8.9	—5.7	—10.5	—8.3	—3.8	0	0
Post 2	0	-------	-------	-------	-------	-------	-------	-------
Post 3	0	—11.4	—7.6	—11.4	—10.8	-------	-------	-------
Post 4	0	—10.2	—6.0	-------	-------	-------	-------	-------
Post 5	0	-------	-------	-------	-------	-------	-------	-------
Post 6	0	—14.0	-------	-------	-------	-------	-------	-------
Post 7	0	—10.8	—6.4	—12.7	—8.3	—10.2	—5.1	0
Post 8	0	—13.0	—5.7	—12.0	—5.7	—8.9	—4.5	+2.5
Post 9	0	—10.8	—5.1	—10.8	—5.7	—8.9	-------	-------
Post 10	0	—12.1	—4.5	—8.9	—3.8	—6.4	—4.4	+3.8
Extra	0	—10.2	—6.4	—9.5	—8.3	—7.6	—3.2	—1.3
Average	0	—11.3	—5.9	—10.8	—7.3	—7.6	—3.4	+1.0

full tide, when the waves were washing over the posts, and at low tide, when the posts were dry. There was always a certain amount of replacement of sand by the ebbing tide, as for instance is shown by the average erosion of 11.3 centimeters (4.5 inches) by the tide on the twenty-ninth, and its replacement to a depth of 5.9 centimeters, as shown by measurements next day. The last tide which reached the posts, as measured on the fourth, had almost entirely replaced the eroded sand. It will be noted that an average of 11.3 centimeters was reached on the twenty-ninth, which was a depth equal to or exceeding the depth to which the

TABLE 4.

Showing Erosion and Replacement of Sand by Successive Tides. Also Contrasting Erosion Shown at Full Tide With That Left After Its Ebbing. Line of Posts Parallel to Beach.

Height of last tide covering posts, ft.	6.2	6.5	6.5	6.6	6.6	6.5	6.2
Date	4/29	4/29	4/30	4/30	5/1	5/2	5/3
State of tide	Low	Full	Low	Full	Low	Low	Low
Average	0	—6.5	—2.9	—7.5	—7.1	—2.8	+0.6
	cm.	cm.	cm.	cm.	cm.	cm.	cm.
Post A	0	—8.9	—3.2	—8.3	—7.6	—2.5	+3.2
Post B	0	—3.2	—3.8	—8.9	—8.9	—2.5	0
Post C	0	—3.2	—1.9	—6.4	—5.7	—2.5	0
Post D	0	—5.1	—1.9	—5.7	—5.1	—1.6	+1.3
Post E	0	—5.7	—1.3	—8.3	—7.0	—2.2	0
Post F	0	—8.3	—6.3	—10.2	—10.2	—5.1	—0.6
Post G	0	—11.4	—1.9	—4.4	—5.1	—3.2	0

eggs had been buried, and if more care had been taken to find the maximum, a very much higher average would have been attained. Unfortunately several of the posts were swept away, although 50 centimeters in length. A similar erosion and replacement was shown by a line of posts parallel to the beach, but distant about fifteen feet from each other. (See Table 4.) These posts, which were remnants of an old breakwater, were not placed low enough on the beach to show the

TABLE 5.

Showing Amount of Erosion at Various Stages of Tide.

	Line of posts parallel to beach		Line of posts across beach	
Date in 1919	4/29	4/30	4/29	4/30
	cm.	cm.	cm.	cm.
Average before tide	0.0	—2.9	0.0	—5.9
Average during tide	—6.5	—7.4	—11.3	—10.8
Average after tide	—2.9	—7.1	—5.9	—7.3
Maximum during tide	—11.4	—10.2	—14.0	—12.7

erosion to its full extent, but the measurements obtained illustrate and corroborate the facts brought out by the other series. To bring these together in a more striking fashion, the data have been placed in Table 5. The phenomenon of erosion and replacement is obvious from these measurements, the extent naturally varying according to the weather and the resultant size of the waves, which were small at the time.

We may therefore conclude with some assurance that the eggs of the grunion are laid at times which allow them to escape the great danger of coming within the area of advancing erosion, and it is clear why the coming of the spawning runs subsequent to the highest tides of a series should serve the survival of the species.

But this does not conclude the matter. For, furthermore, the utilization of the tides during the full of the moon means the use of the lower series of tides which occur during the spawning period. The tides accompanying the dark of the moon are all very much higher (see Figure 6) and carry the area of erosion much farther up the beach, therefore uncovering all the pods left under the deposited sand by the preceding lower series. This occurs two weeks after the spawning at the precise time when the eggs are ready to hatch, and is clearly highly advantageous to the species.

The explanation of the apparent failure to make use of the higher series of tides for spawning may exist in the fact that eggs laid during one of the higher series might not be reached for a month or more by following tides, and then only with the sand depositing lip shown in Figure 7. In Figure 6 the greatest height reached daily is shown for each series of tides, and an alternation of lower series during the full moon with a higher series during the dark of the moon is to be observed during March, April, May, June, and July. Eggs laid during the dark of the moon would therefore risk lying in the sand for the time elapsing between the higher series, at least a month, and in May or June they might be left for a full year because of the habit of laying the eggs very near the upper limit reached by the waves. But in so far as personal observations show, there are no runs during the dark of the moon. Only during the tides when it is full are the eggs laid. The high tides of the last of April were observed with this in view, but not a single grunion was found on the beach, despite the fact that the preceding and the succeeding tides, during the full moon, were characterized by abundant runs. This corroborates in part the words of Mr. Joplin previously quoted: "On the second, third, and fourth nights *after the full moon.*" Hearsay evidence was obtained in one case concerning fish taken at the dark of the moon, but as it would not bear critical examination, it could not well be accepted. In so far as actual observation is concerned, then, the fish run only during the full moon, and until further proof of runs during other periods is obtained this conclusion should hold. It is, as it now stands, a very remarkable one, implying as it does a migration discriminating between series of tides.

There seems no obvious reason, however, why eggs could not be laid during the tides coming in the dark of the moon, if the run should be during a very late part, when the series was receding and the eggs could be laid where the eroding area of the succeeding lower series (ten days later) could reach them. If they were laid at a higher level they would, as has been observed, be compelled to await the tides of a month later, or perhaps, if laid in May or June, fail altogether of liberation. They would thus be exposed to a much greater danger of falling prey to terrestrial enemies, because of the much longer incubating period, and the greater accessibility to those enemies of the higher part of the beach. The temperature changes and the extent of evaporation in the surrounding sand would also be very much greater. But if laid at a low level, there would in reality be but a single tide of the series in which this could be done, because of the greater rapidity of the fall. Although it is not very obvious why this opportunity should not be utilized, nor indeed is there absolute proof on hand that it is not, yet it is plain that the runs occur for the greater part during the most advantageous tides.* These facts, even so far as we are able to state them confidently, are remarkable instances of adaptation, and habits altered to fit the varying character of the tides more closely than is at present obvious would be not less, but more remarkable.

Such adaptations may be possible only during a certain season of the year, for the tides seem capable of accommodating such habits only during the months between March and July inclusive. To show the

---

*Since writing the above, grunion were taken in small numbers during the tides beginning on the first and second of June.

The dark of the moon occurred on the twenty-ninth of May. On that night, and on the thirtieth and thirty-first of May, careful watch was kept over a mile and a half of beach for a period of an hour and a half after the scheduled time of high tide; but no sign of grunion was seen. The tide on the twenty-ninth of May was 6.7 feet at San Diego, the point of reference for our locality; on the thirtieth, 6.7 feet; on the thirty-first, 6.3 feet; on the first and second of June (when the grunion were taken), 5.7 feet and 5.0 feet, respectively. The latter two tides approximated the height of the tides at the full of the moon, which were 5.4 feet on April 16; 5.7 feet on May 14; and are predicted by the United States Coast and Geodetic Survey as 6.0 feet on June 13. The tide on the first of June, of 5.7 feet, was therefore the one during which a run of grunion could have been expected, according to our reasoning, and the eggs laid at that time will probably be freed as larvæ by the next full moon tides.

However, the number of grunion observed was very small. The fish were seen at 12.30 a.m. and at 12.40 a.m. during the tide commencing June 1 on two different parts of the beach, and in each case but a single small school was observed. During the tide commencing June 2, but four scattered individuals were seen at the very height of the tide, despite prolonged search. It is perfectly possible that such small runs could escape attention and caution must therefore be used in concluding that no fish spawn on the other tides during the dark of the moon. This does not seem probable, aside from our direct observations, for during the earlier tides of each series the beach is as a rule occupied by people passing or having beach parties, and any run of fish would be quickly noticed and the news spread. The conclusion that the main run *does* occur during the full of the moon seems therefore entirely probable.

possible reason for this, Figure 6 has been drawn. As has been previously mentioned, the highest tides on each day are shown on the chart, those coming after nightfall being indicated by a very heavy line, those coming during the day by a light line. The grunion, coming in to spawn on wide sandy beaches, would be supposedly the natural prey of many enemies such as sea gulls if the run occurred during the day, and in their utilization of night tides they may have become adapted to meet this circumstance. If so, the tides used must not only be the receding ones in each series, but must also come at night. Such tides are peculiar to the months, March to August, inclusive.

The validity of this reasoning is, however, open to some question, as it is very difficult to prove that daylight spawning is impractical. Even though no such reason exists, the fact that the spring months are the natural spawning months of other species of smelts (Atherinoids) as well as of many other species of fish, predisposes the observer to conclude that such might have been the season for the grunion from the beginning. If so, the use of the receding tides to avoid the eroding of the advancing tides would require use of night tides at that season. That such is probably advantageous is evident enough, however. It will suffice, then, to conclude that the season utilized by the grunion is remarkably well adapted to its purposes.

Reasoning in regard to the mechanism of the adaptation of the grunion to the tidal action is plainly enough a futile task. The migrating impulse does not arise in response to the light of the moon, for during the last spawning run in May, heavy clouds obscured the moon through each night. Beyond this no data are at hand to utilize in speculation.

The perfection of these adaptations, or in other words, the invariable success of the spawning is open to serious question. An example of an imperfection was found in the spawning run of May 16 to 18. The eggs were laid during calm weather, when the waves washed up the beach less than is usually the case, and the spawning therefore occurred low on the beach, especially on the eighteenth. But on the nineteenth to the twenty-second heavy swells prevailed and the area of erosion rose farther up the beach than the area in which the eggs were laid, as was ascertained by measurement from fixed posts. The consequences were what might have been expected, that very few eggs from the spawning on the eighteenth remained in the sand, a fact proved by extensive digging. This did not, of course, mean the actual destruction of the eggs, in so far as present knowledge goes, but did mean that whatever advantage the grunion gained in the protection of its eggs was partially lost, and that they were forced to endure the same rate of mortality that

Imperfections in Adaptation.

fish with less perfected habits but far greater fecundity undergo. The eggs of the grunion, few in number, are obviously far less able to survive the energetic attacks of the multifarious enemies which might prey on them in the open sea than the eggs of such fish as the cod, numbering into the millions.

Such deviations from the average course of events must be disastrous to the eggs of the grunion, the more so the more precise the act of spawning required to fit into the natural phenomena concerned. Thus the spawning of the grunion during the dark of the moon might necessitate a run on a single night, and a marked deviation in the tides either on that night or the following would prove destructive in the extreme. The spawning run at the full of the moon occurs two or three days after the approximately greatest height of the tide has been reached, and just previous to the sharp fall in height of the series. This timing of the runs eliminates, of course, the danger of unusual fluctuations during several tides of nearly equal height, when a strong wind might carry the waves excessively high and hence sweep out to sea all the deposited eggs. In other words, the lateness of the run during the series of tides, in losing the use of the series of nearly equal highest tides (sometimes four or five in number), also avoids the danger that an unusual height in the last of them would destroy the deposited eggs. It is plain that such great natural fluctuations are the greatest enemies the grunion eggs must meet, being to them in the nature of catastrophes, and the timing of the runs appears to be an adaptation to meet them.

But the grunion egg is not without ability to survive one type of natural "catastrophe." That it is subject to being washed out to sea is easily understood, and the reverse of that, a failure to be liberated at the proper time, would seem to be expected occasionally, especially if the eggs are laid at an unusual distance up the beach, and are covered too deeply by the deposited sand. To test their ability to withstand burial for a month, five pods of eggs were dug out of the beach on the twentieth of April, kept until after the spawning runs in May, then washed out of the sand. These eggs were laid in the runs on the sixteenth, seventeenth, and eighteenth of April, hence were five days more than a month old; but a large percentage hatched (respectively, 10 per cent; 30 per cent; 30 per cent; 50 per cent; 20 per cent). These larvæ were all ready for hatching on the twenty-eighth of April, as was ascertained by hatching some from each lot. The larvæ of the grunion, therefore, after reaching the proper stage for hatching, are capable of lying quiescent in their capsules for fully two weeks, awaiting the proper conditions under which they may escape. This is an unusual phenomenon, perhaps something which no other species is known to be capable of.

At the same time, this halt in the normal development is made at great expense. The larvæ which hatched when first ready—in time for the first high tides after spawning—gave a very high percentage of swimming young, while those kept for a prolonged period gave a very much lower one. The delayed larvæ were also very much less vigorous, hatching out slowly during a period of three-quarters of an hour, whereas four minutes sufficed for the normally released larvæ to escape. It is impossible to avoid the conclusion that a prolongation of the egg stage is highly injurious, and surely results in the failure of a great many of the young to survive. In this must be found one of the disastrous results avoided by the habit of spawning during the full moon tides and not during the higher tides of the dark of the moon. Eggs

Fig. 9. Beach showing erosion of sand by gullies, at Long Beach, May 23, 1919. The tide at the dark of the moon has left debris near the foot of the cliff, while that at the full of the moon has left the line of debris along the edge of the dark (moist) sand, above the area of erosion.

laid during the full moon tides would be liberated in two weeks by the succeeding very high tides, whereas those laid during the dark of the moon at the height of the tides, would be compelled to await the tides a month later, with the added danger that even those might not reach them.

The story concluded by the escape of the larvæ into the surf comprises one of the most marvelous of the many strange chapters in the life histories of fish. It is eloquent of exquisite adaptations to a seemingly minor physical phenomenon—the erosion of the beach by one part of a wave, and its upbuilding by another. Those waves are made use of which carry the fish high on the beach; those parts of the tide are utilized which allow the pods of eggs to be laid without risk that further rise of the tide will carry the eroded area over them; the runs occur during those tides in a series which are the last available, because of their height, thus eliminating in so far as possible the danger that unusually rough weather will sweep away the eggs; and the eggs are laid during those series of tides which will allow of the escape of the larvæ two weeks later. The larvæ themselves do not hatch until the sand over them is swept away, even though a month pass by; but when the time comes and the waves of the high tides wash over them, eroding the surrounding sand, they are ready to escape. The eggs are laid in what is nearly dry land, or moist soil, a fact which has no parallel in our knowledge of marine fishes. They are subject to attack by terrestrial enemies, but escape the far more numerous marine enemies. Indeed, so advantageous to the species are these adaptations that the female lays thousands of eggs where millions are laid by other forms. But perhaps the most remarkable fact is the dependence of the existence of a species on a phenomenon so unobtrusive as the shifting of the sand on the seashore.

TABLE 6.

Series of Measurements of Height of Fixed Posts Set at Three-Foot Intervals Across Beach, Parallel to Posts of Table 7, to Show Erosion and Replacement by the Tide. Taken at Fifteen-Minute Intervals During a High Tide of 5.6 Feet on May 13, 1919, at Long Beach. The Deviation of the Tops of the Posts From a Straight Line Given in Table 8.

SERIES A.

	1	2	3	4	5	6	7	8	9	10	11	12
7.15 ------	13.0	13.0	13.0	13.0	13.0	13.0	13.0	13.0	13.0	13.0	13.0	13.0
7.30 ------	13.0	13.0	13.0	13.0	13.0	13.0	13.0	12.4	12.8	13.7	13.2	13.5
7.45 ------	13.0	13.0	13.0	13.0	13.0	13.0	12.9	12.2	12.0	13.6	16.8	16.5
8.00 ------	13.0	13.0	13.0	13.0	13.0	13.0	12.5	11.5	12.5	14.0	14.0	16.4
8.15 ------	13.0	13.0	13.0	13.0	12.7	12.5	11.6	12.0	13.7	14.5	15.5	14.5
8.30 ------	13.0	13.0	12.8	12.9	12.3	11.8	10.6	12.0	13.5	15.3	17.2	17.7
8.45 ------	13.0	13.0	12.1	12.3	11.5	11.6	13.0	14.0	14.1	15.8	17.4	17.3
9.00 ------	13.0	12.9	12.0	11.9	11.0	11.7	13.7	13.3	14.8	17.1	17.2	16.8
9.15 ------	13.0	12.9	12.0	11.5	10.6	11.5	12.9	13.6	15.3	16.8	17.7	17.6
9.30 ------	13.0	12.9	11.5	10.9	10.3	11.6	13.0	14.4	15.6	17.0	16.8	19.7
9.45 ------	13.1	12.7	11.5	11.0	10.3	11.6	12.6	14.0	14.6	16.4	18.0	19.6
10.00 ------	13.1	13.0	11.7	11.2	10.5	11.9	12.7	13.4	14.0	16.7	17.5	18.0
10.15 ------	13.1	13.0	11.8	11.2	10.5	11.8	12.6	12.9	14.4	16.4	17.2	18.1
10.30 ------	13.2	13.0	11.6	11.2	10.5	11.8	12.6	12.5	13.7	15.0	16.5	16.8
10.45 ------	13.2	13.0	11.6	11.1	10.5	11.8	12.6	12.6	13.7	15.1	16.0	16.1
11.00 ------	13.2	13.0	11.6	11.1	10.6	11.8	12.5	12.5	13.6	15.0	16.1	16.3
11.15 ------	13.2	13.0	11.6	11.1	10.6	11.8	12.5	12.5	13.6	15.0	15.9	16.3

TABLE 7.

Series of Measurements of Fixed Posts Similar and Parallel to Those of Table 6. This Differs From That of Table 6 in Presence of Gully From Post 6 Outward.

SERIES B.

	1	2	3	4	5	6	7	8	9	10	11	12
7.15 ------	13.0	13.0	13.0	13.0	13.0	13.0	13.0	13.0	13.0	13.0	13.0	13.0
7.30 ------	13.0	13.0	13.0	13.0	13.0	13.0	13.0	13.0	13.0	13.0	11.8	12.0
7.45 ------	13.0	13.0	13.0	13.0	13.0	12.8	12.3	12.1	12.9	13.0	14.0	15.9
8.00 ------	13.0	13.0	13.0	13.0	13.0	12.7	12.2	11.4	12.0	15.0	14.2	13.8
8.15 ------	13.0	13.0	13.0	12.8	12.9	12.7	11.6	11.6	13.7	15.2	17.0	14.3
8.30 ------	13.0	13.0	13.0	12.5	12.6	12.4	11.0	12.8	15.0	18.8	17.3	17.0
8.45 ------	13.0	13.0	13.0	12.3	12.0	11.6	11.0	13.6	17.3	17.9	17.0	16.0
9.00 ------	13.0	12.8	12.1	11.2	11.0	12.1	12.5	17.2	17.6	19.6	18.2	15.7
9.15 ------	13.0	12.6	11.8	10.6	10.5	12.2	15.8	16.7	17.9	17.3	17.2	20.0
9.30 ------	13.0	12.5	11.5	10.0	10.5	13.3	19.2	19.0	19.3	20.0	20.1	17.3
9.45 ------	13.1	12.4	11.2	9.8	10.5	13.8	17.0	18.1	19.1	20.6	20.9	19.8
10.00 ------	13.2	12.6	11.3	9.9	10.5	13.9	16.8	16.8	18.1	19.2	19.7	18.3
10.15 ------	13.3	12.4	11.3	9.8	10.7	14.8	17.3	16.8	18.5	19.8	20.0	18.0
10.30 ------	13.2	12.6	11.2	9.8	10.6	14.5	17.1	16.5	18.1	19.3	18.8	17.6
10.45 ------	13.3	12.6	11.2	9.8	10.6	14.6	17.0	16.6	18.1	19.0	18.1	17.2
11.00 ------	13.3	12.6	11.2	9.8	10.7	14.8	17.0	16.5	18.2	19.0	18.3	17.5
11.15 ------	13.3	12.6	11.2	9.8	10.7	14.8	17.0	16.5	18.2	18.9	18.3	17.2

TABLE 8.

Deviation of Tops of Posts Utilized in Making of Tables 6 and 7, From a Straight Line. Data to Be Used in Correcting Profile of Beach.

	Series of Table 6 (cm.)	Series of Table 7 (cm.)
Post 1	0.0	0.0
Post 2	—1.0	+1.5
Post 3	—1.0	+1.8
Post 4	+0.8	—0.2
Post 5	+2.9	+2.4
Post 6	+4.0	+5.2
Post 7	+3.3	+3.5
Post 8	+1.2	+1.9
Post 9	+1.2	+2.2
Post 10	+1.2	+2.3
Post 11	+2.3	+1.0
Post 12	0.0	0.0

TABLE 9.

Number of Posts Reached by Waves During Taking of Series Given in Tables 6 and 7. The Average of Six Waves Given Was Utilized in Plate 7 After Smoothing Plotted Curve by Graphical Interpolation. Post 1. Highest.

GREATEST HEIGHT REACHED BY WAVES.

Wave	7.45	8.00	8.15	8.30	8.45	9.00	9.15	9.30	9.45	10.00	10.15	10.30
1	7.5	11.5	8.5	8.5	8.0	6.0	8.5	7.5	8.5	7.0	10.0	11.0
2	9.0	9.0	9.0	3.5	10.5	5.5	7.5	3.5	8.5	7.5	10.5	8.5
3	8.5	9.0	8.0	5.5	2.5	7.5	7.5	10.5	8.0	8.0	9.0	12.0
4	10.5	10.0	9.0	9.0	9.0	7.5	6.0	9.5	11.0	8.5	12.0	11.5
5	7.0	8.5	10.5	7.5	5.5	4.5	9.5	8.0	6.0	10.0	9.5	10.0
6	11.5	10.0	10.5	6.0	3.0	10.5	3.0	8.0	4.0	7.5	7.5	11.0
Av.	9.0	9.7	9.3	6.7	6.4	6.9	7.0	7.8	7.7	8.1	9.8	10.7

3

Reprinted from pp. 1-14 of *Technical Report No. 2,* Beach Erosion Board, U.S. Army Corps of Engineers, 1942, 43 pp.

A SUMMARY OF THE THEORY OF OSCILLATORY WAVES

M. P. O'Brien

LIST OF SYMBOLS

Y = A vertical displacement of water particles.
X = A horizontal displacement of water particles.
t = A specific time.
L = Wave length.
C = Wave velocity.
T = Wave period.
g = Acceleration due to gravity.
d = Water depth.
τ = Surface tension.
ρ = Density of water.
z = Depth below still water level.
x = Displacement in direction of wave travel.
y = Surface ordinate.
h = Wave height from trough to crest.
a_z, b_z = Horizontal and vertical displacements of a water particle
E_p = Potential energy of wave per unit width of crest.
E_k = Kinetic energy of wave per unit width of crest.
w = Weight of water per cubic foot.
a = Half of wave height.
$\bar{U}$ = Velocity of mass transport.
ω = Vorticity.
R = Radius of rolling circle (Gerstner theory).
r = Radius of tracing circle (Gerstner theory), or polar coordinate.
θ = Angular displacement (Gerstner theory), or polar coordinate.
P = Power transmitted by wave.
C_Z = Group velocity.
μ = Absolute viscosity of water.
ν = Kinematic viscosity of water.
ϕ = Potential function.
α = A function of wave height, or an angle.
$m = 2\pi/L$.
n = An integral number.
U = Horizontal component of velocity of water particle.
W = Vertical component of velocity of water particle.
G = Volume of mass transport per unit width of crest.

Section 1. INTRODUCTION

Quantitative predictions of the effect of wave action in shore-line processes requires, first, statistical information on the wave height, length, period, and direction, and water depth, at each locality for each season; and second, a precise knowledge of the behavior of waves approaching a shore line through deep water, into and through shoaling water. This paper deals with the second phase only, presenting a summary of the theoretical treatments of oscillatory wave motion on a free water surface. It is limited to the type of wave encountered in shore protection problems. Much of the theory requires experimental verification, and mention will be made of the extent to which confirmation has been obtained, without, however, quoting experimental results.

The derivations of equations, where not given herein, are to be found in the references quoted.

Section 2. CHARACTERISTICS OF OSCILLATORY WAVES

All wave phenomena have some elements in common and it is important to establish at the outset the approximate boundaries of the subject to be covered by this summary. The term oscillatory has been used to emphasize the periodic character of the phenomenon and to eliminate the ideas associated with the term "wave of translation." The waves under consideration induce a steady state of oscillation which differs in many particulars from the transient state characteristic of a wave of translation. Tides and seiches are also periodic but are of such length in comparison with the water depth as to permit simplifications in analysis which are not valid for the shorter period surface waves to be considered herein. Ripples are periodic and are produced by the wind but are of no consequence in shore-protection problems representing an initial phase in the generation of wind waves. Solitary and seismic waves present an essentially different problem and are not considered here. The waves under consideration are periodic disturbances under the control of gravity and inertia and of such height and period as to break on a sloping shore line.

To be more specific, the waves to be discussed are those which break on the shore line, throw sand into suspension, and possibly generate littoral currents. Periods are usually of the order of 4 to 12 seconds along ocean shore lines but may be much greater. Heights vary widely with wind action, period, and bottom conditions, ranging from a few inches to many feet. Wave crests in deep water may be inclined at an angle with the shore line, parallel to the shore at the plunge point. Several wave trains of different periods and heights may simultaneously approach the shore from different directions, giving rise to a beat frequency in the breakers—this is the basis for the common observation that waves "break in sets." Under some conditions the wave energy may not be completely dissipated in the breakers and the plunge and an appreciable portion may return seaward in the form of a reflected wave. In all cases the observed wave pattern may be considered as the result of superposition of primary and reflected waves of different wave trains, each periodic and of approximately constant amplitude. Accordingly, the first step in the analysis of oscillatory waves in water is to study the behavior of single wave trains of uniform period and amplitude as they progress over horizontal or sloping bottoms.

Although the motion in a breaking or plunging wave is too complicated for present methods of analysis, the essential periodic character of the waves is retained and the motion is oscillatory, in the broad sense of the term, from deep water to the limit of uprush. It should be noted that the plunge of the wave is the final break before the uprush on the beach.

The creation of oscillatory waves is probably the result of wind traction on the water surface. Once disturbed, pressure differences develop on the leeward and windward sides of the crest and permit a direct transfer of energy from the wind to the water through the medium of normal as well as tangential stresses. The details of the mechanism by which a small wave builds up into a large one should be investigated experimentally, but for the present it is sufficient to note that energy is continually transferred from the atmosphere to the surface layers of the ocean and that the waves thus formed provide the mechanism which transfers this energy to the shore line. From theory and experiment it is known that ocean waves in deep water travel great distances without appreciable loss of energy by internal friction. It is evident that on the average the waves observed at any point along the shore will be more controlled by distant than local winds.

The form of oscillatory waves depends upon the ratio of height to length and upon the local wind effect. Very long, low swells from distant storms are almost, but not exactly, sinusoidal in deep water. As they are telescoped by the retarding effect of a shoaling bottom, the crests become sharper while the troughs flatten and the form is then approximately that of a trochoid. Short waves under the influence of local wind are usually assymetrical or may be trochoidal.

At the plunge point the combination of height, bottom slope, depth, and period reaches a limit for a stable wave form and the crest plunges.

Considering the progress of a single wave train of uniform height and length from deep water to the shore over a gradually sloping bottom, it seems reasonable to assume that the same period would be observed at all points along the path. This statement is by no means axiomatic, for waves often coalesce or break up into shorter waves; however, laboratory experiments on small waves moving over limited reaches show that the individual wave crests maintain their identity. The point raised is important in the application of wave equations to field conditions.

From the theoretical standpoint Airy concluded that the conditions of constant pressure and continuity could not be satisfied simultaneously on a sloping bottom, and from experiment it is known that an abrupt change in depth may cause waves to break up into shorter waves. In groups including a finite number of waves, as in the wave pattern around a ship, it is observed that waves at the front of the train disappear while new waves develop at the rear. A frequently observed similar phenomenon, where a wind wave can be followed for a considerable distance and then seems to disappear, probably results from the superposition of wave trains of different periods. The fundamental question is whether the component waves of each wave train maintain their identity. If they do, it follows that their period is the same at every point along the path even though the apparent period of the combined wave trains may vary from point to point.

The question of permanence of oscillatory waves has been the subject of much speculation and it is of interest to review some of the conclusions drawn. Rayleigh (1) stated that:

> * * * I think that the reader who follows the results of the calculations here put forward is likely to be convinced that permanent waves of moderate height do exist.
>
> It may be mentioned that most of the authorities * * * express belief in the existence of permanent waves, even though the water be not deep, provided of course that the bottom be flat.

Stokes came to a similar conclusion as follows:

> In fact it will presently appear that it is only an indefinite series of waves which possesses the property of being propagated with uniform velocity and without change of form; at least this is the case when the waves are such as can be propagated along the surface of a fluid which was previously at rest.
>
> It appears from the above, that of all waves for which the motion is in two dimensions, which are propagated in a fluid of uniform depth, and which are such as could be propagated in a fluid previously at rest, there is only one particular kind, namely, that just considered, which possesses the property of being propagated with a constant velocity and without change of form; so that a solitary wave cannot be propagated in this manner.

The bottom slopes encountered off sandy coasts are usually gradual and, when plotted to true scale, seem almost horizontal. Whether or not the theories applicable to a horizontal bottom may be applied to these very gradual slopes remains to be determined by experiment or direct observations. Though identity of individual wave crests seems to be maintained over a gradually sloping bottom even this feature should be investigated.

The relationship between period, length, and velocity characteristic of all wave phenomena is

$$L = CT \tag{1}$$

This formula is essentially a definition. Independent measurements of length, velocity, and period should agree with equation 1 exactly; and deviations therefrom should be considered as observational errors. Agreement with this equation, however, is not a confirmation of the dynamical theories of oscillatory waves.

Oscillatory waves on the surface of water are characterized by periodic variations in the motion of the water particles, pressure, and surface elevation, each quantity passing through a series of values and returning to its initial value at regular intervals of time. At intervals equal to the wave length measured in the direction of motion, each of these quantities will be equal to and in phase with its former value. From the standpoint of the mathematical analysis, the approximately constant atmospheric pressure on the water surface and the existence of appreciable vertical accelerations are of basic importance. The amplitude of orbital motion decreases exponentially below the surface and becomes very small at a depth equal to half the wave length and for most purposes the orbital velocity is inappreciable at even smaller ratios of depth to length. The effect of decreasing depth is a continuous function. Although some authors give a contrary impression, there is therefore no sharp line of demarcation between shallow-water and deep-water waves. The dividing point is purely arbitrary and is dependent upon the desired accuracy of computation. No abrupt change occurs when the depth is equal to half the wave length and the present arbitrary division of deep-water and shallow-water waves on the basis of the depth-length ratio is meaningless.

Present-day wave theory deals with waves of stable form in which all elements of the wave profile advance with the same velocity relative to the undisturbed water. From a theoretical standpoint, the wave motion is unchanged if the water is imagined to move relative to a fixed point at a velocity equal and opposite to the wave velocity, thus bringing the waves to rest relative to the same point of observation. The analysis is thus changed to one of steady flow in which the surface form, and distribution of horizontal and vertical accelerations must be so distributed as to satisfy the equation of continuity, and the condition of constant surface pressure. In addition, if the water is assumed to be frictionless, which it very nearly is for purposes of wave theory, the flow must be irrotational. The problem is thus reduced to the determination of an analytical expression, or the construction of a flow net, to satisfy these steady conditions. Continuity requires that the flow system in each small element of the wave be so constructed as to indicate no destruction or creation of material. The condition of constant surface pressure is very nearly satisfied for waves moving without wind action, and surface pressure differences due to wind are probably not great as compared with the hydrostatic pressure differences. Waves lose energy so slowly that the assumption of constant energy along each streamline seems reasonable. Irrotational motion occurs in fluids devoid of viscosity and this condition may not be entirely satisfied in natural waves; if the motion is not irrotational, the energy may be different along different streamlines. For a complete discussion of rotational and irrotational motion the reader is referred to Lamb, "Hydrodynamics," or Prandtl-Tietjens, "Fundamentals of Hydro- and Aero-Mechanics."

The aim of the classical analysis of wave motion has been to find the surface form and wave velocity which satisfy the conditions outlined.

Section 3. WAVES OF SMALL AMPLITUDE

The observed permanence of oscillatory waves is utilized for the mathematical analysis of the oscillatory wave problem. The solution then requires the determination of the dynamic conditions required for such permanence of wave form. This concept is fundamental to the analysis of wave motion of the type under consideration and its importance accounts for the preceding discussion of whether or not waves actually do maintain their identity. Quantitatively, permanence of form requires that every element of the wave form advance with the same velocity. Consequently, a wave may be imagined as being brought to rest relative to a fixed reference system by superimposing a mass velocity in the water equal but opposite to the wave velocity. This does not modify the internal flow problem but does simplify the analysis to one of steady flow.

If water of constant depth is disturbed by small periodic impulses, the wave length and velocity are related to the depth by the equation (3).

$$C=\sqrt{\frac{gL}{2\pi}\tanh\frac{2\pi d}{L}} \qquad (2)$$

So long as the surface curvature is everywhere large enough to eliminate the influence of surface tension and the wave height is small as compared with d and L, this equation applies to both deep-water and shallow-water waves. Combining equation 2 with equation 1 gives C as a function of d and T (figure 1) or L as a function of d and T (figures 2 and 2a).

RELATIONSHIP OF WAVE VELOCITY, PERIOD, AND DEPTH

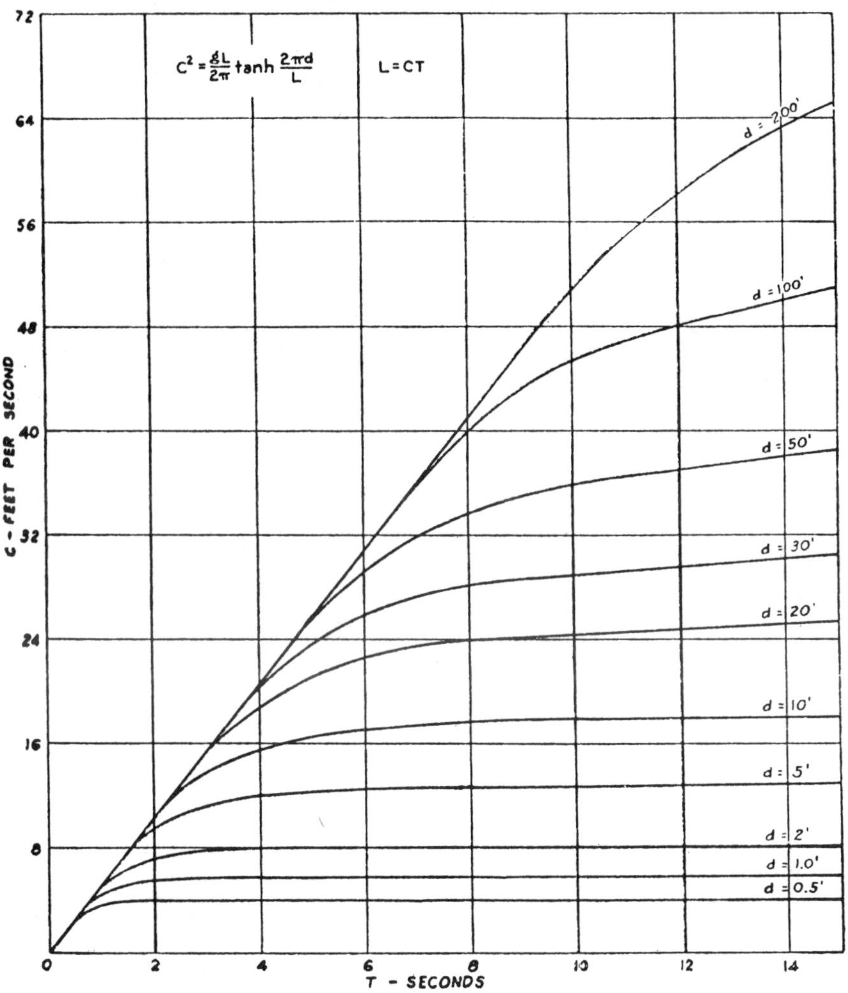

FIGURE 1.

The effect of surface tension is negligible in nature but in model work it may be important, especially near the plunge point where the radius of surface curvature becomes very small. Experimental data will be necessary to establish the lower limit of size at which waves break in a manner similar to natural waves; however, some indication of the wave size for which surface tension is important is given by the equation for ripples (surface tension waves) (4).

$$C=\sqrt{\frac{gL}{2\pi}+\frac{2\tau\pi}{\rho L}} \qquad (3)$$

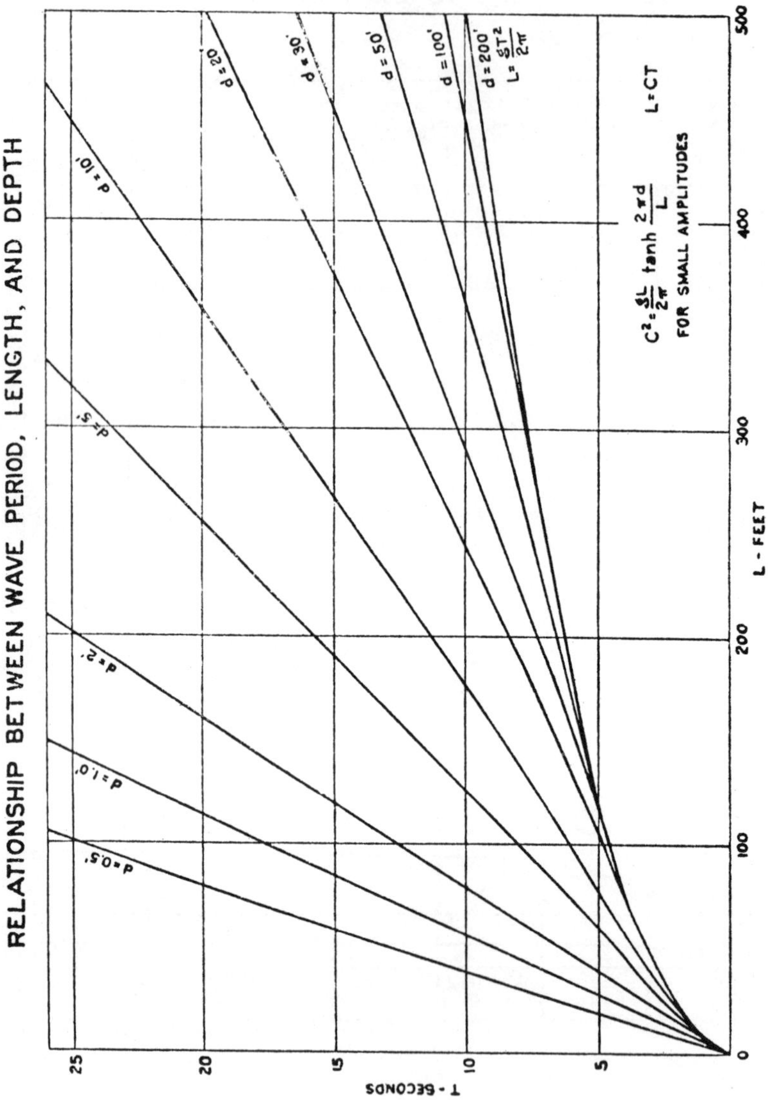

FIGURE 2.

This equation shows a minimum point at

$$C_{min}=\sqrt{2\pi\frac{\tau}{\rho g}}$$

or

$$L_{min}=1.414\sqrt[4]{\frac{\tau g}{\rho}}$$

For water in contact with air at usual atmospheric temperatures, $\tau=0.0051$ pounds per foot; $L_{min}=0.056$ ft.; $C_{min}=0.756$ ft./sec. When $L=3L_{min}$, the wave velocity given by equation 3 differs from that given by equation 2 by less than 5 percent. The equations

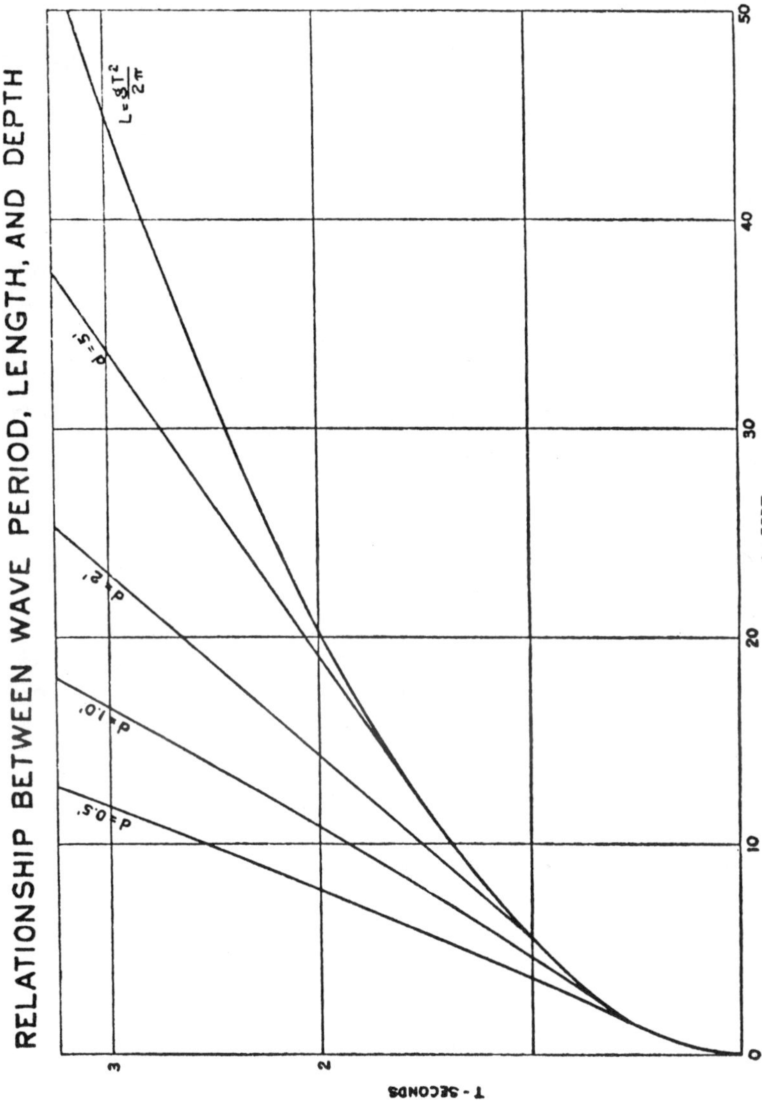

FIGURE 2a.

apply only to ripples of very small amplitude and no quantitative conclusions may be drawn from them regarding the effect of surface tension near the upper limit of the height-length ratio. Since a wave of 1-second period in deep water has a length of 5.1 feet, it is seen that the radius of curvature is so large in comparison to that of a ripple that the effect of surface tension is probably small even for waves of usual laboratory dimensions.

As d becomes very large in equation 2 the term $\tanh \frac{2\pi d}{L}$ approaches unity and the velocity expression becomes

$$C=\sqrt{\frac{gL}{2\pi}} \tag{2a}$$

since

$$L=CT \tag{1}$$

then:

$$C=\frac{gT}{2\pi}=5.12T, \tag{2b}$$

and

$$L=5.12T^2 \tag{2c}$$

Equations $2b$ and $2c$ are the limiting curves in figures 1, 2, and $2a$. As an example of the depth at which the bottom effect becomes negligible, let us assume $d=0.5\ L$. Then $\tanh\ (2\ \pi\times 0.5)=0.9963$ and the wave velocity is 0.9963 times the velocity in an unlimited depth. If an error of 5 percent is permissible, the corresponding $\frac{d}{L}$ is 0.25. Evidently, the limits of applicability of the deep-water wave equations depend on the accuracy desired and $\frac{d}{L}=0.5$ is a purely arbitrary point of division, though frequently mentioned in papers on wave phenomena as the basis of the definition that waves in water having a depth greater than half the wave length are deep-water waves, and waves in lesser depths are shallow-water waves.

If the wave length is very large as compared with the depth, $\tanh \frac{2\pi d}{L}$ approaches $\frac{2\pi d}{L}$ and equation 3 takes the form:

$$C=\sqrt{\frac{gL}{2\pi}\frac{2\pi d}{L}}=\sqrt{gd} \tag{2d}$$

This is the well-known equation for the velocity of long waves of small amplitude in still water.

The horizontal and vertical displacements (3) of the water particles from their mean position at a distance z below the still water level are functions of x and t as shown by:

$$X_z=\frac{h}{2}\cdot\frac{\cosh\frac{2\pi}{L}(d+z)}{\sinh\frac{2\pi d}{L}}\cos 2\pi\left(\frac{x}{L}-\frac{t}{T}\right) \tag{4a}$$

and

$$Y_z=\frac{h}{2}\frac{\sinh\frac{2\pi}{L}(d+z)}{\sinh\frac{2\pi d}{L}}\sin 2\pi\left(\frac{x}{L}-\frac{t}{T}\right) \tag{4b}$$

Here, z is negative measured downward from the still-water surface ($z=-d$ at the bottom). The displacements are out of phase by $T/4$ in time and $L/4$ in the direction of motion. The wave form is sinusoidal and applies to waves of small amplitude only.

The full amplitude of oscillation expressed as a fraction of the height from trough to crest is

Horizontal: $$\frac{a}{h}=\frac{\cosh\frac{2\pi}{L}(d+z)}{\sinh\frac{2\pi d}{L}} \tag{5}$$

Vertical: $$\frac{b}{h}=\frac{\sinh\frac{2\pi}{L}(d+z)}{\sinh\frac{2\pi d}{L}} \tag{6}$$

At the surface, $z=0$, $\frac{a_s}{h}=\coth\frac{2\pi d}{L}$ and $\frac{b_s}{h}=1$.

Figures 3 and 4 show the values of equations 5 and 6 with $\frac{d}{L}$ as the parameter. As $\frac{d}{L}$ approaches zero, the horizontal displacements approach uniformity while the vertical displacements increase linearly from bottom to surface. These circumstances permit a simplification in the theory of long periodic waves which, however, does not alter the fact that they are merely a limiting case of oscillatory wave motion.

Equation 5 can be converted to another form by expressing the hyperbolic functions in terms of exponentials as follows:

$$\frac{a}{h}=\frac{e^{\frac{2\pi d}{L}}\cdot e^{\frac{2\pi z}{L}}+e^{-\frac{2\pi d}{L}}\cdot e^{-\frac{2\pi z}{L}}}{e^{\frac{2\pi d}{L}}-e^{-\frac{2\pi d}{L}}}$$

For very large depths, $d\to\infty$ and $e^{-\frac{2\pi d}{L}}\to 0$ giving

$$\frac{a}{h}=e^{\frac{2\pi z}{L}} \tag{7}$$

The quantity z is again negative measured downward from the surface. The same result is obtained for $\frac{b}{h}$ and the paths traced out by fluid elements are circles.

The ratio of the vertical to horizontal amplitudes of the surface orbits is then:

$$\frac{b_s}{a_s}=\tanh\frac{2\pi d}{L}=\frac{e^{\frac{4\pi d}{L}}+1}{e^{\frac{4\pi d}{L}}-1} \tag{8}$$

Equation 2 for the velocity of a wave may be written then:

$$C=\sqrt{\frac{b_s}{a_s}\cdot\frac{gL}{2\pi}} \tag{8a}$$

The fact that the wave velocity depends, in general, upon the wave length eliminates the possibility of building up the characteristics of

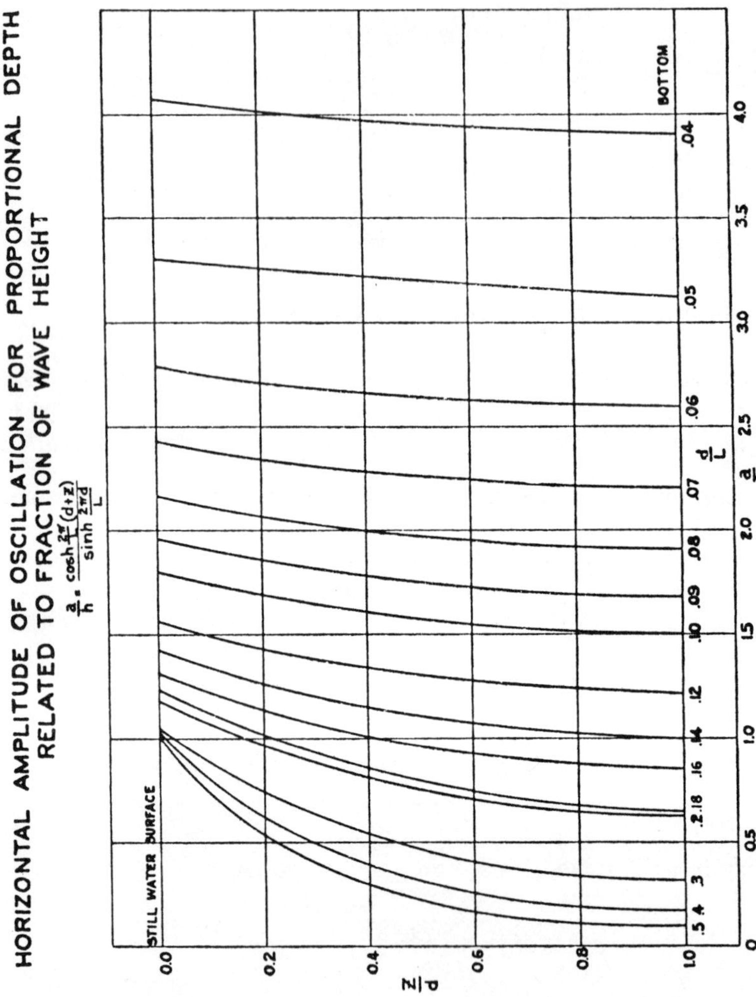

FIGURE 3.

a wave of any form as the summation of the characteristics of waves of different length and amplitude. Each wave length has a characteristic velocity, and superposition of different lengths would give a constantly changing wave form whereas waves observed in nature appear to preserve their identity, at least for distances which are large in comparison to one wave length.

The horizontal and vertical components of the velocity of water

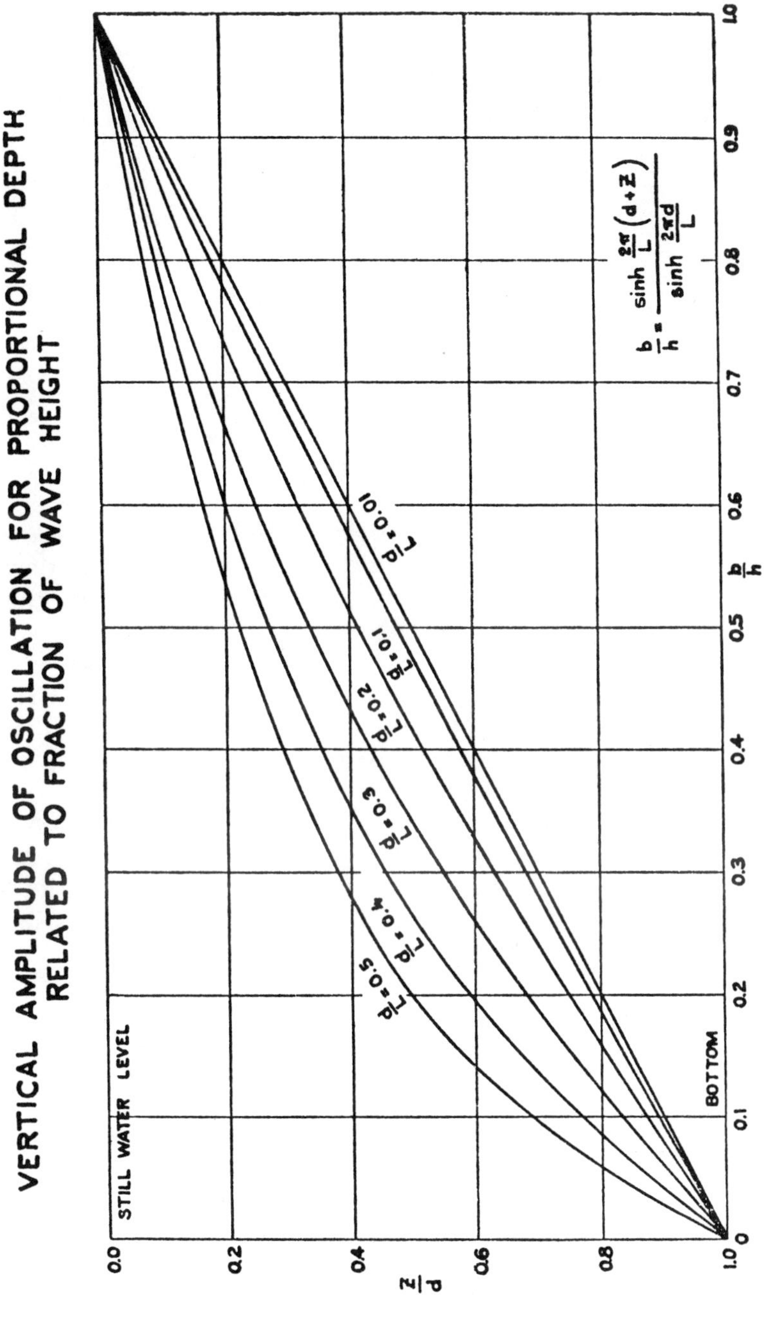
VERTICAL AMPLITUDE OF OSCILLATION FOR PROPORTIONAL DEPTH
RELATED TO FRACTION OF WAVE HEIGHT
STILL WATER LEVEL
BOTTOM
d/L = 0.5
d/L = 0.4
d/L = 0.3
d/L = 0.2
d/L = 0.1
d/L = 0.01
b/h = sinh 2π/L (d + Z) / sinh 2πd/L
Z/d
0.0
0.2
0.4
0.6
0.8
1.0
b/h
0
0.1
0.2
0.3
0.4
0.5
0.6
0.7
0.8
0.9
1.0

FIGURE 4.

particles occupying an average position at distance z below the surface are obtained by differentiating the displacements with respect to t.

$$U_z = \frac{\partial X_z}{\partial t} = \frac{\pi a}{T} \sin 2\pi\left(\frac{x}{L} - \frac{t}{T}\right) \tag{9a}$$

$$W_z = \frac{\partial Y_z}{\partial t} = -\frac{\pi b}{T} \cos 2\pi\left(\frac{x}{L} - \frac{t}{T}\right) \tag{9b}$$

The average velocities over one-half cycle are then:

$$U_{z_{ave}} = \pm\frac{2a}{T} \qquad W_{z_{ave}} = \pm\frac{2b}{T}$$

and the maximum orbital velocities are:

$$U_{z_{max}} = \frac{\pi a}{T} \qquad W_{z_{max}} = \frac{\pi b}{T}$$

In these equations, a and b are the full amplitudes, while T is the period of one full cycle. Computation of these velocities for particular values of h, T, and d, are effected by equations 5 and 6 or figures 3 and 4, using the value of L for the specified d and T from figure 2.

These theoretical equations apply only to waves of small amplitude passing over water of constant depth but "the numerical results show that a variation in the depth will have no appreciable influence, provided that the depth everywhere exceeds (say) half the wave length" (5).

At any instant the total energy per unit width of crest in one wave length is obtained as the sum of the potential and kinetic energies. The kinetic energy is the summation from bottom to surface of the individual kinetic energies corresponding to the orbital velocities as given by equations 9a and 9b.

The potential energy is computed (6) from the elevation or depression above the still water surface as:

$$E_p = w \int_{x=0}^{x=L} \int_{z=0}^{z=-y} z \, dz dx$$

Integrating with respect to z under the surface curve, where y is the particular value of z at the free surface, we obtain:

$$E_p = \frac{w}{2} \int y^2 dx$$

For a small wave of sinusoidal form described by the equation:

$$y = \frac{h}{2} \cos 2\pi\left(\frac{t}{T} - \frac{x}{L}\right)$$

? Nonlinearity in shallow water. ? Solitary waves

the potential energy at any time t per wave length per unit breadth is then:

$$E_p=\frac{wh^2}{8}\int_{x=0}^{x=L}\cos^2\frac{2\pi x}{L}\,dx$$

$$=\frac{wLh^2}{16} \qquad (10)$$

The kinetic energy per unit breadth per wave length is obtained by integration of the equation:

$$E_k=\frac{w}{2g}\int_{x=0}^{x=L}\int_{z=-y}^{z=d}(Uz^2+Wz^2)\,dx\,dz$$

In a deep-water wave of sinusoidal form

$$E_k=\frac{wLh^2}{16}=E_p$$

and the total energy is then:

$$E=E_k+E_p=\frac{wLh^2}{8} \qquad (11)$$

and substituting for L from equation 2c

$$E=0.64\ wh^2T^2 \qquad (11a)$$

[*Editors' Note:* Material has been omitted at this point.]

LIST OF REFERENCES

1. Rayleigh, Lord—On Periodic Irrotational Waves at the Surface of Deep Water—Philosophical Magazine, Vol. XXXIII, 1917, pp. 381–389.
2. Stokes, G. G.—On the Theory of Oscillatory Waves—Transactions, Cambridge Philosophical Society, Vol. VIII, 1847, p. 441, and Supplement, Scient. Papers, Vol. I, p. 314.
3. Lamb, Horace—Hydrodynamics, 6th Ed., 1932, p. 367.
4. Lamb, Horace—Hydrodynamics, 6th Ed., 1932, p. 459.
5. Lamb, Horace—Hydrodynamics, 6th Ed., 1932, p. 368.
6. Lamb, Horace—Hydrodynamics, 6th Ed., 1932, p. 260.
7. Levi-Civita, T.—Détermination rigoureuse des ondes d'ampleur finie—Mathematische Annalen, Vol. XCIII, 1925, pp. 264–314.
8. Struik, D. J.—Détermination rigoureuse des ondes irrotationelles périodiques dans un canal à profondeur finie, Mathametische Annalen, Vol. XCV, 1926, pp. 595–634.
9. Lamb, Horace—Hydrodynamics, 6th Ed., 1932, p. 419.
10. Mitchim, C. F.—Oscillatory Waves in Deep Water, The Military Engineer, March–April, 1940.
11. Michell, J. H.—On the Highest Waves in Water, Philosophical Magazine, (5), Vol. XXXVI, 1893, pp. 430–437.
12. Havelock, E. T.—Periodic Irrotational Waves of Finite Height—Proceedings of the Royal Society, London, Series A, Vol. 95, 1918, pp. 38–51.
13. Rayleigh, Lord—Hydrodynamical Notes—Philosophical Magazine, Vol. XXI, 1911, pp. 177–195.
14. Gerstner, Franz—Theorie der Wellen—Abhandlungen der königlichen böhmischen Gesellschaft der Wissenschaften, Prague, 1802; also Gilbert's Annalen der Physik, Vol. XXXII, 1809, pp. 412–445.

15. Rankine, W. J. M.—On the Exact Form of Waves Near the Surface of Deep Water—Philosophical Transactions of the Royal Society, London, 1863, p. 127.
16. Gaillard, D. D.—Wave Action in Relation to Engineering Structures—1904, reprinted at Engineer School, Fort Belvoir, Va., 1935.
17. Lamb, Horace—Hydrodynamics, 6th Ed., 1932, p. 381.
18. Lamb, Horace—Hydrodynamics, 6th Ed., 1932, p. 237.
19. Reynolds, Osborne—On the Rate of Progression of Groups of Waves and the Rate at which Energy is Transmitted by Waves—Nature, Vol. XVI, 1877, pp. 343-344; also Scientific Papers, Vol. 1, p. 198.
20. Rayleigh, Lord—On Progressive Waves—Proceedings London Mathematical Society, Vol. IX, 1877, pp. 21-26; also Scientific Papers, pp. 322-327.
21. Lamb, Horace—Hydrodynamics, 6th Ed., 1932, p. 382.
22. Wien, W.—Lehrbuch der Hydrodynamik, Leipzig, 1900, Section 3, pp. 201-212.
23. Rayleigh, Lord—Hydrodynamical Notes—Philosophical Magazine, Vol. XXI, 1911, pp. 177-195.
24. Airy, G. B.—On Tides and Waves—Encyclopaedia Metropolitana, London, 1842, p. 289.
25. Ivanov, A. A.—Bulletin, Academy of Sciences, U. S. S. R., Sir. Geogr. Nr. 5/6, pp. 477-490.
26. Lamb, Horace—Hydrodynamics, 6th Ed., 1932, p. 623.
27. Hough, S. S.—On the Influence of Viscosity on Waves and Currents—Proceedings, London Mathematical Society (1), Vol. XXVIII, 1896, pp. 264-288.

ERRATA

Page 9, the second line under equation (2*d*) should read: "amplitude in shallow water."

Page 10, the line under equation (6) should read: "At the surface,

$$z = 0, \frac{a_s}{h} = \coth \frac{2\pi d}{L} \text{ and } \frac{b_s}{h} = 1\text{"}$$

Page 10, equation (8) should read:

$$\text{"}\frac{b_s}{a_s} = \tanh \frac{2\pi d}{L} = \frac{e^{\frac{4\pi d}{L}} - 1}{e^{\frac{4\pi d}{L}} + 1}\text{"}$$

4

Reprinted from *Technical Memorandum No. 3,* Beach Erosion Board, U.S. Army Corps of Engineers, 1944, 47 pp.

SHORE PROCESSES AND BEACH CHARACTERISTICS

W. C. Krumbein

[*Editors' Note:* Figure 1 and Plates I Through IV are not reproduced here.]

Note: The following paragraph was omitted from page 1 of the original publication.

Use of the term "closed system" is not completely rigorous even for isolated bays. Wave energy comes into the bay from the open sea, and materials (sand and silt) may move seaward or be brought in from outside. However, the flux of energy and material through the boundaries could be measured, so that events within the bay proper may be treated as though a closed system were present.

1. Experimental Design for Beach Studies

The study of shore processes in nature is complicated by the presence of a large number of mutually dependent variables, none of which is susceptible to control in the field. As a consequence it has been difficult to discern and largely impossible to specify the physical law or laws controlling the behavior of the environment. It is believed, however, that the study of a relatively simple natural situation by the "closed system" method familiar to the exact sciences will allow the discovery and specification of, at least, the principal fundamental relations governing the behavior of matter in the system.

By definition a closed system is one in which the boundary conditions and the total energy of the system are known, and in which the transformations of matter may be observed and measured. In the study of beach processes isolated bays most nearly satisfy the conditions for closed systems. An isolated bay may be considered as a particular combination of boundary conditions, matter, and energy. All three of these may be measured and expressed to a degree satisfactory for a first approximation to theory. Comparison of a number of bays affords a basis for evaluating boundary effects, so that principles established locally may be applied generally to beach phenomena.

The problem to be studied was then formulated as follows: given a bay with associated headlands, beach, and cliffs, to evaluate the physical processes which occur there. Beaches are composed of particles which must be derived from some source; the particles must be carried from the source to the point of deposition, and the agent which carries them must be energized in some manner or other. The source of the beach material is the adjacent terrain, which may be studied by geological methods. Energy is supplied by waves which strike the shore. The study of these waves, including their role in generating shore currents and carrying material, is a problem in earth physics. The interaction of matter and energy as it affects the land-forms developed (beaches, bars, spits, sea cliffs, etc.) may be studied by geological or geophysical methods; an evaluation of the processes in terms of beach stabilization and control is in the domain of engineering.

This analysis of the problem indicates the kinds of data to be obtained. These include (1) the characteristics of the waves in terms of height and period, as an index of the energy being supplied to the system; (2) the supply of beach and associated material, in terms of the amount available, as well as the size, shape, and other dynamical attributes of the particles; (3) specification of the erosional and depositional land-forms in the environment, and their relation to processes which form them and materials which compose them; (4) definition of the boundary conditions which control the distribution of wave energy in terms of refraction patterns, and which exert an effect on the materials and land-forms adjacent to the boundaries.

The present study is an approach to this experimental design. Limitations arising partly from war conditions prevented the collection of complete data on each of the four factors involved. They are all touched upon, however, to frame the study within its larger background.

The coast of California is ideal from the closed system point of view, in that many of its beaches are found on relatively isolated portions of the coast, each bounded by headlands which to a large extent prevent the migration of matter or wave energy from one system to another. In choosing a bay and beaches for study it is desirable that the beaches be free from artificial structures and the bays represent a type, so that the results may be extended to similar bays. Halfmoon Bay satisfies both of these conditions. It is complicated by very few artificial structures, and in form it is typical of a number of other bays, including Drake Bay, Bolinas Bay, and San Pedro Bay.

By designing a study which repeats certain measurements at intervals, information regarding the stability of the beach in terms of width, slope, sand size, etc. may be obtained. The study should extend throughout at least a year to evaluate seasonal effects. At Halfmoon Bay the study was confined to the spring season, but data are available on winter and summer conditions from previous work by M. P. O'Brien (unpublished reports in Beach Erosion Board files).

2. The Physical Setting of Halfmoon Bay

a. Topography and Hydrography. - Halfmoon Bay lies about 25 miles south of the Golden Gate. It is sheltered on the north by Pillar Point, which rises 181 feet above the sea, and on the south by Point Miramontes, 6.5 miles south-southeast of Pillar Point measured along the curving shore. The curvature of the shoreline is greatest at the northern end, gradually diminishing to the south, in the manner of a logarithmic spiral. The bay is bordered by a sandy beach which terminates against a sea cliff cut into loosely consolidated sand and gravel. Sand dunes occur locally along the shore. The bay is shown in figure 1.

Immediately east of Pillar Point is a belt of low land which extends northwestward toward Point Montara. Although this belt resembles a tidal slough filled with detritus, the San Mateo topographic map indicates that the surface rises to more than 50 feet above sea level within 1½ miles north of the bay. The sediments exposed in this trough are the same as the terrace materials to the east. This low land is therefore part of the general terrace which rims the bay. The surface of the terrace is gently tilted in a north-south direction along an east-west axis; it lies only a few feet above sea level at Pillar Point, but rises to some 60 feet above the sea at Point Miramontes. A gentle warp is superimposed upon the tilt, so that the height of the terrace varies somewhat along its extent, rising to its greatest height at Point Miramontes. The terrace merges inland with the range of hills which parallels the bay to the east and south.

Several small streams enter Halfmoon Bay from the north and east. The total drainage area of the bay is about 35 square miles, but no data are known to the writer on the total stream discharge.

Halfmoon Bay deepens gradually with the regularity interrupted by two reefs which trend to the south-southeast from Pillar Point (see figures 1 and 2). The northern reef is just submerged at low tide and extends about a mile from the Point. The southeast reef has depths of about 20 feet over it at low tide. The greatest depth within the bay is about 10 fathoms, just east of the southeast reef. The bottom is generally sandy, although the trend of the reefs suggests that they are submarine extensions of the bedrock at Pillar Point.

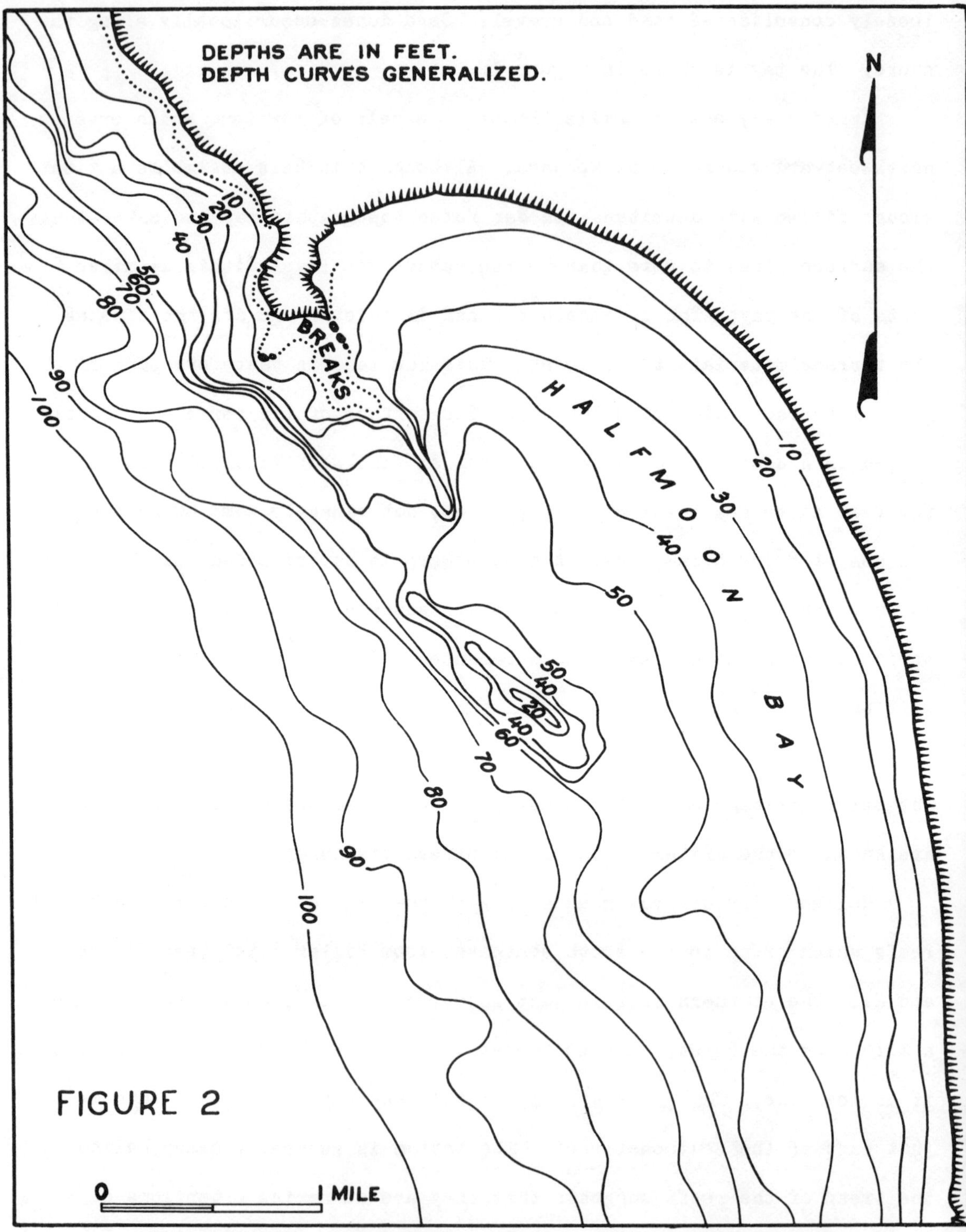

SUBMARINE RELIEF - HALFMOON BAY, CALIF.
(0 - 100 FT. ONLY)

b. Geology. - The terraces along Halfmoon Bay are composed of Pleistocene sand and gravel. The elevation of the terrace and the presence of sea cliffs attests to the relatively recent development of the bay. The large mass of Montara Mountain represents the oldest rock in the vicinity. Pillar Point, as well as the hills which lie to the east of the town of Halfmoon Bay, and an early marine terrace exposed at sea level at Point Miramontes, are all composed of Tertiary sediments. Figure 3 is a geological sketch map of the area.

The only formations which lie along the bay other than Pleistocene and recent deposits are the Purisima and Merced formations. The Purisima is of Miocene-Pliocene age. The particular beds which are exposed at Point Miramontes are dark gray to black fossiliferous shale, with a general northwest-southeast strike, and a moderate dip toward the northeast. The beds are truncated and the Pleistocene deposits rest unconformably upon them. The rocks of Pillar Point belong to the Merced formation, of late Pliocene age. They consist of marine conglomerate, sandstone, and shale, folded into a series of minor synclines and anticlines. The members vary in their resistance to erosion. The structural pattern of the rocks is revealed by recent cutting of Pillar Point.

The geological history of the bay area includes the intrusion of quartz diorite into ancient sedimentary rocks, probably during pre-Jurassic times. The Jurassic and Cretaceous periods left no rock records near the bay, but during this interval uplift and erosion exposed the quartz diorite at the surface. The Tertiary is represented by several formations, which rest unconformably upon the older rocks. The Tertiary sediments are largely marine, and represent an inundation of the region by the sea. At the close of the Tertiary came a marked disturbance, in which the rocks were uplifted, folded, and faulted. Erosion developed marine terraces, a portion of which is exposed on the Purisima at Point Miramontes. Following truncation came depression of the coast, during which Pleistocene sand and gravel were deposited as an apron on the western slopes of the hills. This episode was followed by uplift, warping, and tilting. The present sea cliffs were cut into these elevated Pleistocene deposits, and the modern beach with its fringe of dunes was developed.

c. Meteorology and Oceanography. - The local wind rose shows that northwesterly winds prevail during April and May. At San Francisco winds blow from this direction 38 per cent of the time in April and 51 per cent of the time in May. Seaward of Halfmoon Bay the percentages are 20 and 32 for April and May respectively. These winds have an average force of 4 units on the Beaufort scale. Winds from the north blow less than 10 per cent of the time during both months at San Francisco, but seaward they occur 16 and 19 per cent of the time. The general prevalence of winds from a northwesterly direction is probably explained by a relatively permanent high-pressure area over the Pacific west of the Golden Gate.

The California current in the open sea to the west of the bay has a generally southerly direction, but according to Sverdrup (Oceanography for Meteorologists, p. 203) areas of upwelling north of the Golden Gate and south of Monterey Bay cause swirls during the Spring and early summer months. These swirls develop a northward current along the coast west of Halfmoon Bay. No data are given within the bay itself. The writer knows

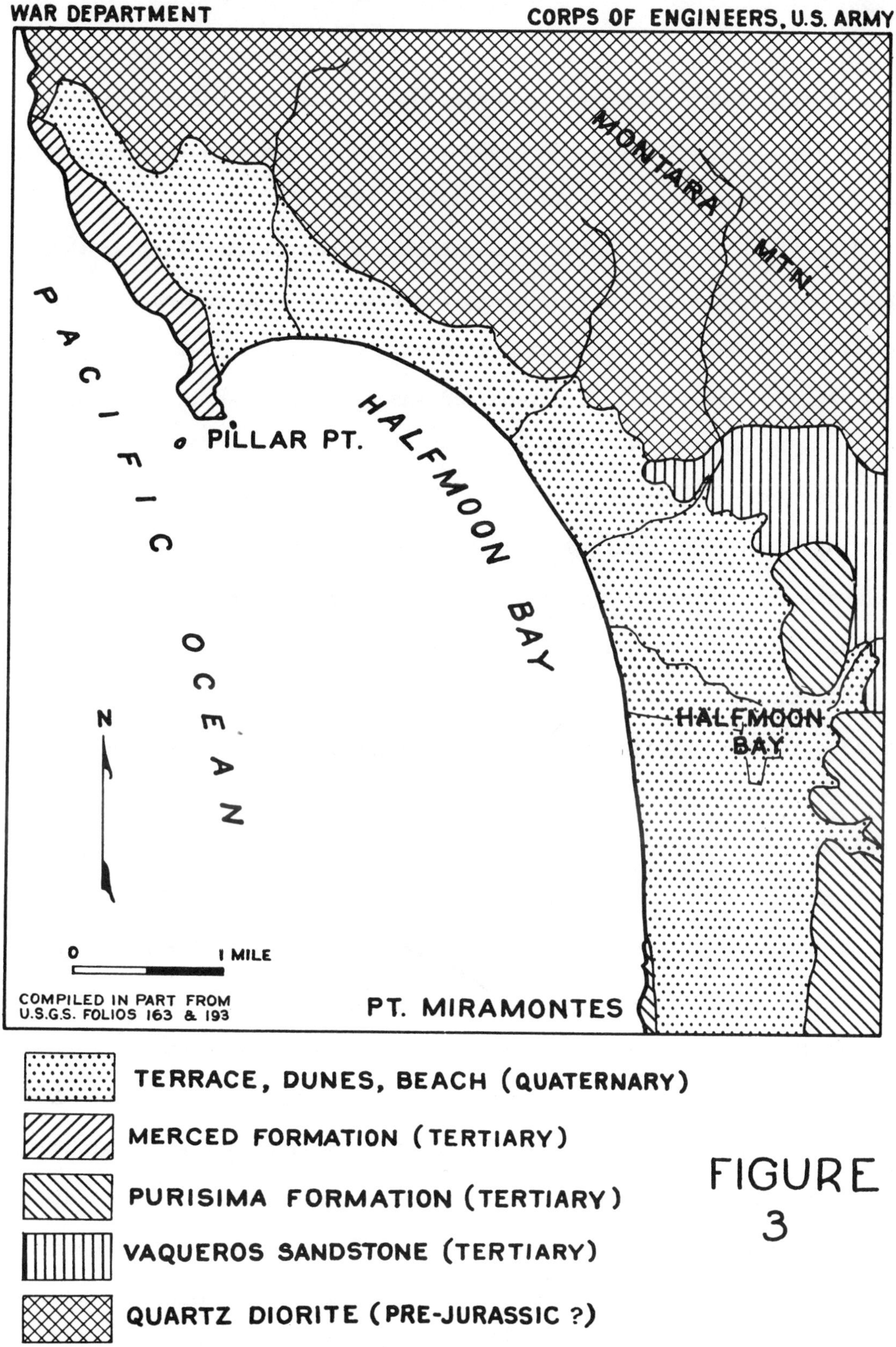

FIGURE 3

AREAL GEOLOGY OF HALFMOON BAY AND VICINITY

of no specific data on waves in the open sea along the coast from the Golden Gate to Santa Cruz. Wave observations were made within the bay during this study; the results are given later.

The tidal cycle in Halfmoon Bay occurs about an hour earlier than at the Golden Gate. The tides are mixed with the long runout following higher high water; the high water interval is 10 hours 30 minutes. The mean range of the tide is 3.8 feet and the diurnal range is 5.5 feet.

3. Observational Data

The present study extended from 4 March to 9 May 1942. The beach was visited each Saturday and except for the first reconnaissance visit, a systematic plan was followed in obtaining data on each visit. Stakes were set along the base of the sea cliff at approximately half mile intervals and numbered 1 to 13 from south to north. The location of these stations is shown on figure 1, and a description is given in table 1. All stations were not visited each Saturday, but an alternation was followed so that about 10 sets of observations were made each time within the tidal limits.

a. <u>The State of the Beach</u>. - The western side of Pillar Point is bordered by numerous reefs. The eastern side is fringed with reefs at its southern end, but develops into a cobble beach to the north. A rock spur separates this pocket beach from the main strand to the east. The spur essentially prevents the migration of cobbles to the east, although some of the fine sand from the bay shore has moved westward. East of the spur a sand beach extends uninterrupted nearly to Point Miramontes. Rock reefs appear at low tide about a mile north of the point and become

Table 1

Description of Stations along Halfmoon Bay

Station	Miles from Pt. Miramontes	General Remarks
1	0.25	Rock reefs abundant, sand very coarse, beach fairly narrow, cliffs 60 feet high, erosion evident.
2	0.80	Occasional rock reefs, sand coarse, cliffs 50-60 feet high, erosion evident.
3	1.20	Well developed berm, cliffs 40 feet high.
4	1.80	Well developed berm, cliffs 20 feet high.
5	2.20	Well developed berm, sand moderately coarse, sand dunes on terrace.
6	2.60	Berm present, stream parallels cliff inland of beach. Cliff 15 feet high.
7	3.20	Well developed berm, prominent dune belt parallels wide beach.
8	3.70	Well developed berm, dune belt ends just north of station.
9	4.10	Partially eroded berm, wide beach, sand moderately fine. Cliffs 10 feet high.
10	4.60	No berm present, wide beach, sand rather fine. Some erosion evident. Cliffs 10 feet high.
11	5.10	No berm present, active erosion along cliffs which are 20 feet high.
12	5.50	Small trace of berm in re-entrant. Wide beach, very fine dark sand. Very active erosion along cliffs which are 15 feet high.
13	6.10	No berm present. Very wide beach, very fine dark sand, always moist. Cliff here about 1 - 2 feet high.

more prominent to the south. At Point Miramontes itself a continuous reef extends outward from the sea cliff. In this vicinity the sand beach becomes discontinuous and patchy.

The width of the beach varies from one end of the bay to the other. On 11 April 1942 it was 250 feet wide east of Pillar Point and diminished somewhat irregularly to 140 feet wide north of Point Miramontes. During subsequent visits the beach widened to some extent along its entire course.

Erosion is active at Point Miramontes and north to station 3, a distance of about 1 mile. Slumped portions of the cliff are common along the beach. Beyond station 3 the cliff is protected by a well developed berm, which is present continuously to station 9 just north of Miramar. Beyond station 9 the foreshore slope extends to the base of the sea cliff, which is awash at high tide. Erosion becomes prominent again at station 11 and continues to Princeton, where extensive cutting is evident.

Several areas of dune sand fringe the beach between stations 4 and 8. These dunes are best developed in the vicinity of station 7, but nowhere do they extend inland farther than a few hundred feet beyond the beach. Plate I includes several photographs of the beach showing the varying character of the beach along its extent.

b. Profiles and Beach Slopes. - Ten beach profiles were measured on April 11. They were located along the crests of beach cusps wherever the latter were present. The profiles are shown in figure 4; they indicate a decreasing slope of the foreshore from south to north and show a well developed berm from stations 3 to 9. A trace of an earlier berm was found at station 12, protected by a re-entrant into the sea cliff. The points marked "S" indicate sand samples taken at the time the profiles were measured.

The berm on April 11 appeared to be largely the result of the previous winter's wave conditions. The height of the berm above MLLW averaged 12.7 feet; the individual values are shown in table 2. It is interesting to note that O'Brien's observations in 1930 indicated an average elevation of 13 feet for the previous winter's berm.

Table 2

Elevation of Berm above MLLW, 11 April 1942

Station	Berm elevation, feet
1	Not present
3	16
4	13
6	11
7	14
8	12
9	13
10	Not present
11	Not present
12	10 (Trace only)
13	Not present

Limitations of time prevented the measurement of complete beach profiles at later dates, but beach slopes were measured during each visit with a Brunton hand compass. The complete data are shown in figure 5, arranged according to dates. O'Brien's observations of June 1930 are included. Despite the scatter of the individual points there is a definite trend to the data. The position of O'Brien's points within the scatter suggests that the slope relations of the beach have not changed significantly within the last 12 years.

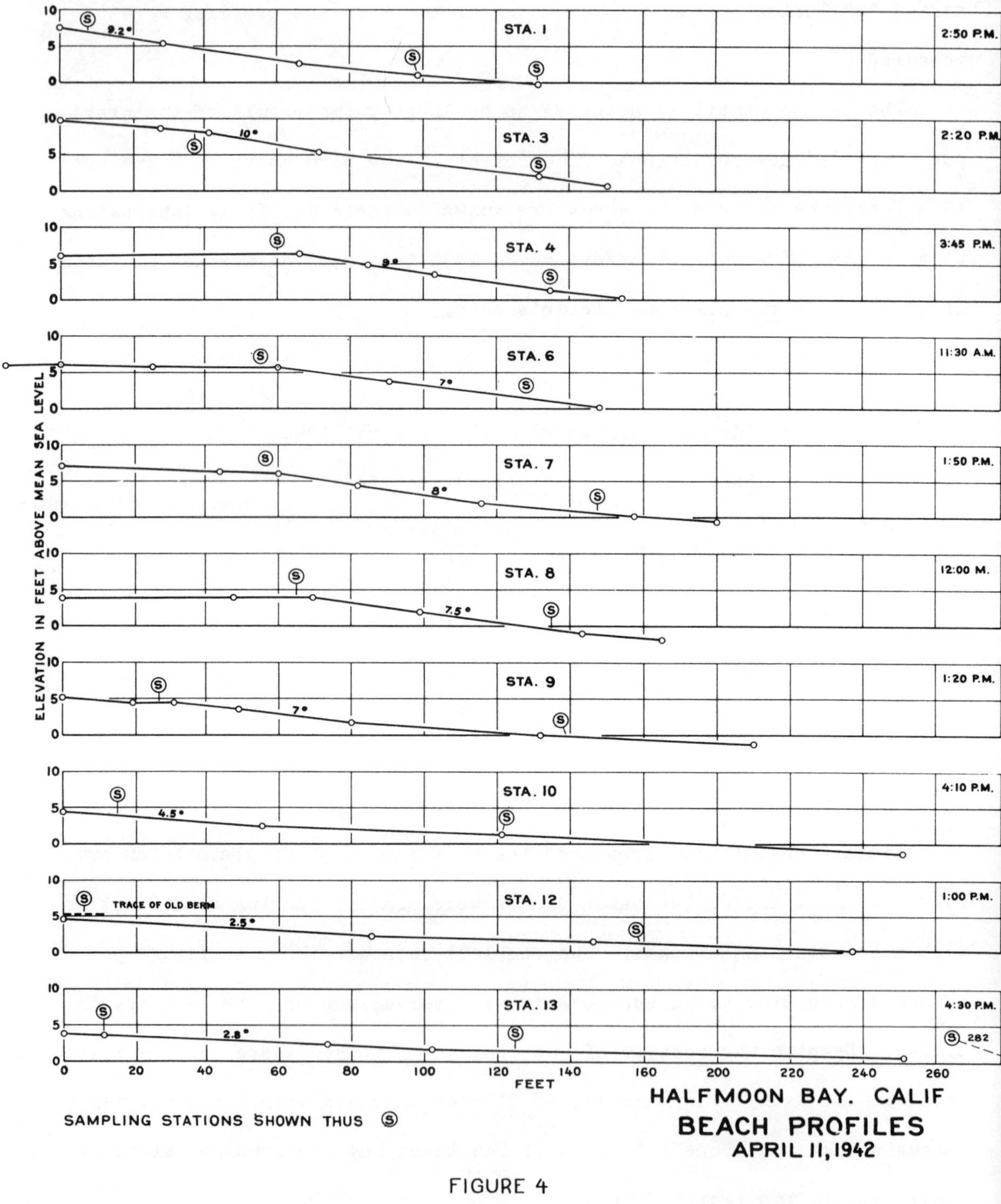

ELEVATION IN FEET ABOVE MEAN SEA LEVEL
STA. 1
2:50 P.M.
9.2°
STA. 3
2:20 P.M.
10°
STA. 4
3:45 P.M.
9°
STA. 6
11:30 A.M.
7°
STA. 7
1:50 P.M.
8°
STA. 8
12:00 M.
7.5°
STA. 9
1:20 P.M.
7°
STA. 10
4:10 P.M.
4.5°
STA. 12
1:00 P.M.
TRACE OF OLD BERM
2.5°
STA. 13
4:30 P.M.
2.8°
282
FEET
SAMPLING STATIONS SHOWN THUS S
HALFMOON BAY. CALIF
BEACH PROFILES
APRIL 11, 1942

FIGURE 4

c. Beach Cusps. - Measurements and observations on cusps were made during each visit to Halfmoon Bay. The number of cusps, their average length, and the standard deviation of length for each station are given in table 3. Standard deviation is a measure of the average scatter of a series of observations about their mean value. Formulas for the calculation of the standard deviation are given in all statistics texts.

On April 11 cusps were well developed and apparently continuous from stations 1 to 9. A moderate sea was running, with waves approaching the bay from the southwest. On April 18 conditions had changed. A southwest storm occurred during the week, and some of the previous cusps were partly eroded; elsewhere entirely new cusps had formed. A striking feature of some of the cusps, especially near station 3, was their asymmetry. The crest lines trended northward of the normal by as much as 32°. The asymmetry was not completely systematic, however, because undistorted crests occurred among them.

On April 25 the aspect of the cusps had nearly completely changed. In the south a new set of shorter cusps was present, with traces of the earlier cusps on a higher berm. At station 5 the old cusps had been completely cut away, and a sea cliff 5 feet high had been cut into the berm. At station 7 new cusps were faintly discernible, but at station 8 no trace of cusps was found. By May 2 the newer cusps were well developed as far north as station 8. The cliff at station 5 was still prominent, and no new cusps had formed there. On May 9 the sea was relatively calm, with waves approaching the bay from the northwest, and cusps were well developed.

Table 3

Observations on Beach Cusps

Date 1942	Station	Number Measured	Average length feet	Standard Deviation per cent	Remarks
4/11	1	11	147	16.7	
"	3	13	199	22.0	Winter cusps?
"	4	13	164	25.9	
"	6	5	148	9.5	
"	7	8	159	4.9	
"	8	10	148	17.1	
"	9	6	175	3.4	Winter cusps
4/18	2-3	11	222	15.4	Winter cusps?
"	4	10	182	15.5	
"	5	9	175	21.6	
"	7	8	174	5.0	
"	8	13	152	23.1	
"	9	6	175	3.4	Same as before
4/25	2-3	11	208	27.0	Storm cusps?
"	4				Old cusps eroded
"	5				Old cusps eroded
"	7	4	126		4 old cusps
"	8				No cusps at all
5/2	2	11	149	18.2	
"	3	12	143	22.2	Also 6 old cusps
"	5-6	10	147	9.0	
"	8	10	153	15.8	

The observations suggest that cusps formed during the winter and early spring are longer and are associated with higher berms than those formed in late spring and summer. The writer assembled all available data, including O'Brien's 1930 observations, and made a tentative classification of the cusps into two main groups, with a possible intermediate group. The results are shown in figure 6. The upper series of points includes long cusps formed during winter or during spring storms. The lowermost line represents short cusps formed in summer or during quiet conditions in the spring. Between the two sets are scattered points which represent a transition between the extremes. If the classification is sound, it suggests two things: (1) cusp length is a function of the average wave and tidal conditions which prevail at the time of their formation, and (2) cusp length within any series decreases systematically from south to north along the beach. The writer is not prepared at present to discuss theories of cusp formation in terms of wave periods, lengths, or heights. Apparently cusps represent a steady state condition along the shore; although abrupt changes in conditions may destroy old cusps by erosion, new ones begin to form almost simultaneously and remain relatively constant in size and form as long as the new conditions prevail. It is observed that cusps are present only where a berm occurs along the beach. The seaward edge of the berm is serrated, but the low water line is nearly straight. The cusps are therefore marginal land-forms associated with the berm.

An oft-mentioned feature of cusps is their regularity in length. When detailed measurements are made, however, it is found that occasionally an unusually long or short cusp occurs within a series. The effect of these irregularities is that the standard deviation of length is larger than expected. In the present instance the average standard deviation of all cusps is of the order of 15 per cent of their length, but this value varies along the beach, as table 4 indicates.

Table 4

Variation in Standard Deviation of Cusp Length

Stations	Standard Deviation
1,2,3	20 per cent
4,5,6	16 " "
7,8,9	11 " "

Thus the variability of the cusps diminishes to the north, perhaps as a function of the degree of shelter from direct wave attack afforded by Pillar Point and its associated reefs.

d. Sand Samples and the Properties of Sediments. - Sand samples were collected at low water, mean water, and high water levels, and from the crest of the berm. All samples were taken from the crest lines of beach cusps, wherever the latter were present. Sampling was repeated at intervals, especially from mean water level, which was sampled on April 11, 25, and May 9.

The sand samples were taken to a depth of one foot, in order to average out nonsystematic variations. They were quartered in the field, and in the laboratory they were washed, dried, and sieved.

The present report includes only the size data of the samples, but other attributes may be equally significant. Work with sediments has shown that six particle properties are important. These are size, shape (sphericity), roundness, mineral composition (density), surface texture, and particle orientation. In addition, sediments have various megascopic mass properties, as porosity, permeability, compactibility, etc. The treatment of the size data in the present discussion is an example of the attack which may be made on each particle property.

The particle attributes may be used in interpreting dynamical processes. Two such processes are selective transportation (sorting action) and particle abrasion. Sorting action is largely a function of settling velocity, and expresses itself in terms of size, shape, and density. Abrasion, on the other hand, changes the roundness and surface texture of the particle's most markedly. Particle orientation (the arrangement of the particles within the deposit) appears to reflect the particular conditions under which deposition occurs.

The size data of the samples were expressed as weight frequency percentages, and plotted as cumulative curves on probability paper. The use of probability paper permits the graphical determination of the median diameter, the geometric mean diameter, and the logarithmic standard deviation. The geometric mean diameter will be used in the following discussion. It is related to the center of gravity of the logarithmic size distribution, and is stated by Rouse to be the most significant average for dynamical studies of particles.

Figure 7 shows the geometric mean diameter of all mean water level samples plotted as a function of distance along the beach. O'Brien's 1930 data are included. A negative exponential function is suggested, which is shown by the semi-log plot of figure 8. The agreement with the straight line is satisfactory and indicates that the functional relationship is of the form

$$r = r_0 e^{-ax} \quad . \ . \ . \ . \ (1)$$

where r is the size of any point x along the beach, r_0 is the initial size at x = 0, and a is the coefficient of size decrease.

One of the questions which arose in the present study was the extent to which repeated samples would give the same size data. This was answered by plotting each set of samples separately on semi-log paper, and determining the parameters of the straight lines. These are the initial size r_0, and the coefficient a of equation 1.

Table 5

Size Parameters of Mean Water Level Samples

Date	Initial size, r_0	Coefficient a
6/12/30	1.18 mm.	0.35 (1/mile)
4/11/42	1.18	0.41
4/24/42	1.19	0.42
5/ 9/42	1.30	0.45

The variation shown by these figures is within the range of scatter of any one set of data, and indicates that on the whole any single set of samples affords a good approximation to the trend, although any one sample may show considerable differences.

The samples collected from other levels were also plotted on semi-log paper, as shown in figures 9 and 10. The parameters (including the average of the mean water set) are given in table 6. The systematic increase in the value of r_0 in the table indicates that average size increases downslope normal to the beach. The systematic increase in the coefficient

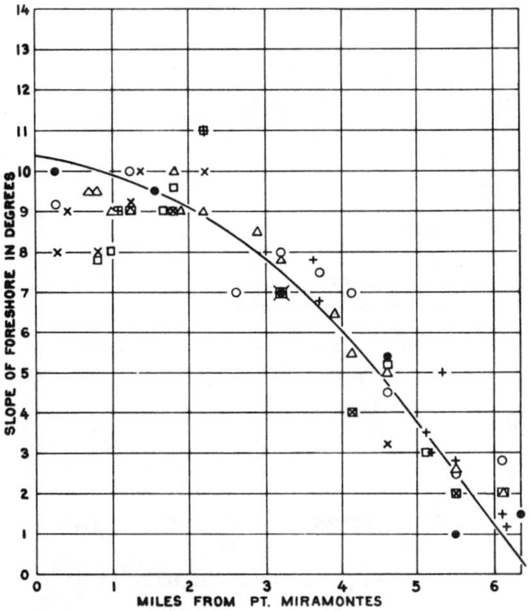

- ● JUNE 21, 1930 O'BRIEN
- ○ APRIL 11, 1942 W.C.K.
- △ APRIL 18, 1942 "
- □ APRIL 25, 1942 "
- + MAY 2, 1942 "
- × MAY 9, 1942 "

FIGURE 5

OBSERVED RELATION BETWEEN FORESHORE SLOPE AND DISTANCE

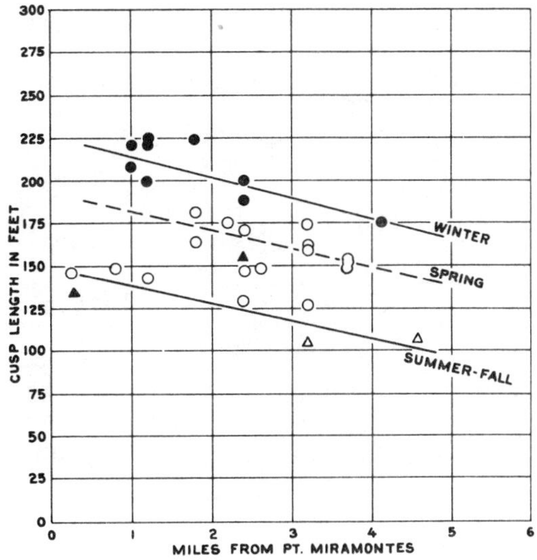

- ● WINTER
- ○ SPRING
- ▲ SUMMER
- △ FALL

FIGURE 6

PROBABLE SEASONAL VARIATIONS OF CUSP LENGTH AND DISTANCE

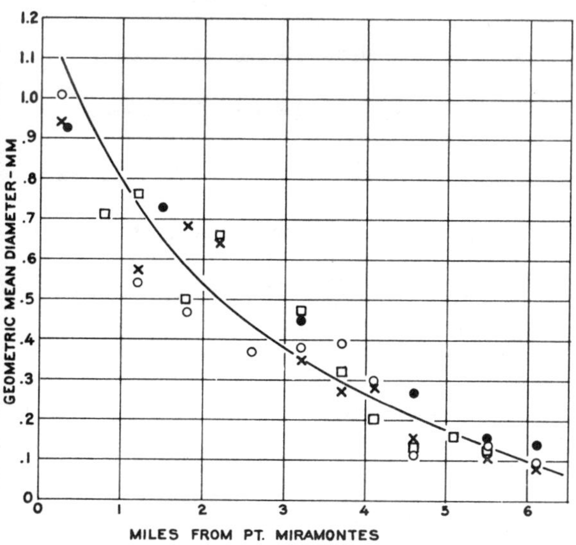

- ● JUNE 21, 1930 O'BRIEN
- ○ APRIL 11, 1942 W.C.K.
- □ APRIL 25, 1942 "
- × MAY 9, 1942 "

FIGURE 7

OBSERVED RELATION BETWEEN SAND SIZE AND DISTANCE OF MEAN SEA LEVEL SAMPLES

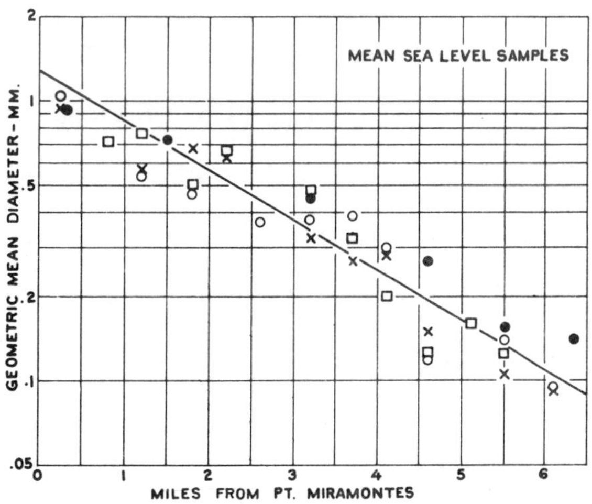

- ● JUNE 21, 1930 O'BRIEN
- ○ APRIL 11, 1942 W.C.K.
- □ APRIL 25, 1942 "
- × MAY 9, 1942 "

FIGURE 8

OBSERVED RELATION BETWEEN SAND SIZE AND DISTANCE OF MEAN SEA LEVEL SAMPLES

(SEMI-LOG PLOT OF FIGURE 7)

a indicates that the lower samples have a relatively greater rate of size decrease along the beach.

Table 6

Size Parameters of all Samples

Samples	Initial size, r_o	Coefficient a
Berm	0.54 mm.	0.17 (1/mile)
High water level	0.88	0.39
Mean water level	1.25	0.41
Low water level	1.50	0.45

The logarithmic standard deviation, σ_ϕ , is a measure of the spread of the size frequency curve about the mean diameter. Large values indicate poor sorting, because σ_ϕ increases with increasing spread of the curves. There is a wide scatter of the individual values of σ_ϕ in the mean water level samples. However, when the data of any one set of samples is averaged, trends appear. Two of these are shown in figure 11. The upper graph represents mean water samples, and shows a marked maximum (representing poorer sorting) at stations 8 and 9. It was observed during sampling that very coarse material occurred beneath several inches of finer sand at the surface. This downward coarsening was encountered on all three sampling occasions. The large value of σ_ϕ apparently arises from a mixture of two size distributions in the samples. The lower material may represent a coarse layer from the winter beach, now exposed at low water.

The lower graph of figure 11 represents the high water sand samples. The absence of the maximum at stations 8 and 9 suggests that the coarse layer has been covered to a greater depth by the spring beach accretion.

The straight line in the graph indicates a gradual improvement in sorting from south to north. The parallelism of this line with the dashed line in the upper figure indicates that the general change in sorting is the same for both levels of samples, if the lag effect is discounted. The relation in the lower graph is typical of many beach sands. Generally the sorting improves in the direction in which size decreases, which is the case here.

e. Relation between Sand Size and Beach Slope. - Figure 12 shows beach slope plotted against the corresponding geometric mean diameter of the mean water level samples collected at the time the slope was measured. The data show a distinct trend, and the finite abcissa at zero slope suggests a positive exponential function. This is shown in figure 13; the fair agreement with the straight line suggests that the function holds at least to a first approximation. From the graph,

$$r = R_o e^{+bS} \quad . \; . \; . \; . \; (2)$$

Here r is the size associated with any slope of S degrees; R_o is the initial size for S = 0; and b is a coefficient of size increase. If size is taken as the independent variable the function becomes logarithmic:

$$S = (1/b)\log_e(r/R_o) \quad . \; . \; . \; . \; (3)$$

Equations 1 and 2 may be used to examine the relation between slope and distance, which was illustrated in figure 5. Those data fail to yield a straight line on either double log or semi-log paper. However, by eliminating r between the two equations, S may be found as a function of x:

$$R_o e^{bS} = r_o e^{-ax}$$

$$S = -(a/b)x + (1/b)\log_e(r_o/R_o) \quad . \; . \; . \; . \; (4)$$

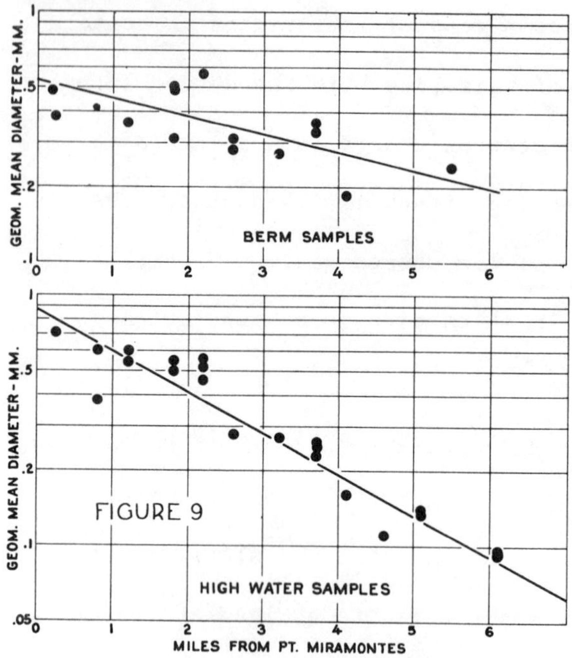

OBSERVED VARIATION OF SAND SIZE WITH DISTANCE FOR BERM AND HIGH WATER SAMPLES

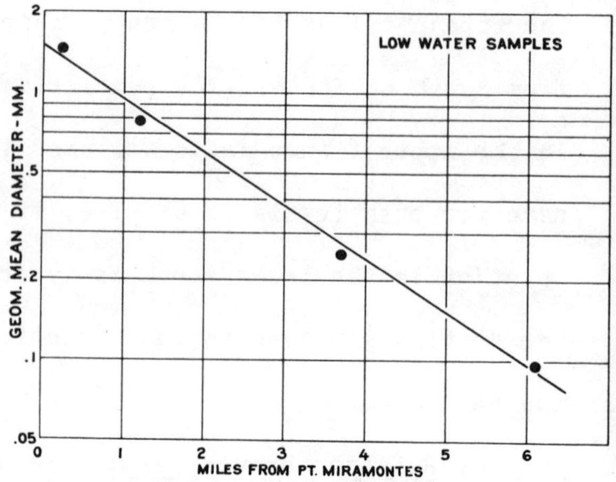

FIGURE 10

OBSERVED RELATION BETWEEN SAND SIZE AND DISTANCE OF LOW WATER SAMPLES

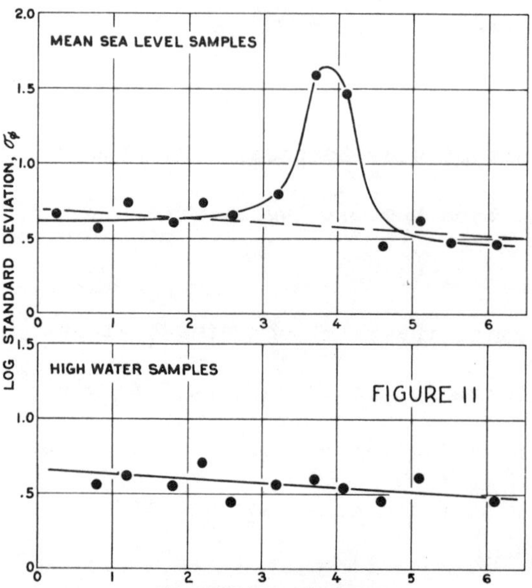

OBSERVED VARIATION OF SAND SORTING WITH DISTANCE FOR MEAN SEA LEVEL AND HIGH WATER SAMPLES

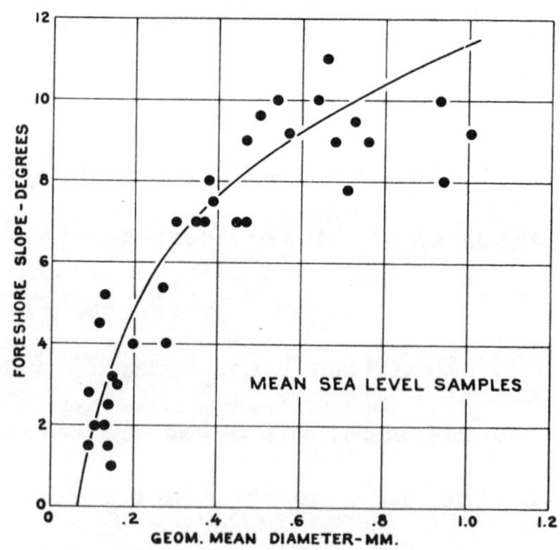

OBSERVED RELATION BETWEEN FORESHORE AND SAND SIZE FOR MEAN SEA LEVEL SAMPLES

FIGURE 12

The predicted relation is linear, in which $-(a/b)$ is the slope, and $(1/b)\log_e(r_o/R_o)$ is the y-intercept. The numerical values of the several constants may be read from figures 8 and 13.

Table 7

Constants from Figures 8 and 13

$r_o = 1.25$ mm.

$R_o = 0.07$ mm.

$a = 0.41$ (1/mile)

$b = 0.23$ (1/mile)

Substituting these values in equation 4, we obtain:

$$S = -1.78x + 12.52$$

The coefficient of x represents the tangent of nearly 61°. The predicted line is shown in figure 14, which is figure 5 redrawn with equal scale units for degrees and miles. The agreement is excellent beyond about the first mile from Point Miramontes. The lower slopes in the initial parts of the beach can be explained as being due in part to boundary conditions. Rock reefs in this stretch, extending outward from shore, act partly as groins, with the result that the normal beach slope is reduced in their vicinity.

f. Waves and Currents. - The waves in Halfmoon Bay are much larger at Point Miramontes than near Pillar Point. This situation is illustrated in Plate II. A visual device was used to measure the relative wave heights from shore. It consisted of a slot cut into a sheet of closely-ruled graph paper, which was held at a fixed distance in front of the eye, and the wave height at the plunge point was measured on the scale. The first measurements were made with some skepticism, but after it was found that rather consistent results were obtained, a systematic series of reading was made. Table 8 includes all the data, based on averages of twelve wave heights at each station. The average wave periods were also indicated.

Observations were also made on April 25 from a boat anchored offshore. The depth of water at the wave crest and trough, and the time of passage of the crests were recorded for 25 successive waves. The sea was relatively rough, but four sets of satisfactory data were obtained. The average height and period were determined for each station, and the plotted curves of figure 15 show the results.

During the observations waves approached the bay from the northwest, and a fresh northwest breeze was blowing. Wave measurements on shore the same day yielded periods shorter than from the boat, but this is attributed in part to the running surf, which rendered it difficult to sort out the individual periods.

The combined wave data, including both shore and boat observations, are shown in figure 16. The arbitrary scales of height were converted to their numerical equivalent of approximately 1 unit = 1.5 feet. The wave height at Point Miramontes was of the order of 7 feet at the plunge point. Each set of observations is indicated separately on the graph. In each instance (except possibly on April 18) the waves approached the bay from the northwest, but the effects of refraction were such that the waves appeared to be essentially normal at all points along the shore.

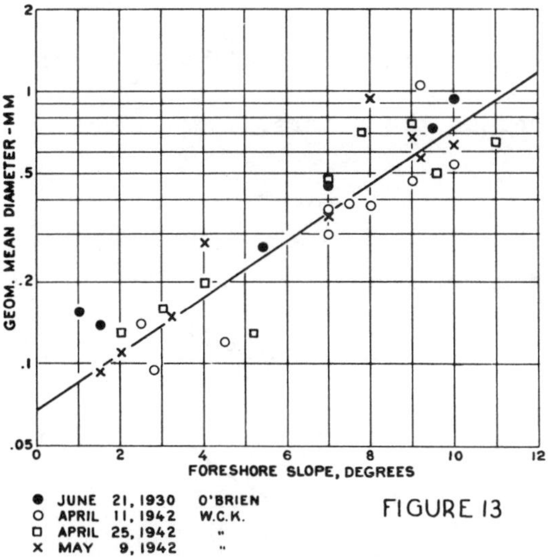

FIGURE 13

OBSERVED RELATION BETWEEN SAND SIZE AND FORESHORE SLOPE

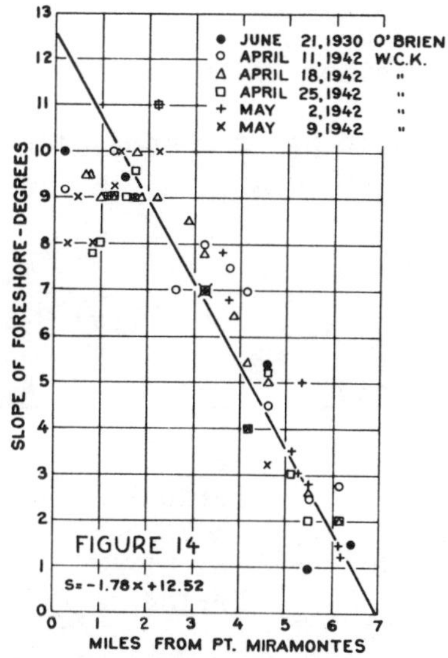

FIGURE 14

PREDICTED RELATION BETWEEN FORESHORE SLOPE (S) AND DISTANCE (x)

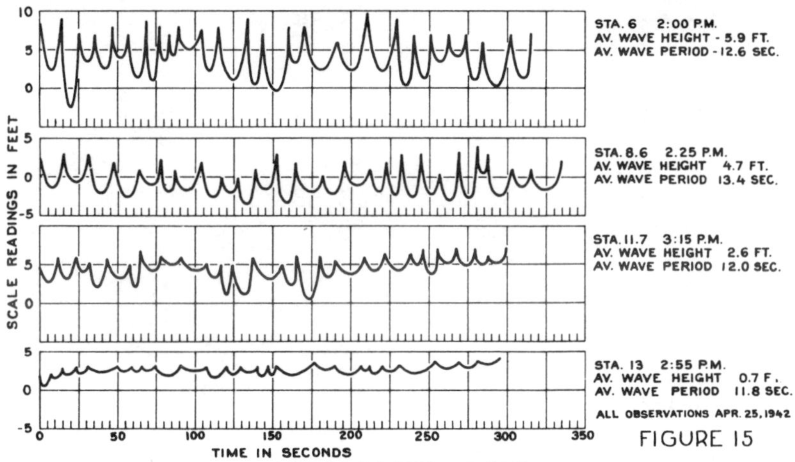

FIGURE 15

WAVE OBSERVATIONS

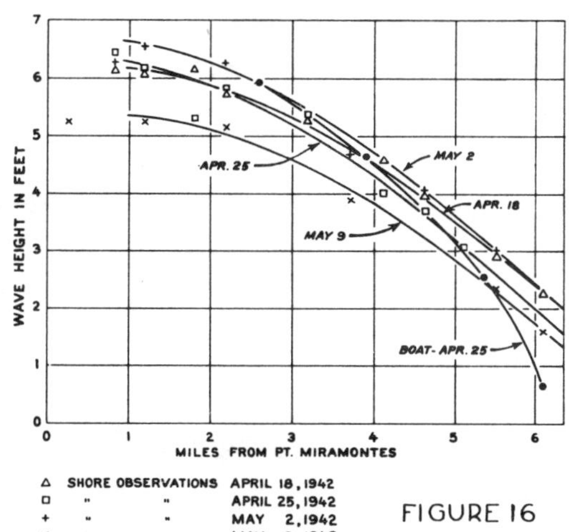

FIGURE 16

OBSERVED VARIATIONS OF WAVE HEIGHTS ALONG BAY SHORE

The weather was much milder on May 9 and the wave height was less, but the curve has the same general trend as the others.

Table 8

Wave Heights and Periods

Station	Data	4/18/42	4/25/42	5/2/42	5/9/42
1	Height*				3.50
	Period**				10.4
2	Height	4.10	4.30	4.18	
	Period	9.0	10.6	12.5	
3	Height	4.05	4.12	4.38	3.50
	Period	9.5	10.7	12.5	11.5
4	Height	4.10	3.54		
	Period	9.8	9.5		
5	Height	3.80	3.88	4.18	3.42
	Period	8.8	10.4	10.7	10.1
7	Height	3.50	3.58		-
	Period	9.9	9.7		12.5
8	Height			3.11	2.58
	Period			11.1	11.8
9	Height	3.04	2.68		
	Period	8.7	-		
10	Height	2.62	2.46	2.70	
	Period	-	-	-	
11	Height		2.04		
	Period		9.9		
12	Height	1.95		2.00	1.57
	Period	9.2		11.9	11.6
13	Height	1.50			1.05
	Period	9.3			11.4

* Arbitrary scale units. 1 unit = approximately 1.5 feet.
** Periods in seconds.

NB- Boat data are summarized in figure 15.

The boat data were transformed to the shoreline scale by adjusting the observed height at station 6 with the shore observations. The remaining points were plotted according to the ratio thus obtained, and are shown as black circles on the graph. The agreement between boat and shore data is fair; the steeper slope of the boat data may be a result of the necessity for anchoring the boat in fairly deep water to avoid surf, so that the waves were not always at corresponding points in their approach to the shore.

The average period of all waves observed during this study is of the order of 11 seconds. This corresponds to an average wave length of 620 feet in deep water.

Observations on currents and beach drift were much less satisfactory than wave observations. No equipment or colored dyes were used. What little data were obtained were contradictory, with marked reversals of direction due partly to conflicting currents set up by the beach cusps. O'Brien's 1930 report indicates a similar situation, with currents reversing their direction abruptly, or leaving the shore at an angle. On May 9, at station 5, a strong southward current along a shallow bar was found to run at 2.5 feet per second. Conversations with local fishermen indicated a complex current system, which varies with the season and the weather. One believed that the strongest currents were toward the north, due to winter storms, but that drift was to the south in quieter weather. Another was of the opinion that circulation within the bay centered about stations 5 and 6, because driftwood tends to collect along this stretch of the beach. It is possible that currents in the open sea to the west may generate slow eddies within the bay, which in turn are modified by local winds and waves.

Indirect evidence of the direction of shore drift is obtained from the beach material itself. The average sand size decreases markedly to the north, and the sorting improves to the north. This suggests drift of the sand from Point Miramontes to Pillar Point. Fine sand is common in the pebbly pocket beach west of the rock spur at Pillar Point, but very seldom are pebbles found to the east of the spur. This also suggests a predominant drift from south to north.

A sand pit was formerly operated at the mouth of Pilarcitos Creek. Operations began about the turn of the century and continued until nearly 10 years ago. As much as 30,000 cubic yards of sand were removed annually, according to O'Brien. Such removal would impoverish the beach in a down-drift direction, and it is possible that the relatively recent erosion between Miramar and Princeton may have arisen from this interference with normal beach drift. The erosional aspects are discussed later, but if the reasoning applies, it indicates predominant northward shore currents in the bay.

Despite the lack of direct observational evidence, the writer agrees with O'Brien's conclusion of 1930 that the net shore drift is from Point Miramontes toward Pillar Point.

g. <u>Wave Refraction and Wave Energy in Halfmoon Bay</u>. - Differences in wave height along the bay shore are evidence that refraction controls the distribution of wave energy along the shore. The refraction pattern is a function of the boundary conditions within the bay, of the direction of wave approach, and of the wave period. A refraction diagram was prepared, using the principles described by O'Brien and Mason in their summary of wave theory. Figure 2 was used as a base map. Waves of 12-second period were brought in from the northwest, to reproduce approximately the conditions on April 25.

Figure 17 shows the refraction pattern. It was assumed for simplicity that refraction does not occur until depths of 100 feet or less are encountered. The projected distances of travel of the wave were taken as the product of the average velocity between the limiting depths and the time of advance (chosen as 36 seconds for convenience). The effect of the southeast reef is probably greater than that shown on the diagram, because the graphical results are in part a function of the time interval chosen. That cross-waves are generated by the reef is supported by observation, however. On May 9 waves of essentially the same period were approaching the shore from two directions at station 5. The shore current inside the plunge line was to the south, suggesting a stronger effect of the southward refracted waves. The interference pattern may also bear on the observations of a local fisherman that drift tends to accumulate in this vicinity.

The principle that adjacent orthogonals to the wave crests inclose regions across which no power flows laterally was used to appraise the energy distributed along the beach. The distances between the orthogonals along the beach in figure 17 were measured, and from their reciprocals the proportional power was computed. The relative wave energy along the beach was also computed from the wave observations on the principle that the

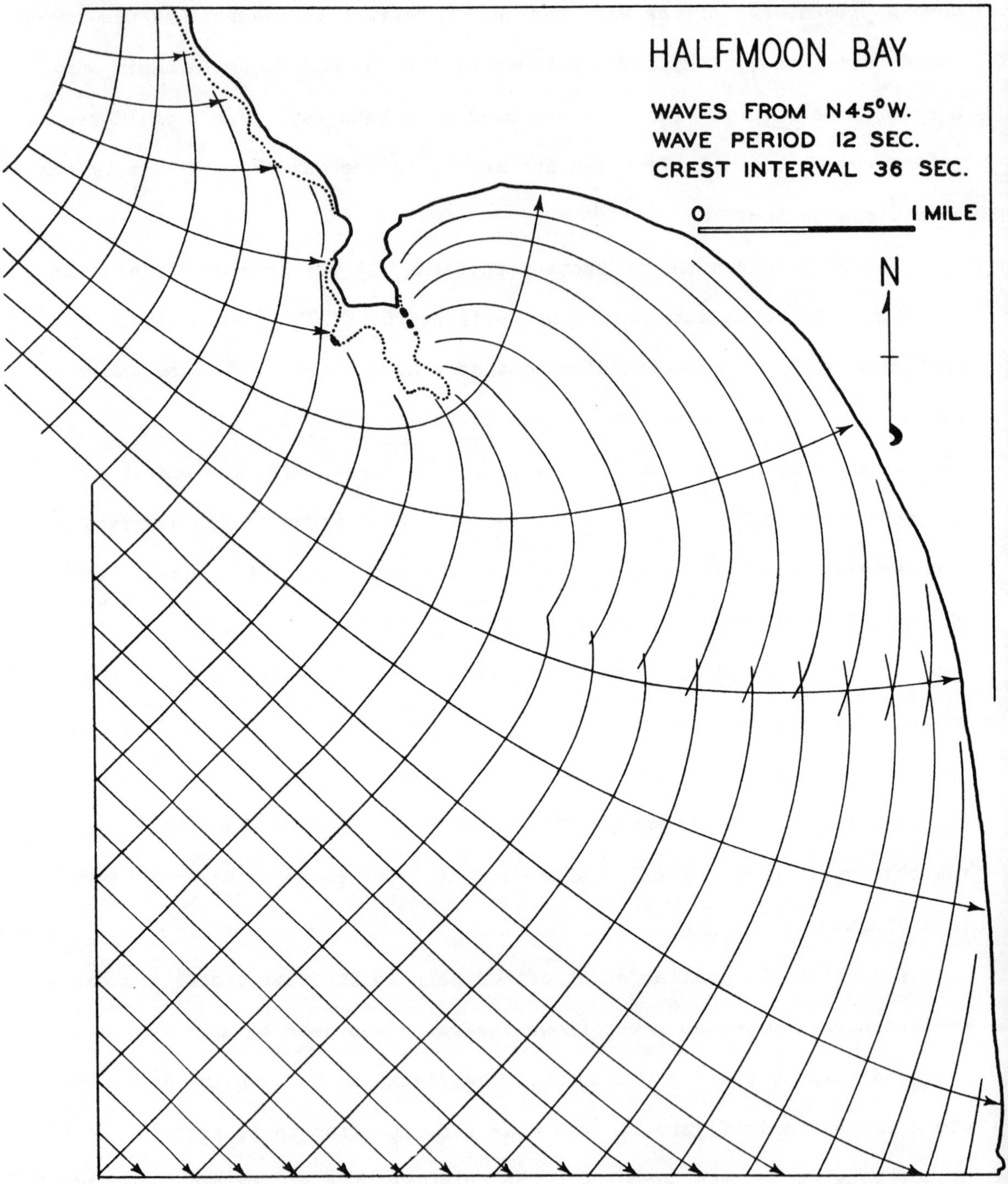
HALFMOON BAY
WAVES FROM N45°W.
WAVE PERIOD 12 SEC.
CREST INTERVAL 36 SEC.
0
1 MILE
N

energy of any wave is proportional to the square of its amplitude. The results of the computations are given in table 9 and are shown in figure 18.

Table 9

Relative Wave Energy, $(h/h_o)^2$

(Here h_o is the greatest h recorded on the given date. Cf. Table 8.)

Station	4/18/42	4/25/42	5/2/42	5/9/42
1				1.00
2	1.00	1.00	0.92	
3	0.98	0.92	1.00	1.00
4	1.00	0.68		
5	0.86	0.82	0.92	0.96
7	0.73	0.70		
8			0.51	0.55
9	0.55	0.39		
10	0.41	0.33	0.38	
11		0.23		
12	0.23		0.21	0.20
13	0.134			0.09

Boat Data

Station	4/25/42
6	0.80
8.6*	0.50
11.7	0.156
13	0.011

Refraction Diagram Data

Station	4/25/42
2	1.00
3.5	0.85
6.6	0.71
10	0.48

* Station 8.6 lies 0.6 of the distance between 8 and 9.

The curved line in figure 18 shows the trend of relative energy per unit crest length per unit time on the basis of the shore observations. The refraction diagram gives results in good agreement with this trend, although the slope of the line joining these points is less steep than that for the wave observations. Nevertheless, the diagram indicates at least the order of energy decrease along the shore, and in the absence of other data would be acceptable as a first approximation.

4. Relations between Wave Energy, Beach Slope, and Sand Size

In the first section of this report it was pointed out that a complete beach study includes information on the boundary conditions, the beach materials, and the energy of the system. It is appropriate to consider how the relative energy along the shore corresponds with the other observed data.

Figure 19 shows a linear relation between beach slope and relative wave energy at all stations where the two were measured at the same time. The equation of the straight line is

$$S = kE + K \quad . \; . \; . \; . \; (5)$$

where E is the relative energy. The constant k has the value 10.0, and K has the value 0.6.

The widest scatter of the points occurs where the slope is most steep. These points also fell below the predicted straight line of equation 4 in figure 14. Thus the boundary conditions, represented by rock reefs, do exert an influence on the beach slope, although the general agreement of the trends in figures 5 and 18 indicates that the energy relation is probably the more important of the two

The relations expressed in the equations previously given permit further predictions, first of the relation between relative energy and sand size, and second between relative energy and distance along the shore.

Inasmuch as beach slope and energy are linear, and beach slope and sand size are logarithmic (equation 3), one may predict that sand size is a positive exponential function of relative energy. This may be shown by eliminating S from equations 3 and 5 and solving for r:

$$(1/b)\log_e(r/R_o) = kE + K$$

$$r = R_o e^{bK} e^{bkE} \quad . \; . \; . \; . \; . \; (6)$$

Equation 6 indicates that if r is plotted against E on semi-log paper, a straight line should be obtained, in which the parameters may be determined from previous knowledge. Thus from table 7 we have $b = 0.23$ and $R_o = 0.07$. From figure 19, $k = 10.0$ and $K = 0.6$. Hence $R_o e^{bK} = 0.08$, and $bk = 2.3$, so that the predicated line is

$$r = 0.08\ e^{2.3E}$$

The straight line of figure 20 is drawn according to the prediction, and indicates a satisfactory agreement.

In similar manner the functional relationship between relative energy and distance along the shore may be found by eliminating S between equations 4 and 5 and solving for E:

$$kE + K = -(a/b)x + (1/b)\log_e(r_o/R_o)$$

$$E = -(a/bk)x + (1/bk)\log_e(r_o/R_o) - (K/k) \quad . \; . \; (7)$$

Again the parameters of the line may be predicted. The slope of the line is $-(a/bk) = -0.178$, and the y-intercept is $1.252 - 0.06 = 1.192$. Figure 21, redrawn from figure 18, shows the predicted line. The slope has been drawn in accordance with the different scale units. The values above 1.0 on the energy axis are imaginary, and indicate that the linear relation begins at the point along the beach where refraction due to Pillar Point and the southeast reef begin to manifest themselves.

The internal consistency of the observational data from Halfmoon Bay strengthens the writer's belief that if relatively simple natural situations are chosen, it is possible to discern the underlying physical laws which control the behavior of matter in the environment. A purely theoretical approach, with a prediction of the exact nature of all the functions, is probably not possible at the present state of knowledge, but a combined observational and analytical attack is feasible. From a theoretical viewpoint, equations 5 and 6 are probably the most fundamental of those given, because they bring in the relations of sand size and beach slope to the relative wave energy. The rates of change of these two equations are:

$$d \quad dE = k$$

$$dr/dE = bkr$$

From these relations it would appear that the slope is a linear function of energy essentially independent of other factors, whereas average sand size is influenced by itself and by the slope relations along the beach. Before the full implications of these equations can be analyzed, additional beaches of the same general type as Halfmoon Bay should be investigated.

5. Erosion and Deposition at Halfmoon Bay

Comparison of earlier maps with the latest chart of Halfmoon Bay shows no noticeable change in the general alignment of the shore. This suggests a stable strand, but field evidence throws some doubt on the subject. A stable beach would have a source of material to replenish the annual down-beach drift of sand, so that a steady state shoreline would occur along the entire stretch. At the down-beach end of the bay would be a trap or other means for disposing of the transported material, such

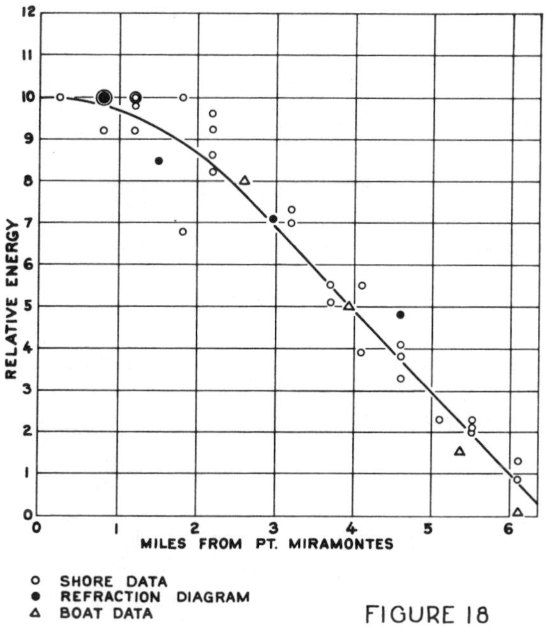

FIGURE 18

OBSERVED RELATION BETWEEN RELATIVE WAVE ENERGY AND DISTANCE

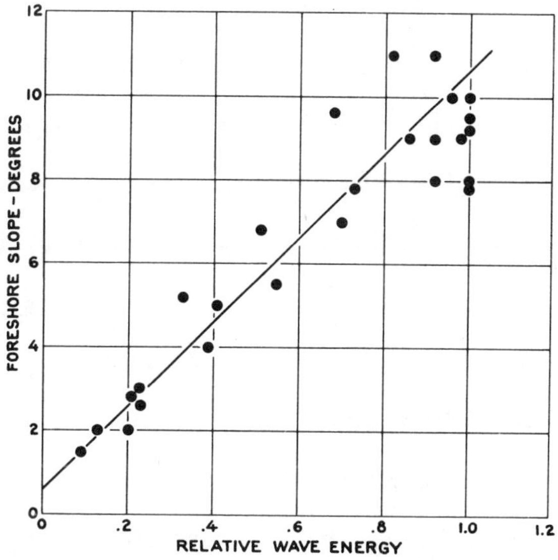

FIGURE 19

OBSERVED RELATION BETWEEN FORESHORE SLOPE AND RELATIVE WAVE ENERGY

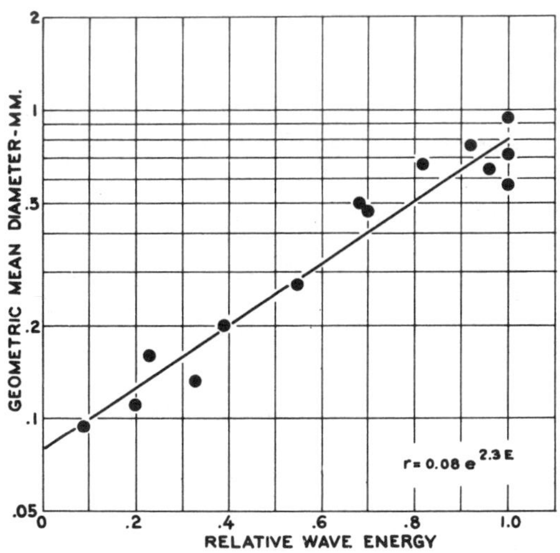

FIGURE 20

PREDICTED RELATION BETWEEN SAND SIZE (r) AND ENERGY (E)

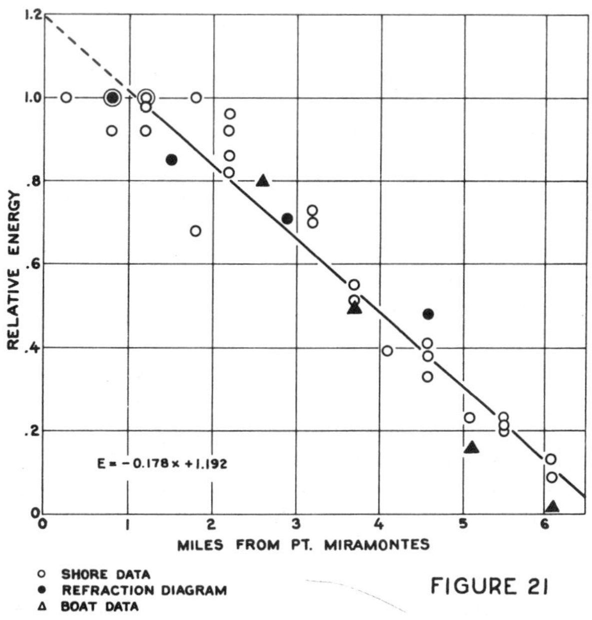

FIGURE 21

PREDICTED RELATION BETWEEN RELATIVE WAVE ENERGY (E) AND DISTANCE (X)

as a belt of sand dunes. Along Halfmoon Bay erosion is active near Point Miramontes, but it is also active near Princeton. Moreover, there are no dunes near Pillar Point, and the dunes which do occur are near the center of the bay shore. The situation along the bay is therefore not simple, and the following sections attempt to explain the observed conditions.

a. Estimates of the Rate of Erosion. - At the base of the cliff near Point Miramontes and for a mile north occur several semi-conical masses of freshly slumped terrace material, which are subject to active wave attack. An estimate of the volume of one of these masses yields the value 70,000 cubic feet, so that some 200,000 cubic feet of sand and gravel had recently slumped from the terraces. (See Plate III.) The material is relatively unconsolidated and probably slumps readily as a result of undercutting. If only 10 such slides occur per year over the mile stretch, approximately 700,000 cubic feet of material is washed away annually. This is equivalent to about 26,000 cubic yards. It is perhaps a coincidence that this figure is nearly the same as O'Brien's estimate of annual sand transport based on operations of the sand pit at Pilarcitos Creek.

A local resident informed the writer that about 25 years ago a road ran along the edge of the terrace south of Princeton. He estimated that its eastern margin was perhaps 40 feet from the edge of the cliff. Since that time erosion has cut away all but small patches of the pavement. The cliff is about 20 feet high, and the eroded stretch is approximately a mile long. Hence the volume removed is equivalent to about 4 million cubic feet of terrace material. Allowing 25 years for the process, the annual removal becomes 6,000 cubic yards. Add this to the 26,000 cubic yards from the southern cliffs, and the total annual removal per year along the bay shore is of the order of 32,000 cubic yards.

b. Estimates of the Rate of Accretion. - If some 30,000 cubic yards of terrace material is annually fed to the beach, there should be evidence of its accumulation somewhere within the closed system, or a mechanism for removing it from the system. Two possibilities are open, (1) the beach material is abraded to fine mud during its movement and is carried out to deep water beyond the bay; (2) the material accumulates partly in the sand dunes along the shore, and partly over the bay bottom.

Recent studies have largely supported the view that abrasion is essentially a second-order term in accounting for size decrease along beaches or in streams. A recent report by the Beach Erosion Board establishes fairly conclusively that abrasion on sand beaches is essentially negligible, a conclusion with which the writer concurs. A study of the sphericity, roundness, and mineral content of the sand samples will shed some light on this question, but from the writer's experience it is most likely to support the present view.

Selective transportation is a much more important process than abrasion in the segregation and removal of material. Selection may be affected by wind or by waves and currents. Sand dunes in general may be considered as traps for beach sand. They occur where deposition exceeds transportation, or where favorably directed winds remove sand from the beach. Along Halfmoon Bay dunes occur most prominently between stations 7 and 8, and less so between stations 5 and 6. The dune belts are very narrow and relatively low. The edge of the terrace projects through the dunes locally, indicating that they are mainly a veneer of sand blown up on the terrace. The photographs of Plate IV illustrate these conditions. If one estimates the dune volume liberally, by assuming them to be 1 mile long, 100 feet wide, and 15 feet high, he arrives at the value 7.5 million cubic feet, which is equivalent to about 280,000 cubic yards. This is less than 10 times the annual sand movement along the beach. Considering that the bay has been present for several thousand years at least, this figure becomes negligible. Hence even under a selective transport assumption, the most reasonable explanation is that the material is carried out into the bay or beyond.

The area of the bay bottom is approximately 5.5 square miles within the zone inclosed between the curve of the bay shore and a straight line from Pillar Point to Point Miramontes. This is equivalent to approximately 17 million square yards. Hence it would require more than 180 years for the annual beach increment to form a layer 1 foot deep over the bay bottom. The corresponding average annual deposition of about 0.006 foot would not be detectable over the time that reliable surveys are available. Even considering that the material would not be uniformly distributed over the bottom, a slight shoaling of water here, and the filling of a depression there, would hardly be detected.

c. Topographic Analysis of the Sand Dunes. - The topography of the dunes between stations 7 and 8 is hillocky, and the configuration of the surface suggests predominant sand movement to the southeast. This suggests that most of the sand is picked up near station 8 and blown up-beach diagonally to station 7, rather than being blown inland normal to the beach at both stations. This was tested by taking sand samples from the dune crest at the two stations. The results show that the sand is finer to the south, in contrast to the beach sand, which is finer to the north. Table 10 gives the size data.

Table 10

Comparison of Dune and Beach Sand at Stations 7 and 8

Station	Beach Sand		Dune Sand	
	Geom. Mean		Geom. Mean	
7	0.40 mm.	0.80	0.19 mm.	0.37
8	0.33	1.60	0.21	0.37

Although two samples are not sufficient to demonstrate the direction of transport, they do support the topographic analysis of the dunes.

The location of the principal dunes near station 7 rathen than near station 13 is probably a function of sand size and of prevailing winds. The fine sand at station 13 remains wet indefinitely, whereas that at station 7 dries rapidly on a relatively high berm. The trough east of Pillar Point may also influence the northwesterly winds so that they are directed against the beach in the vicinity of station 7. In terms of the introductory remarks in this report, a complete energy analysis of the system should include measured data on the winds. This aspect was not included in this study.

6. Concluding Remarks

The present report includes mainly the field observations made during the study. Much more detailed information is available on the geology of the terraces, but this material is deferred until the shape, roundness, and mineral composition of the beach samples are studied.

The writer believes that this report in its present form offers a strong argument in favor of detailed scientific studies of selected beaches as a supplement to the engineering studies now made. The principles developed during such studies will undoubtedly find an immediate application to problems of beach stabilization and control. The role of energy in the system deserves much more study, and reliable and convenient methods should be developed to obtain data on wave heights and periods, both in the open sea and near shore. Methods of estimating currents should also be improved.

The writer is indebted to Professor M. P. O'Brien for many helpful suggestions regarding this study. He is also indebted to Dr. C. M. Gilbert of the Department of Geology at Berkeley for his assistance in the field work. Dr. M. A. Mason, Principal Engineer of the Beach Erosion Board, made helpful suggestions in the manuscript.

Part II

COASTAL HYDRODYNAMICS

Editors' Comments on Papers 5 Through 14

The discipline of coastal hydrodynamics is shared by engineers, applied mathematicians, geologists, oceanographers, and perhaps one or two other academic divisions as well. What these contributors have in common is a fascination with the mechanisms by which waves, tides, and coastal currents interact and move sediments. With separate and distinct backgrounds, each of these groups has focused on a set of problems that has proved to be relatively slow to reveal basic mechanisms. The great progress that has been made—both in field and laboratory experiments, as well as the theoretical analysis—is certainly a result of the range and diverse approaches of these investigators.

The papers presented here (with the exceptions of Johnson's review of the general subject, and Eagleson and Dean's analysis of forces on sediment particles) follow the development of the contemporary models for nearshore circulation. While being only a single element of coastal hydrodynamics, nearshore circulation—and more specifically, longshore currents and sediment transport—is fundamental to this subject. A second element of this model, the onshore–offshore transport, is briefly discussed in Part III ("Beach Processes"). Unfortunately, this latter area is not as well understood today, perhaps in part because of the added complexities of dealing with motions through the zone of breaking waves.

It is interesting to note that most of the authors in this section continue to publish today. The continued refinement of their ideas, and their continued interest and commitment to these subjects is a major reason we have achieved our present level of understanding. We also point out that it is no accident that Inman and Longuet-Higgins have more than a single contribution. Their work over the past two decades, and that of their students and collaborators, has been monumental and internationally recognized.

We anticipate an accelerated and exhaustive study of both onshore and alongshore mechanics during the last quarter of this century. The framework for these studies is in large part provided by the papers presented here.

As an introduction to this section, no discussion is more suitable than this treatise by J. W. Johnson. We placed it here rather than as the first paper because of the historical impact of the preceding selections. Johnson's paper is not a benchmark in that it does not represent a contribution that resulted in a significant change in the thrust of research in the discipline, but it is a classic introduction to the study of the impact of water waves on beaches. Readers unfamiliar with coastal processes, terminology, and dynamics will appreciate the author's clarity and depth of presentation. His detailed review of the literature provides an excellent preparation for the remaining papers in this volume. Johnson discusses several of these references, and his comments provide a historical context as well as a scientific guide for their understanding. His review of the engineering applications of coastal process research is also a valuable contribution to both this volume and the discipline.

A great deal of our literature concerning coastal processes grew out of the need for inshore information for amphibious planning during World War II. Significant among these was the development of analytical tools for the computation of longshore currents. Regardless of whether the concern is for the safe landing of equipment on the beach, or engineering of the beach itself, a sound estimate of the magnitude and direction of the nearshore currents is a necessity.

Putnam, Munk, and Traylor's paper (Paper 6) combines the rigorous analysis of field data with both laboratory observations and theoretical models. This complete approach to the problem exemplifies the thoroughness of their studies. (The review by C. J. Galvin, also included in this volume, contains an excellent discussion of the Putnam et al. paper.) Although there are some errors in their theoretical development of longshore currents based on the conservation of momentum, this study proved to be the foundation for most of the latter derivations. Recent theories for longshore currents, including the work of M. S. Longuet-Higgins (also presented in this volume), have begun with the basic ideas stated in this early work and added the refinements of a more sophisticated analysis of wave mechanics.

The analysis of longshore currents presented by Putnam et al. is primarily concerned with the somewhat simplified problem of currents along a short, straight shoreline. Shepard and Inman extend this discussion to a more regional scale and analyze the interaction of longshore currents with the complex nearshore bathymetry; however, the generalization from a simple longshore current to the concept of a nearshore circulation is not achieved without a considerable loss in theoretical analysis.

The terminology and classification introduced by Shepard and Inman has continued to the present. More recent studies, including several papers presented in this volume—e.g., Bowen and Inman (Paper 11, 1969)—have attempted to derive theoretical analysis for these flow patterns, with increasing degrees of success. Particular attention has been given to the relationship between the bottom contours and the positions and orientation of rip currents. Therefore this earlier study at the Scripps Institution is an invaluable reference to the reader in understanding the relationship of these rip currents to the larger patterns of circulation along the open coast.

No collection of papers dealing with coastal hydrodynamics would be complete without a contribution from the research laboratory at the Massachusetts Institute of Technology. The experimental and theoretical studies from MIT, under the direction of Arthur T. Ippen, have played an important role in the growth and development of our discipline. The paper selected for this volume is no exception.

The mechanics of coastal sediment transport is a subject that has been extremely slow in yielding to theoretical analysis. The paper by Peter Eagleson and Robert Dean (Paper 8) is one of the earliest attempts at applying small-amplitude wave theory to the description of this transport. Their effort, developed in a series of reports to the Beach Erosion Board, with Ippen and others, is summarized in this ASCE contribution. In particular, the authors develop a theoretical model for the initiation of motion of sand particles and a net profile of equilibrium. First considered in Cornaglia's treatise, these topics are also discussed in the paper by Miller and Zeigler, included in Part III.

We are still without a complete theoretical model of unsteady sediment transport. There is little doubt, however, that these early efforts are the building blocks upon which such a model will ultimately be founded.

The contributions of Longuet-Higgins and Stewart to research in coastal processes have been significant from both the theoretical and the applied areas of the discipline. Their concept of the radiation stress has become fundamental to the modern analysis of the physics of water waves. The serious student of the subject will want to read their original papers published in the *Journal of Fluid Mechanics.* The present selection (Paper 9) was considered a more desirable choice for this volume because it attempts to narrow the gap between the theoretical hydrodynamics and the more applied coastal scientist and engineer. Unfortunately, space restrictions have necessitated the editing of this selection. The introduction of the radiation stress concept has been included, as well as its application to wave set-up. The radiation stress analysis has become one of the most important new developments in wave mechanics in recent years. As explained in this selection, many coastal phenomena observed either in the field or laboratory can be understood better from an application of the nonlinear properties of the waves. The concept of wave set-up is useful in the development of nearshore circulation models, as is further discussed by Sonu's paper in this volume (Paper 13). Of course, the explanation of longshore currents from the radiation stress analysis has been well documented in the literature, including the second selection by Longuet-Higgins, also included in this part.

The need for the development of an adequate theory for longshore currents has attracted a large number of contributors since Putnam, Munk, and Traylor began their research in support of the military operations for World War II. C. J. Galvin's review is a comprehensive study of these investigations and includes an examination of the principles that should be fundamental to any derivation of a physical model for the wave generation of this velocity field. As in the Putnam et al. paper, Galvin is concerned with the dynamics of flow along a straight open coast, without the complexities introduced by an irregular bottom topography.

After commenting on the problems associated with all the major theories for longshore currents, Galvin suggests that an empirical approach is the most promising method for developing a predictive model. This conclusion must be reconsidered in light of the more recent advances of both Longuet-Higgins (Paper 12, 1970) and Bowen and Inman (1969) presented in the following selections for this volume. Nonetheless, Galvin's review is an important reference for both the student and the researcher alike.

In addition to this review, C. J. Galvin has made a significant contribution to the understanding of longshore currents, especially with regard to laboratory and analytical studies.

The earlier discussions of nearshore circulation in this part deal primarily with the longshore current. In this next paper Bowen and Inman address the question of rip current dynamics, from both an experimental and a theoretical approach. The rip current was described by Shepard and Inman (Paper 7, 1950), and Bowen and Inman include these observations in their discussion.

The authors' analytical model uses the edge wave as the controlling mechanism for the dimensions of the nearshore circulation cell. Their study is a noteworthy contribution in applying edge waves to the interpretation of these circulation systems. For additional background on edge waves, refer to the papers cited by Bowen and Inman, especially those of Echart [1951] and Ursell (1952). The application of the theory of edge waves has been included in many of the recent studies, of which the present selection is but one example.

Anthony Bowen published a second paper on this subject ("Rip Currents, 1 Theoretical Investigations") in the same issue of the *Journal of Geophysical Research* that is of interest even though we are unable to include it in this volume. In it, Bowen uses the radiation stress and the related wave set-up and set-down to drive a longshore current and rip cell.

The application of the radiation stress to the generation of longshore currents is also discussed in Paper 12, in which Longuet-Higgins notes Bowen's contributions to this subject.

The application of the radiation stress concept to the theory of a longshore current is the subject of this next selection, also by Longuet-Higgins. Recognizing the fundamental work of Putnam, Munk, and Traylor (Paper 6), the author presents a new development based on the application of the conservation of momentum, with the radiation stress analysis.

Although a complete theoretical derivation for longshore currents generated by waves is still unfinished, this more recent approach appears most promising. A companion paper, "Longshore Currents Generated by Obliquely Incident Sea Waves, 2," by Longuet-Higgins also was published in this issue of the *Journal of Geophysical Research.* The further refinement of this approach is certainly interesting, but space limitations prevent its appearance in this volume.

One of the limitations in the development of theoretical models of nearshore circulation is the difficulty in making accurate field measurements. C. J. Sonu has spent considerable time and effort trying to solve this problem. His work provides some of the best data for evaluating the various theories cited in the literature.

Sonu has published several fine papers dealing with the nature of the currents and waves, and their interaction with the inshore bathymetry. This particular paper (Paper 13) was selected for its thorough discussion of the application of the concepts of radiation stress and wave set-up in the explanation of the circulation patterns, as well as reference to the possible influence of edge waves on the observed shoreline configurations. Sonu's comments draw heavily on the ideas presented in the other papers in this book, and thus it provides an interesting forum to test these theories with independent field observations. His conclusions are familiar, but nonetheless significant, restating the need for continued detailed field studies of the nearshore environment.

The earlier papers in this part have developed the relationship between shoaling waves and the resulting longshore currents. To the coastal geologists, explanations of these mechanisms are secondary to the transport of beach sediments by these currents. Komar and Inman, in Paper 14, address this question in an excellent attempt to deal with the transport problems with the same analyses that have been applied to questions concerning currents and waves. As such it is an ideal paper to close this part of the book.

The difficulties in conducting meaningful field studies are magnified when one attempts to quantify the movement of the sediment. Komar and Inman describe these difficulties in detail, and one can appreciate the significance of their theoretical developments in light of these complexities. This problem is the focal point of much of the present research in coastal processes, and their list of references is useful. Additional discussion can be found in *Waves on Beaches, and Resulting Sediment Transport,* edited by Meyers.

REFERENCES

Arthur, R. S. (1962). "A Note on the Dynamics of Rip Currents," *Jour. Geophys. Research* **67**:2777–2779.

Bowen, A. J. (1969). "Rip Currents. 1. Theoretical Investigations," *Jour. Geophys. Research* **74**:5467–5478.

Bruun, P. (1963). "Longshore Currents and Longshore Troughs," *Jour. Geophys. Research* **68**:1065–1078.

Davis, Richard A., Jr., and William T. Fox (1972). "Coastal Processes and Nearshore Sand Bars," *Jour. Sed. Petrology* **42**(2):401-412.

Dean, R. G. (1973). "Heuristic Models of Sand Transport in the Surf Zone," Sidney Australia: Conf. on Eng. Dynam. in the Surf Zone.

Eagleson, P. S., B. Glenne, and J. A. Dracup (1963). "Equilibrium Characteristics of Sand Beaches," *Am. Soc. Civil Engineers Proc., Jour. of the Hydraulics Div.* **HY1**:35–57.

Inman, D. L, and R. A. Bagnold (1963). "Littoral Processes," in *The Sea III,* M. N. Hill, Ed. New York: John Wiley & Sons, 529–553.

Ippen, A. T., and P. S. Eagleson (1955). "A Study of Sediment Sorting by Waves on a Plane Beach," Beach Erosion Board Tech. Memo 63.

King, C. A. M., and M. J. McCulloch (1971). " A Simulation Model of a Complex Recurred Spit," *Jour. Geology* **79**(1):22–36.

Komar, P. D. (1971). "The Mechanics of Transport on Beaches," *Jour. Geophys. Research* **76**(3):713–721.

Longuet-Higgins, M. S. (1953). "Mass Transport in Water Waves." *Phil. Trans. Roy. Soc.* **A245**:535–581.

McKenzie, P. (1958). "Rip Current Systems," *Jour. Geology* **66**(2):103–113.

Munk, W. H. (1949). "Surf Beats," *Am. Geophys. Union Trans.* **30**(6):849–854.

Saville, T. (1950). "Model Study of Sand Transport Along an Infinitely Long Straight Beach," *Am. Geophys. Union Trans.* **31**(4):555-565.

Schiffman, A. (1965). "Energy Measurements in the Swash-surf Zone," *Limnology and Oceanography* **10**:255–260.

Ursell, F. (1952). "Edge Waves on a Sloping Beach," *Proc. Roy. Soc.* (London), Series A **214**:79–97.

5

Reprinted from *Am. Assoc. Petroleum Geologists Bull.* 40(9):2211-2232 (1956)

DYNAMICS OF NEARSHORE SEDIMENT MOVEMENT[1]

J. W. JOHNSON[2]
Berkeley, California

ABSTRACT

Considerable research on the basic mechanics of shoreline processes has been done over the last 15 years by geologists and engineers. Much of this research was done during World War II, but in more recent years numerous investigations have been made in connection with the qualitative and quantitative effects of shoreline structures on nearshore sediment movement. This paper summarizes some of the more important investigations that have been made on the many phases of this phenomenon.

INTRODUCTION

The geological aspects of shoreline processes have been studied actively in the past, and are described in detail in various geology journals. The first comprehensive discussion of these processes was by Gilbert (1890). Johnson (1919) later formulated the concepts and nomenclature that are now generally accepted by workers in this field of study. Over the years, however, certain additions and modifications in Johnson's descriptions of terms and processes have been necessary (Shepard, 1952; Wiegel, 1952). Most of the early publications on shoreline processes have been qualitative in character. In recent years, however, two important events occurred which created conditions under which rather extensive quantitative data on shore processes have been assembled. One of these events was World War II, which created a heavy demand for detailed information on beach and surf characteristics in connection with the War's extensive amphibious operations, particularly in the Pacific Theater. The second factor which has made necessary an extensive program of shoreline studies is the Congressional Act which designated the Corps of Engineers, U. S. Army, directly responsible for evaluating the effects of the construction of a littoral barrier on the adjacent shoreline.

Considerable research on the basic mechanics of shoreline processes has been done in the last 15 years. The solution of many of these problems has necessitated a close collaboration between the geologist and the engineer. The results of many of the investigations sponsored by the Military during World War II have now become available for peacetime use in planning and designing shoreline structures, as well as for evaluating the effect of such structures on the adjacent shorelines. Fortunately, investigations of the basic mechanics of shoreline processes are being continued by various universities and government agencies. For example, The Beach Erosion Board, Corps of Engineers, is actively engaged in a rather comprehensive research program on nearshore sediment movement and related problems.

[1] Manuscript received, February 20, 1956.

[2] Professor of Hydraulic Engineering, Department of Engineering, University of California. The writer prepared this paper while engaged in research made possible by a John Simon Guggenheim Fellowship.

The general character of many of the investigations completed in recent years is briefly summarized. The sediments discussed are the relatively coarse granular materials occurring along our shorelines; that is, the sands which are found on the beaches and cover the ocean bottom near shore and which are affected by the action of the surface waves. The relatively fine material in estuaries and protected bays is not considered.

CHARACTERISTICS OF WAVE ACTION

Over the period of a few years the general configuration of a shoreline appears to be relatively permanent, unless modified by a man-made structure. Superimposed on a shoreline of apparently permanent form, however, are the local changes of the beaches which result from the daily, seasonal, and annual variations in the intensity and direction of wave attack. The most important variables involved in the character of beaches are the waves, the beach material, and the headlands containing the beach. For a particular beach, the waves may change in characteristics from day to day, whereas the beach responds comparatively slowly to their attack except during periods of great storm waves. The position of the water table also may influence the rate of aggradation and degradation of a beach (Grant, 1948). The general character of a particular beach at a particular time is the long-term result of the waves to which it is normally subjected and the supply and type of the sediment available.

The waves most commonly impinging on a shoreline are those generated by wind. Waves may be generated either by local winds or by storms at a relatively great distance from the shoreline (Bigelow and Edmondson, 1947). Locally generated waves generally are referred to as "seas" and waves generated by a distant storm are referred to as "swell." Workable relationships between the characteristics of waves and a generating wind are fairly well established, and convenient graphical methods are available for estimating the wave characteristics (height, period, and direction) from known wind conditions (U. S. Army, 1954a).

The two types of waves (sea and swell) usually exist simultaneously at any time in open water. The swell, however, might be completely obscured by the local wind waves or seas, except near shore where the swell has "peaked up" to a greater height. The swell generally contains much more energy than the local wind waves; consequently, it is the important sediment-moving factor when the energy is released by "breaking" at a shoreline.

When waves approach a coast with crests parallel with the shore, important transformations in the wave characteristics occur. The wave velocity and the wave length decrease, while the period remains the same, and the total energy is slightly reduced by bottom friction. The height first decreases by a small amount as the waves move into shallow water and then increases up to the point of breaking. A maximum wave height occurs at this point. The breaking action is accompanied by relatively high accelerations and velocities of the water particles. No theory has been developed for the complicated motion of these water particles; however, extensive laboratory studies have been made by Iversen (1951) to yield information on the details of the internal motion of breaking waves. The result of breaking is a highly turbulent flow condition that is capable of placing large amounts of sediment into suspension. Field measurements (Mason, 1953) show a relatively large percentage of sediment in suspension near the plunge point of a breaker, as compared with the region on either side of this point.

Waves approaching a shoreline at an angle not only undergo the transformations mentioned, but they also are bent, or refracted, because the inshore portion of the wave front travels at a lower velocity than does the portion in deeper water; consequently, the waves swing around and conform to the bottom contours. The characteristics of the bottom topography, the wave period, and the wave direction in deep water determine the pattern of the wave crests in shallow water. The result of refraction is a change in height and direction of the waves. With very irregular bottom conditions, the heights may differ greatly between closely adjacent points along a coast. The magnitude of the height and direction changes resulting from refraction can be estimated by use of a refraction diagram. The details of the construction of such diagrams are described elsewhere (Johnson, O'Brien, and Isaacs, 1948). The interpretation and use of refraction diagrams by both the geologist (Munk and Traylor, 1947) and the engineer (Johnson, 1953a) also are described elsewhere.

Although waves tend to become parallel with the coast as a result of refraction, they usually break at a slight angle to the shore, with the result that a littoral current is induced and is effective in moving a mass of water (and the sediment placed in suspension by the breaking waves) slowly along the coast. The strength of the littoral current in terms of the wave and beach characteristics for a relatively straight shoreline has been studied in both the laboratory and the field (Putman, Munk, and Traylor, 1949; Shepard, 1950a; Inman and Quinn, 1952). The factors affecting the rate of sand movement by such currents are discussed in the following paragraphs.

EFFECT OF WAVE EXPOSURE ON BEACH TYPE

Beaches adjust themselves to absorb the energy of the attacking waves. Beaches which are subjected to waves from a variety of directions and of a variety of heights and periods, are much more complex in both plan and profile than beaches which are affected by less variable wave conditions. The simplest case is that of a beach in the head of a bay, where the only waves reaching it are those which have come through the entrance and have been refracted and diffracted. The situation results in the screening out of a great number of different periods of waves, and thus the beach is affected only by certain limited types of waves. Beaches so located are adjusted in plan form by the attack of the waves in such a manner that the waves eventually arrive parallel with the beach, unless disturbed by currents caused by freshets or tidal fluctuations.

Beaches which are exposed to waves coming from many directions and waves of varying characteristics are usually of a complex profile. They are affected by currents which may change in direction from day to day, and their profile is likely to be greatly different at closely adjacent points. Such beaches usually have irregular, discontinuous bars of varying number, flat and steep sections, and partial channels or rips throughout the surf zone. For a detailed discussion of the effect of wave exposure on beach characteristics, the reader is referred to the papers of Krumbein (1944), Mason (1950), and Bascom (1951a).

CHARACTERISTICS OF BEACH PROFILES

One of the most important factors in determining the character of a beach profile is the ratio of wave height to wave length—a factor commonly referred to as the "wave steepness" (Johnson, 1949). During extended periods of low wave steepness, beaches may build low berms on the foreshore with very steep profiles on the beach face, and bars and underwater berms tend to disappear or become discontinuous (Fig. 1a). During periods of high wave steepness (storm conditions) the beach face becomes less steep and the offshore bars become more pronounced (Fig. 1b).

Steep beaches become steeper during a period of calm seas (low wave steepness). The berm, typical of summer conditions, then advances seaward, under-

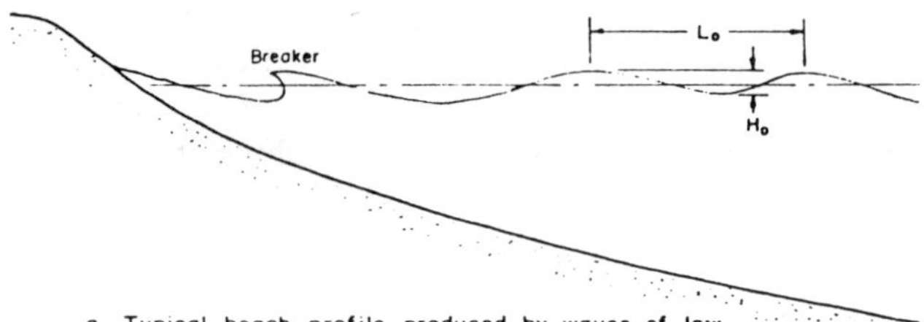

a. Typical beach profile produced by waves of low steepness (H_o/L_o less than 0.025).

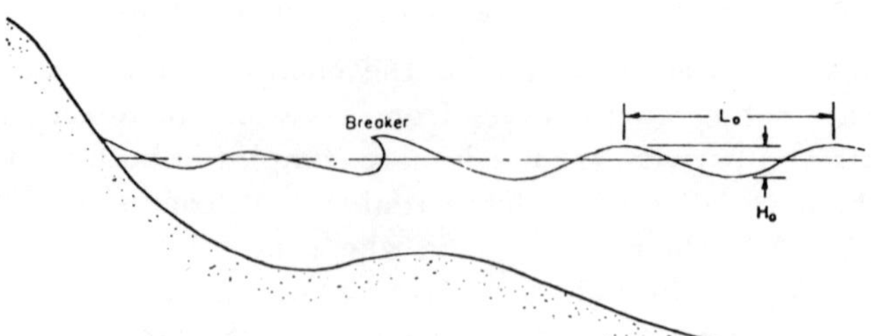

b. Typical beach profile produced by waves of high steepness (H_o/L_o greater than 0.025).

FIG. 1.—Beach profiles produced by calm and storm wave conditions.

water berms of bars tend to disappear, and the beach face may become very steep. Beaches tend to become flatter in profile under the influence of storm waves (high wave steepness) and waves at an angle, and to form a bar or berm offshore at the breaking point.

Beaches of moderate gradient usually have an offshore bar at all seasons of the year, unless they are in partly protected bays. The bar or berm becomes more pronounced when steep waves exist and the beach face becomes flatter. During a period of low wave steepness the beach face becomes steeper and the bar tends to disappear or become discontinuous.

Beaches of flat gradient usually have several offshore bars at all seasons of the year unless they are partially protected from wave action. For example, on the coasts of Washington and northern Oregon, the beaches have three or four bars with summer wave conditions but may have only one during the winter when waves of high steepness occur.

A beach profile resulting from prolonged attack by uniform waves usually is referred to as an "equilibrium profile." In recent years, numerous investigations, both in the field and in the laboratory, have been made of such profiles established under various wave and tide conditions. The most comprehensive field studies are those of Bascom (1951a), Shepard (1950b), Wiegel, Patrick, and Kimberley (1954), and Bruun (1954a). Laboratory studies have been conducted under controlled conditions to investigate equilibrium profiles formed both with and without tide effects (Rector, 1954; Watts and Dearduff, 1954; Bruun, 1954b), as well as with a varying wave period (Watts, 1954).

Also investigated in the laboratory has been the character of the profile change occurring when an equilibrium profile was established under one condition of wave steepness and then the wave steepness was suddenly changed and a new equilibrium profile was established (Scott, 1954). The depth of offshore bars and offshore troughs as related to wave conditions has been investigated both in the laboratory and in the field by Keulegan (1945) and by Shepard (1950c). Of interest in the studies of beach profiles has been the size distribution of sand across a beach and the relation between sand size at about mid-tide and the beach slope. Bascom (1951b) has assembled considerable data from Pacific Coast beaches on both of these factors. Figure 2, which shows Bascom's relationship between sand size and beach slope, clearly demonstrates the effect of wave exposure, a factor which apparently was not considered in previous studies. The Beach Erosion Board (U. S. Army, 1954a) recently has presented a plot similar to Figure 2 but has limited its designation of exposure to the broad classification of "Lake Michigan, Atlantic Coast, and Pacific Coast" beaches. Neglecting the effect of wave exposure, as well as not obtaining sand samples at a specified location (as at Bascom's Reference Point; Bascom, 1951a), probably accounts for the high degree of scatter in the previous attempts to relate sand size and beach slope.

In addition to studies of beach profile changes, investigations have been made of areal and seasonal changes in the configuration and the distribution of sediment of beaches (Bascom, 1951a; Inman, 1953; Trask, Johnson, and Scott, 1955). Other studies have included such problems as beach cusps (Evans, 1938; Kuenen, 1948), ripple marks (Scott, 1954; Bagnold, 1940; Manohar, 1955), sorting (Ippen and Eagleson, 1955), abrasion of beach sands (Mason, 1942), and the movement

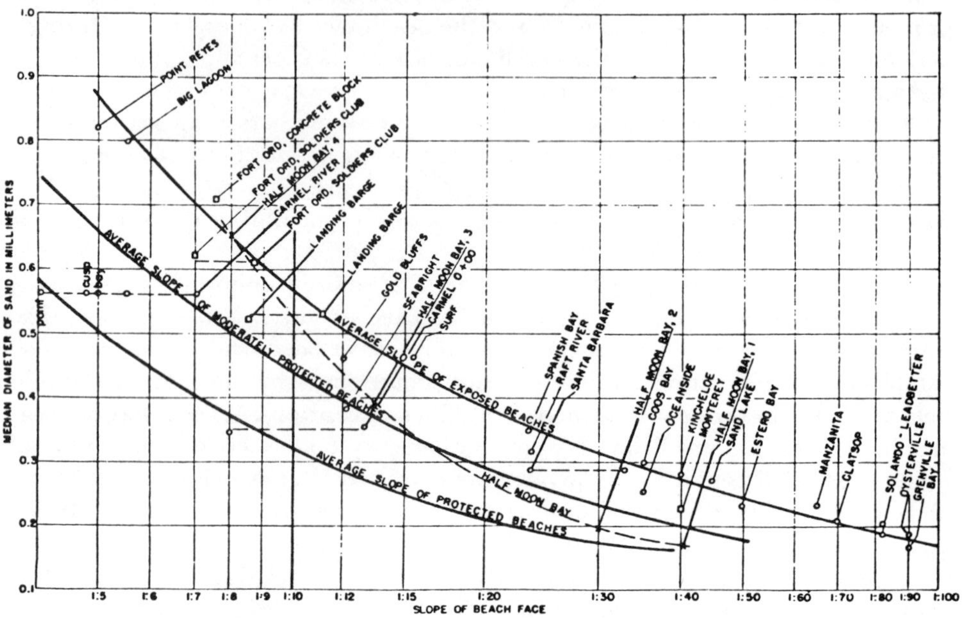

FIG. 2.—Relationship between beach slope and sand size at mid-tide level for Pacific Coast beaches (after Bascom, 1951b).

of sand near the heads of submarine canyons (Inman, 1950; Shepard and Inman, 1951).

SAND TRANSPORT BY LITTORAL CURRENTS

The result of waves breaking at an angle to a shoreline is to generate a longshore or littoral current. It is this current, combined with the agitating action of the breaking waves, that is the primary factor in causing a movement of sand along a coastline (Grant, 1943). This movement takes place in two manners—in suspension and by rolling in a zigzag motion along the beach face (Johnson, 1919; Saville, 1950). For a beach with an equilibrium profile formed by waves of relatively large steepness (Fig. 1b), which is characteristic of storm conditions, the sediment movement is mainly in suspension (Saville, 1950). In the case of an equilibrium beach profile formed by waves of low steepness (Fig. 1a), which is typical of calm summer conditions, the transport appears to be the result of rolling or skipping along the beach face. It is believed that as much as 80 per cent of the material moved by wave action is moved in the area shoreward of the breaking point (Mason, 1953).

The principal factors affecting the rate of littoral transport have been studied in detail in the laboratory (Krumbein, 1944; Saville, 1950; Johnson, 1953b; Sauvage de Saint Marc and Vincent, 1955). Saville (1950) was perhaps the first to show that for a constant wave energy and angle of wave attack there is a particular critical wave steepness at which the littoral transport is a maximum. This maximum appears to occur at a wave steepness between 0.02 and 0.03. Thus, contrary to popular belief, for a given wave energy it is not the relatively steep storm waves that cause a relatively large littoral transport, but rather the intermediate or summer waves which are the major factor in shoreline processes.

Although storm waves may remove large quantities of sand from beaches, they appear merely to move the material offshore into deeper water. This material is later moved back onto the beach during a sustained period of action involving waves of relatively small steepness. With sediment traps located along a profile of a model beach, Saville (1950) showed that during the period of highest transport the larger percentage of the material was moved by drifting along the beach face. The greatest transport of sediment along the beach occurs when the beach profile is in equilibrium with waves whose steepness is approximately 0.025. Previous laboratory experiments (Johnson, 1949) have shown that the beach profile changes from a "storm" profile (with an offshore bar) to an "ordinary" or summer profile (no offshore bar) at this same critical wave steepness of 0.025 (Fig. 1). Apparently the rate of transport for an approach angle increases with increasing wave steepness on a beach possessing an ordinary profile. If an offshore bar starts to form at the critical wave steepness of 0.025, the ability to transport sand is reduced. The increase in the size and the change in the location of the offshore bar with increasing wave steepness apparently is not conducive to an increase in the rate of transport.

Laboratory studies have been made to determine the angle between the shoreline and the wave crests in deep water which gives the maximum rate of littoral transport for a given wave energy. The investigations of Johnson (1953b) indicate the maximum rate of transport occurs when the angle of wave approach is between 30° and 40°. The studies of Sauvage de Saint Marc and Vincent (1955) indicate a value close to 50°.

Because wave characteristics are changing constantly under natural conditions, some study has been made of the effect of a change in wave characteristics on the rate of littoral transport (Johnson, 1953b). In this investigation equilibrium conditions (both beach profile and rate of transport) were first established on a model beach with a wave steepness representing relatively calm wave conditions ($H_0/L_0 = 0.0217$). The wave steepness then was abruptly increased to storm conditions ($H_0/L_0 = 0.039$), and the change in rate of transport with time was observed. As previously mentioned, the largest percentage of the littoral transport occurs shoreward of the breaking point of the waves. These tests showed that although minor variations of the wave steepness occur in this zone, the rate of littoral transport is not materially affected by such variations because the time, or number of waves, required to change the profile from the "storm" or "calm" type, or *vice versa*, is considerable. Major changes in the transport rates occur, not because of slight changes in the beach profile, but because the beach profile is shifted from one distinct type to another.

As yet no general relationship between wave and sediment characteristics is available for estimating the rate of littoral transport that occurs along a given shoreline. The few laboratory experiments already mentioned assist in defining the important variables, but probable rates of transport along natural shorelines can be estimated only from the amount of material trapped by shore structures, either natural or man-made, and from knowledge of the rate of depletion of sources of supply, for example, eroding cliffs or headlands. The known localities where such measurements have been made are presented in Table I. In addition to the calculated average annual rate of littoral transport for each locality, this

TABLE I. SUMMARY OF MEASURED RATES OF LITTORAL DRIFT ALONG COASTS OF UNITED STATES

Location	*Predominant Direction of Drift*	*Rate of Drift (cu. yds. per Year)*	*Method of Measure of Rate of Drift*	*Years of Record*	*Reference*
		Atlantic Coast			
Suffolk Co., N. Y.	W	300,000	Accretion	1946–1955	U. S. Army (1955a)
Sandy Hook, N. J.	N	493,000	Accretion	1885–1933	U. S. Army (1954b)
Sandy Hook, N. J.	N	436,000	Accretion	1933–1951	U. S. Army (1954b)
Asbury Park, N. J.	N	200,000	Accretion	1922–1925	U. S. Army (1954b)
Shark River, N. J.	N	300,000	Accretion	1947–1953	U. S. Army (1954b)
Manasquan, N. J.	N	360,000	Accretion	1930–1931	U. S. Army (1954b)
Barneget Inlet, N. J.	S	250,000	Accretion	1939–1941	U. S. Army (1954b)
Absecon Inlet, N. J.	S	400,000	Erosion	1935–1946	U. S. Army (1954b)
Ocean City, N. J.	S	400,000	Erosion	1935–1946	U. S. Congress (1953a)
Cold Spring Inlet, N. J.	S	200,000	Accretion	—	U. S. Congress (1953b)
Ocean City, Md.	S	150,000	Accretion	1934–1936	U. S. Army (1948a)
Atlantic Beach, N. C.	E	29,500	Accretion	1850–1908	U. S. Congress (1948)
Hillsboro Inlet, Fla.	S	75,000	Accretion	—	U. S. Army (1955b)
Palm Beach, Fla.	S	150,000 to 225,000	Accretion	1925–1939	U. S. Army (1947)
		Gulf of Mexico			
Pinellas Co., Fla.	S	50,000	Accretion	1922–1950	U. S. Congress (1954a)
Perdido Pass, Ala.	W	200,000	Accretion	1934–1953	U. S. Army (1954c)
Galveston, Texas	E	437,500	Accretion	1919–1934	U. S. Congress (1953c)
		Pacific Coast			
Santa Barbara, Calif.	E	280,000	Accretion	1932–1951	Johnson (1953b)
Oxnard Plainshore, Calif.	S	1,000,000	Accretion	1938–1948	U. S. Congress (1953d)
Port Hueneme, Calif.	S	500,000	Accretion	1938–1948	U. S. Congress (1954b)
Santa Monica, Calif.	S	270,000	Accretion	1936–1940	U. S. Army (1948b)
El Segundo, Calif.	S	162,000	Accretion	1936–1940	U. S. Army (1948b)
Redondo Beach, Calif.	S	30,000	Accretion	—	U. S. Army (1948b)
Anaheim Bay, Calif.	E	150,000	Erosion	1937–1948	U. S. Congress (1954c)
Camp Pendleton, Calif.	S	100,000	Accretion	1950–1952	U. S. Army (1953a)
		Great Lakes			
Milwaukee Co., Wis.,	S	8,000	Accretion	1894–1912	U. S. Congress (1946)
Racine Co., Wis.	S	40,000	Accretion	1912–1949	U. S. Congress (1953e)
Kenosha, Wis.	S	15,000	Accretion	1872–1909	U. S. Army (1953b)
Ill. State Line to Waukegan	S	90,000	Accretion	—	U. S. Congress (1953f)
Waukegan to Evanston, Ill.	S	57,000	Accretion	—	U. S. Congress (1953f)
South of Evanston, Ill.	S	40,000	Accretion	—	U. S. Congress (1953f)

table also presents the method of rate determination (whether by accretion or scour), the period of measurement, and the predominant direction of drift. The longer the period of measurement, the more representative is the stated rate of transport. The sources of the various data are presented in the last column of Table I. Although some of these stated rates of transport were based on measurements extending over a long period of years and therefore are representative of the average rates of transport in those localities, unfortunately no simultaneous observations of the wave characteristics were made to permit a possible correlation between the rate of transport and the sediment and wave characteristics.

The observations at Santa Barbara (Johnson, 1953b) are of interest in giving some details on the short period rates of littoral transport. This harbor was formed

FIG. 3.—Sand deposition in lee of breakwater at Santa Barbara, California.

in 1929 by the construction of a breakwater. The prevailing direction of drift along this section of the coast is from west to east, and the drift now is intercepted completely by the breakwater (Fig. 3). This interception results in a deposition of sand in the harbor and in an erosion problem for the beaches east of the harbor. In the few years following the construction of the breakwater, the effects of this erosion have been felt as far as 10 miles east. To maintain the harbor in a navigable condition and to provide nourishment to the easterly beaches it became evident shortly after the breakwater was constructed that a program of sand bypassing was necessary. A summary of the total accretion in the harbor since construction of the breakwater is given in Figure 4 (Johnson, 1953b). Most of these data were obtained from surveys by the Corps of Engineers in connection with a harbor-dredging program. As indicated in Figure 4, the harbor has been dredged every two or three years since 1935. Although the annual rate of transport of sand along this section of the coast (Table II) is the

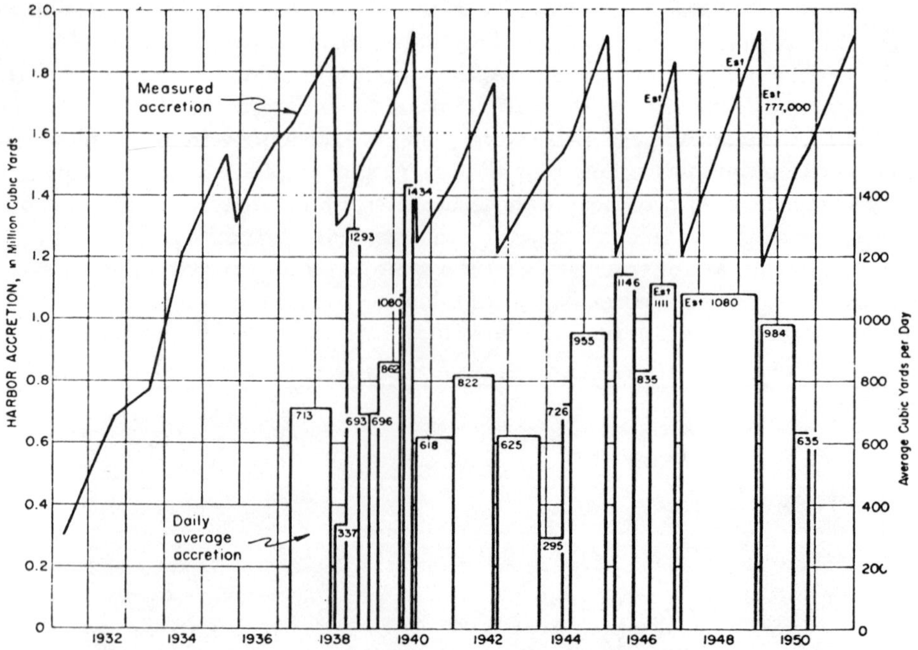

FIG. 4.—Summary of accretion in Santa Barbara Harbor and average rates of littoral transport during periods of harbor surveys.

primary item of interest, the average daily rate is also important. Average daily rates computed for the various periods between harbor surveys are shown in Figure 4.

A critical examination of the Santa Barbara data reveals that sufficient information is available to establish the average and maximum annual rates of sand movement along that section of the California coast. Although the determinations of harbor accretion were made more frequently in later years, observa-

TABLE II. ESTIMATED ANNUAL VOLUMES OF ACCRETION IN SANTA BARBARA HARBOR, 1932–1951, INCLUSIVE

Year	*Accretion Cubic Yards*	*Year*	*Accretion Cubic Yards*
1932	225,000	1942	245,000
1933	265,000	1943	210,000
1934	390,000	1944	235,000
1935	200,000	1945	295,000
1936	225,000	1946	400,000
1937	205,000	1947	330,000
1938	235,000	1948	370,000
1939	280,000	1949	330,000
1940	310,000	1950	300,000
1941	260,000	1951	283,000

Average, 1932–51, incl. = 279,650 cu. yds. per year.

tions at even more frequent intervals are necessary to determine accurately the actual rate of transport for a given wave condition. Harbor surveys should be made as frequently as the reliability of survey methods permits, but in particular they should be made when the wave characteristics are observed to change appreciably. Frequent observations of harbor accretion and wave conditions, however, are not the only factors to be considered. The effect of tide and the general variability of height, period, and direction of the waves are added complications to the formulation of a general equation for transport.

PREDOMINANT DIRECTION OF LITTORAL TRANSPORT

The direction of littoral drift at a particular time is dictated by the direction of the alongshore component of the wave velocity at the breaking point (Fig. 5). On many coastlines important reversals in the direction of littoral drift occur because of the seasonal variation of the direction of wave attack. Usually, however, the intensity of wave attack predominates in one direction, with the

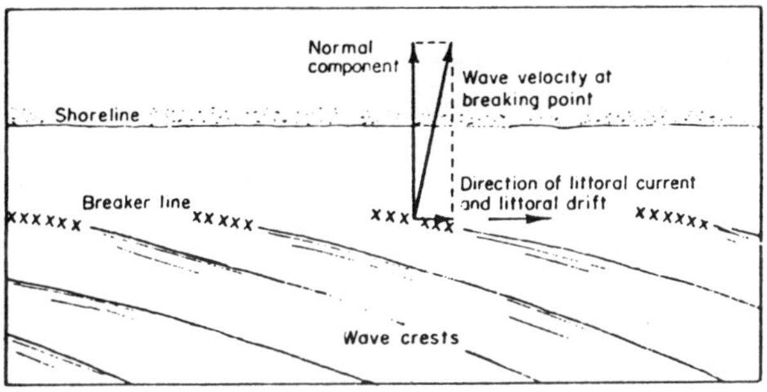

FIG. 5.—Components of wave velocity when waves break at angle to shoreline.

result that a net or predominant direction of drift occurs. Thus, in Table I the predominant direction of drift is stated for the various localities for which rates of littoral transport were computed. Undoubtedly the drift occurs in one direction along these various coastlines at certain times of the year, and in the opposite direction during the remainder of the year; however, a net drift occurs in the direction and with the rate as indicated in the table. For example, along the South Atlantic coast of the United States the littoral drift is northward during the summer season when light winds from the south and southeast prevail, but during the fall and winter strong northeasterly storms accompanied by relatively high seas drive the sand southward (Hall and Herron, 1950). These winter storms are more severe than the summer storms with the result that the predominant drift is southward along the south Atlantic coast.

The determination of the predominant direction of littoral transport has long been a study of interest to the geologist (Gulliver, 1899). In many instances it is necessary to know both the direction of littoral transport at any one time and the predominant direction of littoral transport over a normal climatic cycle. The predominant direction is the more difficult to determine and may involve locating the position of natural and unnatural littoral barriers and those areas called nodal zones in which the net littoral transport changes direction. In these zones the net littoral drift is zero or, in other words, the downdrift components of littoral drift are equal to the updrift components. An excellent example in this respect is the coast of New Jersey where extensive study of sand movement by the Corps of Engineers and others has established that a nodal point occurs at Manasquan, New Jersey (Fig. 6), with the predominant drift being northward north of this point and southward south of this point (U. S. Congress, 1928).

Although the methods used in determining the direction of littoral transport may differ from place to place (U. S. Army, 1954a), determination of the instantaneous and predominant directions of littoral transport and the location of littoral barriers and nodal zones ordinarily is accomplished by consideration of such factors as (a) shore patterns in the vicinity of headlands, (b) the configuration of the banks and beds of inlets and streams, (c) accretion or erosion effects of man-made structures, (d) statistical analysis of wave energy, (e) characteristics of beach and bed materials, and (f) current measurements.

In connection with shoreline studies at existing and proposed coastal works, the Corps of Engineers has accumulated a wealth of information on the rates and predominant direction of littoral transport along the Atlantic, Pacific, Gulf, and Great Lakes coasts of the United States. The known rates of drift are relatively few and are summarized in Table I. More voluminous, however, are the data on the predominant direction of drift. These various data on drift are summarized in Figure 6, wherein the arrows indicate the direction of drift, as compiled primarily from both published and unpublished reports of the Corps of Engineers. The indicated directions of drift are predominant directions but important reversals in drift do occur at many of the localities listed. Also shown in Figure 6 are the rates of littoral drift for the locations listed in Table I.

Another item of importance with respect to currents is the confusion that often occurs between the strength of the littoral currents and the strength of the large-scale ocean currents. These latter currents, as measured by the drift of bottles and floating débris, usually are relatively weak as a sand transporting agency as compared with the wave-induced littoral current. At localities where these two currents are opposed, the wave-induced littoral current usually is the stronger of the two, and therefore determines the predominant direction of littoral drift.

The laboratory experiments on littoral drift were made under conditions representing a straight beach of infinite length. Such conditions are approximated on occasions in nature; however, most natural shorelines are irregular in plan with the alignment often being interrupted by headlands. The general movement of littoral materials along such irregular shorelines is much more complex than with the rather idealized conditions studied in the laboratory by Saville (1950) and others (Johnson, 1953b; Sauvage de Saint Marc and Vincent, 1955).

In these various laboratory studies of littoral transport most of the alongshore movement occurs in the turbulent region of the breaker zone. Along natural shorelines, however, turbulent forces seaward of the breaker zone apparently are great enough to cause a general alongshore movement of sand out to depths as great as 60–80 feet. Although oceanic currents in some localities may be of sufficient strength to transport sand along the sea bottom, the primary mechnism of sediment transport in deep water appears to be the result of turbulence resulting from the oscillatory motion at the sea bed by surface waves. Sediment placed into temporary suspension by such turbulence than may be transported along the sea bottom, and possibly around headlands, as a result of either or both of the following forces: (a) an oceanic current, or (b) when the forward velocity of the water under the wave crests exceeds the backward velocity under the trough, a net transport occurs in the direction of the wave travel (Grant and Shepard, 1939). Two rather comprehensive investigations of the fundamental mechanics of sand movement by oscillatory waves have been made recently (Li, 1954; Manohar, 1955). The first of these studies (Li, 1954) was concerned with the development of a criterion for the change from laminar to turbulent flow. The second phase of this study (Manohar, 1955) was concerned with the character of the bed sediment movement. Because of the action of surface waves, there are initial and general movements of sediment, as well as the initiation, various stages of development, and complete disappearance of bed undulations or ripples. The movement of sediment takes place in a boundary layer that is developed at the sea bottom from the effects of the viscosity of the fluid. The initial and general motion of small sizes of sediment is caused by laminar shear, and similar motions of large sizes of sediment are caused by lift forces in a turbulent boundary layer. Ripples, in general, are not formed unless the flow is turbulent in the boundary layer. All motion in turbulent flow and the various stages of development of ripples were found to be functions of a dimensionless function representing intensity of flow of the fluid near the bottom.

The source and transportation of sediments along natural coastlines have been studied in detail on numerous occasions (Handin, 1951; Trask, 1952; Trask, 1955; and Pincus, 1954). The most difficult problem has been the establishment of the source of the sediments occurring on a certain shoreline. By studying both the textural and petrographic characteristics of sediments it has been established in recent investigations that sediment moves around promontories that previously were considered to be complete littoral barriers (Trask, 1952; Trask, 1955). The mechanism by which sediments move around many promontories is that in the neighborhood of the headland the high local turbulence created by and diffused from the barrier apparently keeps the sediment in suspension. The suspended sediment is finally carried to the downcoast by the current along the barrier, due to the continuity of the flow. Underwater photographs of the sea bottom in depth of 60–80 feet seaward from prominent headlands show con-

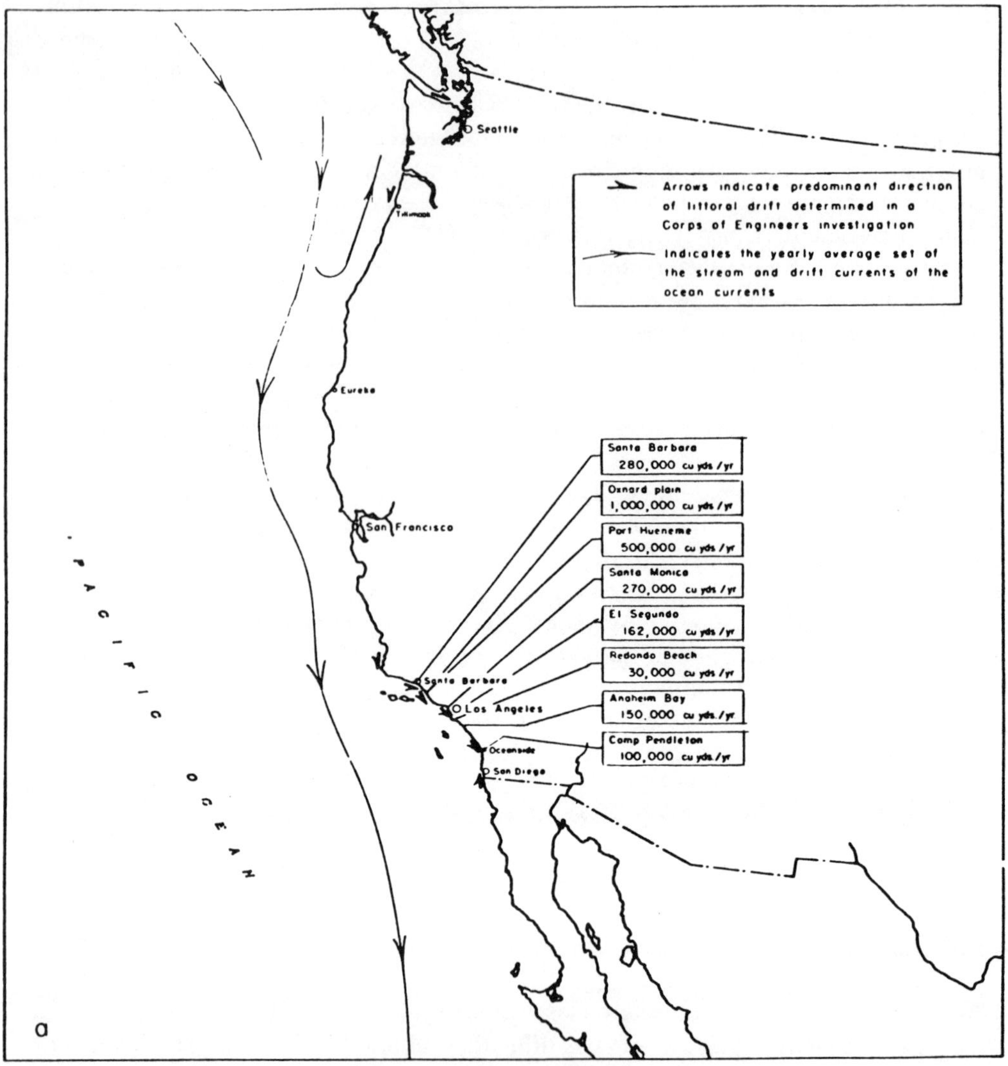

FIG. 6.—Summary of observations on littoral drift and predominant direction of drift in United States.

siderable ripple formations, which indicates that sediments are also passing such headlands in relatively large depths as a result of the forces discussed (Manohar, 1955).

In addition to the problem of sediment movement around headlands, there is the important problem of sediment movement at inlets to tidal bays or estuaries (Eaton, 1951). Such inlets act as littoral barriers and often migrate as a result of deposition on the updrift side and erosion by currents along the downdrift shore. Tidal inlets on the Atlantic coast where the fresh-water discharge is relatively small are subject to ebb and flood tidal flow of about equal intensity. Such

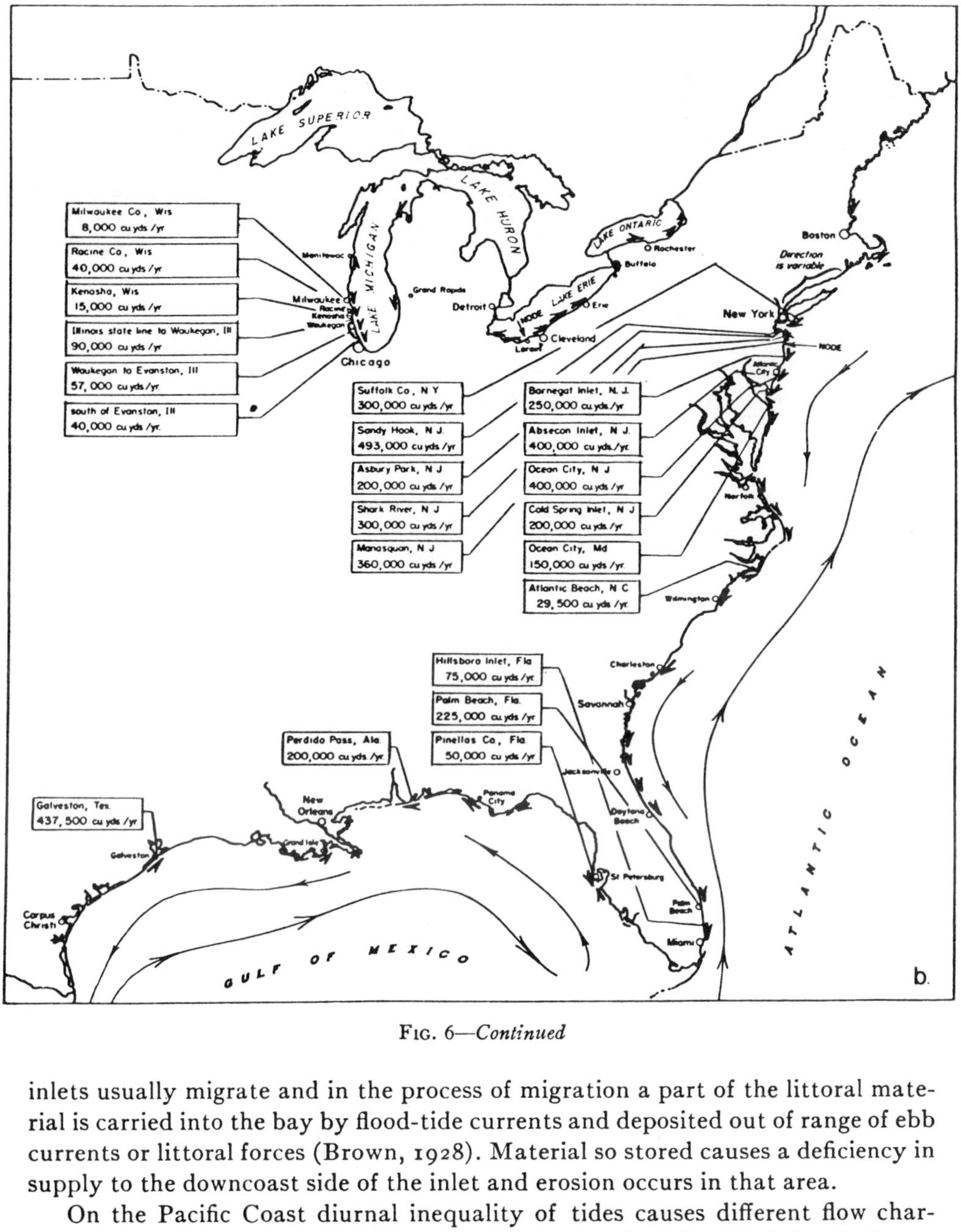

FIG. 6—*Continued*

inlets usually migrate and in the process of migration a part of the littoral material is carried into the bay by flood-tide currents and deposited out of range of ebb currents or littoral forces (Brown, 1928). Material so stored causes a deficiency in supply to the downcoast side of the inlet and erosion occurs in that area.

On the Pacific Coast diurnal inequality of tides causes different flow characteristics at tidal inlets from those previously described for the Atlantic Coast. Maximum velocities are reached in ebb flow. At an unimproved inlet, littoral material entering the inlet during flood tide is swept seaward by the ebb flow and deposited offshore in the form of a crescent-shape bar. The depth profile and alignment of the bar are dependent on the littoral forces, the character of the littoral material, and tidal characteristics of the inlet.

Several attempts have been made to determine the relation between the tidal prism and the sectional area of the entrance channel for natural maintenance of the channel without excessive currents (O'Brien, 1931; Robbins, 1933). For the channel at the throats of the Pacific Coast estuaries, it appears that the cross-sectional area in square feet below mean sea-level is approximately equal to the tidal prism in acre-feet, measured between mean-higher-high-water and mean-lower-low-water (Robbins, 1933). If an inlet is stable rather than migratory, littoral material is being transported across it by natural forces, and the throat dimensions are fixed by the scouring capacity of the ebb current. It is generally believed that the bar is the principal path followed by littoral material crossing an inlet. It appears reasonable to assume that a stable inlet with a mature bar is not necessarily a littoral barrier (Eaton, 1951).

The effect of wave action at stream or tidal entrances is to close the opening; whereas, the stream discharge or tidal flow is to keep the entrance open. Any increase in either of the forces acting is to cause a change in the regimen of the entrance. When the shoreline configuration and submarine topography are such that the effects of wave refraction cause marked differences in wave attack at closely adjacent locations along the shoreline, a stream often will migrate to the point of low wave attack where the entrance can be maintained with a less expenditure of energy by tidal or stream flow than at points where the wave attack is relatively high (Bascom, 1954).

SEDIMENT TRANSPORTATION, DEPOSITION, AND EROSION AT MAN-MADE LITTORAL BARRIERS

There are three basic types of man-made coastal structures which function as littoral barriers. These consist of (a) a dredged channel, (b) a jetty or groin, and (c) an offshore or detached breakwater. The littoral processes in the vicinity of such works are summarized briefly as follows (Eaton, 1951).

Dredged channels.—Harbors are often connected with deep water offshore by means of a dredged channel through the littoral zone (Fig. 7). Such a channel creates greater than normal depths with the result that littoral material accumulates therein. Sediment of small enough size to be moved in the deeper depths seaward from the end of the dredged channel would, of course, not be affected. Measurements indicate that most of the longshore transport of material occurs in the vicinity of the breakers where the available wave energy is converted suddenly from an oscillatory motion into the form of turbulence. For that portion of the wave which moves over a dredged channel, however, breaking does not occur because of the increased depth, and the wave energy passes the normal point of breaking to be spread by refraction and then dissipated further inshore. The degree of turbulence, therefore, is insufficient to transport material across the channel and the material accumulates as indicated in Figure 7. To maintain the channel in a navigable condition, this accumulation of littoral material must be dredged periodically. If this material is removed and redeposited on the downcoast side of the channel, normal littoral transport will occur in that region and the shoreline will remain in an equilibrium position. If, however, the channel deposits are placed elsewhere, then the supply of material to the downcoast beach is reduced and erosion and retreat of the shoreline probably will result (Fig. 7). In a harbor such as shown in Figure 7 the action of the waves is to restore the natural littoral transport of material and thus reduce the area of the entrance to a size compatible with the tidal prism. The equilibrium size of entrance to be ex-

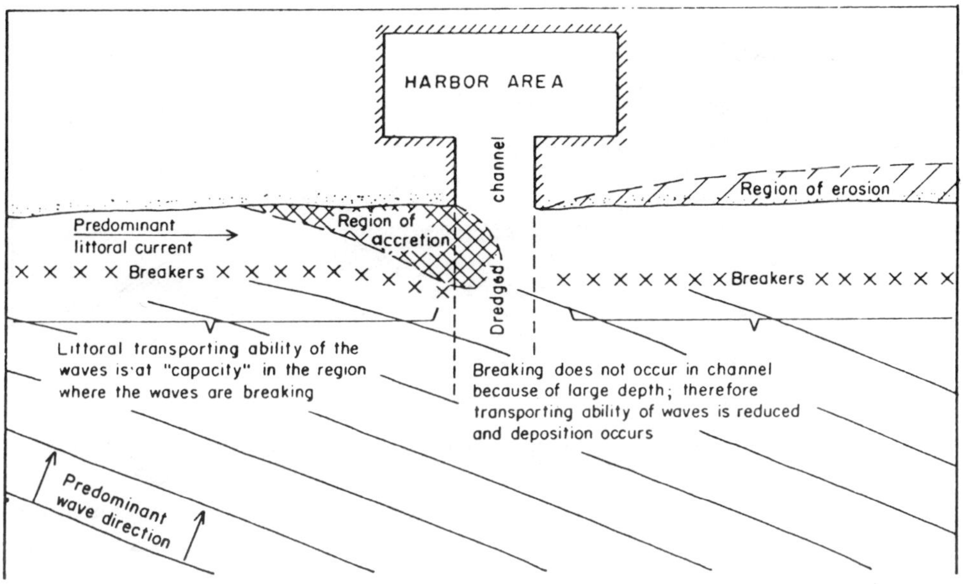

FIG. 7.—Schematic representation of transportation, deposition, and scour of littoral sediments at channel dredged through littoral zone.

pected might be estimated by the relationships between entrance area and tidal prism as given by O'Brien (1931) or Robbins (1933).

Harbors created by shore-connected breakwaters.—The effect of a structure which extends seaward from the shore and across the littoral zone is to act as a dam and trap the littoral drift. The impounding capacity is dependent upon the height of the structure, the bottom slope, and the equilibrium alignment of the shore in that region. The equilibrium alignment is one which is normal to the resultant littoral forces. Thus, in Figure 8, if the original shoreline was stable with respect to the material balance and a breakwater is constructed as shown, accretion will first occur in the form of a fillet on the upcoast side with an alignment tending toward equilibrium. This will create a deficiency in material supplied to the downcoast shoreline, in which erosion probably will occur with the shoreline also tending toward equilibrium. As the upcoast fillet approaches equilibrium, littoral material will move along the outer face of the breakwater and be deposited in the relatively calm water in the lee of the structure. Thus the turbulent character of the wave action upcoast from the breakwater tip is sufficient to transport littoral material at capacity. As the waves reach the tip, however, and are refracted and diffracted into the lee of the structure, the turbulence is insufficient to transport the material and deposition occurs. This deposit continues to grow toward the downcoast shoreline, and when it reaches the shoreline the material balance will be re-established on each side of the barrier.

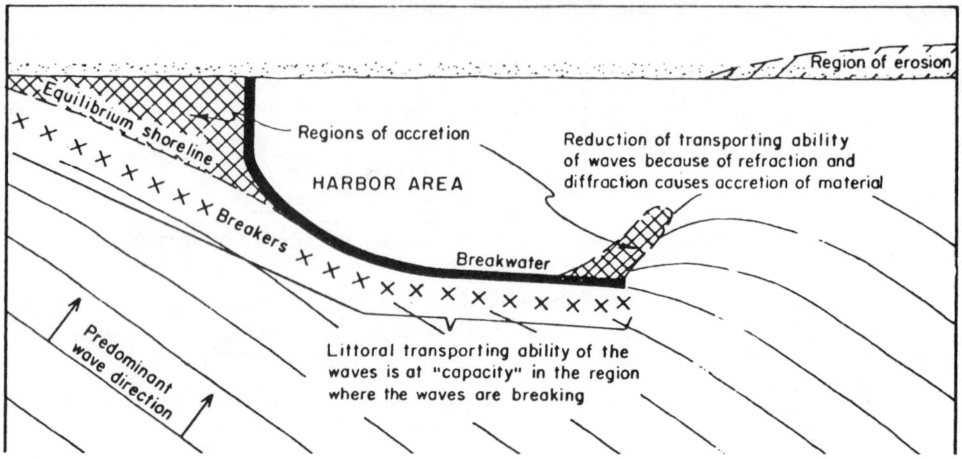

FIG. 8.—Schematic representation of transportation, deposition, and scour of littoral sediments at shore-connected breakwater.

The alignment of the harbor deposit depends primarily on the predominant wave direction. A typical example of such a harbor deposit is that at Santa Barbara, California (Fig. 3).

A variation of a harbor formed by a shore-connected breakwater is the case where two breakwaters must be provided to insure protection from storm waves which may approach the entrance from various directions. Pronounced reversals in the direction of littoral drift usually occur in such instances.

Detached breakwater.—This type of structure intercepts the waves and creates a protected area of relatively calm water. The original theory of such a breakwater location was that the littoral material would move along the coast uninterrupted by the presence of the structure and consequently no maintenance problems from sediment deposition would be involved. This assumption, however, is in error because the result of the refraction and diffraction of the waves behind the structure is to reduce the energy available for littoral transport in the lee of the structure as compared with the energy available on both the upcoast and downcoast shorelines (Fig. 9). The result of this reduction in available energy is that littoral material accumulates in the protected area (Handin and Ludwick, 1950). If this accumulation of material is not removed periodically by dredging, the accretion eventually may extend completely out to the breakwater in the form of a tombolo. For a

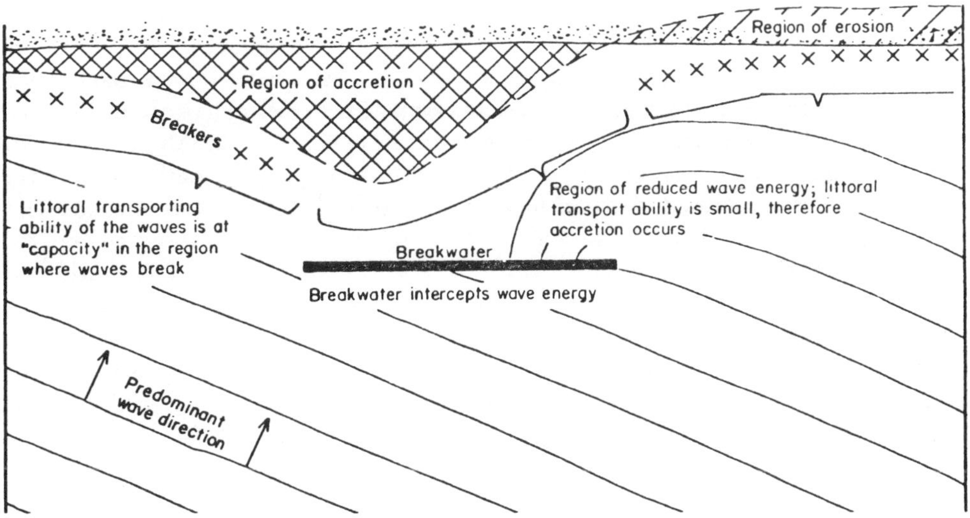

FIG. 9.—Schematic representation of transportation, deposition, and scour of littoral sediments at detached breakwater.

discussion of the phenomenon of tombolo formation the reader is referred to the model studies of Sauvage de Saint Marc and Vincent (1955).

On the upcoast side of a detached breakwater the accretion advances beyond the region of direct effect of the structure itself, and a corresponding erosion occurs on the downcoast side (Fig. 9). A typical example of a harbor of this type is that at Santa Monica, California (Fig. 10).

FIG. 10.—Tombolo in lee of breakwater at Santa Monica, California.

The erosion of shorelines downcoast of man-made littoral barriers has introduced many engineering and economic problems. The construction, improvement and maintenance of beaches by artificial deposition of sand on downcoast shorelines is rapidly gaining prominence where damage by erosion has occurred. Four types of artificial nourishment of beaches have been tried; namely, the offshore dumping method, the stockpiling method, the continuous supply method, and the direct placement method. A list of localities where each of these methods has been tried and an appraisal of the effectiveness of the tests have been presented in detail elsewhere (Hall and Herron, 1950; Hall, 1952; and Harris, 1954). The continuous supply method of passing sand past inlets or harbor entrances which act as littoral barriers appears to be the answer to the maintenance problem where large depths for navigation purposes must be insured. Notable sand by-passing plants have been built at South Lake Worth Inlet, Florida (Watts, 1953), Salina Cruz, Mexico (Rolland, 1952), and Rudee Inlet, Virginia (McDonald, 1955).

SUMMARY

The foregoing discussion summarizes some of the more important investigations of nearshore sediment movement that have been conducted over 15 years by engineers and geologists, primarily in connection with shore and beach protection problems. A rather complete list of references is included to permit a reader to study further into the particular problems of interest.

REFERENCES

BAGNOLD, R. A., 1940, "Beach Formation by Waves; Some Model Experiments in a Wave Tank," *Jour. Inst. Civil Engrs.*, Vol. 1 (1940–41), *Paper 5237*, pp. 27–52.

BASCOM, W. N., 1951a, "The Relationship between Sand Size and Beach-Face Slope," *Trans. Amer. Geophys. Union*, Vol. 32, pp. 866–74.

———, 1951b, *Shoreline and Beach Characteristics; Manual of Amphibious Oceanography*, Office of Naval Research, Washington D. C.

———, 1954, "The Control of Stream Outlets by Wave Refraction," *Jour. Geology*, Vol. 62, No. 6, pp. 600–05.

BIGELOW, H. B., AND EDMONDSON, W. T., 1947, "Wind Waves at Sea, Breakers and Surf," *U. S. Navy H. O. Pub. 602.* 177 pp.

BROWN, E. I., 1928, "Inlets on Sandy Coasts," *Proc. Amer. Soc. Civil Engrs.*, Vol. 54, pp. 505–53.

BRUUN, PER, 1954a, "Coast Erosion and the Development of Beach Profiles," *Beach Erosion Board Tech. Memo. 44*, Corps of Engineers, Washington D. C. 79 pp.

———, 1954b, "Use of Small-Scale Experiments with Equilibrium Profiles in Studying Actual Problems and Developing Plans for Coastal Protection," *Trans. Amer. Geophys. Union*, Vol. 35, No. 3, pp. 445–52.

EATON, R. O., 1951, "Littoral Processes on Sandy Coasts," *Proc. First Conf. Coastal Engin.*, pp. 140–54.

EVANS, O. F., 1938, "Classification and Origin of Beach Cusps," *Jour. Geology*, Vol. 46, pp. 615–27.

GILBERT, G. K., 1890, "Lake Bonneville," *U. S. Geol. Survey Monographs*, Vol. 1. 438 pp.

GRANT, U. S., 1948, "Influence of the Water Table on Beach Aggradation and Degradation," *Sears Foundation: Jour. Marine Research*, Vol. VII, No. 3 (November 15), pp. 655–60.

———, 1943, "Waves as a Sand-Transporting Agent," *Amer. Jour. Sci.*, Vol. 241 (February), pp. 117–23.

———, AND SHEPARD, F. P., 1939, "Shallow-Water Sediment-Shifting Processes Along the Southern California Coast," *Proc. Sixth Pacific Sci. Cong. of Pacific Sci. Assoc.*, Vol. II, pp. 801–05.

GULLIVER, F. P., 1899, "Shoreline Topography," *Proc. Amer. Acad. Arts and Sci.*, Vol. 34, pp. 151–258.

HALL, J. V., JR., 1952, "Artificially Nourished and Constructed Beaches," *Beach Erosion Board Tech. Memo. 29*, Corps of Engineers, Washington D. C. 25 pp.

———, AND HERRON, W. J., 1950, "Test of the Nourishment of the Shore by Offshore Deposition of Sand," *ibid.*, *Tech. Memo. 17.* 32 pp.
HANDIN, J. W., 1951, "The Source, Transportation, and Deposition of Beach Sediment in Southern California," *ibid.*, *Tech. Memo. 22.* 125 pp.
———, AND LUDWICK, J. C., 1950, "Accretion of Beach Sand Behind a Detached Breakwater," *ibid.*, *Tech. Memo. 16.* 13 pp.
HARRIS, R. L., 1954, "Restudy of Test-shore Nourishment by Offshore Deposition of Sand, Long Branch, New Jersey," *ibid.*, *Tech. Memo 62.* 18 pp.
INMAN, D. L., 1950, "Submarine Topography and Sedimentation in the Vicinity of Mugu Submarine Canyon, California," *ibid.*, *Tech. Memo. 19.* 45 pp.
———, 1953, "Areal and Seasonal Variations in Beach and Nearshore Sediments at La Jolla, California," *ibid. Tech. Memo. 39.* 82 pp.
———, AND QUINN, W. H., 1952, "Currents in the Surf Zone," *Proc. 2d Conf. Coastal Engin.*, pp. 24–35.
IPPEN, A. T., AND EAGLESON, P. S., 1955, "A Study of Sediment Sorting by Waves Shoaling on a Plane Beach," *Beach Erosion Board Tech. Memo. 63*, Corps of Engineers, Washington D. C. 83 pp.
IVERSEN, H. W., 1951, "Studies of Wave Transformation in Shoaling Water, Including Breaking," *Symposium on Gravity Waves*, National Bureau of Standards, pp. 9–32.
JOHNSON, D. W., 1919, *Shore Processes and Shoreline Development.* 584 pp. John Wiley and Sons.
JOHNSON, J. W., 1949, "Scale Effects in Hydraulic Models Involving Wave Motion," *Trans. Amer. Geophys. Union*, Vol. 30, pp. 517–25.
——, 1953a, "Engineering Aspects of Diffraction and Refraction," *Trans. Amer. Soc. Civil Engrs.*, Vol. 118, pp. 617–48.
———, 1953b, "Sand Transport by Littoral Currents," *Proc. 5th Hydraulics Conf.*, Bull. 34, State Univ. Iowa Studies in Engin., pp. 89–109.
———, O'BRIEN, M. P., AND ISAACS, J. D., 1948, "Graphical Construction of Wave Refraction Diagrams," *U. S. Navy Hydrographic Office Pub. 605.* 45 pp.
KEULEGAN, G. H., 1945, "Depths of Offshore Bars," *Beach Erosion Board Tech. Memo 8*, Corps of Engineers, Washington D. C. 18 pp.
KRUMBEIN, W. C., 1944, "Shore Currents and Sand Movement on a Model Beach," *ibid.*, *Tech. Memo. 7.* 44 pp.
KUENEN, H. 1948, "The Formation of Beach Cusps," *Jour. Geology*, Vol. 56, pp. 34–40.
LI, H., 1954, "Stability of Oscillatory Laminar Flow Along a Wall," *Beach Erosion Board Tech. Memo. 47*, Corps of Engineers, Washington, D. C., 47 pp.
MANOHAR, W., 1955, "Mechanics of Bottom Sediment Movement Due to Wave Action," *ibid.*, *Tech. Memo. 75.* 121 pp.
MASON, M. A., 1942, "Abrasion of Beach Sand," *ibid.*, *Tech. Memo 2.* 13 pp.
———, 1950, "Geology in Shore-Control Problems," *Applied Sedimentation*, pp. 276–90. John Wiley and Sons.
———, 1953, "Surface Water Wave Theories," *Trans. Amer. Soc. Civil Engrs.*, Vol. 118, p. 569.
McDONALD, T. J., 1955, "Sand Bypassing System at Rudee Inlet for Nourishment of Virginia Beach," *Proc. Amer. Soc. Civil. Engrs.* (in press).
MUNK, W. H., AND TRAYLOR, M. A., 1947, "Refraction of Ocean Waves: a Process Linking Underwater Topography to Beach Erosion," *Jour. Geology*, Vol. 55, pp. 1–26.
O'BREIN, M. P., 1931, "Estuary Tidal Prisms, Related to Entrance Areas," *Civil Engineering*, Vol. 1, No. 8, pp. 738–39.
PINCUS, H. J., 1954, "The Motion of Sediment Along the South Shore of Lake Erie," *Proc. 4th Conf. Coastal Engin.*, pp. 119–46.
PUTNAM, J. A., MUNK, W. H., AND TRAYLOR, M. A., 1949, "The Prediction of Longshore Currents," *Trans. Amer. Geophys. Union.*, Vol. 30, pp. 337–45.
RECTOR, R. L., 1954, "Laboratory Study of Equilibrium Profiles of Beaches," *Beach Erosion Board Tech. Memo. 41*, Corps of Engineers, Washington D. C. 38 pp.
ROBBINS, T. M., 1933, "Maintenance and Improvement of Entrance Channels to the Pacific Coast Ports," *World Ports*, Vol. 21, No. 3, pp. 18–20.
ROLLAND, M. C., 1952, "Sand By-Passing Plant at Salina Cruz, Mexico," *Proc. 2d Conf. Coastal Engin.*, pp. 177–86.
SAUVAGE DE SAINT MARC, M. G., AND VINCENT, M. G., 1955, "Transport Littoral Formation de Fleches et de Tombolos," *Proc. 5th Conf. Coastal Engin.*, pp. 296–328.
SAVILLE, THORNDIKE, JR., 1950, "Model Study of Sand Transport Along an Infinitely Long, Straight Beach," *Trans. Amer. Geophys. Union*, Vol. 31, pp. 555–65.
SCOTT, T., 1954, "Sand Movement by Waves," *Beach Erosion Board Tech. Memo. 48*, Corps of Engineers, Washington D. C. 37 pp.
SHEPARD, F. P., 1950a, "Longshore Current Observations in Southern California," *ibid.*, *Tech. Memo. 13.* 54 pp.

——, 1950b, "Beach Cycles in Southern California," *ibid., Tech. Memo. 20.* 26 pp.
———, 1950c, "Longshore-Bars and Longshore-Troughs," *ibid., Tech. Memo. 15.* 32 pp.
———, 1952, "Revised Nomenclature for Depositional Coastal Features," *Bull. Amer. Assoc. Petrol. Geol.*, Vol. 36, No. 10, pp. 1902–12.
———, AND INMAN, D. L., 1951, "Sand Movement on the Shallow Inter-Canyon Shelf at La Jolla, California," *Beach Erosion Board Tech. Memo. 26*, Corps of Engineers, Washington D. C. 29 pp.
TRASK, P. D., 1952, "Source of Beach Sand at Santa Barbara, California, as Indicated by Mineral Grain Studies," *ibid., Tech. Memo. 28.* 24 pp.
———, 1955, "Movement of Sand Around Southern California Promontories," *ibid., Tech. Memo. 76.* 60 pp.
———, JOHNSON, C. A., AND SCOTT, T., 1955, "Cut and Fill on Point Reyes Beach, California," *Univ. California, Inst. Engin. Research*, Ser. 14, Issue 19 (unpublished).
U. S. ARMY, 1947, "Beach Erosion Report on Co-operative Study at Palm Beach, Florida," *Beach Erosion Board*, Corps of Engineers, Washington, D. C. (unpublished), p. 17.
———, 1948a, "Survey of Ocean City Harbor and Inlet and Sinepuxent Bay, Maryland," Baltimore District, Corps of Engrs., Baltimore, Maryland (unpublished), p. 21.
———, 1948b, "Shoreline Effects, Harbor at Playa Del Rey, Calif.," Enclosure 20, Los Angeles District, Corps of Engrs., Los Angeles, California (unpublished), pp. 39–40.
———, 1953a, "Interim Report on Harbor-Entrance Improvement, Camp Pendelton, California," Los Angeles District, Corps of Engrs. (unpublished), p. 11.
———, 1953b, "Preliminary Analysis of Cooperative Beach Erosion Study, City of Kenosha, Wisconsin," Milwaukee District, Corps of Engrs., Milwaukee, Wisconsin (unpublished), p. 17.
———, 1954a, "Shore Protection Planning and Design," *Beach Erosion Board Tech. Rept. 4*, Corps of Engrs.
———, 1954b, "Atlantic Coast of New Jersey, Sandy Hook to Barnegat Inlet," *Beach Erosion Control Report on Cooperative Study*, New York District, Corps of Engrs., New York, N. Y. (unpublished), pp. 39–40.
———, 1954c, "Beach Erosion Control Report, Cooperative Study of Perdido Pass, Alabama," Mobile District, Corps of Engrs. (unpublished), p. 12.
———, 1955a, "Atlantic Coast of Long Island, N. Y., Five Island Inlet and Shore Westerly to Jones Inlet," *Beach Erosion Control Report on Cooperative Study*, New York District, Corps of Engineers, New York (unpublished), pp. A-16.
———, 1955b, "Sand By-Passing at Hillsboro Inlet, Florida," *Beach Erosion Board Bulletin*, Corps of Engrs., Vol. 9, No. 2, pp. 1–6.
U. S. CONGRESS, 1928, "Manasquan River and Inlet, New Jersey," *House Doc. 482*, 70th Cong., 2d Sess., p. 14.
———, 1946, "Beach Erosion Study, Lake Michigan Shore Line of Milwaukee County, Wisconsin," *House Doc. 526*, 79th Cong., 2d Sess., p. 16.
———, 1948, "North Carolina Shore Line, Beach Erosion Study," *House Doc. 763*, 80th Cong., 2d Sess., p. 20.
———, 1953a, "Ocean City, New Jersey, Beach Erosion Control Study," *House Doc. 184*, 83d Cong., 1st Sess., p. 23.
———, 1953b, "Cold Spring Inlet (Cape May Harbor), New Jersey," *House Doc. 206*, 1st Sess., p. 30.
———, 1953c, "Gulf Shore of Galveston Island, Texas, Beach Erosion Control Study," *House Doc. 218*, 83d Cong., 1st Sess., p. 19.
———, 1953d, "Appendix I, Coast of California, Carpinteria to Point Mugu," *ibid., House Doc. 29*, 83d Cong., 1st Sess., p. 25.
———, 1953e, "Racine County, Wisconsin," *ibid., House Doc. 88*, 83rd Congress, 1st Sess., p. 10.
———, 1953f, "Illinois Shore of Lake Michigan," *ibid., House Doc. 28*, 83d Cong., 1st Sess., p. 39.
———, 1954a, "Pinellas County, Florida," *ibid., House Doc. 380*, 83d Cong., 2d Sess., p. 27.
———, 1954b, "Port Hueneme, California," *House Doc. 362*, 83d Cong., 2d Sess., p. 25.
———, 1954c, "Anaheim Bay Harbor, California," *House Doc. 349*, 83d Cong., 2d Sess., p. 37.
WATTS, G. M., 1953, "A Study of Sand Movement at South Lake Worth Inlet, Florida," *Beach Erosion Board Tech. Memo. 42*, Corps of Engineers, Washington D. C. 24 pp.
———, 1954, "Laboratory Study of Effect of Varying Wave Periods on Beach Profiles," *ibid., Tech. Memo. 53.* 19 pp.
———, AND DEARDUFF, R. F., 1954, "Laboratory Study of Effect of Tidal Action on Wave-Formed Beach Profiles," *ibid., Tech. Memo. 52.* 21 pp.
WIEGEL, R. L., 1952, *Waves, Tides, Currents and Beaches: Glossary of Terms and List of Standard Symbols, Council on Wave Research, Engin. Foundation, Univ. California*, Berkeley, California. 71 pp.
———, PATRICK, D. A., AND KIMBERLEY, H. L., 1954, "Wave, Longshore Current, and Beach Profile Records for Santa Margarita River, Oceanside, California, 1949," *Trans. Amer. Geophys. Union*, Vol. 35, No. 6, pp. 887–96.

6

Reprinted from *Am. Geophys. Union Trans.* 30(3):337-345 (1945)

THE PREDICTION OF LONGSHORE CURRENTS

J. A. Putnam, W. H. Munk, and M. A. Traylor

(Contribution from the Scripps Institution of Oceanography, New Series, No. 422, and Fluid Mechanics Laboratory Publication No. He-116-279)

Abstract--Longshore currents are an important factor in beach erosion. A relationship is derived from theoretical considerations between the velocity of the longshore current and the height and period of the breaking waves, the angle between the breakers and the beach contours (assumed straight and parallel), and the beach slope in the breaker zone. Field observations and laboratory experiments support this relationship.

Introduction

Waves breaking at an angle with the shoreline set up a longshore current, or littoral current, which may reach a velocity of several knots, and which is capable of carrying large amounts of sediment. If the flow of sediments is impeded, either by direct obstructions in the path of the longshore current, or by interfering with the waves that maintain the current, the equilibrium of the beaches is disturbed, and silting or eroding will follow. A knowledge of the relationship between the character of the ocean waves and the direction and magnitude of the longshore current has proven helpful in the design of jetties and breakwaters.

The impetus for this study came, however, from a different source. During the War, considerable difficulty was experienced in landing amphibious craft in the presence of a longshore current. At times the magnitude of this current was sufficient to cause a landing vehicle to broach, even though it had successfully negotiated the breakers. Often this represented a greater hazard than that afforded by the breaking waves, and it was attempted, therefore, to include a forecast of the longshore currents among the forecast of other characteristics of the surf zone [SVERDRUP and MUNK, 1946].

Field and laboratory observations

Observations of longshore currents were initiated along the beach at Oceanside, California, and simultaneously in the Fluid Mechanics Laboratory of the Department of Engineering at the University of California, Berkeley, California. The results from these studies apply only to straight beaches with parallel contours. It has not been possible to evaluate the effect of currents beyond the breakers, nor the effect of irregularities in the shoreline.

Beach observations--The following wave data were collected each day that measurements were made (see Table 1): breaker height H, wave period T, angle of breaking , depth of breaking h, mean beach slope m, and velocity of longshore current V. The angle is difficult to obtain, but it can be measured from a pier by standing directly above the breakers and taking a bearing along the breaking crest. Soundings were taken along the pier, and the position of the breakers noted. This was necessary because the beach did not have a regular slope but was characterized by an outer and inner bar. These bars were constantly shifting position, particularly with heavy surf. The wind speed and direction were recorded when the wind was strong enough to be of importance.

Longshore currents were measured by three different methods. Two of these involved the travel time of a current cross over a known distance. The current cross consisted of a sheet metal cross, suspended from a small float, and held upright by a lead weight (see Fig. 1). The line between the float and the cross could be adjusted to measure the current at any desired depth. The damping effect of the wooden braces shown in Figure 1 held the instrument fairly steady in the surf Two types were used, a large cross 16 inches high and a small cross eight inches high, the latter for measurements in shallow water.

The first method was to take an average of several 100-ft runs. One end of a 100-ft line was attached to the current cross and the other to a block of wood which could be lowered from the pier to the water surface by a short line. The cross was dropped into the current, the block was lowered to the surface and then the time required to draw the 100-ft line taut was measured.

Table 1--Tabulation of field data

No.	Date	Time	H	T	α	m	V
	1945	h m	ft	sec	°		ft/sec
1	Feb. 9	. . . .	5	10	15	0.016	2.5
2	Feb. 21	. . . .	5.5	9	12	0.02	2.2
3	Feb. 24	. . . .	7	9	15	0.023	3.0
4	Mar. 5	. . . .	6	7	7.5	0.017	1.8
5	Mar. 14	10 15	5	10	10	0.031	3.6
6	Mar. 14	13 30	8	10	10	0.022	2.8
7	Mar. 14	14 30	8	10	10	0.023	2.3
8	Mar. 16	10 40	6.5	12	10	0.02	2.4
9	Mar. 16	13 50	4.5	12	10	0.02	2.4
10	Mar. 16	14 45	4.5	12	10	0.019	2.7
11	Mar. 16	15 10	4.5	12	10	0.019	2.1
12	Mar. 22	. . . .	6.5	15	5	0.016	1.7
13	Mar. 23	. . . .	8	7	17.5	0.022	5.2[a]
14	Mar. 24	10 00	5	8	10	0.03	3.3
15	Mar. 24	10 30	8.5	8	12	0.02	2.5
16	Mar. 27	. . . .	5	15	5	0.026	2.4
17	Apr. 9	09 45	9	8	15	0.019	5.5[a]
18	Apr. 9	11 00	9	8	15	0.019	3.9

[a]Following 22 knot wind blowing within 22° of parallel to coast.

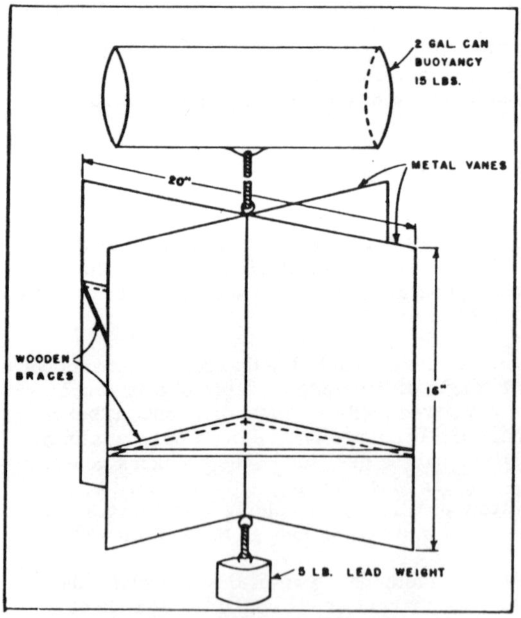

Fig. 1--Current cross used for field measurements of longshore currents

The second method was to drop the cross into the current, either from a boat or from the pier, and to permit it to drift freely. Observers followed along the shore and recorded the time when the cross passed markers at 100-ft intervals until the cross was washed up on the beach, or had drifted out to sea. This method permitted the measurement of currents over long stretches and minimized the effect of errors in timing.

A third method was tried along the beach a few miles north of Oceanside. Fluorescein dye markers were dropped into the current from a low-flying blimp, and were photographed from a plane at five-minute intervals. A party on the beach recorded the wave height and period. On the photographs the progress of the dye could be measured from landmarks along the beach, and the velocity computed. This method has the advantage that breaker angles may be measured directly. However, the diffusion of the dye makes it difficult to determine the exact point of advance. Furthermore, the combination of good flying weather and good waves was so rare that it was more satisfactory to use the other methods.

The surf at Oceanside was usually characterized by a zone of higher breakers on the outer bar, and a zone of lower breakers on the inner bar (see Fig. 2). On days where the majority of waves were breaking on the outer bar, the current was almost constant from the outer line of breakers to shore. During spring tides, however, when only a few of the highest waves were breaking on the outer bar, there would be a weak current between the two lines of breakers, and a strong current between the inner line of breakers and shore. At all times the current was small or negligible beyond the outer breaker.

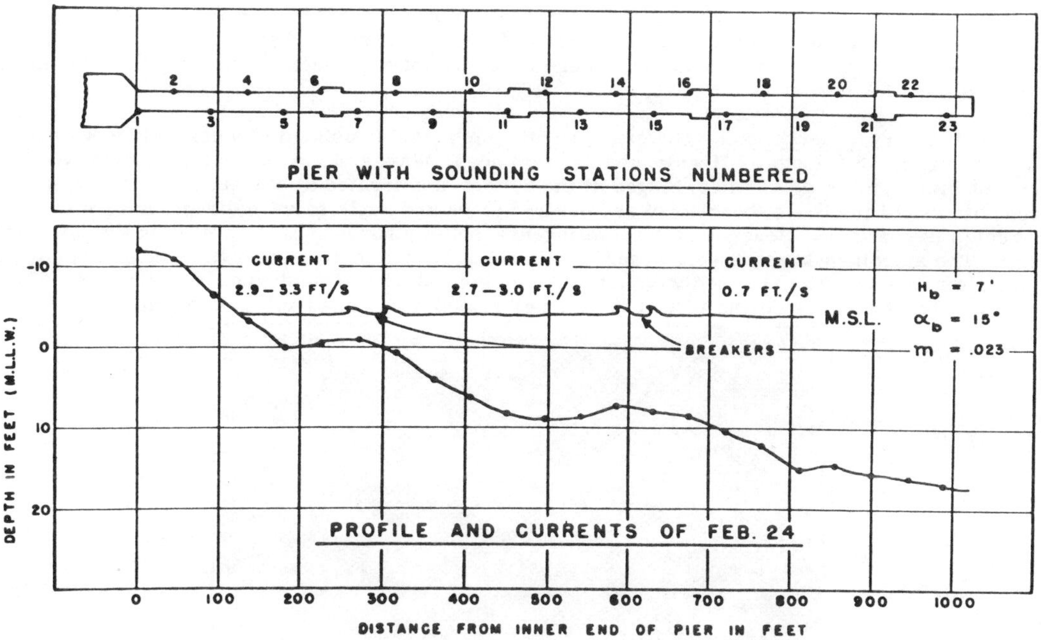

Fig. 2--Typical arrangements for field observations at Oceanside

Along any given stretch of beach the velocity varied as much as 25 per cent either way from the mean value. The observed variations are similar in duration (three to five minutes) to the surges that are observed in rip currents, and are probably associated with the variations in wave height over similar time intervals. Consequently these variations are variations in time, and not in position. This is clearly borne out by observation from a blimp of the movement of six dye markers at different parts of the Oceanside beach (see Fig. 3). In this figure the velocities of markers (1) and (2) were high between runs 1 and 2, and low between runs 2 and 3 (in the case of marker (1) the velocities were low between runs, 2 and 4. The velocities of markers (4), (5), and (6) were high between runs 3 and 4, and low between runs 4 and 5.

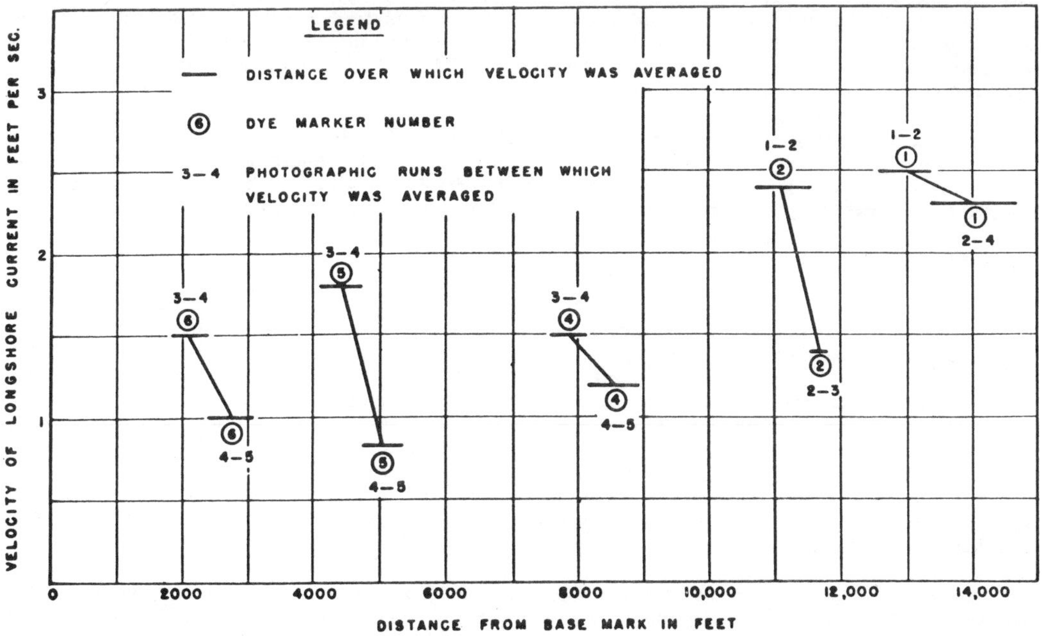

Fig. 3--Movement of dye-markers photographed from a blimp

When waves break nearly parallel to a coast, strong rip currents are set up. These currents run parallel to the beach for a short distance, and then out to sea. They reach velocities up to 1.5 knots. All cases where rip currents were predominant have been eliminated from this investigation.

Laboratory observations--The experimental equipment consisted of a concrete wave basin approximately 58 ft long, 39 ft wide, and two feet deep. Waves of any desired frequency were generated with a motor-driven flap, hinged at the bottom, and located at one end of the tank. At the opposite end of the basin, beaches of any desired slope and angle to wave direction were constructed (see Fig. 4). Beaches of three different absolute roughnesses were used in the investigation. The smoothest beaches were covered with metal sheets or very smooth neat cement. Intermediate roughness was obtained with natural sand bonded with cement to prevent erosion. The roughest beaches were coated with 1/4-inch pea gravel and bonded with a thin grout.

Fig. 4--Laboratory wave basin

Wave heights, breaker heights, and the depth of breaking were measured with electric point gages. Wave periods were obtained by observing the frequency of the wave-generation flap. The angle of the breakers was measured from vertical photographs. The longshore current was measured by timing the travel of KM_nO_4 dye over each five-foot interval marked along the beach. A careful review of the data indicated that measurements of longshore current along different sections of the central position of the beach (five to 15 ft in Figure 4) were consistent and reproducible.

The experimental results are presented in Table 2. All values represent the average of at least three separate determinations. For all runs reported, the length of the beach was ten or more times the width of the surf zone, and all waves broke well inside the toe of the beach slope. Figure 5 illustrates typical surf conditions in the laboratory experiments.

Analysis of the data

The longshore current is a stream which receives energy from the breaking waves. Equilibrium conditions prevail when the fraction of the energy received (available to maintain the longshore motion) equals the rate of dissipation of mechanical energy by friction. An exact analysis is rendered difficult because only a small portion of the wave energy is applied to the maintenance of the current. Although a correlation of the observed data can be obtained by intuitive reasoning based on kinematical considerations or by dimensional analysis, it is believed that relations derived from mechanical principles will yield a better understanding of the phenomenon and also provide a better basis for forecasting longshore currents. Accordingly, two approaches are presented; one based on energy balance, and the other on momentum considerations, both assuming that waves have been breaking long enough to set up a steady state, and that the beach is straight, and the bottom slope uniform.

Energy approach--Consider a volume of water ABCDE (see Fig. 6) extending between shore and the breaking waves, from the surface to the bottom. Over a width of beach dx the total wave energy entering the volume ABCDE equals $C\,E\cos\alpha\,dx$, where C is the wave velocity (almost equal to the

Table 2--Tabulation of laboratory data and results

No.	Aver. K	Slope m	Breaker conditions					Longshore velocity V	
			T	H	h	α	$(mH^2/T) \sin 2\alpha$	Calc.	Obs.
			sec	ft	ft	°		ft/sec	ft/sec
1	0.0397	0.066	1.00	0.47	0.75	18.3	0.205	1.21	0.78
2	0.0397	0.066	1.06	0.32	0.44	13.8	0.143	0.74	0.64
3	0.0397	0.066	1.14	0.40	0.56	14.6	0.166	0.87	0.82
4	0.0397	0.066	1.15	0.31	0.41	12.6	0.133	0.66	0.68
5	0.0397	0.066	1.25	0.30	0.39	11.7	0.124	0.59	0.76
6	0.0397	0.066	1.32	0.32	0.40	11.7	0.127	0.61	0.75
7	0.0397	0.066	1.40	0.29	0.37	10.9	0.114	0.54	0.64
8	0.0397	0.144	1.90	0.16	0.24	17.6	0.104	0.58	0.75
9	0.0397	0.144	2.13	0.15	0.23	17.2	0.095	0.55	0.66
10	0.0397	0.144	2.22	0.15	0.24	17.3	0.094	0.55	0.50
11	0.0397	0.241	0.72	0.28	0.48	18.2	0.250	1.27	1.33
12	0.0397	0.241	0.92	0.35	0.52	16.5	0.260	1.24	1.27
13	0.0397	0.241	1.14	0.22	0.28	10.4	0.154	0.60	0.53
14	0.0397	0.241	1.22	0.22	0.27	10.6	0.151	0.60	0.69
15	0.0070	0.100	0.99	0.24	0.32	28.0	0.169	1.65	1.68
16	0.0070	0.100	1.32	0.22	0.27	22.8	0.138	1.20	1.45
17	0.0070	0.100	1.63	0.16	0.23	18.8	0.099	0.96	0.96
18	0.0070	0.100	1.98	0.16	0.22	18.4	0.092	0.85	0.76
19	0.0070	0.139	0.83	0.28	0.43	56.6	0.234	3.03	2.46
20	0.0070	0.139	0.91	0.23	0.33	45.3	0.201	2.35	2.31
21	0.0070	0.139	1.00	0.22	0.29	38.8	0.187	2.06	2.22
22	0.0070	0.139	1.12	0.20	0.24	33.2	0.166	1.69	1.93
23	0.0070	0.139	1.35	0.20	0.25	31.1	0.154	1.58	1.52
24	0.0070	0.260	0.80	0.34	0.62	57.5	0.324	3.97	3.78
25	0.0070	0.260	0.90	0.29	0.43	52.5	0.286	3.26	3.34
26	0.0070	0.260	0.98	0.28	0.41	47.2	0.275	2.99	3.00
27	0.0070	0.260	1.23	0.20	0.26	32.5	0.197	1.82	1.91
28	0.0070	0.260	1.27	0.22	0.23	31.9	0.207	1.78	1.76
29	0.385	0.098	0.95	0.26	0.36	30.1	0.182	0.52	1.03
30	0.385	0.098	1.33	0.21	0.27	21.4	0.130	0.31	0.46
31	0.385	0.098	1.67	0.16	0.20	18.0	0.096	0.22	0.20
32	0.385	0.098	1.99	0.12	0.19	16.4	0.072	0.17	0.15
33	0.385	0.143	1.08	0.33	0.47	30.4	0.232	0.69	1.32
34	0.385	0.143	1.36	0.29	0.38	24.6	0.188	0.52	0.63
35	0.385	0.143	1.58	0.20	0.27	19.3	0.131	0.33	0.36
36	0.385	0.143	1.91	0.20	0.26	18.4	0.121	0.30	0.32
37	0.385	0.143	2.32	0.22	0.30	19.1	0.122	0.30	0.18

1-14 (Lab. 1), natural-sand beach; 15-28 (Lab. 2), beach covered with sheet metal or smooth cement; 29-37 (Lab. 3), beach covered with 1/4-inch pea gravel.

group velocity for breaking waves), E is the mean energy per unit surface area of the breaking waves, and α is the angle formed between the breaking crest and the shoreline. Of the energy that advances parallel to shore ($C\ E \cos \alpha dx$) $\sin \alpha$, a small fraction s is responsible for setting up longshore currents. The greater part of the wave energy entering this volume is destroyed in the breaker and turned into heat, or may be responsible for piling water against the shore and setting up rip currents.

Energy is dissipated in a layer of frictional resistance along the bottom. The frictional force per unit width of beach equals

$$k \rho V^2 \ell' dx \quad \text{(1)}$$

and the rate at which energy is dissipated in the volume ABCDE equals

$$k\rho V^3 \ell' dx \quad \ldots\ldots (2)$$

where k is a friction parameter depending on the hydraulic roughness of the bottom, V the velocity of the longshore current, and ℓ' the distance along the bottom from shore to breaker line. With good approximation $\ell' = \ell$ (see Fig. 6). If m denotes the average beach slope, and h the depth at which the waves break, then

$$m = h/\ell \quad \ldots\ldots (3)$$

Assuming a steady state, all the energy available for longshore currents entering the volume ABCDE due to wave action must equal the loss of energy due to friction

$$sCE \cos\alpha \sin\alpha \, dx = k\rho V^3 \, dx \, h/m \quad \ldots\ldots (4)$$

The following relationships derived by the solitary wave theory have been found to apply to breaking waves with good approximation [SVERDRUP and MUNK, 1946; MUNK, 1949].

$$E = (\rho g/L)(4h\,H/3)^{3/2} \quad \ldots\ldots (5)$$

$$C = [g(h + H)]^{1/2} = L/T \quad \ldots\ldots (6)$$

$$h = 1.28\,H \quad \ldots\ldots (7)$$

$$Q = 4\,h^2 \sqrt{H/3h} \quad \ldots\ldots (8)$$

where ρ is the density of water, g the acceleration due to gravity, H, L, and T the height, length, and period of the breaking wave, h the depth of breaking, and, for later reference, Q is the cross-sectional area beneath the profile of a breaking wave crest. Combining (4), (5), (6), and (7) gives

$$V = K\left[(m\,H^2/T)\sin 2\alpha\right]^{1/3} \quad \ldots\ldots (9)$$

where

$$K^3 = 0.871\,g\,(s/k) \quad \ldots\ldots (10)$$

According to (9) the velocity of the longshore current depends upon the breaker height, the wave period, the angle formed by the breakers, the beach slope, and the hydraulic roughness of the beach. The first three parameters can be predicted from weather maps [SVERDRUP and MUNK, 1946]; the latter two remain relatively constant over a given beach.

The above solution applies to a situation in which the waves have been breaking long enough to set up a steady state, and in which the region chosen is far enough from any obstruction on the beach so that a full-strength current is reached. A more general solution has been derived which is not subject to these limitations. From these solutions numerical values have been computed for the duration of the transient state, and for the distance from a breakwater or discontinuity in the beach over which a current must flow to reach its full velocity. Both these values are surprisingly small. A current will build up to an essentially steady state within five minutes of the time a train of waves starts to break on a beach, and will have practically its full value within 200 yards of a breakwater or other obstruction. The short duration time is consistent with observations of three to five minute variations in longshore currents due to the variation of wave height over similar time intervals (see Fig. 3).

Momentum approach--With the breaking of every wave a certain water mass is thrown into motion in the direction of wave propagation. The longshore component of this motion provides the driving force for the longshore current.

Consider again the volume ABCDE (see Fig. 6). Let Q represent the cross-sectional area of a breaking wave crest moving with velocity C, L the length of the breaking wave, α the angle formed by the breaking crest with the shoreline, and ρ the density of water. Then $\rho Q\,C/L$ is the average momentum per unit surface area, and $C\,(\rho Q\,C/L)\cos\alpha dx$ the mean flux of momentum into ABCDE. The component of momentum flux parallel to shore equals

$$C\sin\alpha\,(\rho Q\,C/L)\cos\alpha dx \quad \ldots\ldots (11)$$

The broken water is quickly slowed down by turbulent friction to the mean velocity V of the longshore current, and eventually expelled from the surf zone, giving rise to a momentum flux outward of

$$V\,(\rho Q\,C/L)\cos\alpha dx \quad \ldots\ldots (12)$$

Fig. 5--Surf-zone on smooth laboratory beach

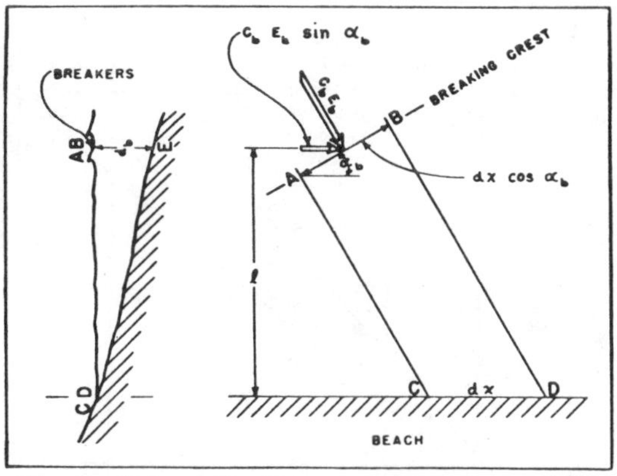

Fig. 6--Schematic diagram for energy and momentum considerations of longshore currents

The difference between (11) and (12)

$$(C \sin \alpha - V)(\rho Q\, C/L) \cos \alpha dx \quad \ldots\ldots (13)$$

represents the net flux of momentum, or equivalent force, applied by the breakers upon the water mass in the surf zone.

Since the longshore current is maintained at an essentially constant rate, this force is balanced by frictional force along the bottom (1). Equations (1), (3), (6), and (13) give

$$V^2 = a\,(C \sin \alpha - V) \quad \ldots\ldots (14)$$

where

$$a = mQ \cos \alpha / khT$$

Solving for V results in

$$V = (a/2)\left[\sqrt{1 + 4C \sin (\alpha/a)} - 1\right] \quad \ldots\ldots (15)$$

where according to (6) and (7) $C = (2.28\ gH)^{1/2}$. Substituting for Q from (8) and using (7) leads to

$$a = (2.61\ mH \cos \alpha)/(kT) \quad \ldots\ldots (16)$$

where, according to (14)

$$k = (2.61\ mH \cos \alpha)(C \sin \alpha - V)/TV^2 \quad \ldots\ldots (17)$$

Equation (15) gives the velocity of the longshore current as a function of the same variables that appeared in (9), and the solution is subject to the same limitations that were discussed when dealing with the energy approach.

Comparison of theory with observed data

In Figure 7 the observed longshore currents are plotted against the wave characteristics as they appear on the right side of (9). Average values of the friction parameter k were determined by substitution of the observed data in (17) for the various types of beach surface investigated. Except for the roughest beach, it was possible to draw a straight line through each set of data, the slope of which is the proportionality constant K of (9). Table 3 illustrates the order of magnitude of the energy fractions available for maintenance of longshore currents, as obtained from (10).

The beach at Oceanside and the natural-sand beach in the laboratory (Lab. 1) have values of s that are comparable. The sand-surface textures of these two beaches were comparable, but the values of the friction parameter k differ by five-fold. A part of this difference may be ascribed to the fact that the beach at Oceanside possessed bars whereas the laboratory beach was a plane,

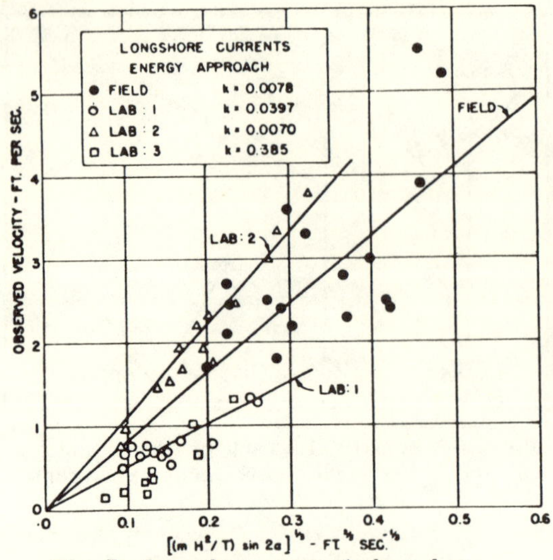

Fig. 7--Longshore currents based on energy approach

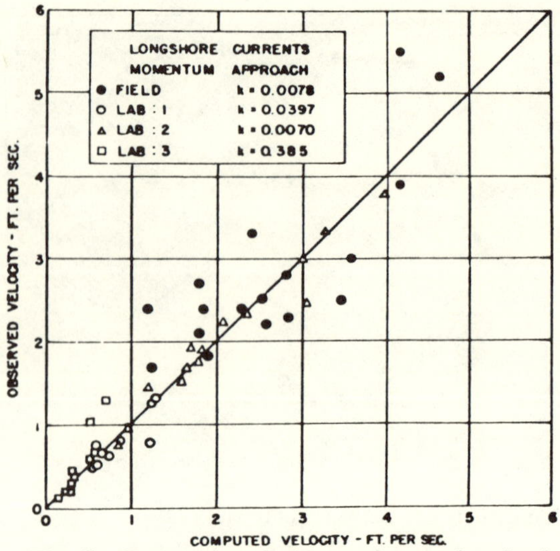

Fig. 8--Comparison between observed and calculated velocities of longshore currents based on momentum approach

Table 3--Values of parameters for field and laboratory data

Beach	k	K	s
Field	0.0078	8.2	0.15
Lab 1	0.0397	5.12	0.19
Lab 2	0.0070	11.02	0.33
Lab 3	0.385	. . . .	. . .

undisturbed surface. However, k depends upon the hydraulic roughness of the beach rather than the absolute roughness of the beach surface. The experiments with the Lab 1 beach were characterized by relatively weak longshore velocities so that viscous forces were relatively large. Thus the higher values of k might be ascribed to boundary layer resistance in contrast to resistance governed solely by roughness. On the other hand, the Lab 2 beach (sheet metal or smooth concrete) was the smoothest one investigated and the values of k are correspondingly small. However, the longshore velocities were in general larger even than those encountered at Oceanside so that it offered less resistance to flow and was therefore able to absorb a larger fraction of the available energy as indicated by the large value of s.

The scatter of the data shown in Figure 7 results primarily from the difficulty of measuring the breaker angle α, both in the laboratory and the field. The very rough beach (Lab 3, 1/4-inch pea gravel) showed a particularly wide range in the individual values of k, from 0.07 to 1.2, and these depended in a systematic fashion upon the longshore velocity. In this case the theory cannot assume a constant average value of k.

Figure 8 shows the observed longshore current plotted against values of the current computed according to (15) using the above average values of k. The agreement of these results indicates that the theory employed is essentially correct.

Conclusion

The laboratory investigation corroborates the field data and the theory, but indicates the need for systematic work, such as has been done in the case of pipe flow, in order to correlate the friction parameter with measureable beach and wave characteristics. Meanwhile, reasonable computations of longshore currents may be obtained by using either (9) or (15) with a value of k taken equal to that obtained from the Oceanside data. Equation (15) has the advantage that no assumption has to be made regarding the selection of the energy fraction available.

The relationships derived here may find application in two ways: (a) in the prediction of longshore currents on the basis of the predicted values for the height, period, and direction of the breaking waves; (b) in the computation of the seasonal variations in the magnitude and direction of longshore currents at chosen localities, on the basis of climatological wave data [**BURT** and **SAUR**, 1948]. The restriction to straight beaches with uniform bottom topography greatly limits the usefulness of these applications.

Acknowledgments--This work represents results of research carried out for the Office of Naval Research, the Hydrographic Office, and the Bureau of Ships of the Navy Department under contract with the University of California. We are indebted to Ralph E. Jentoft, Jr. for his assistance in carrying out the observations while on wartime duty at the Scripps Institution.

References

BURT, W. V., and SAUR, J. F. T., JR., Hindcasting technique provides statistical wave data, Civ. Eng., v. 18, pp. 47-49, 1948.

MUNK, WALTER H., The solitary wave theory and its application to surf problems, N. Y. Acad. Sci., v. 51, pp. 376-424, 1949.

SVERDRUP, H. U., and MUNK, W. H., Theoretical and empirical relations in forecasting breakers and surf, Trans. Amer. Geophys. Union, v. 27, pp. 828-836, 1946.

Department of Engineering (J. A. P.),
University of California,
Berkeley, California
Scripps Institution of Oceanography, and Institute of Geophysics (W. H. M.),
University of California,
La Jolla, California
Museum of Natural History (M. A. T.),
Chicago, Illinois

7

Reprinted from *First Coastal Eng. Conf., Univ. Calif., 1950, Proc.*, pp. 50-59

NEARSHORE CIRCULATION*

F. P. Shepard and D. L. Inman
Scripps Institution of Oceanography
University of California
La Jolla, California

INTRODUCTION

Studies of nearshore circulation were initiated at Scripps Institution during World War II. A method of estimating the velocity of longshore currents from known wave conditions on straight beaches with parallel contours was devised by Munk and Traylor (1945) and later revised by Putnam, Munk and Traylor (1949). Their methods were based on energy and momentum considerations which were applied to the following two types of observations: (1) field observations of longshore currents along the straight beach at Oceanside, California made by Munk and Traylor (1945), and (2) laboratory measurements conducted at the Department of Engineering, University of California.

In 1945 a program of field observations was initiated to study the nearshore currents in relation to a variety of coastal types and submarine configurations. Operations extending over a period of one year involved measurement of currents inside the breakers at 63 stations from the United States -- Mexican boundary to Newport, California (Shepard, 1950) (Fig. 1). The observations were repeated approximately every 12 days. Subsequently the shallow waters adjacent to individual beaches representative of various types of environment have been studied intensively. This work has included a beach with adjacent submerged ridge and canyon topography (Shepard and Inman, 1950), two straight beaches with parallel bottom contours, and one beach at the head of a crescentic bay. In addition the effects of jetties, piers, and points have been investigated. During this work currents inside and outside the breaker zone were investigated. Most of the observations were made in southern California, but studies along many other coasts of the United States and in the Hawaiian Islands indicate that the results have a general application.

All devices used in these investigations were of a free drifting type, the velocity being determined by the distance of travel in a given period of time. Currents outside of the surf zone were measured by surface floats, dye, and triplanes (current crosses) submerged at a variety of depths. Locations were obtained by multiple sextant angles from accompanying boats. Inside the breaker line bottom currents were measured by volley balls given slight negative buoyancy. Surface currents were observed by free floating kelp and dye. Details of the methods are given by Shepard and Inman (1950). Wave recordings were obtained simultaneously with many of the measurements. Angles of wave approach were determined in part by the transit-sighting bar method devised by Forrest (1950).

TERMINOLOGY AND GENERAL PRINCIPLES OF CIRCULATION

Observations of nearshore circulation show that there are certain basic principles which apply to most environments, including in varying degree straight beaches with parallel offshore contours and beaches adjoined by irregular submarine topography. There appear to be at least two interrelated current systems (Fig. 2):

1. The coastal currents which flow roughly parallel to the shore, and constitute a relatively uniform drift in the deeper water adjacent to the surf zone. These currents may be tidal currents, transient wind-driven currents, or currents associated with the distribution of mass in local waters.

*Contribution from the Scripps Institution of Oceanography, New Series No. 532. This work represents in part research carried out for the Beach Erosion Board and the Office of Naval Research under contracts with the University of California.

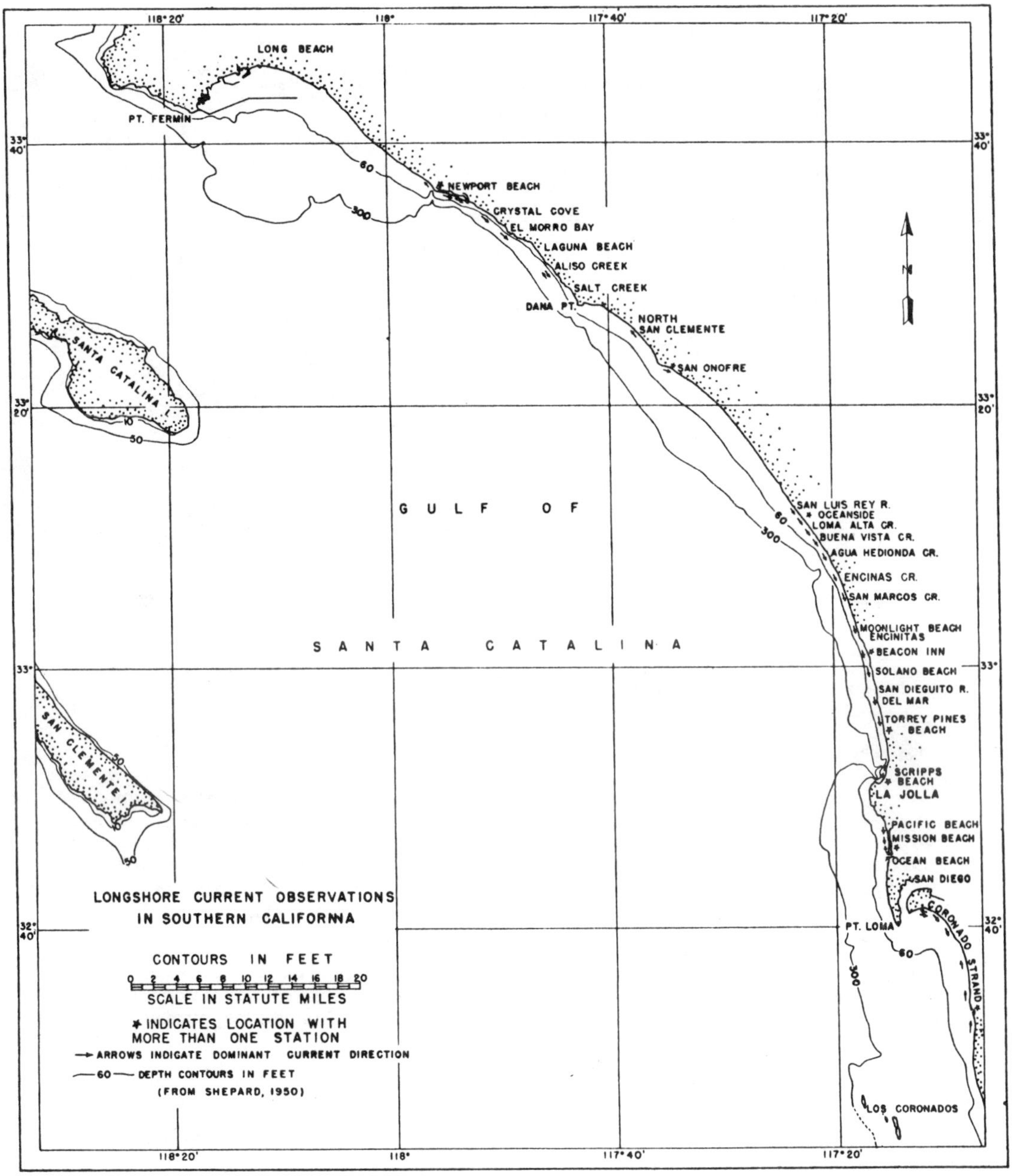

Fig. 1
Showing areas where nearshore current measurements have been made in southern Calif.

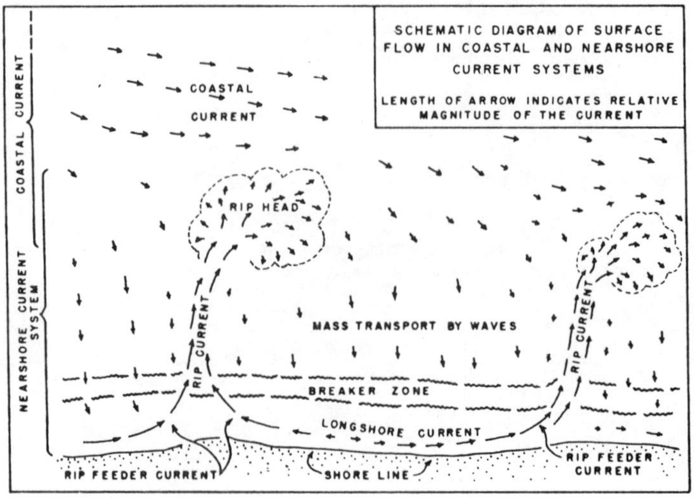

Fig. 2

2. A nearshore system which may be superimposed on the inner portion of the coastal current or in the absence of a coastal current may exist indepently. The nearshore system is associated with wave action in and near the breaker zone and consists of: (a) shoreward mass transport of water due to wave motion, which carries water through the breaker zone in the direction of wave propagation, (b) movement of this water parallel to the coast as longshore currents, (c) seaward return flows, such as flow along concentrated lanes known as rip currents, and (d) longshore movement of the expanding head of the rip current.

Since the rip currents have relatively high offshore velocities, the shoreward movement is restricted to wide lanes in between rip currents. As a result, the circulation pattern takes the form of an eddy or cell with a vertical axis (Fig. 2). The positions of shoreward motion and seaward return are much dependent on the submarine topography, the configuration of the shoreline, and the height and period of the waves. Periodicity or fluctuation of current velocity and direction is a characteristic of flow in the nearshore system (Shepard and Inman, 1950). This variability is primarily due to the grouping of high waves followed by low waves, a phenomenon which gives rise to surf beat (Munk, 1949b).

The direction of longshore current is primarily dependent on two factors: (1) the direction of wave propagation, and (2) the rise in water level due to the shoreward mass transport of the waves, which is greatest in the zones of highest breakers along a beach (wave convergence zones). The longshore currents commonly flow away from these zones of highest waves.

There may be processes other than rip currents by which water may be returned seaward (Munk, 1949a). Our observations indicate the importance of a net seaward drift along the bottom inside the breaker zone and a net shoreward movement at the surface. Comparisons of the offshore and onshore components of surface and bottom currents in the surf zone are given in Table I. This table shows that there is a definite tendency for bottom longshore currents to have a small offshore component. The lack of a pronounced onshore component for the surface currents is probably due to the method of observation (see footnote to Table I on following page).

To date we have not found any indication that these differential net movements between top and bottom extend any distance outside of the breaker zone, but rather that the current moves shoreward from top to bottom in one area and seaward from top to bottom in another outside of the breaker zone.

TABLE 1

Comparison of the offshore and onshore tendencies between surface and bottom currents in the surf zone. Based on all observations, irrespective of magnitude.

Location	Surface Currents[1]			Bottom Currents[2]		
	No. of Obs.	% Offshore	% Onshore	No. of Obs.	% Offshore	% Onshore
Torrey Pines Beach	139	54.7	45.3	512	81.4	18.6
Pacific Beach	162	47.5	52.5	220	81.8	18.2
Mission Beach	224	50.0	50.0	383	76.3	23.7
Total No. of Obs.	525			1115		
Average %		50.5	49.5		79.7	20.3

[1]Surface currents were measured by free floating kelp and dye and thus represent measurements of a relatively thick surface layer and not the surface itself which is known to have a greater onshore tendency (Shepard and Inman, 1950, p. 200).

[2]Bottom currents were measured with volley balls given slight negative buoyancy.

STRAIGHT BEACHES WITH PARALLEL CONTOURS

Most beaches have some curvature and the adjacent bottom topography is usually somewhat irregular. However, there are a sufficient number of beaches which are essentially straight with parallel bottom contours to warrant consideration. This type of beach is ideal for theoretical treatment and was therefore chosen for the investigations of longshore currents by Putnam, Munk and Traylor (1949). Their momentum approach has been widely used. It relates the velocity V, of the longshore currents to the wave height H, period T, angle of approach α, and slope i of the beach, according to the equation:

$$V = \frac{a}{2}\left[\sqrt{1 + \frac{4C \sin\alpha}{a}} - 1\right] \qquad (1)$$

where $a = (2.61\ i\ H \cos\alpha) / (kT)$

and, $C = \sqrt{2.28\ g\ H}$ is the wave velocity, k is the beach friction coefficient (hydraulic roughness), and g is the acceleration of gravity.

Field measurements partly reported by Shepard (1950) have indicated only partial agreement with current velocities predicted from the equation. The discrepancies appear to be due to:

1. Rip currents which normally increase the flow in the dominant direction on the up current side of the rip zone and decrease or reverse the current on the other side (Fig. 3).

2. Cell-like circulation of the nearshore system, which exists even under conditions of normal wave approach (breaker crests essentially parallel to the beach). According to the formula there should be no current under conditions of normal approach. However, with large breakers longshore currents of velocity of one or more knots have been observed for limited distances despite this normal approach.

3. The fluctuating nature of longshore currents which requires measurements over a long interval of time in order to obtain an average that is suitable for comparison with the predicted value.

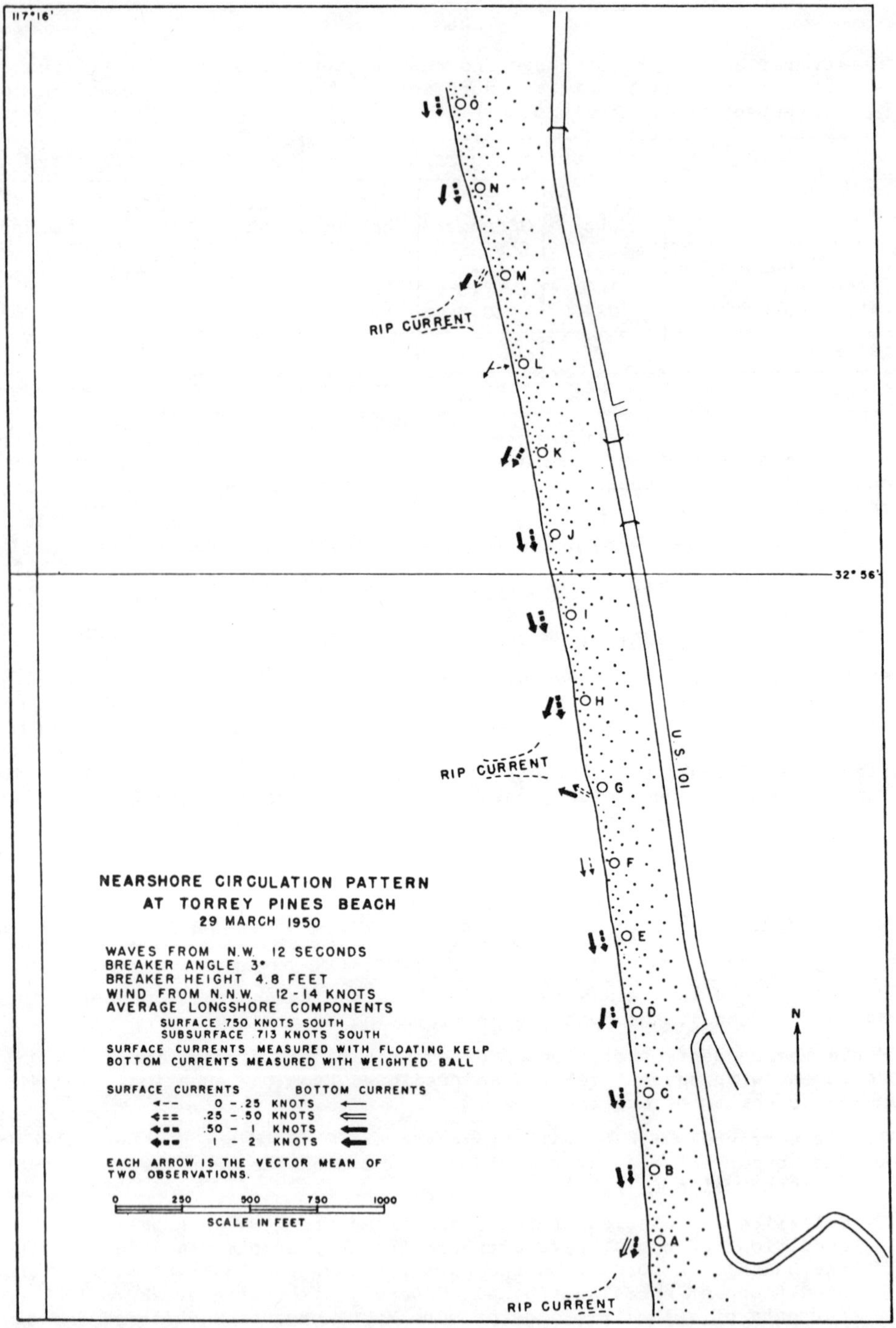

Fig. 3
Showing the influence of circulation pattern and rip currents on the direction and magnitude of longshore currents on a straight beach. Note that in the lee of rip current zones the longshore current is much reduced (station L) and sometimes reversed in direction (station G).

4. The fact that beach friction coefficient (k) was assumed to be a constant for a given beach. Further measurements indicated that it is a function of current velocity (Fig. 4).

5. Many beach approaches are terraced, making application of the formula impossible.

The nearshore circulation is frequently the result of waves approaching the coast from more than one direction, making current predictions extremely difficult.

Despite these discrepancies the longshore currents outside straight beaches with parallel submarine contours appear to be essentially the result of the factors considered by Putnam, Munk and Traylor (1949).

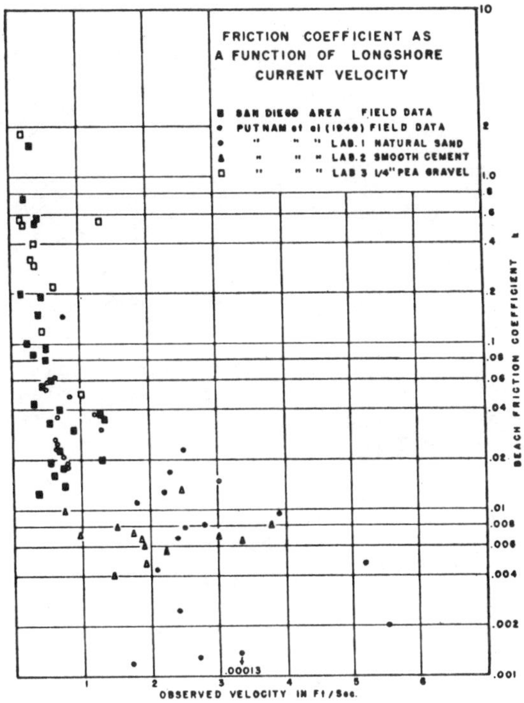

Fig. 4

The average longshore component of currents observed along two relatively straight beaches with parallel bottom contours in the San Diego area were compared with the currents predicted from equation 1. Preliminary plotting of the observed versus the computed velocity showed a rather large scatter, the amount of disagreement apparently being a function of the velocity of the current. This suggested the possibility that the friction coefficient, k, is not a constant for a given beach, but varies with the velocity.

Using the observed value of the longshore current, the value of k was computed for all of the field and laboratory observations listed in Putnam et al (1949) and for those recently obtained in the San Diego area. The coefficient k is plotted as a function of the observed velocity in Fig. 4. Inspection of this figure strengthens the contention that the coefficient k is a function of velocity, particularly for values of current of the order of two feet per second and less. Predicted values of longshore current based on k values from Fig. 4 have shown good agreement with field observations.

The regular spacing between rip channels cutting through the longshore bars along the straight, fine-sand beaches of northern Oregon and southern Washington indicates that the general features of the nearshore circulation pattern described in this report also apply to these northern beaches. At low tide the distance between rip channels can easily be ascertained from car speedometer readings. Observations of this type were made along the straight beaches of Clatsop Spit, Oregon, and Leadbetter Spit, Washington, on the 17th and 18th of June, 1950. The mean distance between rip channels was found to be 0.25 statute miles. The standard deviation of the distance between channels was found to be 0.09 miles for a ten mile section of beach along Clatsop Spit. These observations were made during a period of low waves.

BEACHES BORDERED BY IRREGULAR SUBMARINE TOPOGRAPHY

As yet there has been no quantitative approach to the prediction of currents or circulation along beaches bordered by irregular submarine topography. The situation differs from that existing along straight beaches with parallel contours in that the direction of longshore currents is dependent not only on the angle of wave approach but on the localized piling up of water on the beach at points of wave convergence. This local rise of sea level, the degree of which is still un-

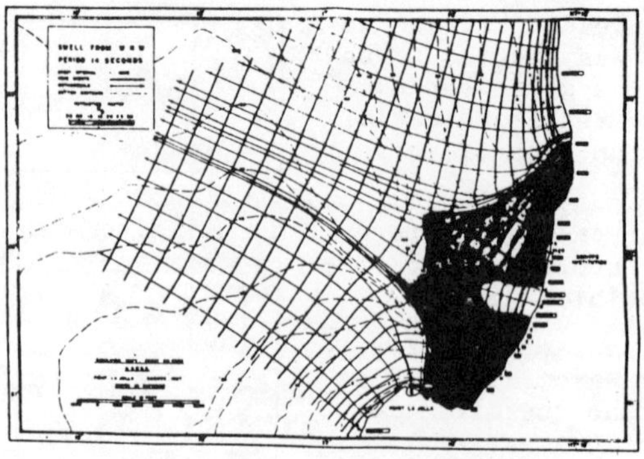

Fig. 5. Wave refraction diagram showing areas of wave convergence and divergence near La Jolla. The relative wave height is given by the length of the bar at various points along the beach. The letters A through H locate the same beach stations shown in Figs. 6 and 7. (From Munk and Traylor, 1947.)

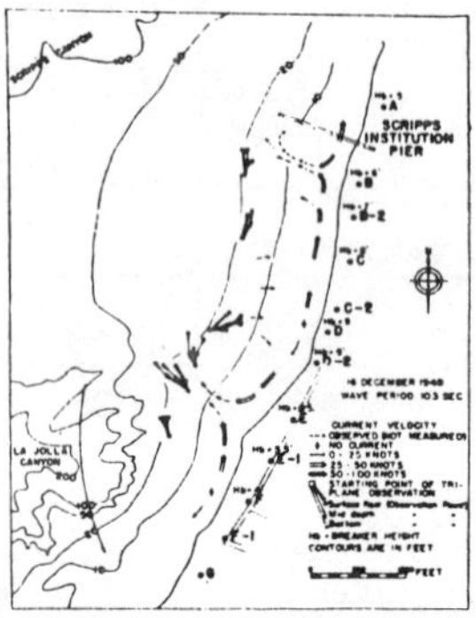

Fig. 6. The cell-like features of the nearshore circulation at Scripps Beach, La Jolla. The point of wave convergence is at station D, as shown by breaker height notation. (From Shepard and Inman, 1950.)

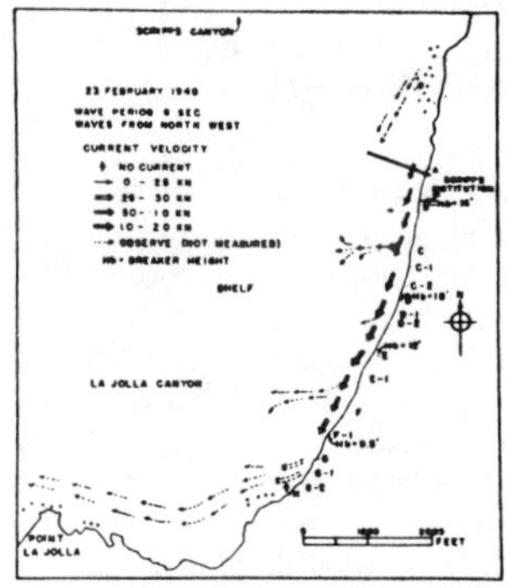

Fig. 7. For short period waves the wave convergence is not as pronounced, and the direction of longshore current is controlled by the direction of wave approach. (From Shepard and Inman, 1950.)

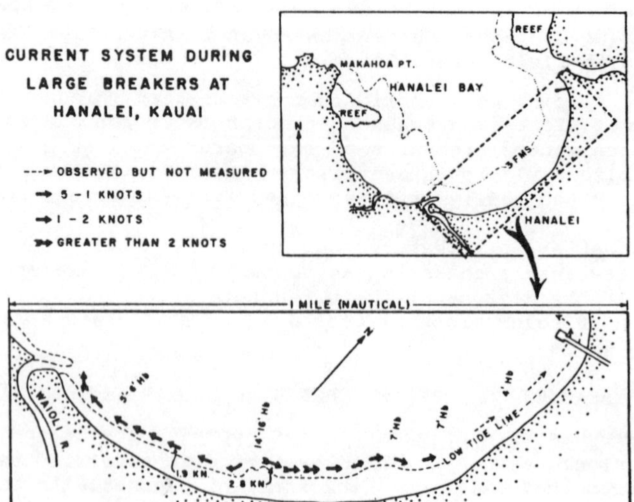

Fig. 8

known, results in a gradient along the beach and accompanying flow in both directions away from the convergence. These currents may be opposed in direction to the angle of wave approach.

Submarine canyons. Near those places where submarine canyon heads approach closely to the coast there are striking examples of wave convergence. At the canyon head the wave orthogonals diverge, decreasing the wave height, whereas they converge on either side, increasing the wave height (Fig. 5). In the La Jolla area it is not uncommon for the breaker height to be 10 times as high north of the two canyon heads as it is inside the heads. Current measurements outside the breakers at the convergence north of La Jolla Canyon indicate a movement towards the beach from the top to the bottom (Fig. 6). Inside the breakers the longshore currents flow away from the convergence in either direction. Rip currents commonly develop at points between the convergence and the divergence zone, where the waves are intermediate in height.

The current systems have been studied in two areas where canyon heads closely approach the coast: at La Jolla (Shepard and Inman, 1950) and at Mugu (Inman, 1950). In both areas the circulation pattern was found to be primarily dependent on the wave period. The longer period waves resulted in a sufficient degree of convergence so that the longshore currents flowed away from the zone of convergence. In this case the circulation cell is well developed and essentially fixed in position (Fig. 6) except as shifted slightly by changes in direction of wave approach. For shorter period waves the convergence is not as pronounced and the direction of wave approach becomes the controlling factor in determining the direction of longshore currents (Fig. 7). In this case the circulation cells are less stable and the position of the rips along the beach is variable. At La Jolla a strong blow accompanied by short period waves approaching diagonal to the coast produces constant current directions with the most pronounced rip at the bend in the coast in the down current direction (Fig. 7).

Crescentic Bay. On the north coast of Kauai in the Hawaiian Islands, Hanalei Bay forms a large crescentic bight into the main trend of the coast. During periods of high waves a large variation in wave heights was observed in a short distance along this bay (Fig. 8). In about a quarter of a mile breaker heights were observed to decrease from as much as 18 feet to 5 feet in one direction and at about half that rate in the other direction. This contrast is only partially explained by the small reefs which fringe both sides of the entrance. The highest breakers were comparable to those on the open coast so that they do not represent a convergence except relative to the adjacent portions of the bay. The longshore currents respond to this relative convergence in the same way as observed in areas adjacent to canyon heads. Currents as high as three knots were measured. At the point of greatest breaker height there is a low ridge extending seaward and the wave crests are bent around this ridge so that they come in directly against the strong outflowing currents. The flow is so strong as to produce a pronounced trough along the shore. In places these troughs are bordered by escarpments in the sand. As can be seen in Fig. 8, the flow away from the zone of high waves extends along the shore as much as three-quarters of a mile.

It is felt that these two cases including submarine canyons and crescentic bay by no means exhaust the possibilities of developing a local convergence with resulting divergent flow of longshore currents.

Small crescentic bays often show currents flowing in at one side and out on the other. During a 60 mile an hour gale at Nahant, Mass., currents were observed flowing in on one side of a small bay with velocities averaging about 4 or 5 knots. The outflow on the other side was much slower, indicating that there was a subsurface return flow. The surface inflow was continuous although considerably retarded during wave troughs.

CIRCULATION IN RELATION TO OBSTRUCTIONS

Points, breakwaters, and piers all influence the circulation pattern and alter the direction of the currents flowing along the shore. In general these obstructions determine the position of one side of the circulation cell. In places where relatively straight beaches are terminated on the down current side by points or other obstructions, a pronounced rip extends seaward. During periods of large waves having strong diagonal approach these rips can be traced seaward for one or more miles.

At the north jetty of the Mission Bay Breakwater a seaward flow was frequently observed in the current lee of the jetty. This same seaward flow north of the jetty exists under most other observed directions of wave approach (Fig. 9).

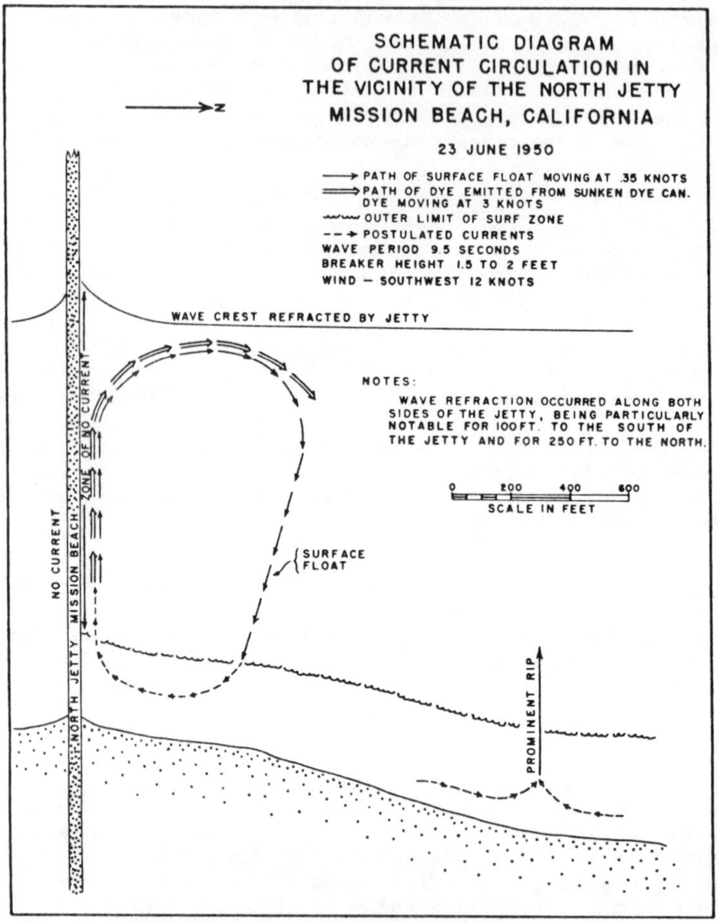

Fig. 9

Where prominent points of land interrupt the predominant longshore current flow, currents opposite in direction to that of the coast in general are likely to develop in the current lee of the point. Examples of this reverse flow have been observed in the lee of Dana Point (See Fig. 1 for location). Indirect evidence of currents flowing in this reversed direction are found where spits extend north in the current lee of obstructions such as at Morro Rock along the central California coast.

ACKNOWLEDGMENTS

The writers were assisted in field operations by Capt. W. H. Quinn, USAF, Lt. A. E. May, USN, Lt. Comdr. E. V. Mohl, USN, D. B. Sayner, Ruth Young, Comdr. R. Panzarini, Argentine Navy, J. D. Frautschy, H. R. Gould, Ruth Inman, R. L. Wisner, A. A. Allanson, and W. Jewell. Helpful suggestions in the course of the work came from W. H. Munk and R. S. Arthur.

REFERENCES

Forrest, D.R. (1950). A method of estimating wave direction: Bulletin of the Beach Erosion Board, vol. 4, No. 2, pp. 31-37, Corps of Engineers, Wash., D.C.

Inman, D.L. (1950). Beach study in the vicinity of Mugu Lagoon, California: Tech. Memorandum No. 14, Beach Erosion Board, Corps of Engineers, Wash., D.C.

Munk, W.H. (1949a). The solitary wave theory and its application to surf problems: Annals New York Acad. Sciences, vol. 51, pp. 376-424.

Munk, W.H. (1949b). Surf beats: Trans. Amer. Geophys. Union, vol. 30, pp. 849-854.

Munk, W.H., and M.A. Traylor (1945). Forecasting longshore currents: Scripps Institution of Oceanography, University of California, Wave Project Report No. 46 (unpublished).

Munk, W.H., and M.A. Traylor (1947). Refraction of ocean waves: a process linking underwater topography to beach erosion: The Jour. of Geology, vol. 55, pp. 1-26.

Putnam, J.A., W.H. Munk, and M.A. Traylor (1949). The prediction of longshore currents: Trans. Amer. Geophys. Union, vol. 30, pp. 337-345.

Shepard, F.P. (1950). Longshore current observations in southern California: Tech. Memorandum No. 13, Beach Erosion Board, Corps of Engineers, Wash., D.C.

Shepard, F.P., and D.L. Inman (1950). Nearshore water circulation related to bottom topography and wave refraction: Trans. Amer. Geophys. Union, vol. 31, pp. 196-212.

8

Reprinted from *Am. Soc. Civil Engineers Proc., Jour. Hydraulics Div.*
85(HY10):53-79 (1959)

WAVE-INDUCED MOTION OF BOTTOM SEDIMENT PARTICLES[a]

P. S. Eagleson,[1] A.M. ASCE and R. G. Dean,[2] A.M. ASCE

ABSTRACT

Theoretical analyses are presented yielding expressions for incipient motion and net velocity of single spherical bottom sediment particles under oscillatory waves.

These relationships are verified by experiment and are extended to the prediction of equilibrium relationships between local wave characteristics, bottom slope and median particle diameter found on natural beaches.

SYNOPSIS

The problem under study is the mechanics of the motion of sedimentary beach materials, particularly as it applied to their sorting due to selective onshore, offshore transport by shallow-water waves.

A theoretical analysis is presented which yields an equation for the net velocity of a single spherical sediment particle as a function of a resistance coefficient, C_D, the fluid mass transport velocity near the bottom and the beach slope. The equation is verified by laboratory measurements.

Measurements of the incipient motion condition are correlated with Reynolds number by means of a derived resistance coefficient.

By means of existing boundary layer theory both the net sediment velocity and incipient motion expressions are extended separately to the prediction of the equilibrium relationships between local wave characteristics, bottom slope and median particle diameter found on natural beaches. Agreement

Note: Discussion open until March 1, 1960. To extend the closing date one month, a written request must be filed with the Executive Secretary, ASCE. Paper 2202 is part of the copyrighted Journal of the Hydraulics Division, Proceedings of the American Society of Civil Engineers, Vol. 85, No. HY 10, October, 1959.

a. Presented at meeting of Hydraulics Div., ASCE, Massachusetts Institute of Technology, Cambridge, Mass., August, 1957.

1. Assistant Prof. of Hydr. Eng., Hydrodynamics Lab., Dept. of Civ. and Sanitary Eng., Massachusetts Institute of Technology, Cambridge, Mass.

2. Research Assistant, Hydrodynamics Lab., Dept. of Civ. and Sanitary Eng., Massachusetts Institute of Technology, Cambridge, Mass.

with natural beach data was found to be good in both cases but no conclusion was reached with respect to the conditions in nature under which each is applicable.

List of Notations

a	Instantaneous acceleration of fluid or sediment particle, ft./sec.^{2}
A, B, E, P, Q	Coefficients associated with kinematic wave properties, as specified
C	Wave celerity, ft./sec.
C_D	Coefficient of drag
C_L	Coefficient of lift
C_M	Coefficient of virtual mass
C_{Rx}	Coefficient of rolling friction
d	Still water depth, ft.
D	Diameter of sediment particle, mm. or ft., as specified
D_e	Sediment diameter as determined from "equilibrium" criterion, ft. or mm.
D_i	Sediment diameter as determined from "incipient" criterion, ft. or mm.
e	Base of natural logarithms, 2.718. . . . .
$—_f$	Subscript "f" referring to effective kinematic fluid quantity or to physical property of fluid
f	Coefficient associated with kinematic wave properties
F_D	Drag force on sediment particle, lbs.
F_F	Fluid force on sediment particle, resultant of viscous resistance forces, lbs.
F_G	Gravitational force on sediment particle, lbs. (buoyant weight)
F_L	Lift force on sediment particle, lbs. (non-viscous)
F_P	Pressure force on sediment particle, lbs.
F_R	Reaction force on sediment particle, lbs.
F_{RX}	Rolling frictional force on sediment particle, lbs.
F_{VM}	Virtual mass force on sediment particle, lbs.
g	Gravitational constant, 32.2 ft./sec.^{2}
H	Wave height, crest to trough, ft.
$—_j$	Subscript "j" referring to property at elevation, j
j	Distance normal to channel bottom at which maximum fluid mass transport velocity occurs, origin at plane through top of bed roughness particles, mm, or ft. as specified.

k	Mean diameter of roughness particle, mm or ft., as specified
L	Wave length, ft.
M	Moment of forces on sediment particle, ft. lbs.; particle mass, slugs.
$—_o$	Subscript "o" referring either to kinematic conditions in region of potential flow or to deep water wave geometry.
R_D	Reynolds number based on net relative fluid-sediment particle velocity.
R_I	Reynolds number, based on instantaneous relative fluid-sediment particle velocity.
$—_s$	Subscript "s" referring to sediment
s	Specific gravity of fluid or sediment, as indicated.
t	Time coordinate, sec.
T	Wave period, sec.
$\overline{U}(y)$	Longuet Higgins' fluid mass transport velocity at elevation y ft./sec. or ft./min. as specified.
$\overline{U}_f$	Effective fluid mass transport velocity, ft./sec. or ft./min. as specified.
$\overline{U}_o$	Stokes' theoretical fluid mass transport velocity evaluated near the bottom of the channel, ft./sec. or ft./min. as specified.
v	Instantaneous velocity, ft./sec.
$\overline{V}_s$	Net sediment velocity, ft./sec. or ft./min., as specified.
$—_x$	Subscript "x" referring to coordinate direction.
x	Coordinate parallel to beach, positive in direction of wave propagation, origin arbitrary, ft.
$—_y$	Subscript "y" referring to coordinate normal to beach.
y	Coordinate normal to channel bottom, positive upwards origin at upper edge of roughness particle, ft.
α	Angle of beach with horizontal, radians
β	A measure of the extent of viscous effects under a wave $(ft.)^{-1}$
γ	Specific weight, lbs./ft.^{3}
ϵ	Coefficient of rolling frictional resistance
θ	Wave phase angle, radians, origin 90° behind wave crest
ν	Kinematic viscosity of fluid, ft.^{2}/sec. or cm.^{2}/sec.
π	3.1415 . . . .
ρ	Mass density, slugs/ft.^{3}
σ	Angular frequency, rad./sec.
Σ	Symbol indicating summation of quantities.

ORDER NUMBER

PO NYAATP08

BOOK NUMBER

9718133

BAY NUMBER

2576-14

TITLE	Beach Processes and Coastal Hydrodynamics (Volume 39)

DESCRIPTION

Volume 39. This is an ex-library book and may have the usual library/used-book markings inside.This book has hardback covers. Clean from markings. In fair condition, suitable as a study copy.

ϕ	Angle between normal to beach and element of sediment particle reaction cone in plane of wave particle motion, radians.
Φ	Functional symbol
λ	Angle between resultant viscous force and face of beach, radians.

INTRODUCTION

As an oscillatory wave advances into shoaling water, a depth is eventually reached at which the unsteady fluid particle motion caused by the wave reaches the bottom. Shoreward of this point, as the depth decreases and the wave steepens, the particle displacement and velocity increase in magnitude. This unsteady fluid motion at the bottom produces hydrodynamic forces on the sediment particles comprising the beach. Under the proper conditions, these forces may be of sufficient magnitude to initiate and maintain motion of the sediment particles. For sediment densities near that of the fluid, the net sediment motion approaches the mass transport of the fluid.

It is recognized that other wave-induced phenomena such as rip and littoral current, breakers and bed fluidization can cause local motion of bottom sediments, but it is the general motion due to wave passage that is examined here as a possible explanation of the observed banding of sediment properties parallel to many natural beaches.

Ippen and Eagleson(1) made extensive laboratory observations of the behavior of discrete spherical sediment particles under wave action. It is the aim of the present study to expand the theoretical basis of the previous investigation through re-evaluation of existing data and through additional related experiments in order to present the data in terms of familiar physical relationships which may be justifiably extrapolated to natural beach conditions.

Ippen and Eagleson(1) divided the observed motion into two classifications:

1. Incipient sediment motion - an instantaneous condition reached when the resultant of all the active forces on the particle intersects the line connecting the bed-particle contact points.

2. Established sediment motion - an oscillatory or quasi-oscillatory condition of motion reached when for some portion of each wave cycle the moment of the instantaneous active forces exceeds that value necessary to initiate motion. The instantaneous hydrodynamic forces are oscillatory in nature and due to wave asymmetry, their net value (temporal average over one wave period) is always in the onshore direction. Gravity, however, provides an unidirectional offshore force. Thus, the particular local sediment, wave and beach characteristics will determine an offshore gravity force and a net onshore hydrodynamic force, the difference between which will lead to onshore, offshore or no net sediment motion.

This paper will take up these two classifications in the order listed and finally will apply the results of both investigations to natural beaches.

Theoretical Considerations

The details of this development are presented in Reference (2): only the briefest outline of steps is included here.

A. Forces Acting

Figure (1) shows the instantaneous forces acting on a discrete spherical sediment particle on a plane beach roughened with spherical particles of uniform diameter.

The individual forces are defined under LIST OF NOTATIONS.

B. Incipient Motion

Incipient motion has been described above as a condition of unstable equilibrium in which the obliquity of the resultant active force must equal the slope of an element (lying in a plane normal to the wave crests) of the polyhedron bounded by the reaction forces.

For a bed composed of irregular sand grains, the number, inclination and angular orientation of the reaction forces may have infinite variation, the average of which defines a mean, right circular "stability cone" the elements of which make angle ϕ with the cone's axis.

If the viscous resistance and apparent mass forces are assumed to act through the upper edge of the particle and all others through the particle center (see Fig. 1), and if moments are taken about the point of bed particle contact, the conditions of incipient motion may be written:

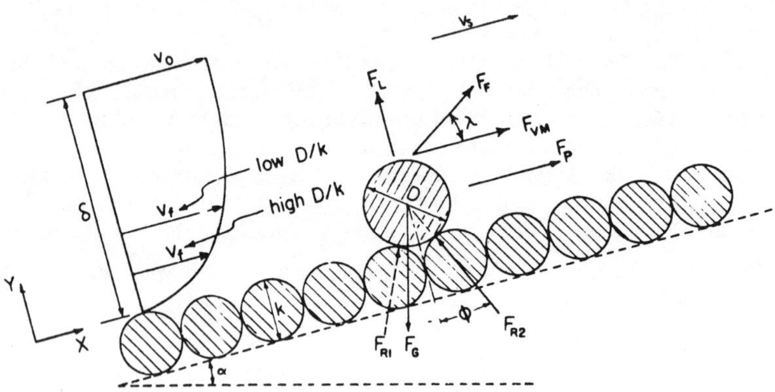

FIG. I FORCES ACTING AND IDEALIZED BED-PARTICLE GEOMETRY

$$\Sigma M = 0 = \underbrace{\rho_f \frac{\pi D^3}{16} v_f^2 [C_D(1+\cos\phi) + C_L \sin\phi]}_{\text{Viscous Resistance and Lift}} \tag{1}$$

$$+ \underbrace{\rho_f \frac{\pi D^4}{12}[C_M a_f(1+\cos\phi) + a_o \cos\phi]}_{\text{Apparent Mass and Pressure}} \underbrace{- \rho_f \frac{\pi D^4}{12} g\left(\frac{S_s}{S_f} - 1\right)\sin(\alpha+\phi)}_{\text{Gravity}}$$

in which the combined viscous and non-viscous normal forces ($F_F \sin\lambda$ and F_L respectively) are expressed in terms of a lift coefficient, C_L.

C. Established Sediment Motion

Equation of Motion

Assuming that the sediment particle is in motion throughout the entire wave cycle, the established motion condition is expressed by summing all forces parallel to the bottom. For waves shoaling with normal incidence, this is

$$\underbrace{M_s a_s}_{\text{Inertia}} = \underbrace{M_f a_o}_{\text{Pressure}} + \underbrace{C_M M_f (a_f - a_s)}_{\text{Apparent Mass}} + \underbrace{C_D \rho_f \frac{\pi D^2}{8}(v_f - v_s)|v_f - v_s|}_{\text{Drag}} \tag{2}$$

$$\underbrace{- g(M_s - M_f)\sin\alpha}_{\text{Gravity}} \underbrace{- \epsilon[F_G \cos\alpha - F_F \sin\lambda - F_L]\frac{v_s}{|v_s|}}_{\text{Bottom Resistance}}$$

The following assumptions are now made:

1. The drag force and the bottom frictional force are both linear functions of the relative instantaneous velocity, $(v_f - v_s)$.

2. Convective accelerations of the sediment particles are negligible.

3. The elevation at which the instantaneous velocity and acceleration are applied is independent of time.

4. The instantaneous fluid particle velocities at the edge of and within the bottom boundary layer are given respectively by expression of the form

$$v_o = \sigma A_o \sin\sigma t + 2\sigma B_o \sin 2\sigma t + E_o \tag{3}$$

$$v = A_f \sin\sigma t + B_f \sin 2\sigma t + P_f \cos\sigma t + Q_f \cos 2\sigma t + E_f \tag{4}$$

where the quantities A, B, P, Q, and E are independent of time and $\sigma t = \pi/2$ under the wave crest.

5. The coefficient of apparent mass is constant with time.

Net Sediment Motion

The net sediment velocity may now be obtained by integration of Eq. (2) over one complete wave cycle to yield:

$$\underbrace{C_D \rho_f \frac{\pi D^2}{8} (\overline{U}_f - \overline{V}_s)^2}_{\text{DRAG}} + \underbrace{C_{Rx} \rho_f \frac{\pi D^2}{8} (\overline{U}_f - \overline{V}_s)^2}_{\text{BOTTOM RESISTANCE}} = \underbrace{g (M_s - M_f) \sin\alpha}_{\text{GRAVITY}} \quad (5)$$

where C_D and C_{Rx} are inversely proportional to the relative velocity, $\overline{U}_f - \overline{V}_s$. A detailed derivation of Eq. (5) is presented in Reference (2).

As was pointed out in Reference (1) both $\overline{U}_f$ and $\overline{V}_s$ are the residue of a cyclic integration of much higher instantaneous velocities. $\overline{U}_f$ and $\overline{V}_s$ are thus to be considered as indicative of the remainder of intense instantaneous action and not, in general, as capable of significant dynamic action of their own. Nevertheless, in a region of linear drag (as in the assumption here) the relative net velocity as determined by the difference between these two residues is identical with the average relative instantaneous velocity. The difference between the two net velocities, therefore, does have dynamical significance.

In order to verify Equation (5), it was necessary to determine, experimentally, values of $\overline{V}_s$, $\overline{U}_f$, C_D and C_{Rx} for given beach, wave and sediment properties.

Experimental Equipment and Procedures

In keeping with the aim of the study, simplifications were made which allowed concentration on primary variables.

The experimental facilities consisted of a 100 foot long, 2-1/2 foot wide and 3 foot deep glass-walled wave channel in one end of which an artificially roughened, fixed plane beach of variable slope was installed. Waves were generated at the other end by a horizontally reciprocating piston actuated by a hydraulic servomechanism with continuously variable speed and stroke.

The range of variables tested was:

Incident Waves

Characteristics of the incident waves in the constant depth portion of the channel are given in Table I.

Beach Slope

Two slopes were used:

$$\tan \alpha_1 = 1/14.8 = 0.0675$$

$$\tan \alpha_2 = 1/22.4 = 0.0447$$

Beach Roughness

The beach surfaces were coated with a uniform layer of immovable sand

TABLE I. Characteristics of Test Waves Before Transformation

Wave No.	H Ft.	L Ft.	T Sec.	d Ft.	d/L	H/L	H_o/L_o Approx.
1	0.186	8.92	1.43	1.75	0.196	0.0209	0.018
2	0.265	11.16	1.68	1.75	0.157	0.0237	0.020
3	0.354	8.42	1.39	1.75	0.208	0.0420	0.039
4	0.234	5.91	1.10	1.75	0.296	0.0396	0.041
5	0.230	4.36	0.94	1.75	0.401	0.0528	0.054
6	0.440	7.20	1.24	1.75	0.243	0.0611	0.061
7	0.357	5.99	1.11	1.75	0.293	0.0598	0.060
8	0.114	30.16	4.04	1.50	0.050	0.0038	0.001
9	0.235	16.22	2.33	1.50	0.093	0.0145	0.009

grains whose mean diameter, as determined by sieving, was:

$$k_1 = 1.83 \text{ mm.} \qquad k_2 = 0.79 \text{ mm.}$$

In addition, a smooth, uncoated surface was used yielding a roughness of assumed mean height:

$$k_3 = 0 \text{ mm.}$$

Spherical Sediment Particles

Characteristics of the movable sediments are given in Table II.

Wave Characteristics

Local wave characteristics were determined by means of capacitance type wave profile wires in conjunction with a recording oscillograph.

Incipient Sediment Motion

Definition of incipient motion was accomplished by gradually varying the local wave characteristics until stationary sediment particles tipped out of their roughness hollows and continued to oscillate.

Unfortunately, it was not possible to obtain a reliable indication of the wave phase at which this condition was reached but note was made of the direction of initial motion.

Established Sediment Motion

The net velocity of single spherical particles was determined by timing their net movement over successive six inch increments on the beach.

TABLE II. Characteristics of Movable Sediments

No.	Nominal Size D mm. (Ft.)	Specific Gravity through Fall Velocity	D/k	
			k = 0.79 mm.	k = 1.83 mm
1	2 (.0066)	1.29	2.62	1.09
2	3 (.0098)	1.19	3.93	1.64
3	3.17 (.0104)	1.28	4.17	1.73
4	5 (.0164)	1.24	6.56	2.74
5	2.3 (.0075)	2.63	3.00	1.25
6	3 (.0098)	2.44	3.93	1.64
7	4 (.0131)	2.17	5.24	2.19
8	5 (.0164)	2.13	6.56	2.74
9	6 (.0197)	2.12	7.88	3.29
10	4.37 (.0143)	7.00	5.72	2.39

In order to get a statistically valid picture of net particle velocity at any beach position this single particle technique was repeated 10 times which gave a 95% confidence interval of $\pm$ 10%.

Fluid Mass Transport Velocities

In order to determine $\overline{U}_f$, the fluid mass transport velocity effective in the motion of a given size sediment under a given wave, globules of a fluid immiscible with and slightly heavier than water were introduced on the smooth beach surface. Measurements of the mean net globule velocity were then made under the same test waves as described above for sediment particles.

To define the distribution of net mass transport velocity through the boundary layer, measurements of the net velocity of a range of globule sizes were made under each of several wave conditions. The net globule velocity was assumed to be that of the fluid at its center.

It is recognized that bottom roughness will modify the near-bottom distribution of instantaneous fluid velocity however its effect on the net (mass transport) velocities was assumed negligible.

Resistance Coefficients

Because of the difficulty of evaluating the coefficient of rolling friction, C_{Rx}, its effect was approximated by including it implicitly in the definition of the drag coefficient, C_D, as was done by Carty(3) in an experimental study of the resistance coefficients of spheres on a plane boundary. The study was performed in the manner of a "fall velocity" determination by timing the steady-state rolling velocity of various spheres on an inclined smooth plane boundary submerged in fluids of different viscosity.

Presentation and Discussion of Results

Net Sediment Velocities, $\overline{V}_s$

A complete tabulation of the net sediment motion data has been published elsewhere(2) thus only a sample will be given here.

In Fig. 2 the average net velocities obtained for various sediment particles under waves No. 2 and 3 are plotted with respect to position on the beach. Following the motion of the 3 mm. glass particle on the smooth beach a helpful summary of general observation is obtained:

A point of zero net velocity may be seen at approximately Sta. 48 where the sediment is in oscillating equilibrium. This same particle introduced offshore of the equilibrium point will move offshore under the predominant driving force of gravity. As the particle moves offshore it moves continually into a weaker fluid velocity field. Eventually the maximum local effective instantaneous fluid velocity under the crest may fall below that value required for incipient particle motion and a quasi-oscillatory sediment motion containing only the offshore cycle may result. With progress into still deeper water, even this motion will vanish and the sediment will come to complete rest.

Onshore of the point of oscillatory equilibrium, the effective fluid velocities provide the predominant driving force and since they increase sharply in the direction of motion, they produce a considerable net convective particle acceleration.

In the vicinity of the breaker, the onshore motion is more and more opposed by the backrush from the preceding waves and the sediments come to another point of oscillatory equilibrium.

Effective Fluid Mass Transport Velocity, $\overline{U}_f$

Longuet-Higgins(4) presents a theoretical solution for the mass transport velocity distribution in a closed channel which satisfies continuity. Within the bottom boundary layer this solution is given by:

$$2\frac{\overline{U}(y)}{\overline{U}_o} = 5 - 8\, e^{-\beta y}\cos\beta y + 3\, e^{-2\beta y} = \Phi \qquad (6)$$

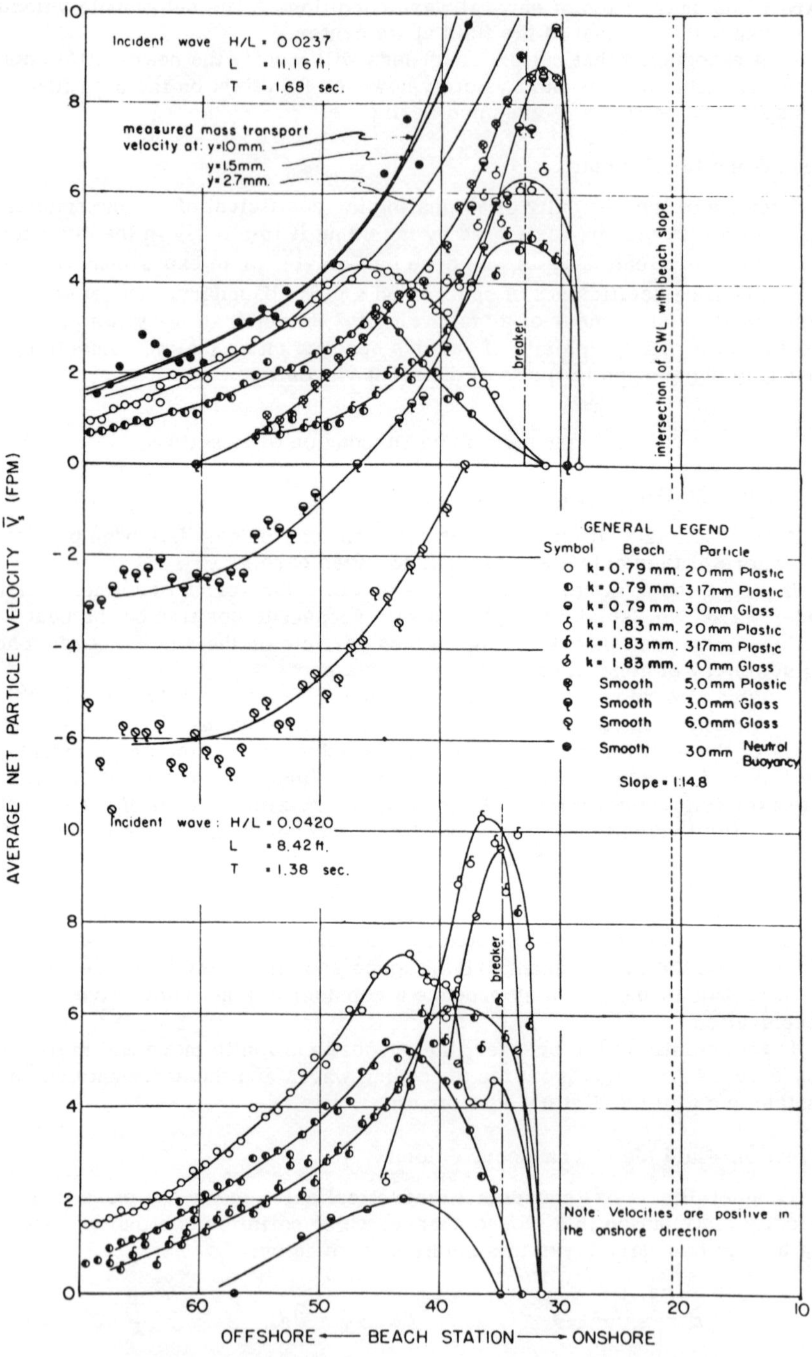

FIG. 2 TYPICAL VARIATIONS IN NET SEDIMENT AND WATER PARTICLE VELOCITY WITH POSITION ON BEACH

in which

$$\bar{U}_o = \frac{1}{2}\left(\pi \frac{H}{L}\right)^2 \frac{C}{\sinh^2 \frac{2\pi d}{L}} \qquad (7)$$

and

$$\beta^{-1} = \sqrt{\frac{2\nu}{\sigma}} \qquad (8)$$

A sample set of measured distributions of net velocity at three separate points on the smooth, 1 to 15 slope under a given wave is shown in Fig. 3. Because of the difficulty of determining the true net boundary layer limit due to the presence of the return flow, the more readily distinguishable point of maximum net velocity was chosen as the reference dimension, j, for dimensionless comparison of the velocity distributions.

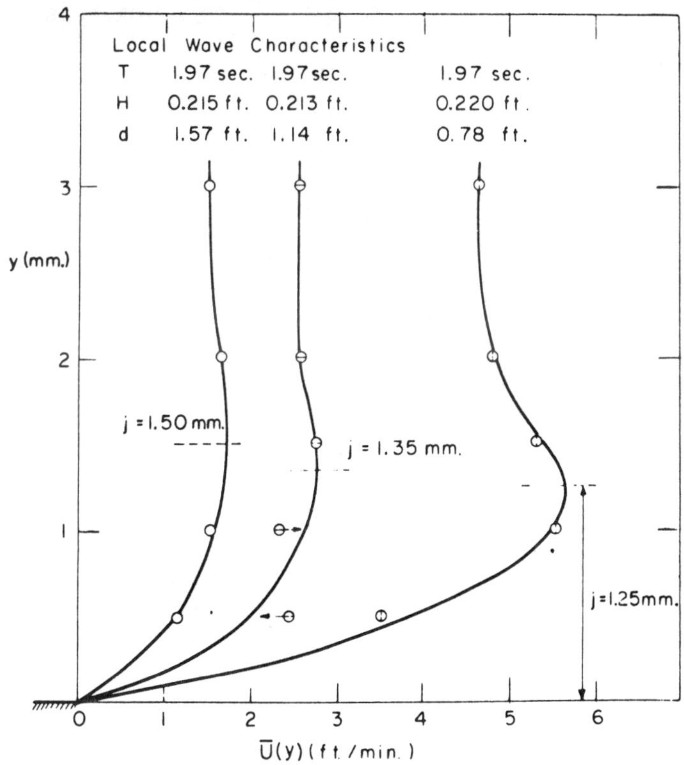

FIG. 3 SAMPLE DISTRIBUTIONS OF NET FLUID VELOCITY WITHIN THE BOUNDARY LAYER ON A SMOOTH BOTTOM

All the experimental velocity distributions are presented on this basis in Fig. 4 where they are compared with the Longuet-Higgins theory Eq. (6).

Measured values of the quantity, j, are compared in Fig. 5 with the theory which is given by the first maximum of Eq. (6) to be:

$$j_{theor.} = 2.306 \sqrt{\frac{2\nu}{\sigma}} = 1.30\sqrt{\nu T} \tag{9}$$

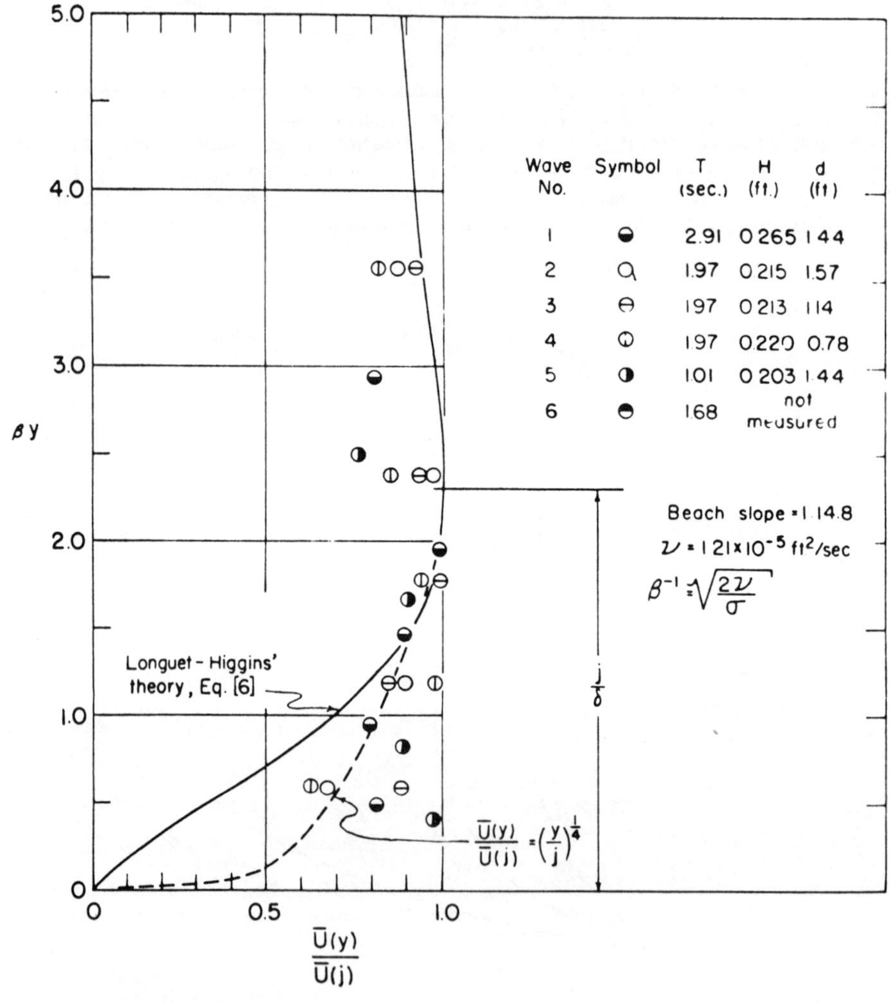

FIG. 4 DISTRIBUTION OF NET FLUID VELOCITY WITHIN THE BOUNDARY LAYER ON A SMOOTH BOTTOM

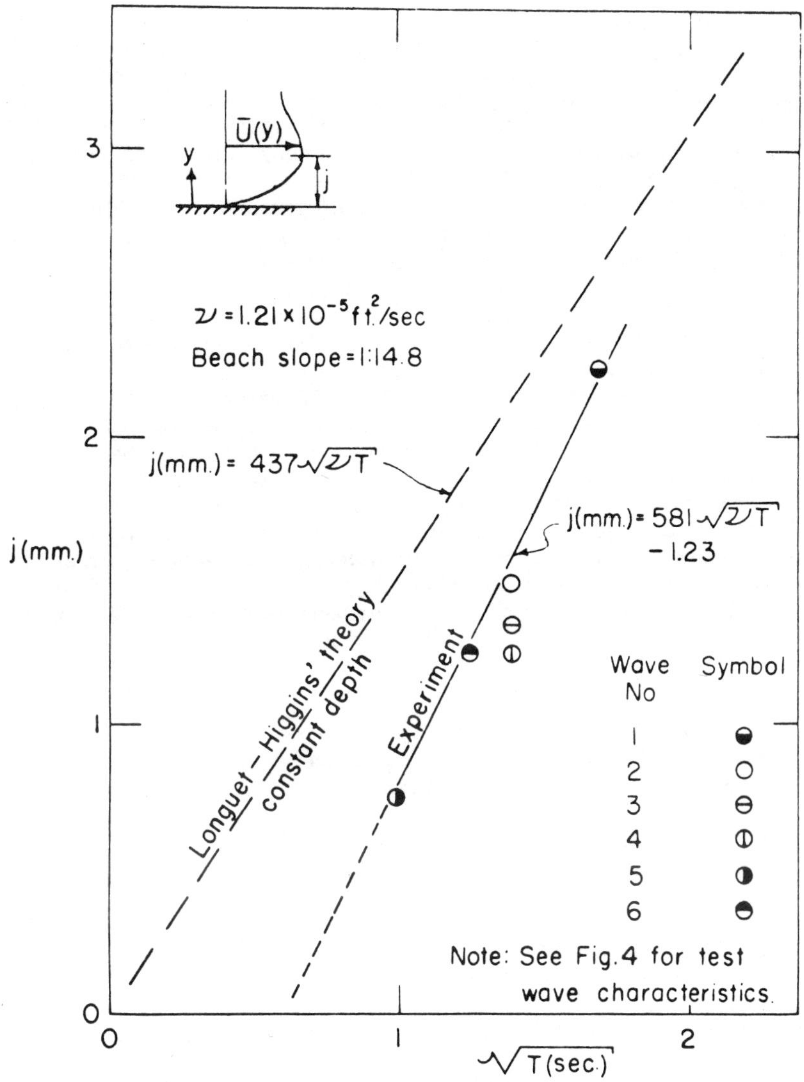

FIG. 5 ELEVATION OF MAXIMUM MASS TRANSPORT VELOCITY WITHIN THE BOUNDARY LAYER ON A SMOOTH BOTTOM

It should be remembered when studying the comparisons of Figs. 4 and 5 that the theory was developed for a constant depth while the experimental bottom had a 1 to 15 slope. For waves 1, 2 and 5, velocity distributions were measured close to the toe of the beach where the wave possessed essentially its constant depth characteristics. The extent of the variation caused by depth change is indicated by the distributions for waves 2, 3 and 4 which were measured at three progressively shallower depths in the transformation of a single wave. It would seem that any effect of bottom slope would increase with decreasing depth to wave length ratio d/L. However, experiment approaches theory with increasing wave period (decreasing d/L in this case) as may be seen from the measurements of j in Fig. 5.

It is thus reasoned that the discrepancies noted between theory and experiment are not primarily a result of the sloping bottom but may be due to some inadequacy of the theory for high frequency waves. For these waves a simple "power law" distribution such as the 1/4th power curve indicated in Fig. 4 may provide a more accurate description of velocity variation close to the bottom. It is interesting that recent unpublished experiments by the senior author have indicated instantaneous bottom shearing stresses to be several times greater than theory for waves comparable in frequency to these.

Because of this apparent inaccuracy in the theory when applied to laboratory waves the mass transport velocity effective in the motion of a given sediment was measured as described under Experimental Procedures. Some of these data are shown in the sample data presentation (Fig. 2). The similarity in the curves of net sediment and net fluid velocity indicates the analogous nature of the two phenomena.

Resistance Coefficients, C_D and C_{Rx}

By evaluating the drag coefficient for spheres rolling down an inclined, smooth plane boundary from the relation:

$$C_D \rho_f \frac{\pi D^2}{8} v_s^2 = F_G \sin\alpha \tag{10}$$

Carty[3] has included implicitly in C_D at least a partial effect of rolling friction. His coefficients are presented as a function of Reynolds Number, R_D, in Fig. (6) where they may be compared with those for spheres in an infinite fluid.

The conventional reference velocity, "v_o," used in drag computations is the root mean square freestream velocity. In this case the component of peripheral velocity in the direction of translation (relative free stream velocity) varies linearly from zero at the bed to twice the translational velocity, v_s, at the upper edge of the particle. The root mean square relative free stream (evaluated over the projected area of the particle) velocity may be shown to be 1.12 times the translational velocity which was used by Carty to define his parameters.

The resistance curve for spheres rolling on a plane boundary was corrected for this velocity effect as shown in Fig. (6).

As mentioned above the experimental resistance coefficients include implicity, the effect of rolling friction. Since this frictional force is a function of F_G, the degree to which Fig. (6) determines C_D uniquely as a function of R_D defines the importance of this neglected force. The small scatter of

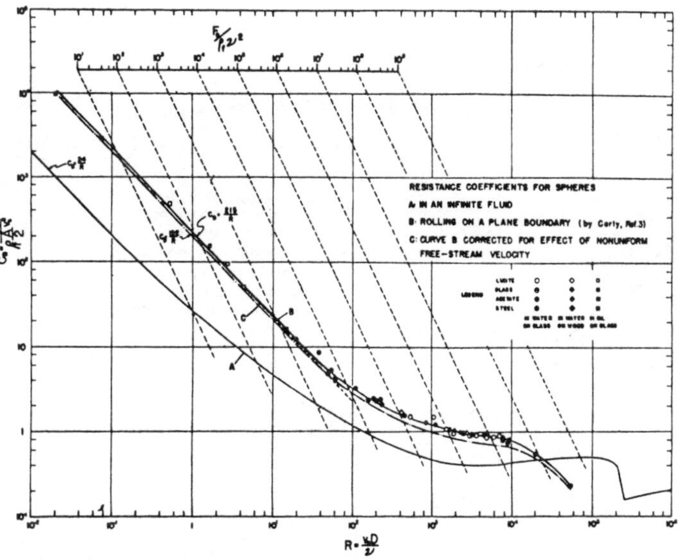

FIG. 6 RESISTANCE COEFFICIENTS FOR SPHERES

the experimental points thus indicates the effect of rolling friction to be quite small.

It should also be noted that there is a basic hydrodynamic difference between the gravity-induced motion studied by Carty and the fluid-induced net sediment motion which is of prime concern here. This difference lies in the sense of particle rotation with respect to the direction of the relative translational velocity. The effect of this reversed flow picture on the resistance coefficient is zero in an infinite fluid and has not been determined in the presence of a solid boundary.

Absorbing the effect of bottom friction in the resistance coefficient, Equation (5) may be written:

$$C_D \rho_f \frac{\pi D^2}{8} (\overline{U}_f - \overline{V}_s)^2 = F_G \sin \alpha \tag{11}$$

In order to demonstrate the validity of Equation (11) it was used to compute resistance coefficients using measured relative net velocities. Data from shoreward of the point of maximum net convective acceleration (i.e. inflection point of the curves such as in Fig. 2) were excluded due to the action of forces not included in Equation (11). All other data for a given value of the drag force, $F_G \sin \alpha$ and for each boundary geometry D/k, were averaged and the resulting coefficients are compared with the Carty curve in Fig. 7. Each plotted point represents the average of from 2 to 9 individual sets of data, each under different wave conditions. Excellent agreement is obtained where the boundary geometries are similar, that is at high D/k. A scattering of points from rough surfaces shows a trend to consistently lower coefficients at the same Reynolds number as the ratio D/k decreases. The effect of low D/k is two-fold. The first is to modify the exposed particle area and

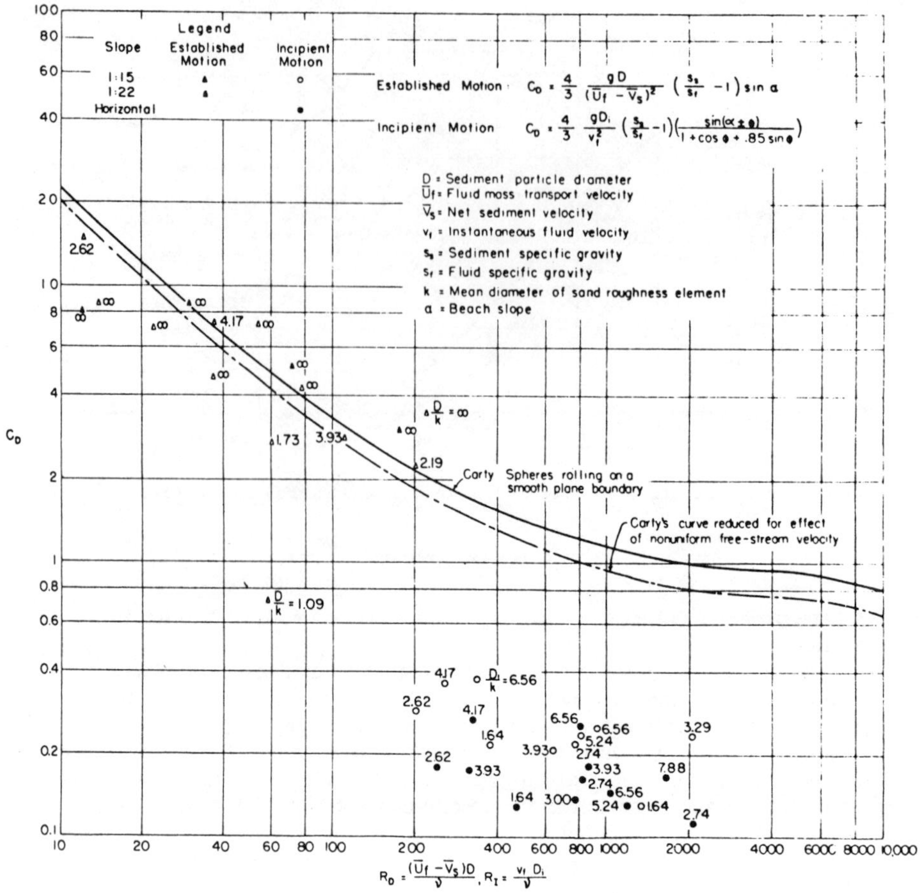

FIG. 7 RESISTANCE COEFFICIENTS DEFINING INCIPIENT AND ESTABLISHED MOTION

consequently the streamline configuration. The second is to introduce a larger bottom resistance which leads to smaller net sediment velocities. Under the definition of Equation (11) both effects should lead to lower coefficients at the same $F_G \sin \alpha$.

If the percentage difference between resistance coefficients for different D/k's is assumed independent of Reynolds number, the effect of bed-particle geometry may be shown as in Fig. 8, for established motion.

Incipient Motion

Equation (1) describes the instantaneous moment acting on a stationary sediment particle. It will be assumed that incipient motion occurs for a given D/k, W_b and α at the wave phase angle at which the hydrodynamic portion of

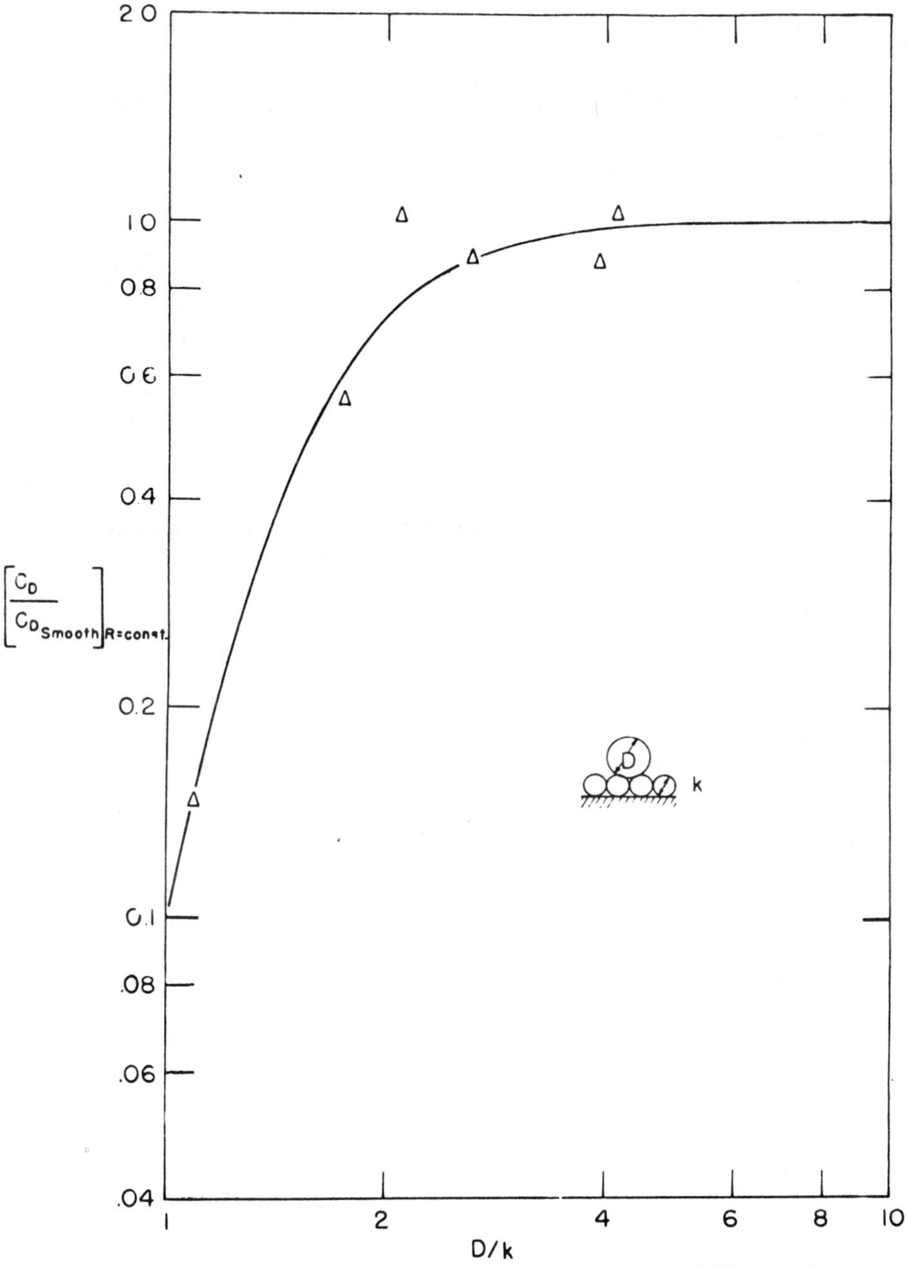

FIG. 8 EFFECT OF BED-PARTICLE GEOMETRY ON THE RESISTANCE COEFFICIENTS OF SPHERES

this moment has its maximum value. Because the moment is composed of drag and inertial components which are 90° out of phase, the value of this phase angle is not immediately obvious.

For long waves Lamb(5) gives the near-bottom velocity distribution as:

$$v_f = \frac{f}{\sigma}\left[\sin\theta - e^{-\beta y}\sin(\theta - \beta y)\right] \qquad (12)$$

where

$$\frac{f}{\sigma} = \frac{H}{2}\,\frac{\sigma}{\sinh\frac{2\pi d}{L}} \qquad (13)$$

Equation (12) and its first time derivative are now substituted into Eq. (1) with the additional assumptions.

$$1. \quad v = v_f \quad \text{when } y = D \qquad (14)$$

$$2 \quad a = a_f \quad \text{when } y = D \qquad (15)$$

Differentiation of the first two terms of Eq. (1) in order to obtain the location of the maximum hydrodynamic moment then yields:

$$\frac{d\Sigma M}{d\theta} = 0 = \frac{d}{d\theta}\left[\frac{3}{4}\frac{v_f^2}{\sigma D}\left[C_D(1+\cos\phi) + C_L\sin\phi\right] + C_M\frac{dv_f}{d\theta}(1+\cos\phi) + \frac{dv_0}{d\theta}\cos\phi\right] \qquad (16)$$

C_M is assumed equal to 0.6 as given by Streeter(6) for a sphere on a smooth plane boundary in potential flow and C_L is assumed equal to 0.85 C_D as given by Chepil.(7) Then:

1. If C_D is assumed constant at the appropriate value determined for established motion at high R - then, as might be expected, the drag and lift forces predominate for all βD and the phase angle for incipient motion is close to 90°.

2. If C_D is assumed inversely proportional to R with the constant of proportionality taking on the appropriate value determined for established motion at low R - then for a large βD (these laboratory data) inertia forces will predominate and the phase angle θ of maximum force will be close to 0° while for small βD (most natural beaches) drag and lift forces will govern as in case 1 above.

Multiple regression statistical techniques were applied to the laboratory data in an effort to obtain independent determinations of C_D, C_M and incipient θ. Due presumably to the scatter of the data, wildly varying results including imaginary and negative coefficients were obtained and the technique was abandoned.

An approximate evaluation of these data may be obtained under the assumption that all accelerative effects may be ignored. With this assumption the condition of incipient motion is given by Eq. (1) rewritten to yield:

$$C_D = \frac{4}{3}\frac{gD}{v_f^2}\left(\frac{S_s}{S_f} - 1\right)\left(\frac{\sin(\alpha \pm \phi)}{1 + \cos\phi + .85\sin\phi}\right) \qquad (17)$$

The effective velocity, v_f, is given by Eqs. (12) and (14) evaluated at $\theta = 270°$ or 90° depending upon whether the initial motion was observed to be in the offshore or onshore direction respectively.

The tangent of the angle of repose ($\tan \phi$) for the various bed-particle combinations was determined experimentally by tilting an immersed roughened plane supporting the particular particle until motion ensued. These data are presented in Fig. 9 in comparison with the theoretical relationship for an idealized bed-particle geometry. It should be remembered that the values of ϕ determined in this manner apply to incipient motion in a moving fluid only when the resultant active force acts through the particle center. Due to the effect of bottom roughness in elevating the point of application of the drag force this assumption becomes poor as D/k becomes small.

The resistance coefficients calculated on the basis of these velocities and repose angles are presented in Fig. 7 as a function of the instantaneous Reynolds number, R_I. Each point shown represents the average of from 1 to 10 separate determinations (each under different wave conditions) of incipient motion for each separate combination of F_G, D/k and α.

Since in all probability the critical phase angle is somewhat less than 90° the drag and lift forces and Reynolds number will be less and the resistance coefficient correspondingly higher.

These data should not be expected to agree completely with those for established motion however since for low D/k at least the latter include an appreciable effect of rolling friction.

It is possible to conclude only that there is no qualitative disagreement between the theory and experiment.

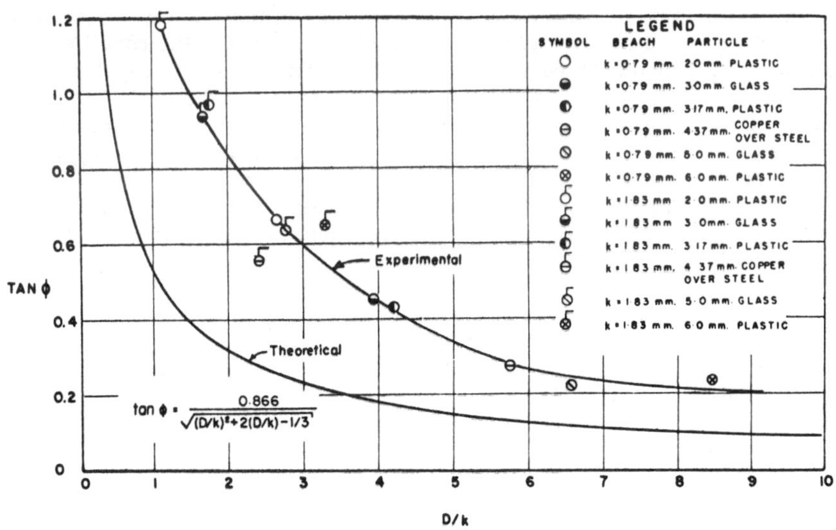

FIG. 9 ANGLE OF REPOSE AS A FUNCTION OF BED-PARTICLE GEOMETRY

Extrapolation of Results to Natural Beaches

Although the type of sediment motion dealt with here may be obscured by any one of several others on natural beaches, certain phenomena of sediment distribution may be explained qualitatively on the basis of the findings reported.

Observations of many natural beaches show the median sediment size to be distributed from coarse to fine in the offshore direction and the skewness of the local size distribution to increase consistently in the same direction.

Two of the laboratory-observed phenomena reported yield a qualitative explanation of these variations. These phenomena are:

1. The equilibrium condition of established sediment motion.

2. The condition of incipient sediment motion.

These conditions will be discussed separately.

Equilibrated Established Motion

As long as instantaneous fluid velocities exceed the incipient value at some time during each wave cycle, the sediment will be in the condition herein referred to as established motion.

In the laboratory, due to the constant experimental slope, the condition of oscillating equilibrium producing zero net motion is a condition of unstable equilibrium as far as any single particle is concerned. This is true with any wave train of essentially constant form since any irregularity which forces the sediment particle off its equilibrium location moves it into a zone of net motion leading away from the equilibrium position.

A typical natural beach, on the other hand, has a slope which varies from mild to steep in the onshore direction, a fact which causes the condition of oscillating equilibrium to be one of stable equilibrium. Any irregularity in the wave train will produce a net hydrodynamic force on the particle which tends to move it away from its equilibrium position. Due to the slope variation this particle displacement brings into play a modified gravity force which acts to restore the particle to its equilibrium position.

Considering the condition D/k = 1 to most nearly represent the bed-particle geometry of equilibrium natural beaches, the resistance coefficient may be written for the range of interest as:

$$C_D = \frac{19.2}{R_D} \tag{18}$$

from Figures (6), (7) and (8).

Substituting Eq. (18) into Eq. (11) and simplifying for the case of interest, i.e. $\overline{V}_s = 0$;

$$\frac{gD^2}{\nu \overline{U}_f}\left(\frac{s_s}{s_f} - 1\right)\sin\alpha = 14.4 \tag{19}$$

For sand in sea water:

a. $s_s = 2.65$

b. $s_f = 1.03$

c. $\nu = 1.45 \times 10^{-5}$ ft/sec.

thus

$$D^2 \sin \alpha = 4.13 \times 10^{-6}\, \overline{U}_f \qquad (20)$$

In order to define $\overline{U}_f$ for natural beaches the velocity distribution of Longuet-Higgins,(4) Eq. (6), is used.

Near the bottom Eq. (6) may be approximated by:

$$\beta y = 0.32\, \Phi^{7/8} \qquad (21)$$

thus

$$\Phi = 2\, \frac{\overline{U}(y)}{\overline{U}_o} = 3.68\, (\beta y)^{8/7} \qquad (22)$$

On a very rough bottom (D/k = 1) the sediment particle is partially embedded in the beach and the fluid velocities acting on the remainder of the lower portion are so weak that their contribution to dynamic pressures is probably quite small. For this reason the fluid velocity effective in motion of these natural sediments will be assumed to act at the upper edge of the particle.

$$\overline{U}(y) = \overline{U}_f \quad \text{when} \quad y = D \qquad (23)$$

The effect of this shielding on the particle area exposed to drag is implicitly in the drag coefficients given in Fig. 8.

Using (22) and (23), (20) becomes:

$$D_e^2 \sin \alpha = 7.60 \times 10^{-6} (\beta D_e)^{8/7}\, \overline{U}_o \qquad (24)$$

in which D is in feet and $\overline{U}_0$ is in ft. per sec.

Eq. (24) describes the equilibrium relationship between particle size, beach slope and local wave characteristics under the assumption that the equilibrium condition of established sediment motion is the controlling mechanism.

Incipient Motion

If, on the other hand, the sediment size which satisfies the conditions for oscillating equilibrium at all locations cannot be moved by the maximum local instantaneous fluid velocity, then the median remaining size will be governed by incipient motion conditions.

Suppose a beach which initially had the same normal distribution of sand sizes at each location is subjected to a given incident wave until the sorting process has reached equilibrium. In the deeper water, only the smallest of the sediment particles will be put into motion with some moving onshore and some offshore. At stations further onshore the size just capable of being moved by the maximum local instantaneous fluid velocity (called the incipient particle size) increases thereby producing an absence of a larger and larger percentage of the smaller sizes. This effect will move the median remaining size to increasingly larger diameters with distance in the onshore direction.

As before considering the conditions D/k = 1 to be most representative of the beach conditions of interest the resistance coefficient will be given by Eq. (18).

As stated above, for small βD the maximum moment is in phase with the velocity. Under these conditions the effective velocity, v_f, as given by Eqs. (12) and (14) may be approximated by:

$$v_f = 0.86 \frac{f}{\sigma} \beta D \qquad .01 < \beta D < 1.0 \tag{25}$$

Eqs. (18) and (25) may be substituted into Eq. (17) to obtain

$$\frac{g D \sigma}{\nu f \beta}\left(\frac{S_s}{S_f} - 1\right)\left(\frac{\sin(\alpha \pm \phi)}{1 + \cos\phi + .85 \sin\phi}\right) = 12.4 \tag{26}$$

Again for sand in sea water, Eq. (26) becomes:

$$\frac{D_i}{\beta} \sin(\alpha \pm .92) = 8.1 \times 10^{-6} \frac{f}{\sigma} \tag{27}$$

where tan ϕ is obtained from Fig. (9) for D/k = 1.

Eq. (27) describes the equilibrium relationship between particle size, beach slope and local wave characteristics under the assumption that the condition of incipient sediment motion is the controlling mechanism.

An additional important simplification, the quantitative effect of which is indeterminate at this time, results from considering the bed to be permeable. Flow into the permeable natural bed under the wave crest will increase particle stability, but flow out of the bed under the wave trough will reduce the effective particle weight and may in fact put the whole surface of the bed in a "quick" of fluidized condition.

This would normally lead to smaller incipient sizes in a zone of onshore motion and larger incipient sizes in a zone of offshore motion.

It should be remembered that the degree to which the "incipient sediment size" represents the median remaining size depends to a great degree upon the initial distribution of sizes at the locality.

At this time it is not possible to state which of the above criteria governs the formation of equilibrium beaches. It is recognized, however, that wherever the incipient size, D_i, (that maximum size which the local instantaneous fluid velocities are just able to move) is smaller than the equilibrium sediment size, D_e, (that size which would be in oscillating equilibrium under the local fluid velocities) the equilibrium criterion cannot govern the median remaining size. For small beach slopes Eqs. (24) and (27) may be combined to yield a definition of this limiting condition:

$$\frac{D_e}{D_i} = .082 \beta^{1/3} \frac{\sigma}{f} \bar{U}_o^{7/6} \left(\sin\alpha^{-1/6} + 1.3 \sin\alpha^{-7/6} \right) = 1 \tag{28}$$

Comparison with Natural Beach Data

It can be seen from Eqs. (24) and (27) that for a given wave system there is a relationship between the local sediment diameter and the local beach slope at equilibrium. For a measured set of natural beach and wave characteristics it is possible to test the applicability of this relationship.

Data concerning local depth and median particle size along fourteen ranges of three natural beaches on the Eastern coast of the United States were supplied by the Beach Erosion Board. A sample of the information is shown in Fig. 10.

The significant wave estimated by the Beach Erosion Board to apply to these beaches has an 8 second period and a deep-water height, H_o = 1.5 ft. (H_o/L_o = 0.0046).

Using small amplitude wave transformation theory wave characteristics bracketing those of the estimated wave were introduced into Eqs. (24) and (27) and the resulting expressions were plotted as the solid lines of Fig. (11).

There is considerable variability in the natural beach data and consequently the arithmetic mean of all measurements at constant H_o/d was determined and plotted along with arrows to show the extent of the variation. Excluded from the averages in Fig. 11-a were all data points for which the theoretical diameter ratio, D_e/D_i, was greater than unity as determined from Eq. (28). The small numbers by each plotted point indicate the number of observations determining that mean.

The variations are due in some part to the difficulty of scaling small local slopes accurately from a plotted beach profile.

Agreement of the mean data with theory is remarkable for both theoretical models.

Thus, for known predominant deep water wave characteristics it is possible, using these criteria, to:

1. Predict median sand size distribution knowing the equilibrium beach profile, or

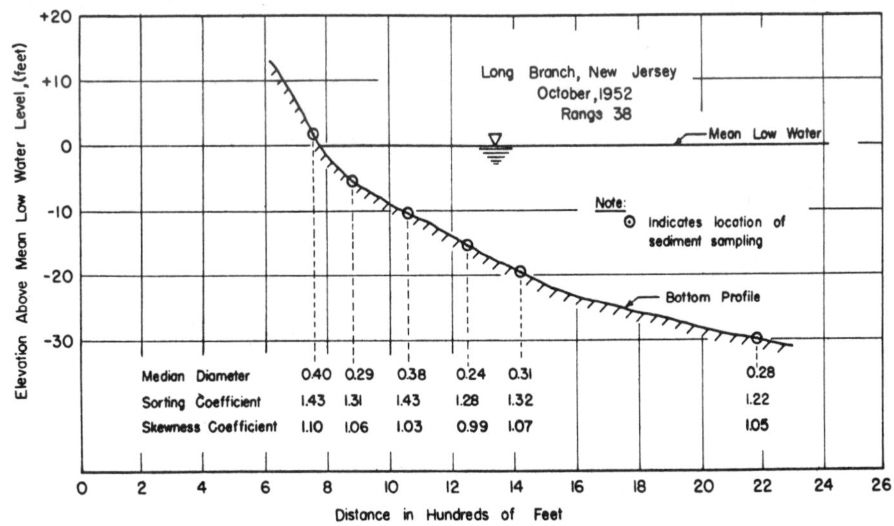

FIG. 10 SAMPLE NATURAL BEACH DATA AS FURNISHED BY THE BEACH EROSION BOARD

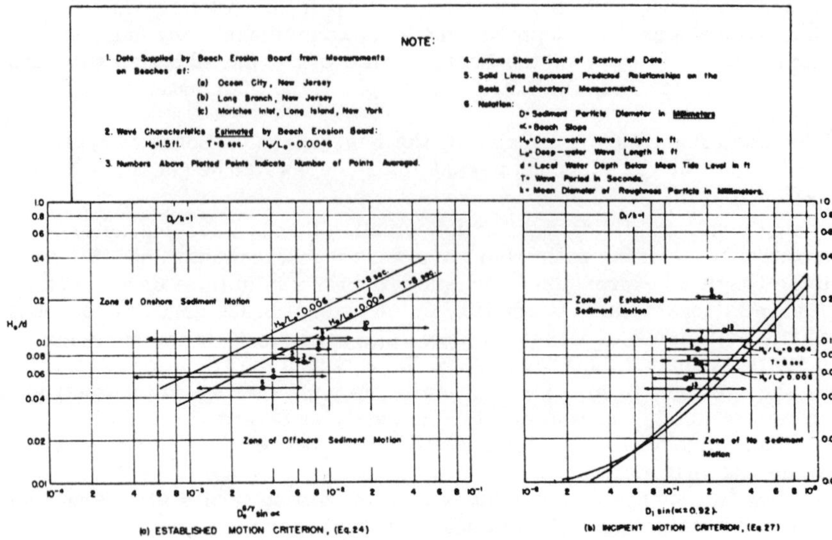

FIG. 11 PREDICTED AND OBSERVED EQUILIBRIUM CHARACTERISTICS OF NATURAL BEACHES

2. Predict equilibrium beach profile knowing the distribution of median sand size.

Use of Eq. (11) to describe the rate of sand motion should permit simultaneous determination of both equilibrium slope and sand size distribution starting from known initial, unstable values of these variables. Current investigations are proceeding in this direction.

A complementary laboratory investigation of two-dimensional natural sand beaches is currently underway in the Hydrodynamics Laboratory at M.I.T. under the sponsorship of the Beach Erosion Board. The aims of this study are:

1. To determine to what extent either of the above criteria may be applied or can be modified to apply to the prediction of equilibrium profiles of natural beaches.

2. To determine to what extent the derived equation of net sediment motion can be utilized in the prediction of transient beach behavior following a change in incident wave form.

SUMMARY AND CONCLUSIONS

A series of experiments was made concerning two facets of the wave induced motion of discrete spherical sediment particles: (1) Established net motion on both smooth and artificially roughened beaches (2) Incipient motion on artificially roughened beaches.

An equation of net sediment motion was derived and verified by the laboratory measurements.

Two criteria for the local mean sediment size - beach slope relationship for equilibrium beach profiles were developed and compared with natural beach data.

The following conclusions may be drawn from this set of experiments:

1. The bottom boundary layer under high frequency ($\sigma > 2$) waves is considerably thinner than predicted by available theory.

2. The instantaneous drag force is linearly related to the instantaneous relative fluid-sediment velocity in the usual range of sediment particle Reynolds number (as defined by the instantaneous relative fluid-sediment velocity) encountered in the zone of established motion in both the laboratory and on natural beaches.

3. All phases of established sediment motion exclusive of the effects of the breaker may be described in terms of the relative net fluid-sediment velocity by the equation:

$$C_D \rho_f \frac{\pi D^2}{8} (\overline{U}_f - \overline{V}_s)^2 = F_G \sin \alpha$$

in which C_D includes the effect of bed resistance and is, by conclusion 2, a linear function of the relative net fluid sediment velocity.

4. Two rational equations have been developed for predicting the relationship between the local characteristics of slope and median sediment diameter on two-dimensional equilibrium natural beaches:

a. On the basis that the median remaining size is indicated by that size which can be maintained in oscillating equilibrium at the point -

$$D_e^{6/7} \sin \alpha = 7.60 \times 10^{-6} \beta^{8/7} \overline{U}_o$$

b. On the basis that the largest particle just capable of instantaneous motion will indicate the median remaining size at the point -

$$D_i \sin(\alpha \pm .92) = 8.0 \times 10^{-6} \frac{f}{\sigma} \beta$$

Comparison of both of these equations with the limited available prototype data is encouraging but it is not yet possible to say with assurance which mechanism controls the natural beach configuration.

ACKNOWLEDGMENT

This investigation was carried out at the Hydrodynamics Laboratory in the Department of Civil and Sanitary Engineering at the Massachusetts Institute of Technology under the sponsorship of the Beach Erosion Board, Corps of Engineers, U. S. Army.

Particular credit is due Mr. Luis A. Peralta who performed much of the early experimental work and data analysis.

1. Ippen, A. T. and Eagleson, P. S., "A Study of Sediment Sorting by Waves Shoaling on a Plane Beach," M.I.T. Hydrodynamics Laboratory Technical Report No. 18, 1955, also published as Beach Erosion Board Technical Memorandum No. 63, Washington, 1955.

2. Eagleson, P. S., Dean, R. G. and Peralta, L. A., "The Mechanics of the Motion of Discrete Spherical Bottom Sediment Particles Due to Shoaling Waves," M.I.T. Hydrodynamics Laboratory Technical Report No. 26, 1957, also published as Beach Erosion Board Technical Memorandum No. 104, Washington, 1958.

3. Carty, J. J., Jr., "Resistance Coefficients for Spheres on a Plane Boundary," Thesis submitted to the Massachusetts Institute of Technology for the degree of B.S., M.I.T., Cambridge, Massachusetts, May, 1957. (Unpublished).

4. Longuet-Higgins, "Mass Transport in Water Waves," Phil. Trans. Roy. Lond. Series A, No. 903, Vol. 245, pp. 535-581, 31 March 1953.

5. Lamb, H., "Hydrodynamics," Dover Publications, New York Sixth Ed., 1932, p. 622.

6. Streeter, V. L., "Fluid Dynamics," McGraw-Hill Book Co. Inc., First Edition, 1948, p. 80.

7. Chepil, W. S., "The use of Evenly Spaced Hemispheres to Evaluate Aerodynamic Forces on a Soil Surface," Trans. A. G. U. Vol. 39 No. 3, June 1958, pp. 397 - 404.

9

Reprinted from pp. 529–540, 561–562 of *Deep-Sea Research* **11**:529–562 (1964)
with the permission of Microforms International Marketing Corporation as
exclusive copyright licensee of Pergamon Press journal back files.

Radiation stresses in water waves ; a physical discussion, with applications

M. S. LONGUET-HIGGINS* and R. W. STEWART†

(*Received* 17 *June* 1964)

Abstract—The radiation stresses in water waves play an important role in a variety of oceanographic phenomena, for example in the change in mean sea level due to storm waves (wave " set-up "); the generation of " surf-beats "; the interaction of waves with steady currents; and the steepening of short gravity waves on the crests of longer waves. In previous papers these effects have been discussed rigorously by detailed perturbation analysis. In the present paper a simplified exposition is given of the radiation stresses and some of their consequencies. Physical reasoning, though less rigorous, is used wherever possible. The influence of capillarity on the radiation stresses is fully described for the first time.

INTRODUCTION

In a series of recent papers (1960, 1961, 1962, see also TAYLOR 1962, WHITHAM 1962) we have attempted to elucidate some of the non-linear properties of surface gravity waves in terms of what we have called the " radiation stress." Some of these non-linear properties have turned out to be unexpected (or at least to differ from properties widely assumed previously in the literature). For this reason a major part of the above mentioned papers has been used for a careful check of the results obtained by using the radiation stress concept, by means of detailed perturbation analysis to the required order of approximation.

One effect of this approach (which we believe to have been necessary) has been that the papers are somewhat analytical, and the straightforward simplicity of the concept may have been partly obscured for some readers. It is the purpose of the present paper to attempt a simple exposition, setting forth the nature and uses of the radiation stress. In many cases results will be stated without proof; readers dissatisfied with any of these are referred to the previous papers. (We shall refer to LONGUET-HIGGINS and STEWART, 1960, 1961, and 1962 as I, II and III). At the same time we shall extend some of our previous results for pure gravity waves so as to include effects of capillarity.

In the first sections of the paper we describe a simple derivation of the radiation stress, both for gravity waves and for capillary waves. In the second part we shall describe the application of these results to a number of interesting phenomena observed in the oceans; in particular to wave " set-up," " surf beats," the steepening of short waves on adverse currents or tidal streams, and the generation of capillary waves by steep gravity waves.

*National Institute of Oceanography Wormley, Godalming, Surrey.
†Institute of Oceanography, University of British Columbia, Vancouver, British Columbia, Canada.

PART I—THE RADIATION STRESSES; A PHYSICAL DISCUSSION

It is well known that electromagnetic radiation impinging on a surface, or originating on a surface, produces a force which is referred to as the " radiation pressure." It is perhaps less well known that a similar phenomenon occurs in the case of acoustic waves and of waves on the surface of fluids (or indeed of internal waves in a stratified fluid). In each case the force is principally in the direction of wave propagation. It is therefore not an isotropic one unless the waves themselves are isotropically distributed (as is the case for electromagnetic waves in an isothermal enclosure). In fluid mechanics it has become customary to use the term " pressure " for the isotropic stress which figures in the equation of state. We have therefore considered it wise to coin the term radiation *stress* as a more general one which does not carry any implication of isotropy‡.

Qualitatively§, the following argument may serve to introduce the concept :

It is well known (LAMB, 1932, Section 250) that surface waves possess momentum which is directed parallel to the direction of propagation and is proportional to

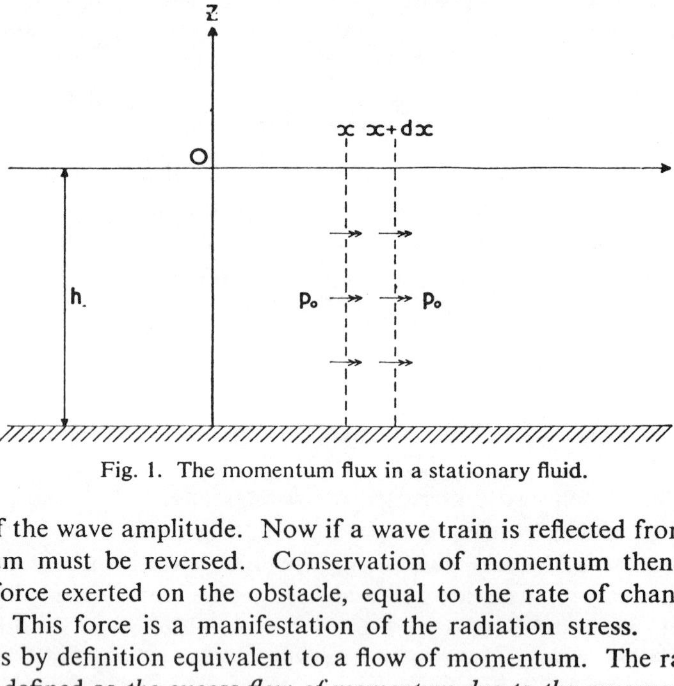

Fig. 1. The momentum flux in a stationary fluid.

the square of the wave amplitude. Now if a wave train is reflected from an obstacle, its momentum must be reversed. Conservation of momentum then requires that there be a force exerted on the obstacle, equal to the rate of change of a wave momentum. This force is a manifestation of the radiation stress.

A stress is by definition equivalent to a flow of momentum. The radiation stress may thus be defined as *the excess flow of momentum due to the presence of the waves.* Let us consider in detail how this stress arises in gravity waves.

1. *Progressive waves in water of uniform depth*

Consider first an undisturbed body of water of uniform depth h (Fig. 1). The pressure p at any point is equal to the hydrostatic pressure :

‡It might be argued that " radiative stress " would be grammatically more correct, but we wish to retain the implied analogy to the well established term " radiation pressure "—and in any case the use of nouns as adjectives is widespread in English.

§Quantitatively, in some special cases, it leads to difficulties and to errors, because some phenomena are incompletely described by the discussion in this paragraph.

$$p = -\rho g z, \tag{1}$$

where ρ, g and z denote density, gravity, and distance measured upwards from the mean surface. If we denote the above expression by p_0 then the flux of horizontal momentum across any vertical plane $x =$ constant is simply p_0 per unit vertical distance. Thus the total flux of horizontal momentum between surface and bottom $(z = -h)$ is

$$\int_{-h}^{0} p_0 \, dz \tag{2}$$

Since this quantity is independent of x, the flux of momentum across an adjacent plane $(x + dx)$ just balances the flux across the first plane, and there is no net change of momentum between the two planes (Fig. 1).

Now consider the momentum flux in the presence of a simple progressive wave motion (Fig. 2). The surface elevation $z = \zeta$ is given approximately by

$$\zeta = a \cos (kx - \sigma t) \tag{3}$$

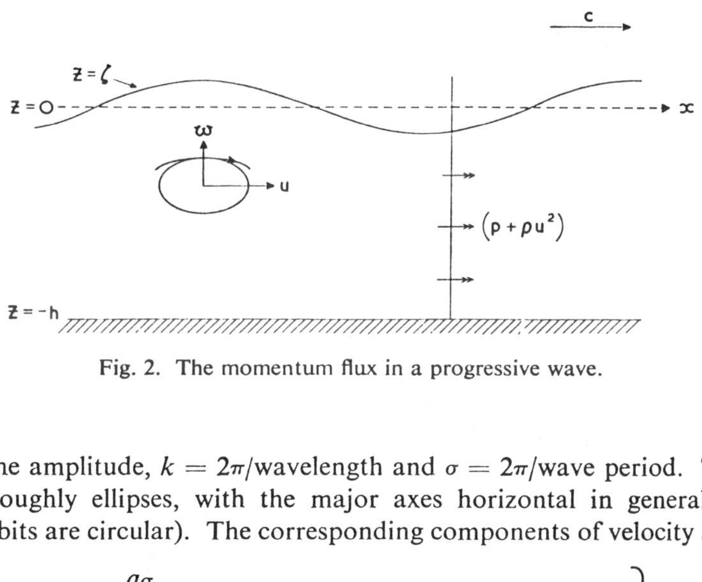

Fig. 2. The momentum flux in a progressive wave.

where a is the amplitude, $k = 2\pi$/wavelength and $\sigma = 2\pi$/wave period. The particle orbits are roughly ellipses, with the major axes horizontal in general. (In deep water the orbits are circular). The corresponding components of velocity are given by

$$\left. \begin{aligned} u &= \frac{a\sigma}{\sinh kh} \cosh k\,(z + h) \cos (kx - \sigma t) \\ w &= \frac{a\sigma}{\sinh kh} \sinh k\,(z + h) \sin (kx - \sigma t). \end{aligned} \right\} \tag{4}$$

A quite general expression for the instantaneous flux of horizontal momentum across unit area of a vertical plane in the fluid is

$$p + \rho u^2. \tag{5}$$

In this expression the second term ρu^2 represents a bodily transfer of momentum ρu (per unit volume) at a rate u per unit time (Fig. 3). The term ρu^2 is evidently analogous to a pressure. Even if the mean value of u itself is zero, the mean value u^2 is of course generally positive.

(Similarly, fluid crossing the plane $x =$ constant possesses z-momentum associated with the velocity component w. A mean product such as $\rho \overline{uw}$, which represents the mean transport of z-momentum across a plane $x =$ constant—or vice versa—is equivalent to a shear stress. In turbulence theory, such mean values are known collectively as Reynolds stresses, and it will be appreciated that the above

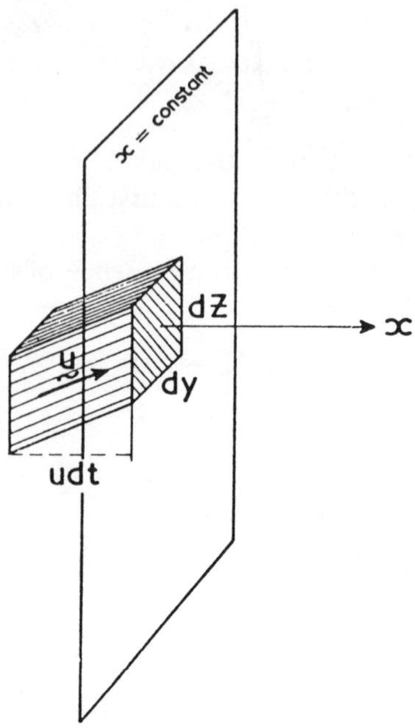

Fig. 3. Bodily transport of momentum across a plane $x =$ constant. In time t a volume udt per unit area has been transported across the plane. The momentum transported is thus $\rho \underline{u}\, udt$.

concepts are also similar physically to those used to explain the origin of pressure and viscosity in the kinetic theory of gasses. If due to turbulent fluctuations, the Reynolds shear stress is frequently parameterised by the concept of "eddy viscosity." Reynolds stresses also occur in waves, but in this case we must seek a different kind of description. For a discussion of the Reynolds stresses particularly appropriate in the present context, see STEWART, (1956)).

To find the total flux of horizontal momentum across a plane $x =$ constant we have to integrate (5) between the bottom $z = -h$ and the free surface $z = \zeta$:

$$\int_{-h}^{\zeta} (p + \rho u^2)\, dz. \tag{6}$$

We now define the principal component S_{xx} of the radiation stress as the mean value of the function (6) with respect to time, minus the mean flux in the absence of the waves, that is to say

$$S_{xx} = \overline{\int_{-h}^{\zeta} (p + \rho u^2)\, dz} - \int_{-h}^{0} p_0\, dz. \tag{7}$$

Special care must be taken to take the mean value *after* integration, since the fluctuation of the free surface itself contributes to the momentum flux. We can see this best by separating the right-hand side of (7) into three parts, that is by writing

$$S_{xx} = S_{xx}^{(1)} + S_{xx}^{(2)} + S_{xx}^{(3)} \tag{8}$$

where

$$\left.\begin{aligned} S_{xx}^{(1)} &= \overline{\int_{-h}^{\zeta} \rho u^2 \, dz} \\ S_{xx}^{(2)} &= \overline{\int_{-h}^{0} (p - p_0) \, dz} \\ S_{xx}^{(3)} &= \overline{\int_{0}^{\zeta} p \, dz.} \end{aligned}\right\} \tag{9}$$

Consider the first contribution $S_{xx}^{(1)}$. Since the integrand is of second order, the upper limit $z = \zeta$ may be replaced effectively by the mean level $z = 0$, since the additional range $0 < z < \zeta$ contributes only a third-order term. Now, both the limits of integration 0, h, being constant, we can transfer the mean value to the integrand. Thus

$$S_{xx}^{(1)} = \overline{\int_{-h}^{0} \rho u^2 \, dz} = \int_{-h}^{0} \rho \overline{u^2} \, dz. \tag{10}$$

The contribution $S_{xx}^{(1)}$, then, is effectively *the Reynolds stress* $\rho \overline{u^2}$ *integrated from the bottom up to the free surface.* It is obviously positive in general.

Consider now the contribution $S_{xx}^{(2)}$. As in equation (10), we may take the mean value inside the limits of integration :

$$S_{xx}^{(2)} = \int_{-h}^{0} (\overline{p} - p_0) \, dz. \tag{11}$$

In other words $S_{xx}^{(2)}$ arises from *the change in mean pressure within the fluid.* Now the pressure $\overline{p}$ generally contains terms proportional to a^2, which can be found by a second order analysis. However, we do not need to calculate all the second-order terms explicitly since $\overline{p}$ may be found directly from a consideration of the *vertical* flux of vertical momentum as follows.

We know that the mean flux of vertical momentum across any horizontal plane, which is $\overline{p + \rho w^2}$, must be just sufficient to support the weight of the water above it. Since the mean level of the water is at $z = 0$, we have then

$$\overline{p + \rho w^2} = - \rho g z = p_0 \tag{12}$$

or

$$\overline{p} - p_0 = - \rho \overline{w^2} \tag{13}$$

Thus $\overline{p}$ is generally less than the hydrostatic pressure p_0. Substitution in equation (11) gives

$$S_{xx}^{(2)} = \int_{-h}^{0} (-\rho \overline{w^2})\, dz. \tag{14}$$

This contribution is then negative in general.

Combining equations (10) and (14) we have

$$S_{xx}^{(1)} + S_{xx}^{(2)} = \int_{-h}^{0} \rho\, (\overline{u^2 - w^2})\, dz \geqslant 0. \tag{15}$$

For, since the major axes of the particle orbits are horizontal we have $\overline{u^2} \geqslant \overline{w^2}$. After substituting for the velocities from equations (4) and carrying out the integration we find in fact*

$$S_{xx}^{(1)} + S_{xx}^{(2)} = \tfrac{1}{2}\, \frac{\rho a^2\, \sigma^2\, h}{\sinh^2 kh} = \frac{\rho g a^2\, kh}{\sinh 2kh} \tag{16}$$

having used in the last step the frequency relation

$$\sigma^2 = gk \tanh ph \tag{17}$$

for waves of small amplitude.

In deep water the particle orbits are circles, and $\overline{u^2}$ equals $\overline{w^2}$; the Reynolds stresses are isotropic in x and z. The positive contribution $\rho\overline{u^2}$ from the horizontal Reynolds stress is then exactly cancelled by the pressure defect $-\rho\overline{w^2}$ arising from the vertical Reynolds stress. On the other hand in shallow water the particle orbits are elongated horizontally, and $\overline{w^2}$ becomes small compared with $\overline{u^2}$. Then $\rho\, (\overline{u^2 - w^2})$ becomes simply $\rho\overline{u^2}$. Since, for the same reason, the kinetic energy is then just $\frac{1}{2}\rho\overline{u^2}$ per unit volume, we see that $S_{xx}^{(1)} + S_{xx}^{(2)}$ is then twice the kinetic energy density, that is, the total energy density of the waves.

There remains the important contribution $S_{xx}^{(3)}$. This is equal to the pressure p integrated† between 0 and ζ and then averaged with respect to time. It is easily evaluated, for near to the free surface p is nearly equal to the hydrostatic pressure below the free surface :

$$p = \rho g\, (\zeta - z). \tag{18}$$

Thus the pressure at any point near the surface fluctuates in phase with the surface elevation ζ. Substitution in the integral gives

$$S_{xx}^{(3)} = \tfrac{1}{2}\rho g \overline{\zeta^2}. \tag{19}$$

So $S_{xx}^{(3)}$ is generally positive and is in fact equal to the density of potential energy, that is to say half the total energy density E :

$$S_{xx}^{(3)} = \tfrac{1}{4}\, \rho g a^2 = \tfrac{1}{2} E \tag{20}$$

*It may be noted that $(\overline{u^2 - w^2})$ is independent of z, for using the incompressible, irrotational and progressive character of the motion we have :

$$\frac{\partial}{\partial z}\, (\overline{u^2 - w^2}) = 2\, \overline{\left(u\frac{\partial u}{\partial z} - w\frac{\partial w}{\partial z}\right)} = 2\, \overline{\left(u\frac{\partial w}{\partial x} + w\frac{\partial u}{\partial x}\right)} = 2\, \overline{\frac{\partial}{\partial x}\, (uw)} = 0.$$

†When the free surface is below the mean level, the velocity field is assumed to be extended analytically up to the mean level. This device leads to the simplest algebra. If preferred, however, the upper limit of integration may be taken at any arbitrary fixed level in the fluid, within a distance of order a from the mean level; the final result is the same.

where

$$E = \tfrac{1}{2}\rho g a^2. \tag{21}$$

Altogether we have from equations (15) and (19)

$$S_{xx} = S_{xx}^{(1)} + S_{xx}^{(2)} + S_{xx}^{(3)} \geqslant 0, \tag{22}$$

or on inserting the values found in equations (16) and (20)

$$S_{xx} = E\left(\frac{2kh}{\sinh 2kh} + \tfrac{1}{2}\right). \tag{23}$$

The ratio $2kh/\sinh 2kh$ lies always between 0 and 1. In deep water ($kh \gg 1$) the ratio tends to 0 and so

$$S_{xx} = \tfrac{1}{2}E, \tag{24}$$

while the shallow water ($kh \ll 1$) it tends to 1 and so

$$S_{xx} = \tfrac{3}{2}E. \tag{25}$$

The transverse components of the radiation stress. Now let us consider in a similar way the flow of y-momentum (momentum parallel to the wave crests) across a plane $y =$ constant. Denoting this by S_{yy} we have corresponding to equation (7) the relation

$$S_{yy} = \overline{\int_{-h}^{\zeta} (p + \rho v^2)\, dz} - \int_{-h}^{0} p_0\, dz \tag{26}$$

where v is the transverse component of velocity. Just as for S_{xx} we can consider S_{yy} as the sum of three parts :

$$S_{yy} = S_{yy}^{(1)} + S_{yy}^{(2)} + S_{yy}^{(3)} \tag{27}$$

where

$$\left.\begin{aligned} S_{yy}^{(1)} &= \overline{\int_{-h}^{\zeta} \rho v^2\, dz} \\ S_{yy}^{(2)} &= \overline{\int_{-h}^{0} (p - p_0)\, dz} \\ S_{yy}^{(3)} &= \overline{\int_{0}^{\zeta} p dz} \end{aligned}\right\} \tag{28}$$

In gravity waves the transverse velocity vanishes everywhere and so

$$S_{yy}^{(1)} = 0, \tag{29}$$

while $S_{yy}^{(2)}$ and $S_{yy}^{(3)}$ are equal to $S_{xx}^{(2)}$ and $S_{xx}^{(3)}$ respectively. Thus

$$S_{yy}^{(2)} = \int_{-h}^{0} (-\rho\overline{w^2})\, dz \leqslant 0, \tag{30}$$

$$S_{yy}^{(3)} = \tfrac{1}{2}\rho g\, \overline{\zeta^2} \geqslant 0. \tag{31}$$

Substitution for w from equations (4) and use of the frequency relation equation (17) leads to

$$S_{yy} = E \times \frac{kh}{\sinh 2kh}. \tag{32}$$

In deep water, $\overline{w^2} = \overline{u^2} = \frac{1}{2}\,\overline{(u^2 + v^2 + w^2)}$; then $S_{yy}{}^{(2)}$ is equal to minus the density of kinetic energy, which is $-\frac{1}{2}E$. Thus $S_{yy}{}^{(2)}$ just cancels $S_{yy}{}^{(3)}$ and S_{yy} vanishes :

$$S_{yy} = 0 \qquad (kh \gg 1) \tag{33}$$

In other words, the deficiency in the mean pressure $\overline{p}$ arising from the Reynolds stress $\rho\overline{w^2}$ is exactly cancelled, in deep water, by the surface deformation effect. In shallow water, on the other hand, the mean square vertical velocity $\overline{w^2}$ is small. Hence $S_{yy}{}^{(2)}$ is negligible and

$$S_{yy} = S_{yy}{}^{(3)} = \tfrac{1}{2}E. \tag{34}$$

Lastly the flow of x-momentum across the plane $y =$ constant is given by

$$S_{xy} = \overline{\int_{-h}^{\zeta} \rho uv \, dz};$$

there is no contribution from the mean pressure. Since $\overline{uv}$ vanishes identically we have always

$$S_{xy} = 0 \tag{35}$$

provided, of course, that the x-direction is the direction of wave propagation. If for some reason we choose a co-ordinate system at an angle (other than a right angle), then there *will* be a non-zero shear stress S_{xy}. Its magnitude may be calculated by the ordinary tensor transformation rules from the two-dimensional tensor S, which in diagonal form is given by

$$S = E \begin{pmatrix} \dfrac{2kh}{\sinh 2kh} + \frac{1}{2} & 0 \\ 0 & \dfrac{kh}{\sinh 2kh} \end{pmatrix}. \tag{36}$$

2. *Standing gravity waves*

Let us combine two progressive waves of equal amplitude a and wavelength $2\pi/k$ so as to produce a standing wave. The free surface is then described by

$$\zeta = 2a \cos kx \cos \sigma t \tag{1}$$

and the components of velocity by

$$\left.\begin{aligned} u &= \frac{2a\,\sigma}{\sinh 2kh} \cosh k(z+h) \sin kx \sin \sigma t \\ w &= \frac{2a\,\sigma}{\sinh 2kh} \sinh k(z+h) \cos kx \sin \sigma t. \end{aligned}\right\} \tag{2}$$

The surface elevation has antinodes at $kx = n\pi$ (where n is an integer) and nodes ta $kx = (n + \frac{1}{2})\pi$, as in Fig. 4. The two components of velocity fluctuate in-phase, proportionally to sin σt, so that the particle orbits are straight lines. Beneath the antinodes the orbits are vertical, beneath the nodes they are horizontal, and at intermediate positions the orbits are inclined generally to the horizontal. The mean product $\overline{uw}$ and also the shearing stress $\rho\overline{uw}$ do not vanish in general, and are functions of the horizontal co-ordinate x. The horizontal variation of $\rho\overline{uw}$ supports a difference in mean surface level between node and antinode. We can use the radiation stress to calculate this difference.

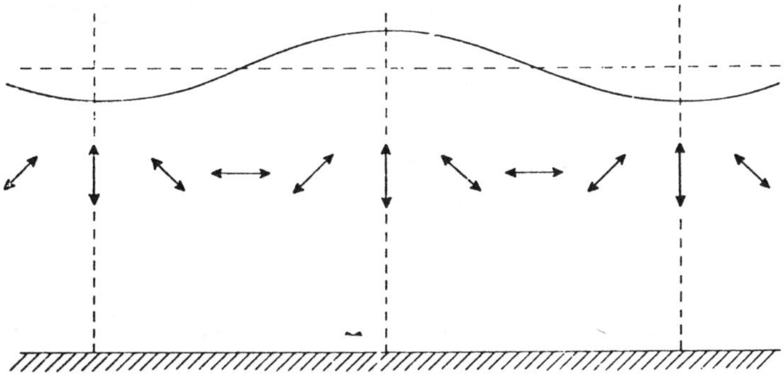

Fig. 4. Particle velocities in a standing wave. The components u and w fluctuate in-phase, and the mean product $\overline{uw}$ is non-zero in general.

Consider the stress S_{xx}, representing the flux of horizontal momentum parallel to the x-axis. This is given by the general relation in equation (7), Section 1, (in which an overbar denotes the average with respect to *time*). As before we may consider the right-hand side as the sum of three parts $S_{xx}^{(1)}$, $S_{xx}^{(2)}$, $S_{xx}^{(3)}$ given approximately by

$$\left.\begin{aligned} S_{xx}^{(1)} &= \int_{-h}^{0} \rho\overline{u^2}\,dz. \\ S_{xx}^{(2)} &= \int_{-h}^{0} (\overline{p} - p_0)\,dz \\ S_{xx}^{(3)} &= \overline{\int_{0}^{\zeta} p dz}. \end{aligned}\right\} \tag{3}$$

where $p_0 = -\rho g z$. The third component $S_{xx}^{(3)}$ is found to be

$$S_{xx}^{(3)} = \tfrac{1}{2}\rho g\overline{\zeta^2} \tag{4}$$

as before. In the second component $S_{xx}^{(2)}$, the time-mean pressure $\overline{p}$ cannot be deduced quite so simply as in the progressive wave, being no longer independent of x. However, it can be found from the more general relation for the vertical flux of vertical momentum :

$$\overline{p + \rho w^2} - \frac{\partial}{\partial x}\int_{z}^{0} \rho\overline{uw}\,dz = p_0 + \rho g\,\zeta, \tag{5}$$

in which the terms on the right represent the total weight (per unit cross-section) of a vertical column of water from z to ζ (Fig. 5); the terms on the left show how this weight is supported : the first two terms represent the mean flux of vertical momentum through the base of the column, while the third term is the resultant of the fluxes of vertical momentum through the vertical sides of the column. Taking

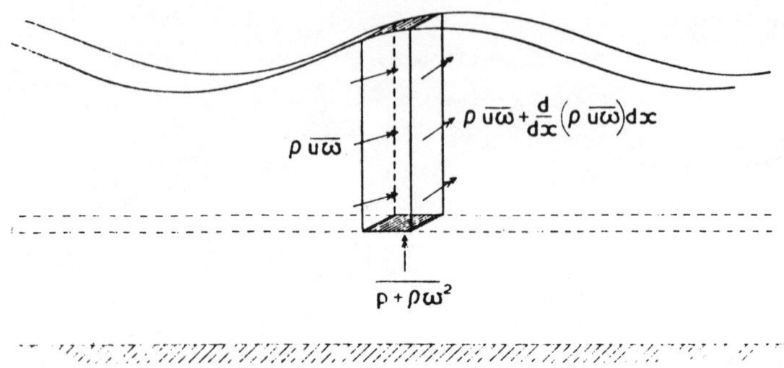

Fig. 5. The balance of momentum in a vertical column of unit cross-section.

$\bar{p} - p_0$ to the left-hand side and the other terms to the right, we have after integration with respect to z :

$$S_{xx}{}^{(2)} = \rho gh\,\bar{\zeta} - \int_{-h}^{0} \rho\overline{w^2} + \frac{\partial}{\partial x}\int_{-h}^{0}\int_{z'}^{0} \rho\overline{uw}\,dz\,dz'. \tag{6}$$

Adding this to $S_{xx}{}^{(1)}$ and $S_{xx}{}^{(2)}$ we deduce

$$S_{xx} = \rho gh\,\zeta + \int_{-h}^{0} \rho\,\overline{(u^2 - w^2)}\,dz + \frac{\partial}{\partial x}\int_{-h}^{0}\int_{z'}^{0} \rho\overline{uw}\,dz\,dz' + \tfrac{1}{2}\rho g\,\overline{\zeta^2}. \tag{7}$$

Now clearly S_{xx} must be a constant, independent of x, for otherwise horizontal momentum would accumulate at some parts of the wave*. Therefore S_{xx} is equal to its horizontal average, that is to say its average with respect to x over a wavelength. Among the terms on the right of equation (7), the horizontal average of ζ is identically zero, while the horizontal average of the third term also vanishes by the periodicity (the momentum fluxes across two vertical walls a wavelenth apart just cancel). So we have simply

$$S_{xx} = \int_{-h}^{0} \rho\,\underline{\overline{(u^2 - w^2)}}\,dz + \tfrac{1}{2}\rho g\,\underline{\overline{\zeta^2}} \tag{8}$$

where an underbar denotes the horizontal mean value. Substituting from equations (1) and (2) we find

$$S_{xx} = \rho g a^2\left(\frac{2kh}{\sinh 2kh} + \tfrac{1}{2}\right). \tag{9}$$

Comparison with equation (23) Section 1 shows that the radiation stress in a

*This follows from the conservation equation for x- momentum: $\partial S_{xx}/\partial x + \partial S_{xy}/\partial y = 0$, and the fact that $S_{xy} = 0$ in these co-ordinates.

standing wave is exactly twice the value in each progressive wave; it represents the sum of the stresses in the incident and reflected waves, as we should expect.

The *local* mean level $\bar{\zeta}$ can now be found from equation (7) since all other terms in the equation are known to the required approximation. In this way we find†

$$\bar{\zeta} = a^2 k \coth 2kh \cos 2kx. \tag{10}$$

This shows that the mean surface level is slightly raised at the antinodes and correspondingly lowered at the nodes.

The various terms on the right of equation (7) do not all give contributions in the same sense; some tend to raise the level at the antinodes and others to lower it. A simpler way to estimate $\bar{\zeta}$ is to return to the momentum flux equation (5) and set $z = 0$. This gives us

$$\overline{(p + \rho w^2)}_{z=0} = \rho g\, \bar{\zeta} \tag{11}$$

(the flux of vertical momentum across the vertical sides of the column is of third order only). But on taking time averages in the Bernoulli equation

$$\left[p + \tfrac{1}{2}\rho\,(u^2 + w^2) + \frac{\partial \phi}{\partial t}\right]_{z=0} = f(t)$$

we have also

$$\left[\overline{p + \tfrac{1}{2}\rho\,(u^2 + w^2)}\right]_{z=0} = C, \tag{12}$$

where C is a constant, not necessarily zero. From equations (11) and (12) we deduce

$$g\,\bar{\zeta} = -\tfrac{1}{2}\,\overline{(u^2 - w^2)}_{z=0} - C. \tag{13}$$

The constant C is determined by the condition that $\underline{\xi} = 0$. Substitution for u and w now gives us equation (10) as before. We note in particular that in deep water ($kh \gg 1$) equation (10) becomes

$$\bar{\zeta} = a^2 k \cos 2kx \tag{14}$$

and in shallow water ($kh \ll 1$)

$$\bar{\zeta} = \frac{a^2}{2h} \cos 2kx. \tag{15}$$

As the depth h diminishes, a and k being fixed, the changes in mean level which are represented by $\bar{\zeta}$ become accentuated.

The evaluation of the transverse stress S_{yy} follows exactly similar lines; it is necessary only to replace $\overline{u^2}$ by $\overline{v^2}$, $= 0$ throughout. Hence

$$\begin{aligned} S_{yy} &= S_{xx} - \int_{-h}^{0} \rho \overline{u^2}\, dz \\ &= S_{xx} - \rho g a^2 \left(\frac{2kh}{\sinh 2kh} + 1\right) \sin^2 kx \\ &= \tfrac{1}{2}\rho g a^2 \left[\frac{2kh}{\sinh 2kh} + \left(\frac{2kh}{\sinh 2kh} + 1\right) \cos 2kx\right]. \end{aligned} \tag{16}$$

Hence S_{yy}, unlike S_{xx}, is a function of x in a standing wave. Perhaps surprisingly,

†This result is in agreement with TADJBAKHSH and KELLER (1960) provided that account is taken of a misprinted sign in their equation (30).

it will be noted that the maximum values of S_{yy} occur at the *nodes* of the surface elevation. The mean value of S_{yy} is given by

$$\underline{S_{yy}} = \rho g a^2 \frac{kh}{\sinh 2kh} \tag{17}$$

which is just twice the value for the progressive wave (equation (32) Section 1) as we should expect.

The radiation shear stress is given by

$$S_{xy} = S_{yx} = 0 \tag{18}$$

as in the progressive wave.

[*Editors' Note:* Material has been omitted at this point.]

REFERENCES

Cox C. S. (1958) Measurements of slopes of high-frequency wind waves. *J. Mar. Res.* **16**, 199–225.

Dorrestein R. (1962) Wave set-up on a beach. *Proc. 2nd. Tech. Conf. on Hurricanes*, Miami Beach, Fla. pp. 230–241. U.S. Dept. of Commerce, Nat. Hurricane Res. Proj. Rep. No. 50.

Fortak H. (1962) On the mathematical formulation of the storm surge. *Proc. 2nd Tech. Conf. on Hurricanes*, Miami Beach, Fla. pp. 196–210. U.S. Dept. of Commerce, Nat. Hurricane Res. Proj. Rep. No. 50.

Harris D. L. (1963) Characteristics of the hurricane storm surge. *U.S. Weather Bur. Tech. Pap. No.* 48. 139 pp.

Hasselmann K. (1962) On the non-linear energy transfer in a gravity-wave spectrum. Part 1. General theory. *J. Fl. Mech.* **12**, 481–500.

Hasselmann K. (1963) On the non-linear energy transfer in a gravity-wave spectrum. Part 2. Conservation theorems; wave-particle analogy; irreversibility. *J. Fl. Mech.* **15**, 273–281.

Hasselmann K. (1963) On the non-linear energy transfer in a gravity-wave spectrum. Part 3. Evaluation of the energy flux and sea-swell interaction for a Neumann spectrum. *J. Fl. Mech.* **15**, 385–398.

Hughes B. A. and Stewart R. W. (1961) Interaction between gravity waves and a shear flow. *J. Fl. Mech.* **10**, 385–400.

Lamb H. (1932) *Hydrodynamics, 6th ed.* Cambrdige U. Press.

LONGUET-HIGGINS M. S. and STEWART R. W. (1960) Changes in the form of short gravity waves on long waves and tidal currents. *J. Fl. Mech.* **8**, 565–583.

LONGUET-HIGGINS M. S. and STEWART R. W. (1961) The changes in amplitude of short gravity waves on steady non-uniform currents. *J. Fl. Mech.* **10**, 529–549.

LONGUET-HIGGINS M. S. and STEWART R. W. (1962) Radiation stress and mass transport in gravity waves, with application to "surf-beats." *J. Fl. Mech.* **13**, 481–504.

LONGUET-HIGGINS M. S. and STEWART R. W. (1963) A note on wave set-up. *J. Mar. Res.* **21**, 4–10.

LONGUET-HIGGINS M. S. (1963) The generation of capillary waves by steep gravity waves. *J. Fl. Mech.* **16**, 138–159.

LUNDGREN H. (1963) Wave thrust and energy level. *Proc. Int. Ass. Hydr. Res. Congr.* London, pp. 147–151.

MILES J. W. (1962) On the generation of waves by shear flows. Part 4. *J. Fl. Mech.* **13**, 433–448.

MUNK W. H. (1949a) The solitary wave and its application to surf problems. *Ann. N. Y. Acad. Sci.* **51**, 376–424.

MUNK W. H. (1949b) Surf beats. *Trans. Amer. geophys. Un.* **30**, 849–854.

PHILLIPS O. M. (1958) The equilibrium range in the spectrum of wind-generated waves. *J. Fl. Mech.* **4**, 426–434.

PHILLIPS O. M. (1960) On the dynamics of unsteady gravity waves of finite amplitude. Part 1. The elementary interactions. *J. Fl. Mech.* **9**, 193–217.

PHILLIPS O. M. (1963) On the attenuation of long gravity waves by short breaking waves. *J. Fl. Mech.* **16**, 321–332.

SAVILLE T. (1961) Experimental determination of wave set-up. *Proc. 2nd Tech. Conf. on Hurricanes*, Miami Beach, Fla. pp. 242–252. U.S. Dept of Commerce, Nat. Hurricane Res. Proj. Rep. No. 50.

STEWART R. W. (1956) A new look at the Reynolds stresses. *Canad. J. Phys.* **34**, 722–725.

TADJBAKHSH I. and KELLER J. B. (1960) Standing surface waves of finite amplitude. *J. Fl. Mech.* **8**, 442–451.

TAYLOR, SIR GEOFFREY (1962) Standing waves on a contracting or expanding current. *J. Fl. Mech.* **13**, 182–192.

TUCKER M. J. (1950) Surf beats : sea waves of 1 to 5 minutes' period. *Proc. Roy. Soc.* A **207**, 565–73.

URSELL F. (1956) Wave generation by wind. *Surveys in Mechanics*. G. I. Taylor. 70th Anniversary Vol. Cambridge Monog. on Mech. and Appl. Math. ed. G. K. Batchelor and R. M. Davies. Cambridge U. Press pp. 216–249.

WHITHAM G. B. (1962) Mass, momentum and energy flux in water waves. *J. Fl. Mech.* **12**, 135–147.

10

Reprinted from *Rev. Geophysics* 5(3):287-304 (1967)

Longshore Current Velocity: A Review of Theory and Data

CYRIL J. GALVIN, JR.

Research Division

U. S. Army Coastal Engineering Research Center, Washington, D. C. 20016

Abstract. A proven prediction of longshore current velocity is not available, and reliable data on longshore currents are lacking over a significant range of possible flows. Theoretical studies have been based on over-simplified models, and empirical predictions have been hampered by lack of data. The empirically modified, momentum-flux theory now accepted as the best prediction is based on an untenable assumption and supported by inappropriate data. Regardless of their validity, however, all six of the testable equations agree fairly well with at least one of six sets of published data, and two agree with both of the better sets of data. These two equations may be used as empirical guides for velocity prediction in the absence of a proven theory. The best prospect for a generally valid velocity prediction appears to be an empirical correlation based on reliable data.

INTRODUCTION

Longshore currents flow parallel to the shoreline and reach their highest velocities between the point of wave breaking and the shoreline. These currents are assumed to be generated by the longshore component of motion in waves that approach the shoreline obliquely.

The effect of longshore currents on landing craft during World War II led to the comprehensive field, laboratory, and theoretical study of *Putnam et al.* [1949]. That study, as amended by the work of *Inman and Quinn* [1951], remains the accepted treatment of longshore currents. More recently, the ability of longshore currents to erode, transport, and deposit beach sediment has prompted further studies by engineers, who must predict the future effect of these currents on coastal structures, and by geologists, who wish to interpret the past importance of these currents in distributing sediment [*Bruun,* 1963; *Brebner and Kamphuis,* 1963; *Inman and Bagnold,* 1963; *Harrison and Krumbein,* 1964; *Galvin and Eagleson,* 1965].

The principal aim of published longshore current studies has been to develop a method for predicting the longshore current velocity. As a result of these studies, there are now at least 12 different equations to predict velocity and at least 352 published field and laboratory measurements of longshore currents. This paper reviews published field and laboratory observations that permit a description of longshore current flow, and it evaluates the theories proposed to predict longshore current velocity. The review covers papers appearing in North American sources; it is selective rather than exhaustive, emphasizing recent results and omitting data known to exist but unpublished [*Galvin and Nelson,* 1967, p. 3].

DATA

The variables. A review of eight published sets of field and laboratory data indicates that only five variables (Figure 1) have been measured in all studies:

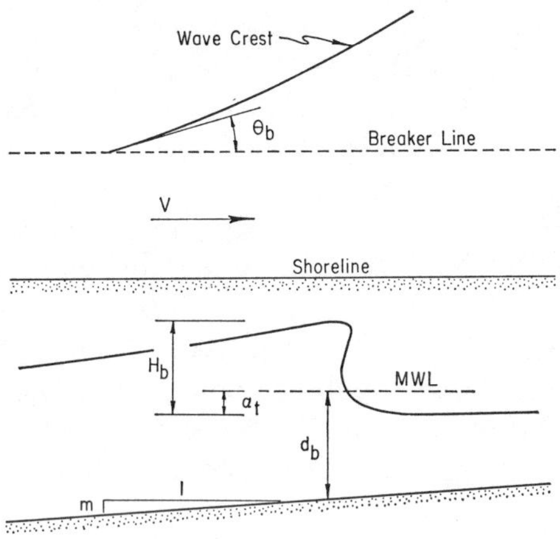

Fig. 1. Definition of variables ($\beta = d_b/H_b$, $\sigma = a_t/H_b$). In field studies, slope is approximated from bottom profiles. MWL is mean water level.

a mean longshore velocity (V), a wave height (H), direction (θ), period (T), and a beach profile or slope (m). In these studies, other variables such as breaker depth, breaker position, and wave length were measured, but none of these additional variables was measured in more than two of the eight studies. Although it would be useful to relate longshore current velocity to deep-water conditions, as *Brebner and Kamphuis* [1963] point out, it is better to attempt to relate the current to the nearest condition of the breaking wave. The choice of the breaker position eliminates the effects of shoaling and refraction over the usually unknown offshore hydrography, the field problem of obtaining simultaneous deep-water and surf zone measurements, and the necessity in laboratory studies of computing hypothetical deep-water directions and wave heights. Since most investigators have measured the height (H_b) and direction (θ_b) at the breaker point, data from the studies that include only deep-water height (H_0) and direction (θ_0) [*Brebner and Kamphuis,* 1963; *Saville,* 1950] have been recomputed for the breaker position [*Galvin and Nelson,* 1967].

Laboratory data. The term 'observation' denotes all the measurements made during given wave conditions. The four laboratory studies under discussion contain 225 observations [*Putnam et al.,* 1949; *Saville,* 1950; *Brebner and Kamphuis,* 1963; *Galvin and Eagleson,* 1965]. (The last-named study will be referred to hereafter as *TM 10.*) Seven other observations have been made by *Eagleson* [1965] on the growth of the current immediately downstream of an obstacle. Twelve of the 38 observations in *TM 10* do not include information on all five variables.

The laboratory basins used in these studies had test beaches that measured, between training walls, from 6 meters (20 feet, *TM 10*) to 18 meters long (60 feet)

[*Saville,* 1950]. Beach slopes in most tests were about 0.1, but they varied from 0.05 to 0.26. Most beach surfaces were of smooth concrete, but anything from sheet metal to bonded gravel has been used [*Putnam et al.,* 1949]. Saville's beach consisted of loose sand in the surf zone and a plane concrete surface off shore. In the four studies, off-beach water depth ranged between 0.3 and 0.6 meter.

There has been no systematic test of the relation between basin geometry and longshore currents, but it is known that wave heights vary irregularly with time and position in wave basins [*Savage,* 1962]. Basin geometry also affects the return of the fluid in the surf zone to the offshore part of the basin and, thus, the development of the currents known to be non-uniform in the longshore direction. Finally, basin geometry may affect the refraction pattern of waves on the beach, unless the training walls are curved to follow the wave orthogonals.

In the laboratory, the wave period (T) can be measured accurately to 0.01 sec, usually sufficient for 1% accuracy. Beach slopes (m) for typical plane concrete laboratory beaches are also accurately measurable, but the slopes of sand beaches involve some approximation. Typical laboratory waves about 5 cm high can be measured to within 5% by resistance gages or ordinary point gages [*TM 10; Ursell et al.,* 1958]. This accuracy is all that is warranted, since wave height can vary by greater amounts along the beach, probably owing to the effects of basin geometry (see Table 1).

The breaker angle (θ_b) is both the most difficult of the variables to measure accurately and probably the most important for determining the strength of the current. *Putnam et al.* [1949] measured θ_b from vertical photographs, and *Galvin and Eagleson* [1965] used a tedious method of visual observation with a specially designed protractor. The protractor gave results agreeing well with the theoretical values obtained from standard small amplitude refraction theory (see Table 7 in *TM 10*).

Longshore current velocity on laboratory beaches has been measured by tim-

TABLE 1. Wave Height Variation on a 6-Meter Laboratory Beach
Breaker height, in centimeters, at distance x from upstream training wall.*

Test Number	T, sec	x, meters 0.9	3.1	5.2
II 3	1.13	5.9	7.0	6.0
II 4	1.25	4.0	5.2	4.0
II 5	1.38	5.9	6.6	4.7
II 7	1.25	0.7	1.4	1.0
II 8	1.25	2.3	4.8	3.1
II 9	1.25	4.3	6.4	5.7
II 10	1.25	4.7	6.0	5.4
II 11	1.25	5.2	6.8	5.4
IV 4	1.25	5.5	5.6	4.9
IV 9	1.25	4.7	4.6	4.0

* Data from *Galvin and Eagleson* [1965]. Each height is average of twenty consecutive waves measured with wire resistance gage.

ing the travel of dye [*Putnam et al.,* 1949; *Saville,* 1950], immiscible fluids [*Brebner and Kamphuis,* 1963] or wooden floats [*TM 10*]. The per cent difference between float velocity and the velocity of the midpoint of a dye patch ranged from less than 1 to 10% in five comparisons (Table 5 in *TM 10*), the float velocities tending to be higher than the dye velocities.

Field data. If measurements are made of a uniform train of waves in the ocean, it is possible to measure repeatedly an average wave period to within about 5% of a mean value, either with a stop watch or from an oscillograph of the water surface (Table 4 in *Galvin and Savage* [1966]). Under favorable conditions, wave height can be measured with similar accuracy from the oscillographs, but accuracy can be significantly poorer.

Wave direction is less accurately measurable by ordinary field techniques. Single measurements of wave direction are probably only rarely accurate to within about 2°, and such an error is relatively important for the small values of θ_b that commonly occur. Often θ_b is measured only to the nearest 5° [*Moore and Scholl,* 1961; *Putnam et al.,* 1949]. Longshore current velocity has been measured by timing the travel of dye [*Putnam et al.,* 1949; *Moore and Scholl,* 1961], seaweed, a weighted soccer ball [*Inman and Quinn,* 1951], and water-filled balloons [*Galvin and Savage,* 1966]. With care and repetition, most such measurements can be within 0.25 ft/sec of the mean of the data. Beach slopes are obtained from profiles through the surf zones and seem to be subject to less variation than may be initially suspected. Thirteen carefully controlled profiles obtained at Nags Head, North Carolina, during five consecutive days of changing wave conditions had nearshore slopes ranging only from 0.025 to 0.030 (profiles supplied by Louisiana State University Coastal Studies Institute, March 1964).

Although measurements can produce numbers that are repeatable within reasonable limits, field data are uncertain because of the uncontrolled variables that influence the longshore current velocity. The two principal unknowns in field work are the effect of variations in nearshore hydrography [*Sonu et al.,* 1967] and the effect of substituting a mean value for the distribution of values of each measured variable. Another field problem is the difficulty of measuring all variables simultaneously, so that they refer to the same physical condition.

Flow changes. When non-uniformity in the alongshore direction has been measured (only in the laboratory [*Berbner and Kamphuis,* 1963; *Eagleson,* 1965; and *TM 10*]), the current typically had a low velocity near the upstream wall, increased to a maximum on the downstream half of the beach, and then fell off as the downstream training wall was neared. By using minature current meters calibrated against the travel of floats, it was possible to measure the one- and two-dimensional velocity distributions in the surf zone. Figure 2 shows the one-dimensional longshore current growth for three tests which differed only in the direction of wave approach. These results are typical of much data (Figure 27 and Table A7 in *TM 10*) indicating that the current requires more distance to reach its maximum velocity at low θ_b than at high θ_b.

Measurements of typical two-dimensional distributions of velocity in laboratory longshore currents have also been made (Figure 24 and Table A7 in *TM 10*). The maximum velocities appeared between the still water line on the beach and

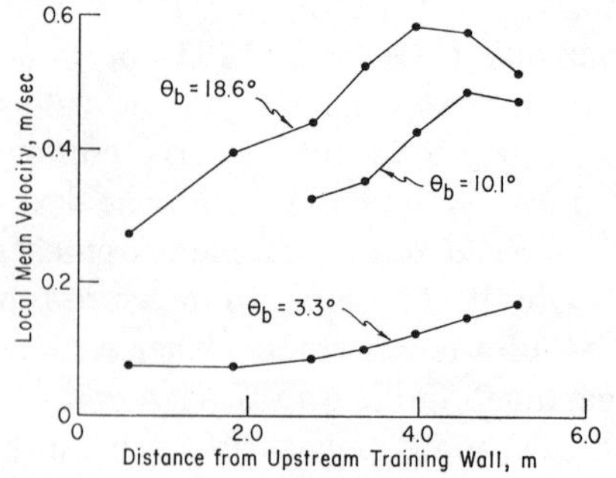

Fig. 2. Velocity distributions along a laboratory beach as a function of mean breaker angle θ_b. All other conditions constant: wave period, 1.25 sec.; wave height, 4.25 cm at 35-cm depth; beach, concrete with 0.11 slope.

the breaker position, usually initially closer to the still water line than to the breaker position. As the current flowed down the beach, the local maxima had a tendency to migrate toward the breaker position. Under certain laboratory conditions, longshore currents may turn seaward into a rip-like current, but such laboratory rip currents are not known to have been observed during any of the laboratory studies discussed here.

On natural beaches, rip currents may cause unsteadiness in longshore current flow [*Putnam et al.*, 1949], but such changes may not always be present [*Galvin and Savage*, 1966]. The non-uniformity of longshore currents on natural beaches has not been studied quantitatively, although the results of *Inman and Quinn* [1951] indicate non-uniform flow may have been present during their observations.

EVALUATION OF DATA

Overlapping laboratory data. There is only one laboratory condition that is nearly repeated in three studies and only about four conditions that appear in two laboratory studies. The common laboratory conditions must be drawn from data taken on smooth, plane beaches of approximately 1 on 10 slope. Observations numbered 15, 138, and 213 satisfy these restrictions, are drawn from three different laboratory studies, and have approximately the same wave conditions (Table 2).

TABLE 2. Comparison of Overlapping Laboratory Data

Observation Number*	H_b, cm	T, sec	θ_b, deg	m	V, m/sec	References
15	7.3	0.99	28.0	0.10	0.51	*Putnam et al.*
138	5.5	1.00	26.0	0.10	0.39	*Brebner and Kamphuis*
213	5.5	1.00	28.0	0.11	0.66	*Galvin and Eagleson*
17	4.9	1.63	18.8	0.10	0.29	*Putnam et al.*
205	4.9	1.50	6.9	0.11	0.37	*Galvin and Eagleson*
48	3.0	1.00	7.5	0.10	0.14	*Brebner and Kamphuis*
199	2.7	1.00	2.9	0.11	0.14	*Galvin and Eagleson*
72	4.9	1.00	14.0	0.10	0.34	*Brebner and Kamphuis*
201	5.2	1.00	14.1	0.11	0.54	*Galvin and Eagleson*
18	4.9	1.98	18.4	0.10	0.23	*Putnam et al.*
88	4.6	1.00	18.0	0.10	0.36	*Brebner and Kamphuis*

* Observation number refers to the numbers assigned to the data in the compilation of *Galvin and Nelson* [1967]. All data are for smooth cement laboratory beaches.

The only important difference in wave variables is that observation 15 has a wave height of 7.3 cm and the other two observations have heights of 5.5 cm. Despite the similar values of wave variables, the observed velocities differ greatly. The velocity in observation 213 (*TM 10*) is 70% larger than the velocity in observation 138 (Brebner and Kamphuis) and 28% larger than the velocity in observation 15 (Putnam et al.). Assuming velocity increases with wave height, the discrepancy is even greater because of the difference in heights. On the other hand, the difference in velocity between observations 15 and 138 is in the direction to be expected from the lower value of wave height and angle in 138. The four pairs of observations, representing somewhat similar wave conditions, also show the same trend (Table 2). The velocities in *TM 10* are significantly greater than those measured by *Putnam et al.* [1949] and by *Brebner and Kamphuis* [1963].

Causes of velocity differences. The size of the differences is considerably larger than is to be expected from differences between float and dye methods of velocity measurement. The consistent sign of the differences makes it unlikely that they are due to random error. The velocities were measured at approximately the same position on the beach, thus ruling out large errors due to non-uniformity of the current. (Putnam et al., Brebner and Kamphuis, and Galvin and Eagleson measured velocities at, respectively, 6.1, 4.6–6.1, and 4.9 meters, from the start of the current.) In the order just listed, lengths of the test beaches in these studies were 12 meters, about 11 meters, and 6–9 meters. Since backwater effects are less on the longer beaches, the higher velocities on the shorter beach (*TM 10*) remain unexplained.

A possible explanation of the difference in measured velocities is that the wave heights in the three studies are not comparable because of different types of breaking or different definitions of the breaking point. Although it does not seem to have been remarked before, the data of Putnam et al. (Figure 3) agree as well with Miche's theory for the maximum stable wave as do the experiments usually quoted as evidence in favor of this theory [*Danel*, 1952]. Miche's theory appears to be valid under conditions producing spilling breakers, implying that the currents measured by Putnam et al. were produced by spilling waves. Since the data

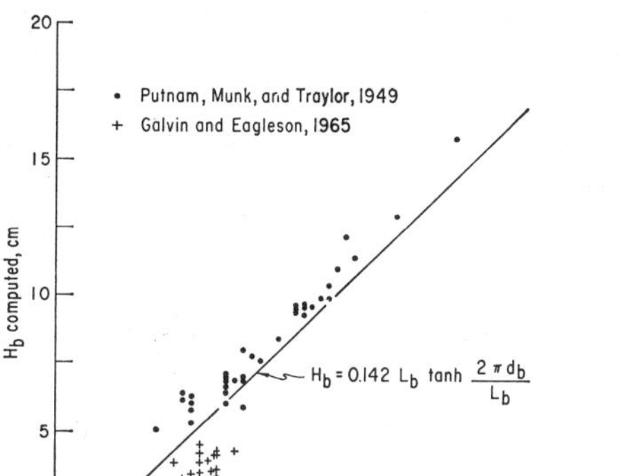

Fig. 3. Comparison of laboratory breaker heights with the theory of *Miche* [1944]. L_b was computed from $T\sqrt{gd_b}$. Data of *Galvin and Eagleson* [1965] are the average heights of 20 consecutive waves.

of Putnam et al. followed Miche's theory, their agreement with the data of Brebner and Kamphuis, in the few cases where it is possible to compare (Table 2), is to be expected because the wave heights in Brebner and Kamphuis' data were computed from a graph derived from Miche's theory [*Le Méhauté*, 1961]. The data in *TM 10,* which do not agree with Miche's theory (Figure 3), were for currents produced by plunging breakers.

Further evidence that the experiments of Putnam et al. involved spilling breakers is provided by the breaker depth-to-height ratio, β. Data on breaker depth are given by Putnam et al. and in *TM 10,* and the β ratios are summarized in Table 3. In general, wave steepness and beach slope determine both β [*Iversen,* 1953] and type of breaking [*Patrick and Wiegel,* 1955]. By comparing the laboratory data with the curves of Iversen and of Patrick and Wiegel (see *Wiegel* [1964], p. 175), it appears that the data of Putnam et al. are in the spilling range and that the data of *TM 10,* obtained with plunging breakers, are in the plunging range of the graph of Patrick and Wiegel. But even though the wave data of Putnam et al. imply spilling, spilling rarely occurs on the steep slopes used in their tests, and photographs accompanying the paper of Putnam et al. clearly indicate breaking by plunging. This suggests that breaker position may have been differently defined, but information on this point is not available.

TABLE 3. Laboratory Breaking Depth-to-Height Ratios

Slope	Number of Observations	Average β*	Reference
0.10 smooth slope	4	1.34	*Putnam et al.* [1949]
All slopes (0.07 to 0.26)	37	1.39	*Putnam et al.* [1949]
0.11 smooth slope	54	0.85	*Galvin and Eagleson* [1965] (observations of 12 wave conditions)
0.10 smooth slope	163	0.86	CERC (unpublished data from films of 12 wave conditions)

* $\beta = d_b/H_b$, where d_b is the mean water depth at the breaker point. $\beta = 1.28$ in solitary wave theory.

Evaluation of laboratory studies. It is necessary to determine if any of the four diverse sets of laboratory data are useful for evaluating theories of longshore current flow. Because of the uncertainties brought out in the paragraph above (see also discussion of Figure 4 in later section), it is assumed that the laboratory data of Putnam et al. are not useful for the purpose of evaluating the theories. Since the breaker data were not measured by Saville or by Brebner and Kamphuis, only the laboratory data of *TM 10* remain. In this review these data will be assumed to supply the most useful test of theory for laboratory conditions, although they have recognized experimental uncertainties (Figure 2 and Table 1).

Evaluation of field studies. The available data do not permit comparisons between field studies, even to the limited extent possible for the laboratory studies. The data of *Inman and Quinn* [1951] are already averages of measurements from 15 positions spaced at 90-meter intervals along the beach. They demonstrate convincingly that local conditions are not necessarily representative of the average conditions along a stretch of beach, since the standard deviation of the average velocity in this study usually exceeds the mean. However, since velocity predictions developed from physical models are all based on local conditions of the breaker, the data of Inman and Quinn, being the average of differing local conditions, are not appropriate for evaluating these models.

The data of *Moore and Scholl* [1961] are measurements of generally low velocities accompanying low waves on steep beaches, a condition not represented in any other study. However, the precision of measurement was low: H_b averaged about 0.4 meter and was measured to the nearest 0.1 meter; θ_b was measured to the nearest 5°, with the result that 36 to 71 observations were either 0° or 5°. This study includes many observations of low flows, several of which flowed against the direction of the breaking wave. It is possible that, for such low velocity conditions, other physical factors exert relatively more influence on the velocity than does the wave direction.

The five observations of *Galvin and Savage* [1966] include useful supplementary information on variability encountered under field conditions but are themselves too few to be useful. Of the four sets of field data available, only the 18 observations of Putnam et al. appear to be useful for evaluating predictions of longshore current velocity.

VELOCITY PREDICTION

Equations to predict longshore current velocity can be grouped into three classes, according as they are based predominately on: (1) conservation of energy of momentum [*Putnam et al.*, 1949; equation 28 of *TM 10; Eagleson*, 1965]; (2) conservation of mass [*Bruun*, 1963; *Inman and Bagnold*, 1963; equation 35 of *TM 10*]; (3) empirical correlation of data [*Inman and Quinn*, 1951; *Brebner and Kamphuis*, 1963; *Harrison and Krumbein*, 1964].

Energy and momentum equations. The first application of physical principles to derive equations for predicting longshore current velocity was made by *Putnam et al.* [1949]. They (and most subsequent investigators) used the solitary wave description of the breaker [*Munk*, 1949] to obtain relations for the energy and momentum flux into a unit length of the surf zone, and they assumed a simple partition of the energy and momentum flux in the surf zone. This led to an energy equation for the mean longshore current velocity:

$$V = [6.97g(s/f)(mH_b^2/T) \sin 2\theta_b]^{1/3} \qquad (1)$$

For details of typical derivations, see the original analysis of *Putnam et al.* [1949] or the more elaborate treatment of *Eagleson* [1965].

The wave variables were defined in Figure 1. Equation 1 involves two unknown constants: *s*, the fraction of incoming energy flux available to the longshore current; and *f*, the Darcy-Weisbach friction factor. (The original coefficient *k* has been replaced by the more commonly used *f*.) The acceleration due to gravity is represented by *g*. Momentum conservation was used to obtain a second equation for the mean longshore current velocity:

$$V = \frac{a}{2}\left[\left(1 + \frac{4}{a}\sqrt{2.28gH_b}\sin\theta_b\right)^{1/2} - 1\right] \tag{2}$$

where

$$a = 20.88(m\,H_b/fT)\cos\theta_b \tag{3}$$

(*Castanho* [1964, equation 10] has recently rederived (2) in a slightly different form.) There are then two equations, (1) and (2), involving three unknowns, V, s, and f. No solution is possible for V unless one of the constants, s or f, is known. Equation 2 was used together with the field and laboratory data to obtain an f value for the field and each of the laboratory beaches. This value of f was then put back into (2) to predict the velocity [*Putnam et al.,* 1949].

This procedure may be criticized because the measurements are used to obtain f and then f is used to predict the measurements; thus it is clear that no independent evaluation of the equation has been made.

There is, however, a more basic objection to the derivation of the equation itself. In deriving (1) and (2), it was assumed that the longshore current is driven by the alongshore component of motion in the breaking wave and that this component is simply the alongshore component of the velocity of the breaker crest given by $C_b \sin\theta_b$. C_b is the speed of the breaking wave crest according to solitary wave theory. Since energy losses accompany the current generation, the assumption implies the inequality (equation 14 of *Putnam et al.* [1949]):

$$C_b \sin\theta_b > V \tag{4}$$

It is a well observed fact, however, that water in the breaking wave is largely withdrawn from the surf zone just prior to breaking. If a longshore current is flowing in the surf zone, then the water particles in the breaking wave will have an alongshore component of motion in addition to $C_b \sin\theta_b$. Furthermore, Figure 4 shows that this inequality (4) does not hold for many observations. For the data from *TM 10* (the bottom histogram of Figure 4), (4) does not hold in 25 of 26 cases; for the data of *Brebner and Kamphuis* [1963] (one-third of which are plotted on the middle histogram), (4) does not hold in 65 of 94 cases on the 1:10 slope.

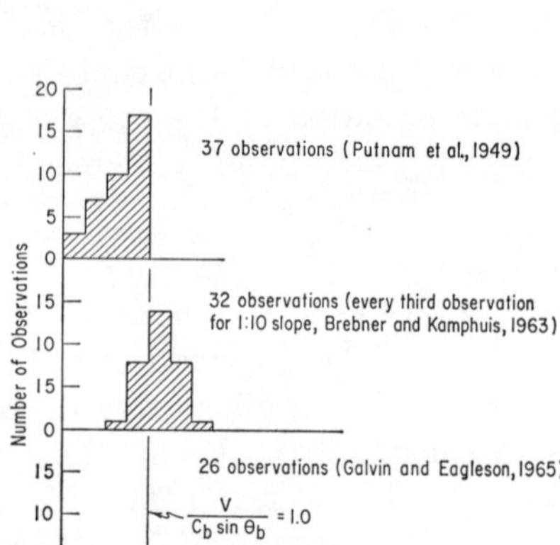

Fig. 4. Laboratory tests of the hypothesis of *Putnam et al.* [1949] that the ratio $V/(C_b \sin\theta_b)$ is less than 1.0.

In constructing Figure 4, the assumptions of Putnam et al. were used. C_b was obtained from the solitary wave equation

$$C_b = \sqrt{gh} \tag{5}$$

where h is the elevation of the wave crest above the bottom, given in terms of wave height as

$$h = H_b(1 + \beta - \sigma) \tag{6}$$

β and σ are defined in Figure 1. Putnam et al. assume $\beta = 1.28$ and $\sigma = 0$. If the values $\beta = 0.85$ and $\sigma = 0.3$ are used instead (*TM 10*), C_b would be 18% smaller, thus increasing the disagreement with the hypothesis (4).

There are at least three reasons why Putnam et al. did not discover that the inequality (4) could be violated: (1) their data were for high values of θ_b (only 3 of 55 observations had $\theta_b < 10°$, and 12 had $\theta_b > 30°$; rough refraction calculations indicate that in some laboratory experiments the wave generator must have made about 90° angle with the beach, a condition that introduces serious reflection problems); (2) the derivation, in effect, sets $\sigma = 0$ by equating trough elevation with mean water level; and (3) their experimental conditions were such that β was much higher than has been found for plunging waves in other studies (Table 3 and accompanying discussion). All three of these factors artificially increase the left side of (4), so that (4) was not violated by their data, but at the time their data were the only available.

More space than would otherwise be appropriate has been devoted to this discussion because equation 2, as modified by *Inman and Quinn* [1951], is currently the accepted description of longshore current velocity [*Wiegel,* 1964; *Bruun,* 1963]; and the reasons for rejecting it should be made clear. Inman and Quinn's subsequent modification is discussed below (see equation 17).

The fact that longshore velocity in the surf zone affects the alongshore component of fluid motion in the breaker was used in another momentum equation (p. 52, *TM 10*) in which the mean alongshore velocity in the breaker, $V_b(x)$, was distinguished from the mean velocity in the surf zone, $V(x)$. The equation agrees qualitatively with the dependence of velocity profile on angle (Figure 2), but it is not useful for quantitative prediction because both V and V_b are unknown.

Recently, *Eagleson* [1965] developed an elegant model of longshore current growth in which longshore current velocity is the only unknown:

$$\frac{V^2(x)}{A} = 1 - \left[1 - \frac{V^2(0)}{A}\right]e^{-Bx} \tag{7}$$

where, for usual beach slopes

$$A = \frac{3}{8}\frac{gH_b{}^2n_b}{d_bf} m \sin\theta_b \sin 2\theta_b \tag{8}$$

$$B = \frac{2}{5}\left[\frac{f}{md_b \sin\theta_b}\right] \tag{9}$$

The f in (8) and (9) is evaluated from the Karman-Prandtl resistance equation for steady, uniform, turbulent flow in pipes of given roughness. According to (7), the predicted velocity profiles are always concave downward, and steady state is always approached faster for lower values of θ_b. These two predictions do not agree with data now available (e.g. Figure 2).

Conservation of mass equations. Bruun [1963] suggests two theories of longshore current flow based mostly on conservation of mass in the surf zone: a rip current theory and a continuity theory using the Chezy equation. Basic to both is a factor Q_b, the cross-sectional area of the significant wave (the average of the highest third of the waves). Bruun assumes that only those waves whose heights equal or exceed the significant wave height, i.e., the highest one-sixth of all waves, break on the bar and contribute to the current. Q_b is given as an integral [*Bruun,* 1963, p. 1070], which *Brebner and Kamphuis* [1963, p. 10] integrate. This integrated result can be written, using Bruun's assumptions, as

$$Q_b = 0.38H_{bs}^2 \tag{10}$$

H_{bs} refers to the breaking height of the significant wave. Assuming that the laboratory case represents a spectrum so narrow that all the waves are nearly the significant wave height, the right side of (10) will be multiplied by 6 in the laboratory case.

In essence, the continuity theory [*Bruun,* 1963, p. 1074] is derived from the following assumptions. A wave breaks at an angle to the beach contributing mass to the surf zone and locally raising the mean water level in the surf zone. This gives a slope to the water surface which creates a longshore current whose velocity is controlled by the frictional head loss in the current (thus introducing energy considerations). A value for the slope S can be derived by dividing the rate of rise of the mean water level at a point on the beach ($Q_b \cos \theta_b/bT$) by the speed at which the plunge point of the breaker travels down the beach ($C_b/\sin \theta_b$) so that

$$S = (Q_b \sin 2\theta_b)/(2bC_bT) \tag{11}$$

where b is the distance between the breaker point and the shoreline. This slope is then used in the Chezy equation for open channel flow.

In order to test it here by available data, the equation has been modified by the assumption that $b = m/d_b$ and, further, by the assumption that the hydraulic radius in the Chezy equation is expressible as kd_b, where $k = 0.8$ in the field and 0.5 in the laboratory. This results in

$$V = C_f'[H_b^{3/2}m(\sin 2\theta_b)/T]^{1/2} \tag{12}$$

where C_f' is a constant (0.25 in laboratory, 0.13 in field) times the 'Chezy' coefficient C_f.

The rip current theory supposes that the breaking waves dump their water across the longshore bar into the trough and that this water runs along the trough until reaching a low point in a bar, where it flows seaward as a rip current. This leads to an equation for non-uniform longshore current which *Bruun* [1963, p. 1072] averages and formulates in terms of shear stress. However, using the same assumptions (including $\beta = 1.0$), the rip current theory can be arranged in a simpler form without including shear:

$$V = 2.31\ \lambda m(\cos\theta_b)/T \tag{13}$$

Equations 12 and 13 are expressions of the ideas presented in Bruun's paper but not stated in this form.

Another rip current theory, developed independently by *Inman and Bagnold* [1963], can be written

$$V = 2.31\ \lambda m(\cos\theta_b \sin\theta_b)/T \tag{14}$$

again employing the assumption that $\beta = 1.0$. Both rip current theories have the same form except for the addition of $\sin\theta_b$ *to* (14), arising from the assumption of Inman and Bagnold that only the alongshore motion in the breaking wave contributes to the current. Since no data on λ collected simultaneously with longshore current data have been published, neither equation 13 nor 14 can be tested, but for reasonable choices of λ, T, and m, it appears that velocity predicted by (13) is unreasonably high and that velocity given by Inman and Bagnold (14) is not impossible. The difficulty with (13) comes from the assumption that all return flow is by rip currents.

The final equation suggested by conservation of mass considerations (*TM 10*) equates the flow rate of the longshore current to an alongshore flow through the wave area, $A_w C_b \sin\theta_b$, so that after solving for V

$$V = KgmT \sin 2\theta_b \tag{15}$$

where

$$K = (1 + \beta - \sigma)/2\beta^2 \tag{16}$$

K can range from about 0.6 to about 1.1 for appropriate choices of β and σ. Because of the approximation involved in (15), $K = 1$ is used throughout this discussion. The form of (15) is unusual in that V is independent of H_b. This independence of H_b, also found in (13) and (14), seems intuitively implausible at first sight, but the mass rate of flow will still depend strongly on H_b, since H_b will determine the cross-sectional area of the current.

Empirical correlation. The equation of *Inman and Quinn* [1951], a modification of (2) obtained by introducing an empirical expression for f, is given as

$$V = \{[(1/4R^2) + \sqrt{2.28gH_b}\sin\theta_b]^{1/2} - (1/2R)\}^2 \tag{17}$$

where $R = 108.3\ mH_b \cos\theta_b/T$. The expression for f was obtained from a least-squares fit of three sets of data: the field data of Inman and Quinn, and the field data and laboratory data of Putnam et al. For reasons discussed in the previous section, only the field data of Putnam et al. seem appropriate to the correlation (see discussion of data).

Brebner and Kamphuis [1963] correctly question the assumptions necessary to obtain equations for velocity from conservation laws, and they give good reasons in favor of developing empirical equations. It is unfortunate that the two equations they developed require data for deep-water conditions, which are available in only two studies [*Saville*, 1950; *Brebner and Kamphuis*, 1963] and in practice have been computed rather than measured.

In another study involving offshore data, *Harrison and Krumbein* [1964] performed a linear multiregression analysis on unpublished data from Virginia Beach, Virginia, and found, for the 11 variables of their study, that the combination of 6 variables that causes the greatest reduction in variability is T, H_0, m, onshore wind speed, offshore wind speed, and wave direction, in that order. Other combinations of variables do very nearly as well, however, and, surprisingly, only 1 or 2% of the variability is due to wave direction. This may be explained by the fact that wave direction was measured at a point 1000–1300 feet offshore, where the relative depth varied from deep water to shallow water depending on wave period. Under such conditions the relation between measured velocity and measured direction is complicated by the unevaluated refraction effects which will vary with deep-water wave direction, wave period, and bottom topographic irregularities.

EVALUATION

Theoretical evaluation. From the previous section it is apparent that, in attempting to arrive at some solution, theoretical treatments of longshore currents have considered models that are drastic simplifications of longshore currents. It has been necessary to assume relations for wave speed, wave shape, water particle velocity, cross-sectional area of the current, velocity distribution and energy distribution within the current, and alongshore variation of the waves and current. Any one of the approximations might be plausible under proper conditions, but the necessity to use them all has produced theories that only occasionally resemble longshore currents as they actually flow.

Basic descriptive data necessary for an adequate explanation of longshore currents are still lacking. Waves on beaches may break by spilling, plunging, or surging, but the prediction of breaker types is not possible except in general terms, and the control volume approach of the theories does not distinguish among them. The magnitude and velocity distribution of the current should be affected by breaker type, since the relative amount and location of energy dissipation depends on breaker type and since the energy available for the current is determined by how much is lost in breaking.

Data are also lacking on how longshore currents fit into the nearshore current system. Most theories assume rip currents are absent, but some assume them present at a known spacing [*Bruun,* 1963; *Inman and Bagnold,* 1963]; yet the existence of rip currents and their spacing are dependent variables of the system. It has already been pointed out that the neglect of the longshore current already present in the breaking wave makes the original theory of *Putnam et al.* [1949] internally inconsistent and that an attempt to account for this velocity [*TM 10*] merely adds another dependent variable to the theory. *Sonu et al.* [1967] have recently pointed out that the whole nearshore current system, and particularly the longshore current, depends sensitively on variations in nearshore hydrography.

A theory based on conservation of energy (or momentum) flux appears to be impractical because (1) the flux assignable to the longshore current is such a small fraction of the flux in the incident waves [*TM 10*] (usually less than a tenth, often less than a hundredth); (2) energy flux in the current is not a constant fraction of the incident flux; and (3) computed flux in the unsteady, non-uniform current depends on coefficients analogous to the empirical coefficients measured in steady, uniform, open-channel flow [*Putnam et al.,* 1949; *Bruun,* 1963; *TM 10; Eagleson,* 1965].

A continuity theory might be more practical than a momentum or energy theory, since the mass flow in the current is the dominant mass flow in the surf zone. This is particularly true for low velocity flows, where observation indicates that the current velocity is merely the averaged alongshore component of the surge-induced water particle velocity. However the approximations necessary in the theories developed so far [*Inman and Bagnold,* 1963; *Brunn,* 1963; *Galvin,* 1963] are extreme.

Because of the inadequacies of physical theory, empirical correlation offers the most promise [*Brebner and Kamphuis,* 1963], but as yet such correlations have not been developed from useful data. The acquisition of reliable data is a limiting factor, and there is increasing evidence that current depends on the offshore and alongshore boundaries to such an extent that each set of data is applicable only to the peculiar conditions producing it (*Sonu et al.* [1961]; also experimental work in progress at the Coastal Engineering Research Center).

Empirical evaluation. Most theories have been put forward with experimental evidence supporting their validity, yet this review indicates that no available theory is adequate and that most available data are inappropriate tests of existing theories.

To examine this paradox, all the testable equations were evaluated with each of the eight sets of data. The six testable equations, functions only of the variables that have been measured, include the energy (1) and momentum (2) equations of Putnam et al., the asymptotic version (7) of Eagleson's equation, the modified version (12) of Bruun's 'Chezy' equation, the continuity equation (15) of *TM 10,* and the correlation equation (17) of Inman and Quinn. Velocities predicted by these equations were calculated on the CDC 1604 at the Computer Center of the University of Wisconsin. Output was obtained both in numerical form and as Calcomp plotter graphs of predicted versus measured velocity for each combination of equation and data set. Agreement between observed and predicted velocity was judged qualitatively by how closely the plotted data clustered around the 45° line between the axes.

It was necessary to supply constants for the two equations of Putnam et al. and the equation of Bruun. Ten values of the constants were chosen over a range from unreasonably low to unreasonably high values, based on the open channel analogy. The computation of these three equations was then iterated for the ten values of the constants, and best values of the constants were identified as those producing the best correlation on the plots.

It was found that every equation gave at least fair agreement with some set of data, and, conversely, except for the field data of Moore and Scholl and of Inman and Quinn, each set of data was predicted at least fairly well by some one of the equations. Best predictions come from the continuity theories of Bruun (equation 12, Figure 5) and *TM 10* (equation 15, Figure 6). These are the only theories to produce fair correlations with both the better sets of data (the field data of *Putnam et al.* [1949] and the laboratory data of *TM 10*). The laboratory data for $\theta_d = 51°$ on Figure 6 closely parallels the predicted curve but is displaced from it.

Each of the four other equations was able to correlate some set of data, although equation 17, derived empirically from three sets of data, agrees fairly well with only one of them, in the qualitative sense of this discussion. Figure 7 shows some of the better correlations that can be obtained for the other four equations. Good correlation of the Brebner-Kamphuis data can be produced by most of the equations, probably because the range of experiments was limited and θ_b and H_b were computed.

Figures 5, 6, and 7 are empirical evidence that apparent correlation between

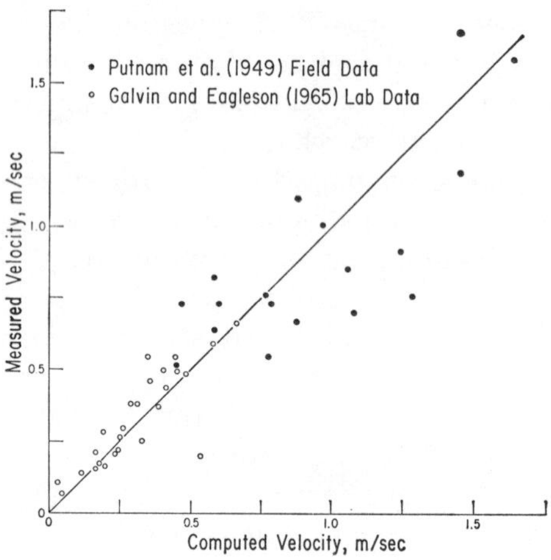

Fig. 5. Best observed agreement between a testable theory and the two usable sets of data. Theory is equation 12 with $C_f = 110\ m^{1/2}/sec$.

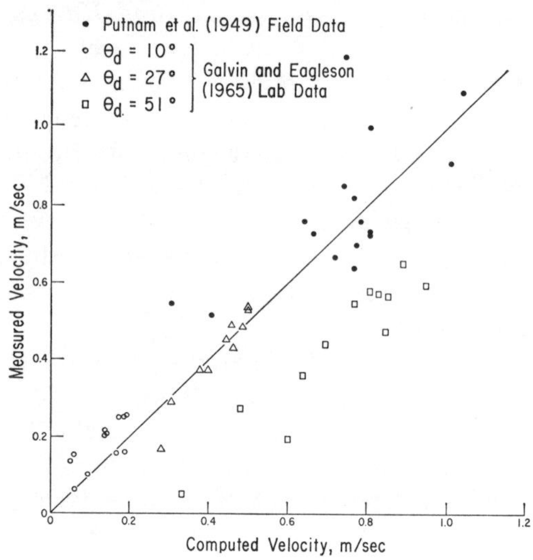

Fig. 6. Comparison between measured velocities and velocities predicted by equation 15. θ_d is the angle between the wave generator and the beach.

theory and data is no guarantee that the theory accurately describes the phenomenon. Such a correlation for some sets of data is possible because (1) there are eight sets of data each covering a limited range of possible conditions, (2) laboratory data are internally correlated, owing to systematic changes in the experimental variables, and (3) the free constant in many equations gives an added possibility of adjustment.

CONCLUSIONS

1. In spite of many studies since the pioneer paper of *Putnam et al.* [1949], there is still no adequate theoretical prediction of longshore current velocity. Previous studies include nine papers containing twelve velocity equations, of which only six can be evaluated with available data.

2. Most available data are not useful. Reliability of experimental techniques and statistics of variability are usually unknown. Available published data

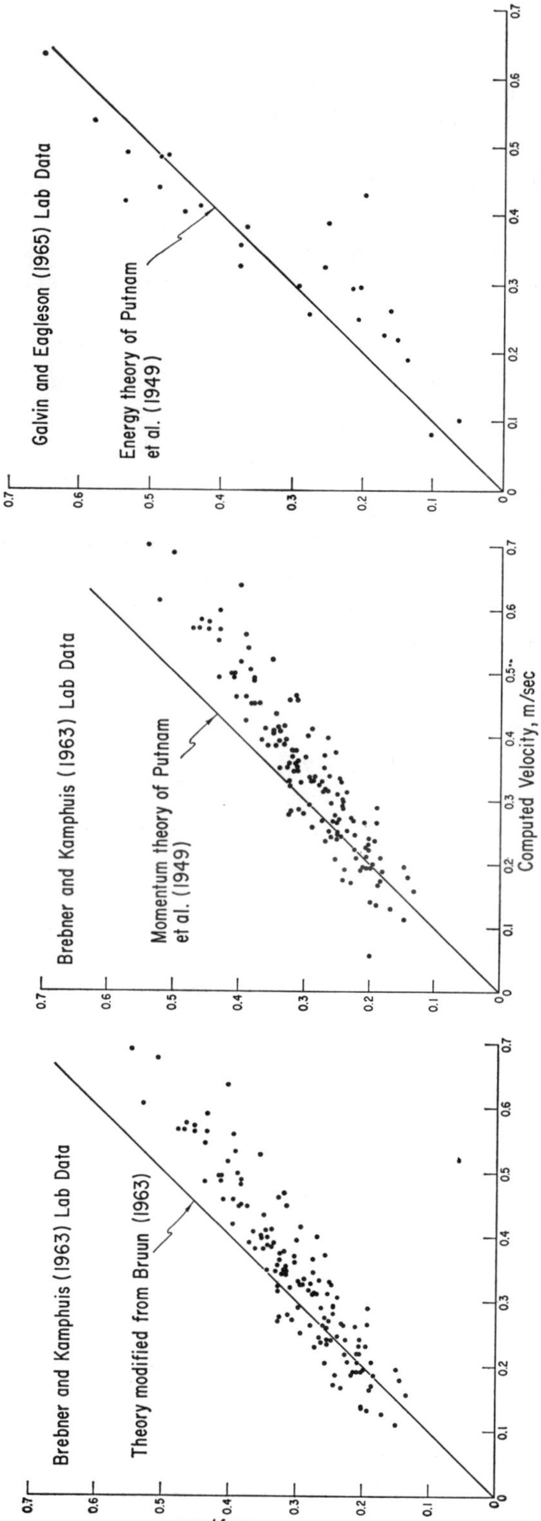

Fig. 7. Examples of misleading correlation between theory and data. *Left:* equation 12 with $C_f = 110\ m^{1/2}/sec$. *Center:* equation 2 with $f = 3.0$. *Right:* equation 1 with $s/f = 25$.

include four sets of laboratory measurements and four sets of field measurements totaling 352 observations. The better sets of data are the field data of *Putnam et al.* [1949] and the laboratory data of *TM 10*.

3. Longshore currents are considerably more complicated than the models on which theoretical studies have been based. Important factors not accounted for in the theories include breaker type, nearshore hydrography, and the nearshore current system.

4. The presently accepted theory of longshore current motion is an empirical modification [*Inman and Quinn,* 1951] of the original momentum flux equation of *Putnam et al.* [1949]. The momentum equation is shown, both conceptually and experimentally (Figure 4), to be based on an erroneous assumption, and the empirical modification, derived from three sets of data, is a fair predictor for only one of these three sets of data.

5. Demonstration of agreement between a proposed theory and a given set of data is no proof that the theory is valid. Such agreement could be found between each of the six testable equations and some one of the four laboratory sets of data, independently of the validity of the theory or the reliability of the data. Two theories based on conservation of mass (equations 12 and 15) were the only ones to produce fair agreement with both of the two better sets of data. These two equations may be taken as empirical guides to velocity prediction in lieu of a general theory.

6. On the basis of energy or momentum flux, longshore currents are a secondary feature of the surf zone, but on the basis of net mass flow they are a primary feature. This suggests that conservation of mass may be better than conservation of momentum or energy as a basis for longshore current theories.

7. At present the best approach to a meaningful prediction of longshore current velocity is through empirical correlation of reliable data.

Acknowledgments. I thank T. Saville, Jr., R. P. Savage, W. H. Munk, and the reviewers for comments on the manuscript, and R. A. Nelson for assistance in compiling the data. M. C. McClellan programmed the testing of the equations at the Mathematics Research Center, University of Wisconsin. This review developed from several investigations at the U. S. Army Coastal Engineering Research Center and an earlier experimental study done under P. S. Eagleson at the M. I. T. Hydrodynamics Laboratory. Permission of the Chief of Engineers to publish this work is acknowledged.

REFERENCES

Brebner, A., and J. W. Kamphuis, Model tests on relationship between deep-water wave characteristics and longshore currents, *Queen's Univ. Civil Eng. Res. Rept. 31,* 1–25, 1963.

Brunn, P., Longshore currents and longshore troughs, *J. Geophys. Res., 68,* 1065–1078, 1963.

Castanho, José, Wave energy and littoral transport, *Proc. Ninth Conf. Coastal Eng.,* 259–266, 1964.

Danel, P., On the limiting clapotis, *Natl. Bur. Standards Circ. 521,* 35–38, 1952.

Eagleson, P. S., Theoretical study of longshore currents on a plane beach, *M. I. T. Hydrodynamics Lab. Tech. Rept. 82,* 1–31, 1965.

Galvin, C. J., Experimental and theoretical study of longshore currents on a plane beach, doctoral thesis, M. I. T. Department of Geology and Geophysics, 1963.

Galvin, C. J., and P. S. Eagleson, Experimental study of longshore currents on a plane beach, *U. S. Army Coastal Eng. Res. Center Tech. Mem. 10,* 1–80, 1965.

Galvin, C. J., and R. A. Nelson, Compilation of longshore current data, *U. S. Army Coastal Eng. Res. Center Misc. Paper 2-67,* 1-19, 1967.

Galvin, C. J., and R. P. Savage, Longshore currents at Nags Head, North Carolina, *U. S. Army Coastal Eng. Res. Center, Bull. 2,* 1966.

Harrison, W., and W. C. Krumbein, Interactions of the beach-ocean-atmosphere system at Virginia Beach, Virginia, *U. S. Army Coastal Eng. Res. Center Tech. Mem. 7,* 1-102, 1964.

Inman, D. L., and R. A. Bagnold, Littoral processes, in *The Sea,* vol. 3, pp. 529-553, Interscience Publishers, New York, 1963.

Inman, D. L., and W. H. Quinn, Currents in the surf zone, *Proc. Second Conf. Coastal Engineering,* 24-36, Council on Wave Research, Richmond, Calif., 1951.

Iversen, H. W., Waves and breakers in shoaling water, *Proc. Third Conf. Coastal Eng.,* 1-12, Council on Wave Research, Richmond, Calif., 1953.

Le Méhauté, B., A theoretical study of waves breaking at an angle with a shoreline, *J. Geophys. Res., 66,* 495-499, 1961.

Miche, R., Movements ondulatoires de la mer, *Ann. Ponts Chaussées, 114,* 25-78, 131-164, 270-292, 369-406, 1944.

Moore, G. W., and D. W. Scholl, Coastal sedimentation in northwestern Alaska, *U. S. Geol. Surv. Trace Elements Invest. 779,* 43-65, 1961.

Munk, W. H., The solitary wave and its application to surf problems, *Ann. N. Y. Acad. Sci., 51,* 376-424, 1949.

Patrick, D. A., and R. L. Wiegel, Amphibious tractors in the surf, *Proc. First Conf. Ships and Waves,* 397-422, Council on Wave Research, Richmond, Calif. 1955.

Putnam, J. A., W. H. Munk, and M. A. Traylor, The prediction of longshore currents, *Trans. Am. Geophys. Union, 30,* 337-345, 1949.

Savage, R. P., Laboratory determination of littoral transport-rates, *J. Waterways Harbors Div., Am. Soc. Civil Engrs., 88,* 69-92, 1962.

Saville, T., Model study of sand transport along an infinitely long straight beach, *Trans. Am. Geophys. Union, 31,* 555-565, 1950.

Sonu, C. J., J. M. McCloy, and D. S. McArthur, Longshore currents and nearshore topographies, *Proc. Tenth Conf. Coastal Eng.,* pp. 525-549, Am. Soc. Civil Engrs., New York, 1967.

Ursell, F., R. G. Dean, and Y. S. Yu, Forced small-amplitude water waves: A comparison of theory and experiment, *M. I. T. Hydrodynamics Lab. Tech. Rept. 29,* 1-33, 1959.

Wiegel, R. L., *Oceanographic Engineering,* pp. 175, 325, Prentice-Hall, Englewood Cliffs, N. J., 1964.

11

Reprinted from *Jour. Geophys. Research* 74(23):5479–5490 (1969)

Rip Currents

2. Laboratory and Field Observations

ANTHONY J. BOWEN

Tidal Institute, University of Liverpool, Birkenhead, Cheshire, England

DOUGLAS L. INMAN

Scripps Institution of Oceanography, University of California
La Jolla, California 92037

The nearshore circulation of water on a plane beach exposed to a uniform wave train, normally incident on the beach, was investigated experimentally in the laboratory. The incident waves generated standing edge waves on the beach of the same frequency as the incoming waves. The interaction between these edge waves and the incident waves gave rise to steady flow patterns (nearshore circulation cells) consisting of an onshore flow toward the breakers, a longshore current in the surf zone, and an offshore flow in relatively strong, narrow rip currents. The rip currents were found to occur at alternate antinodes of the edge waves, and the spacing of the rip current was therefore equal to the longshore wavelength of the edge waves. Although the incoming wave may interact with all the possible edge wave modes of the same frequency, it was found that the interaction with one particular mode is often dominant. A useful estimate of the relative importance of the modes is given by the parameter $\omega^2 x_b/(g \tan \beta)$, where ω is the radian frequency of the edge waves, $\tan \beta$ is the beach slope, and x_b is the width of the surf zone. Field observations made in the Gulf of California strongly suggest that this mechanism is important on real beaches.

INTRODUCTION

The nearshore circulation of water caused by a longshore variation in the height of the breaking waves was investigated theoretically by Bowen (see part 1, which precedes this paper). The radiation stress arising from the excess flux of momentum due to the presence of the waves [*Longuet-Higgins and Stewart*, 1964] was found to provide driving terms for the steady flow only inside the surf zone. The circulation patterns could therefore be produced by longshore variations in the radiation stress in the surf zone. In shallow water, the radiation stress is proportional to the square of the wave height and the nearshore circulation is therefore directly related to longshore variation in breaker height, currents flowing seaward where the breaker height is low.

However, the conditions that were considered by Bowen (part 1) are those in which a sinusoidal longshore variation in the wave height has been assumed; the results obtained from the analysis were then in general agreement with field observations. There is no doubt that regular systems of rip currents are found on real beaches where there are no obvious regular variations in either the topography or the wave field. In these circumstances, the criterion that governs the spacing of the rip currents is of considerable interest.

To investigate this problem under controlled conditions, a series of observations were made in the 50 × 60 foot wave basin at the Hydraulics Laboratory of the Scripps Institution of Oceanography. The experimental arrangement is shown schematically in Figure 1. The wave generator produced a regular wave train, 52 feet wide, which approached the plane beach normally. The beach was of smooth concrete, with slope $\tan \beta = 0.075$. The working section of the beach was 24 feet wide and bounded by smooth barriers that extended seaward from the top of the beach for 16 feet. As end effects are inevitable in this type of experiment, it was decided to have well defined limits whose effect might reasonably be understood rather than to

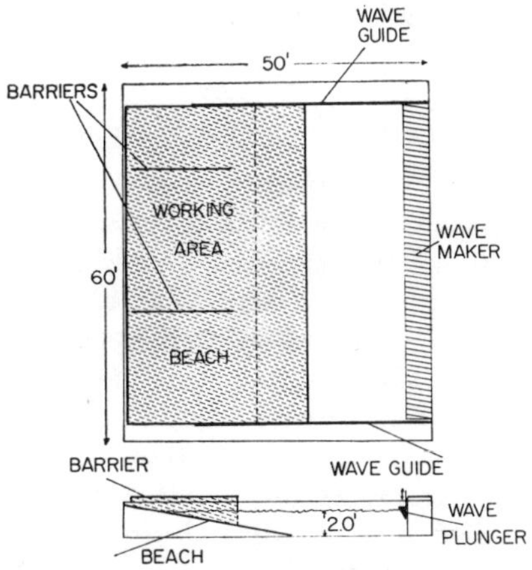

Fig. 1. Schematic diagram of the experimental arrangement in the wave tank.

use baffles and wave absorbers that create complex boundary conditions.

In theory, the set-up at the beach can be in equilibrium with the momentum input of a uniform wave train over any width of beach, provided that the conditions are uniform in the longshore direction. However, as this condition is not observed in nature and not observed even in the laboratory unless the beach width is quite small, it has been suggested that the equilibrium might not be stable. That is, a small temporary disturbance might cause a complete breakdown of the two-dimensional equilibrium. In the absence of an incoming wave train, a small disturbance in the nearshore region will presumably excite the normal modes of oscillation of the region. The normal modes of longshore oscillation, the edge waves, are therefore of interest.

Edge Waves

The theory of edge waves is well established. *Eckart* [1951], using shallow-water theory, showed that an edge wave of longshore wave number $\lambda = 2\pi/(\text{wavelength})$ and radian frequency $\omega = 2\pi/(\text{wave period})$ had a velocity potential ϕ, which is given by

$$\phi = \frac{ga_n}{\omega} \cos \lambda y \cdot L_n(2\lambda x) e^{-\lambda x} \cos \omega t \qquad (1)$$

where $\omega^2 = g\lambda(2n + 1) \tan \beta$ and where x and y are the horizontal coordinates in the offshore and longshore directions, $L_n\ (2\lambda x)$ is the Laguerre polynomial of order n, and a_n is the wave amplitude at the shoreline ($x = 0$). The offshore dependence of the wave form for low values of n is shown in Figure 2. As n is the number of zeros of the function, for higher n the curves are more oscillatory and their amplitude decreases seaward more slowly.

Ursell [1952] solved Laplace's equation in three dimensions to obtain a complicated expression for the normal modes

$$\phi = \frac{ga_n}{\omega} \cos \lambda y \cos \omega t \cdot \Big\{ \exp\left(-\lambda(x \cos \beta + z \sin \beta)\right) + \sum_{r=1}^{n} A_{rn}[\exp\left(-\lambda(x \cos (2r-1)\beta - z \sin (2r-1)\beta)\right) + \exp\left(-\lambda(x \cos (2r+1)\beta + z \sin (2r+1)\beta)\right)] \Big\} \qquad (2)$$

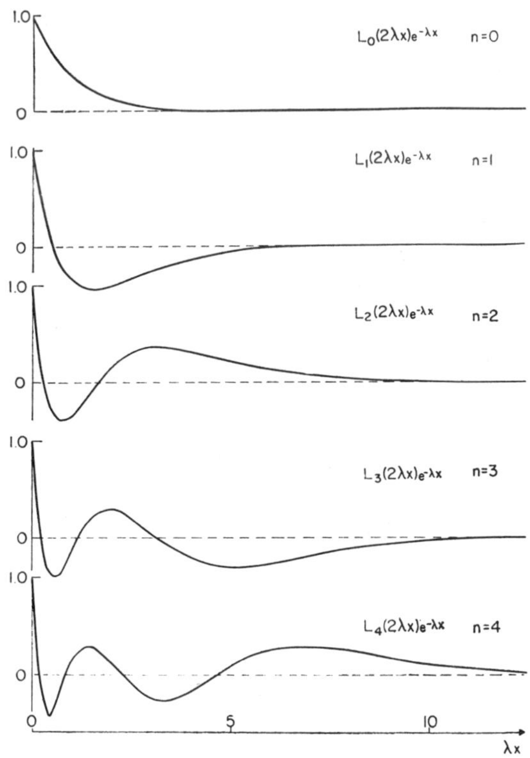

Fig. 2. The offshore dependence of the shallow water solution for the edge waves [after *Eckart*, 1951].

where

$$A_{rn} = (-1)^r \prod_{s=1}^{r} \frac{\tan (n - s + 1)\beta}{\tan (n + s)\beta}$$

and where $\omega^2 = g\lambda \sin (2n + 1)\beta$. In equation 2, z is the vertical coordinate (positive upward). When β and n are small, the two expressions for the edge wave give equivalent results. As β increases, however, the assumptions of shallow-water theory become less applicable and Ursell's formulation might be expected to give a more accurate result. Ursell's theory also suggests a cutoff at $(2n + 1)\beta = \pi/2$, and, if $\omega^2 > g\lambda$, there is a continuous spectrum of waves that are not trapped at the shoreline.

In the laboratory experiments the solid walls provide an extra boundary condition. At each wall the longshore orbital velocity v must be zero, where v is given by

$$v(x, y) = (\partial\phi/\partial y) \sim \sin \lambda y$$

If one barrier is at $y = 0$ and the other barrier is at $y = b$, then

$$v(x, 0) = 0$$

$$v(x, b) \sim \sin \lambda b = 0$$

therefore

$$\lambda b = m\pi \qquad (3)$$

where m is a positive integer. The possible longshore wavelengths are given by $L = 2\pi/\lambda = 2b/m$; the edge waves that can occur in this region therefore have two modal numbers, an offshore modal number n and a longshore modal number m. The possible resonant periods for small values of m, calculated from Ursell's dispersion relation (equation 2), are shown in Table 1.

Experimental Observations

The first experiments run in the hydraulics laboratory were ones in which a wave of long period surged up the beach rather than broke upon it. Under these conditions it was found that standing edge waves were generated that were clearly visible at the shoreline (Figure 3). The rip currents were well formed and obviously correlated with the edge waves, occurring at alternate antinodes of displacement of the waves. The edge waves and rip currents appeared to grow simultaneously and quite rapidly, the pattern being established a few wave periods after the first waves reached the beach. There was no indication of the theoretically possible two-dimensional equilibrium.

Since the edge wave was visible at the water line, the longshore modal number m could be identified. The edge wave was found to have exactly the same period as the input wave and the offshore modal number could therefore be identified by using values in Table 1. Figure 3 shows the response at 6.40 seconds. To obtain the coloration, dye was introduced close to the shore along the whole length of the working section and the successive photographs show this dye moving seaward in the rip currents. The predominant mode is $m = 3$ (and therefore $n = 0$) with the two strong rip currents at alternate antinodes. Careful measurements also showed the existence of the mode $m = 1$ (and therefore $n = 1$), which produced a small rip current flowing along the barrier at the right-hand side of the picture. The amplitude

TABLE 1. Resonant Periods of the Edge Waves ($b = 24$ feet)
From Ursell's theory; period in seconds.

	$n=0$	$n=1$	$n=2$	$n=3$	$n=4$	$n=5$	$n=6$	$n=7$	$n=8$	$n=9$	Cutoff
$m = 1$	11.1	6.45	5.04	4.31	3.86	3.59	3.36	3.22	3.14	3.29	3.06
$m = 2$	7.85	4.56	3.57	3.05	2.73	2.54	2.38	2.27			2.16
$m = 3$	6.40	3.71	2.91	2.48	2.23	2.07	1.94	1.85			1.77
$m = 4$	5.55	3.22	2.52	2.15	1.93	1.79					
$m = 5$	4.96	2.88	2.25	1.93							
$m = 6$	4.53	2.64	2.06								
$m = 7$	4.20	2.44	1.91								
$m = 8$	3.92	2.28									
$m = 9$	3.51	2.04									
$m = 11$	3.34										

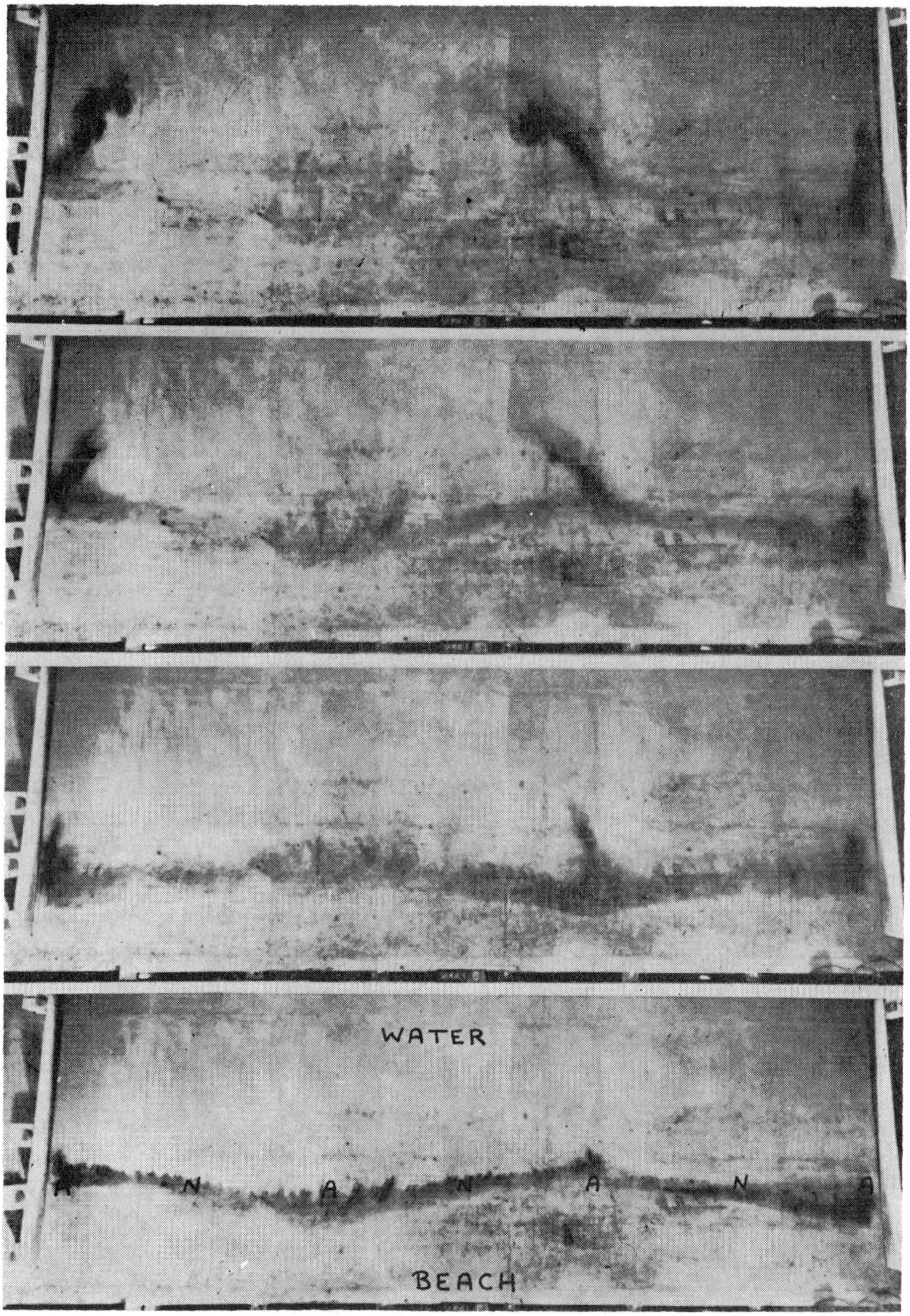

Fig. 3. Edge waves and rip currents in the wave tank for an input wave of period 6.40 seconds. The antinodes *A* and nodes *N* of the edge wave are visible at the water line on the beach.

of this mode was considerably less than that of mode $m = 3$.

The edge waves also occurred when the period of the input waves was not particularly close to one of the resonant periods given in Table 1. In fact, a driving period of 6.0 seconds produced the modes ($m = 3$, $n = 0$), ($m = 4$, $n = 0$), and ($m = 1$, $n = 1$) simultaneously, creating a very complicated flow pattern.

The next series of experiments dealt with shorter wave periods. Well formed, plunging breakers were produced on the beach. In these circumstances no edge waves were visible in the run-up, although the rip currents occurred at the theoretical positions of the alternate antinodes of edge waves having the input frequency. Measurements outside the breakers showed that edge waves were present there, but they were not visible inside the breakers.

The measurements made at the break point seem to give some insight into the mechanism of the interaction. Table 2 gives the data for an experiment at 5.00 seconds, in which the waves broke nearly uniformly except for a small edge wave that was visible in the surf. The position of the breakers is given by their distance from the still water line (SWL) on the beach. The rip current occurs at an antinode of vertical displacement of the edge wave that has a maximum negative value at the time a wave crest arrives. The next antinode along the beach is 180° out of phase; it therefore has a maximum positive value at the time of arrival of a wave crest and augments the wave height (Figure 4). As the input wave and the edge wave have the same period the situation recurs with the passage of each crest. The breakers at the antinodes are alternately large and small. The two components are approximately linearly additive only outside the break point; once the

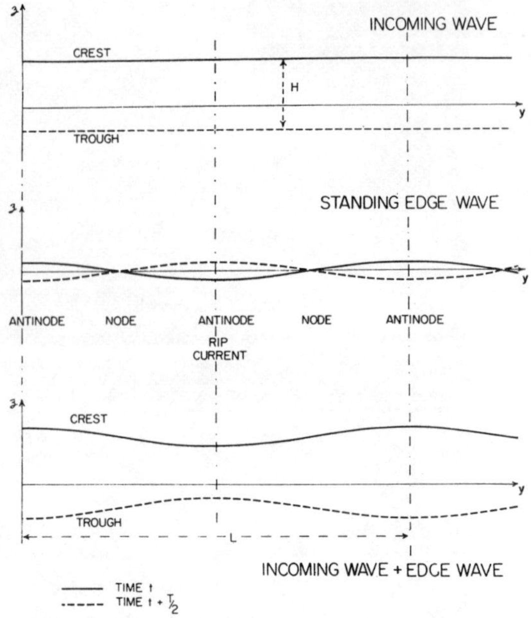

Fig. 4. The interaction near the break point between the incoming wave and a standing edge wave of the same period T.

wave has broken, the bore exists as a unit and consequently the edge wave may not be visible in the surf zone.

The edge wave thus creates a longshore variation in the breaker height that is steady in time. Thus the mechanism proposed by Bowen (part 1) can be applied directly to the interaction of incident and edge waves along a beach. There will be a flow in the surf zone from the region of high breakers to the region of low breakers, and rip currents will occur where the breakers are lowest. It can be seen in Table 2 that the wave height does not actually have longshore symmetry about the node, the wave height in the rip current being increased by the seaward flow; in this particular case the maximum rip current velocity was 5 cm/sec.

In a few of the experiments, an edge wave appeared several minutes after the start of the run and grew slowly until it was very large, usually overtopping the beach. Such an experiment, with incident wave period $T = 3.95$ seconds, is shown in Figure 5. The longshore mode was $m = 2$ with period $T = 7.90$ seconds, exactly twice that of the input period. From Table 1, this can be identified as the mode ($m = 2$, $n = 0$). It was possible to produce

TABLE 2. Observations of the Longshore Variation of Breaker Height, Incident Wave Period 5.00 Second

	Antinode Rip		Node		Antinode No Rip
Longshore distance, cm	0	72	144	216	288
Breaker height, cm	2.05	2.22	2.41	2.93	3.18
Distance of breaker from SWL, cm	34	34	34	37	40

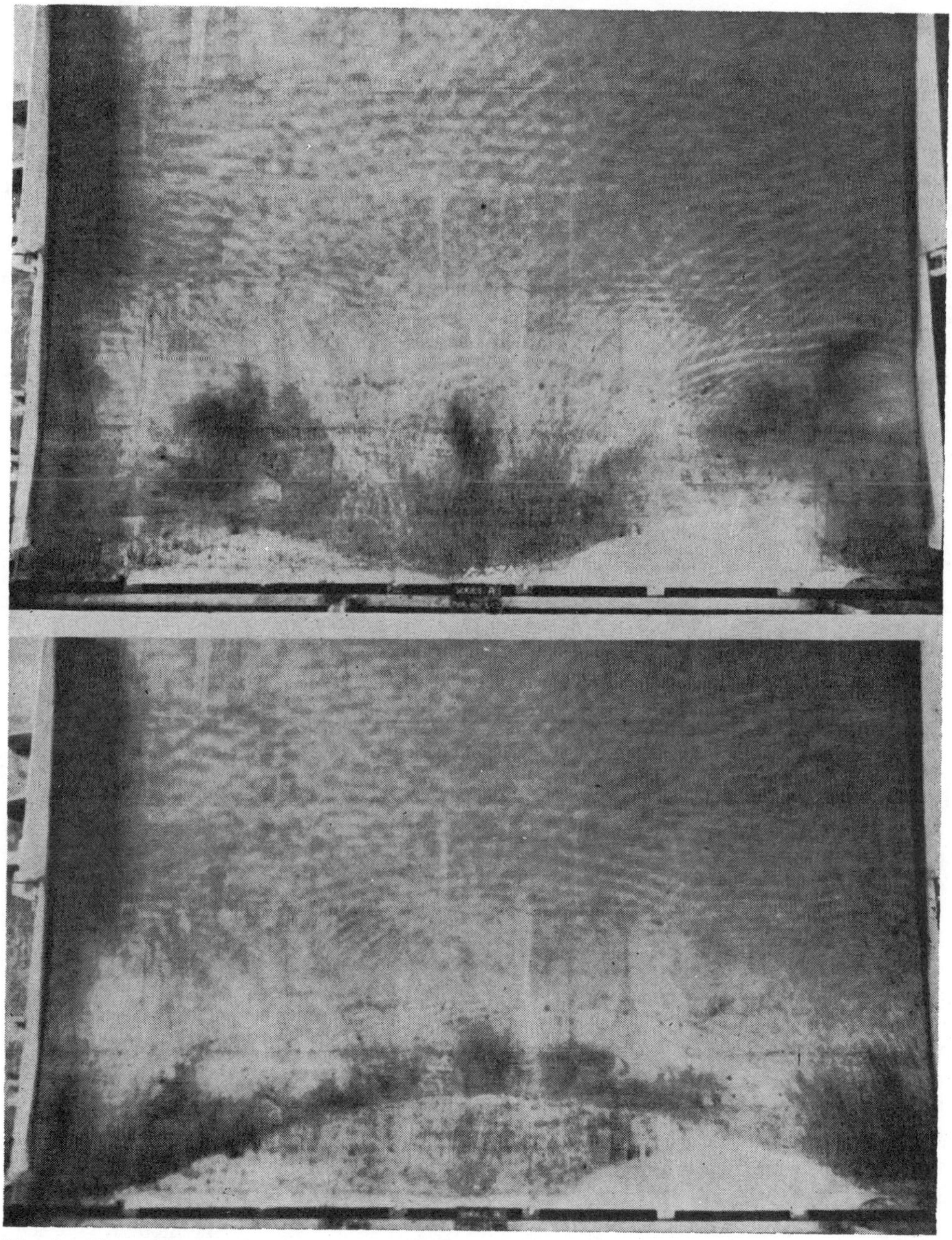

Fig. 5. The subharmonic edge wave resonance; input period 3.95 seconds, edge wave period 7.90 seconds.

other modes provided that the driving period was close to half of the resonant period. This subharmonic edge wave has been previously observed by *Galvin* [1965]. Although the amplitude of these waves grew very large, there were no associated rip currents.

When the edge wave has twice the input period, the crests of the incoming wave are alternately reinforced and reduced at a given antinode of the edge wave because the edge wave only completes half its cycle in the time between the passage of two crest. The next antinode along the shore produces the same effect, 180° out of phase with the first. Although this process may lead to an oscillation at the edge wave frequency, it does not produce any steady terms that could drive the nearshore circulation. This is in agreement with the observations, which showed that there were no circulation patterns associated with these edge waves.

If the important interaction between the incident waves and the edge waves occurs near the breakers, the edge wave amplitude at the break point must be an important parameter; this is given from (1) as

$$a_n \cos \lambda y L_n(2\lambda x_b) e^{-\lambda x_b}$$

where x_b is the width of the surf zone. The shallow-water expression is used, because Ursell's formulation (equation 2) is algebraically intractable, even for low values of n. From equation 29 of Bowen (part 1) it is possible to derive a forcing function

$$B \sin \lambda y = \frac{1}{4} g\gamma \frac{a_n \lambda}{x_b} L_n(2\lambda x_b) e^{-\lambda x_b} \sin \lambda y \qquad (4)$$

where $\gamma = H/(\bar{\eta} + h)$ is the relative wave height defined by Bowen (part 1, equation 10). The direct effect of the height of the incoming wave is to determine the width of the surf zone.

It has been observed that the spacing of the rip currents on real beaches can be roughly correlated with the width of the surf zone [e.g., *McKenzie,* 1958]. *Longuet-Higgins and Parkin* [1962] observed that the spacing of beach cusps could also be correlated with the width of the surf zone and could not be correlated with the longshore wavelength of the edge wave of mode $n = 0$. It is therefore interesting to look at circulation patterns to see how the flow varies as a function of λx_b, the ratio between the width of the surf zone and the longshore wavelength of the edge wave. The importance of edge waves in determining the spacing of rip currents was noted independently by *Bowen* [1967] and *Harris* [1967].

As it is not practical to look at the whole flow pattern as a function of λx_b, the variation of the magnitude of maximum longshore current, v_{max}, in the surf zone was calculated by using the bottom friction equations described in part 1. From the boundary conditions at the breaker line,

$$P = -\frac{B \tan^2\beta}{c\lambda^4}[2 - (2 + \lambda x_b)e^{-\lambda x_b}]$$

where c is the linearized bottom friction velocity. To simplify the problem, the set-up inside the surf zone has been neglected. The maximum longshore velocity occurs at $\sin \lambda y = 1$ and at a distance from the shoreline x_{max}. From equation 35 of Bowen (part 1)

$$\tanh \lambda x_{max} = \frac{P\lambda^4 c}{2B \tan^2\beta} = \tfrac{1}{2}[2 - (2 + \lambda x_b)e^{-\lambda x_b}] \qquad (5)$$

then, as shown in Figure 6, x_{max} is always just less than $x_b/2$, and

$$v_{max} = \frac{B \tan \beta}{c\lambda^2}\{\sinh \lambda x_{max}\,[(2 + \lambda x_b)e^{-\lambda x_b} - 2] + 2[\cosh \lambda x_{max} - 1]\}$$

Substituting for B from (4), we have

$$v_{max} = g\gamma \frac{\tan \beta}{4c} a_n L_n(2\lambda x_b)e^{-\lambda x_b} F(\lambda x_b) \qquad (6)$$

where

$$F(\lambda x_b) = \frac{1}{\lambda x_b}\{\sinh \lambda x_{max}\,[2 - (2 + \lambda x_b)e^{-\lambda x_b}] - 2[\cosh \lambda x_{max} - 1]\} \qquad (7)$$

$F(\lambda x_b)$ and λx_{max} are shown in Figure 6.

Figure 7 shows the nondimensional velocity maximum given by

$$V = \frac{4cv_{max}}{g\gamma \tan \beta a_n} = L_n(2\lambda x_b)e^{-\lambda x_b} F(\lambda x_b) \qquad (8)$$

as a function of λx_b, for the three lowest modes, $n = 0, 1, 2$. For the mode $n = 0$ the maximum value of V occurs when $\lambda x_b \simeq 1$, $(L/x_b \sim 6)$; for mode $n = 1$ the maximum is at $\lambda x_b \simeq 2$, $(L/x_b \sim 3)$; and for the mode $n = 2$ the maximum value occurs when $\lambda x_b \simeq 3$, $(L/x_b \sim 2)$. Field observations of rip current spacing have generally shown $L/x_b \sim 4$ but with a wide scatter from 1.5 to 8.0. As the measurement of spacing tends to emphasize the shorter wavelengths and therefore the lower modes, e.g., from (1)

$$L = (gT^2/2\pi)(2n + 1) \tan \beta$$

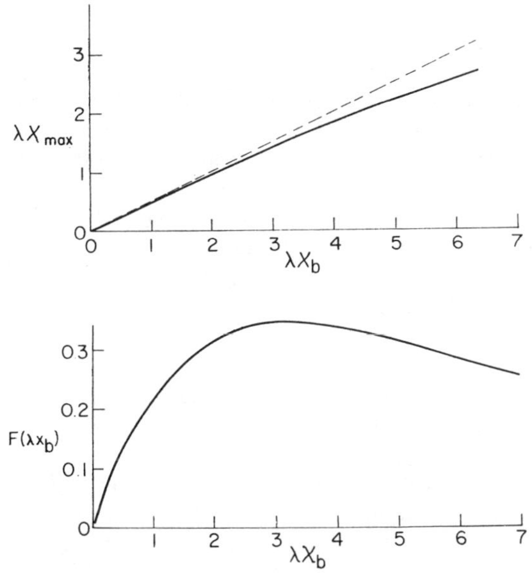

Fig. 6. x_{max}, the position of the maximum longshore current, and F (λx_b) as functions of λx_b.

where T is the edge wave period, the theory and observations are in general agreement. The nondimensional velocity V is dependent on the particular mode selected, because the term a_n, the amplitude of the edge wave at the shoreline, varies with the mode of the edge wave. Considering all possible modes, the maximum value of V is found to be least for mode $n = 0$ and largest for mode $n = 1$, the curve $F(\lambda x_b)$ peaking at $\lambda x_b = 3$ and decaying for larger λx_b (Figures 6 and 7). This prevents the higher modes from achieving a large velocity at some large value of λx_b.

If one considers the effect of variations in wave height with time when the period is kept constant, some interesting possibilities appear for determining the most likely edge wave mode. From the dispersion relation for the edge waves,

$$\lambda x_b = \omega^2 x_b/[g(2n + 1) \tan \beta]$$

where $\omega^2/(g \tan \beta)$ is a constant for a particular experiment. The offshore length scales in Figure 7 are then deceptive as λ is a function of the offshore modal number n. In Figure 8 the same curves have been replotted as a function of the nondimensional width of the surf zone, $\omega^2 x_b/(g \tan \beta)$.

There seems to be a possibility of determining which mode an incoming wave of given height will excite. The smallest waves, giving a low value of x_b, will produce a flow dominated by the mode $n = 0$; for example, the flow at $T = 6.40$ seconds shown in Figure 3 has a value of $\omega^2 x_b/(g \tan \beta)$ of 0.42. The transition from mode $n = 0$ to $n = 1$ with increasing $\omega^2 x_b/(g \tan \beta)$ was investigated in the laboratory. Table 3 shows the pertinent results, which are in good agreement with the suggested criterion. An interesting result occurred for $T = 3.55$ seconds, which was run for the case $\omega^2 x_b/(g \tan \beta) = 2.40$. The dominant mode was ($m = 2$, $n = 2$), occurring presumably because there is no resonance of mode $n = 1$ near this period (see Table 1). For higher values of $\omega^2 x_b/(g \tan \beta)$ the mode $n = 0$ was never visible. However, the other modes tended to occur simultaneously; the dominance of one mode was rarely clear and the rip current patterns were rather confused.

It is encouraging that some interesting results appear in the last section without reference to the value of the amplitude, a_n, of the edge wave at the shoreline. It was tacitly assumed that a_n was roughly constant for all modes; however, the observed amplitudes of the edge waves were not equal, the stronger

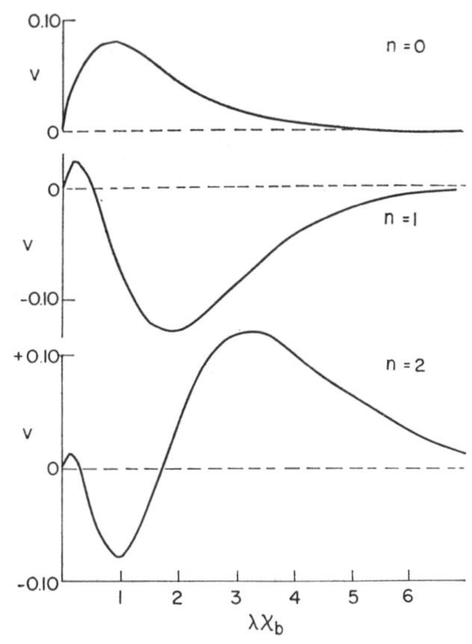

Fig. 7. The nondimensional maximum longshore currents V as a function of λx_b for edge wave modes $n = 0, 1, 2$.

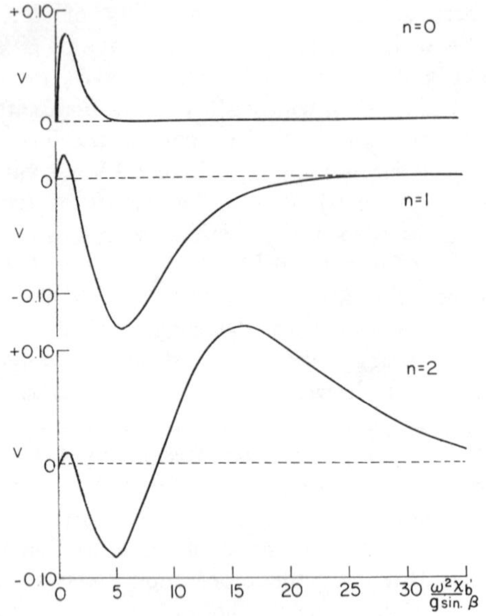

Fig. 8. The nondimensional maximum longshore current V as a function of $\omega^2 x_b/(g \tan \beta)$, for edge wave modes $n = 0, 1, 2$.

flows being associated with larger edge waves. As the estimate of which flow pattern will be dominant seems to be correct, this suggests that the edge wave is generated by a resonant interaction involving the steady flow. Since the flow pattern that would be dominant if all the possible edge wave modes had the same amplitude will actually be associated with the edge wave of largest amplitude, this flow pattern will be even more dominant than the preceding theory suggests. The resulting equilibrium amplitude a_n is presumably a complex function of the intensity of the steady currents.

Although it seems reasonable that a normally incident wave train should excite a longshore mode that has a resonance near the input period, the precise mechanism of the interaction is not clear. There is an apparent correlation between the amplitude of the edge wave at the shore and the magnitude of the standing wave caused by the wave reflection at the beach. When the waves surge up the beach and there is a large reflected wave, large edge waves are visible at the water line. As soon as the wave begins to break with a clear plunge, however, the edge waves can no longer be observed in the surf. When the wave breaks cleanly, not only is the reflection coefficient reduced, but also a considerable part of the reflection takes place at the break point.

This may be the same problem as that of the generation of transverse waves in a wave flume with a flat bottom. However, solutions for even this, apparently simpler, case are not available. The correlation of the amplitude of the standing edge wave with the standing wave due to the reflection does again suggest the possibility of a resonant interaction. The reflected wave has frequency ω and zero longshore wave number and the edge wave has a frequency ω and longshore wave number λ; together they produce a steady flow, frequency zero, and longshore wave number, λ. If the steady flow or a steady change in the sea surface can interact with the incoming wave to drive the edge wave, the steady effect provides the term $\cos \lambda y$; the reflected wave, the term $\cos \omega t$. The difficulty occurs in finding a composite function of x that resembles $L_n(2\lambda x)e^{-\lambda x}$. The question remains as to how well the x dependencies must match before energy can be effectively transferred into an edge wave mode.

Even if the growth mechanism of the edge waves were more fully understood, the problem of calculating the equilibrium amplitude would still be formidable. Unfortunately, the equilibrium amplitude a_n is a most important parameter.

TABLE 3. Observations of the Relative Importance of the Possible Modes

Wave Period, sec	Dominant Mode	Minor Mode	$\dfrac{\omega^2 x_b}{g \tan \beta}$
3.71	$(m = 9, n = 0)$	$(m = 3, n = 1)$	2.00
3.25	$(m = 4, n = 1)$	$(m = 12, n = 0)$	2.85
2.92	$(m = 5, n = 1)$	None	3.85

Field Observations

Several field observations that give good qualitative descriptions of the nearshore circulation patterns were mentioned by Bowen (part 1). These observations have provided some information about the important processes taking place in the nearshore region and have been used to check the general characteristics of the flow patterns derived from the theoretical models. However, few of these observations were accompanied by detailed measurements. To determine whether the conditions observed in the wave basin ever occur in the field, one needs good observations in a region of simple topography. Unfortunately, the measurements of *Shepard and Inman* [1950, 1951] were made at La Jolla, California, where there are rather extreme variations in the offshore topography.

Therefore, a series of field trips were made to El Moreno Beach on the northwest coast of the Gulf of California. This beach is on a sand spit that extends north from Point San Felipe for many miles. The beach has a steep face ($\tan \beta = 0.148$), rising above a nearly horizontal low tide terrace.

The power spectra of the incoming waves were measured from an array of digital wave sensors placed in and near the surf zone following a procedure outlined by *Koontz and Inman* [1967]. The breaker height and depth, the width of the surf zone, and the spacing of the rip currents were also measured. The results given in Table 4 include three cases for which the rip current spacing was rather regular. In each of these cases (DAS 73, 76, 79) the spacing is in good agreement with the calculated longshore wavelength of an edge wave of the input period of mode $n = 1$. The other experiment, DAS 77, is one in which the spacing is not regular, but the mean spacing is of the order of the wavelength of the mode $n = 1$.

The power spectra at two staffs, 40 feet from the beach and 20 feet apart, are shown in Figure 9. The wind wave peak and the tidal peak are clearly shown. The fact that the peak energy is the same at both staffs is not surprising as the whole rip current system moves along the shore on a net longshore current, and differences in wave height due to the presence of edge waves cannot be detected.

As the beach at El Moreno is very long in comparison with the spacing of the rip currents, there are presumably no end effects, and the situation may differ in two ways from the laboratory experiment. First, if an edge wave of the input frequency is generated, all offshore

TABLE 4. Field Measurement at El Moreno, October 1966

	DAS 73	DAS 76	DAS 77	DAS 79
Wave period, sec	3.50	3.25	3.20	3.00
Breaker height, feet	1.2	1.0	0.6	1.0
Breaker angle	16°	21°	15°	10°
Width of surf zone, feet	10	8.0	6	8.0
Rip current spacing, feet	29.5	28.5	26.0	23.5
	29.5	22.0	21.5	22.0
	33.5	26.5	33.5	22.5
	34.0	21.5	16.0	18.5
	24.0	25.5	20.0	21.0
	33.0	21.0	36.5	20.5
	28.5	23.0	21.0	20.5
		26.0	24.5	20.0
		25.0	18.0	
			18.3	
Mean spacing	30.5	24.5	23.5	21.0
Theoretical Wavelength of Edge Wave				
Mode $n = 0$, feet	9.4	8.0	7.8	7.0
Mode $n = 1$, feet	28.1	24.0	23.3	20.8
Mode $n = 2$, feet	46.8	40.0	39.0	34.8
$\omega^2 x_b/(g \sin \beta)$	6.7	6.2	4.8	7.3

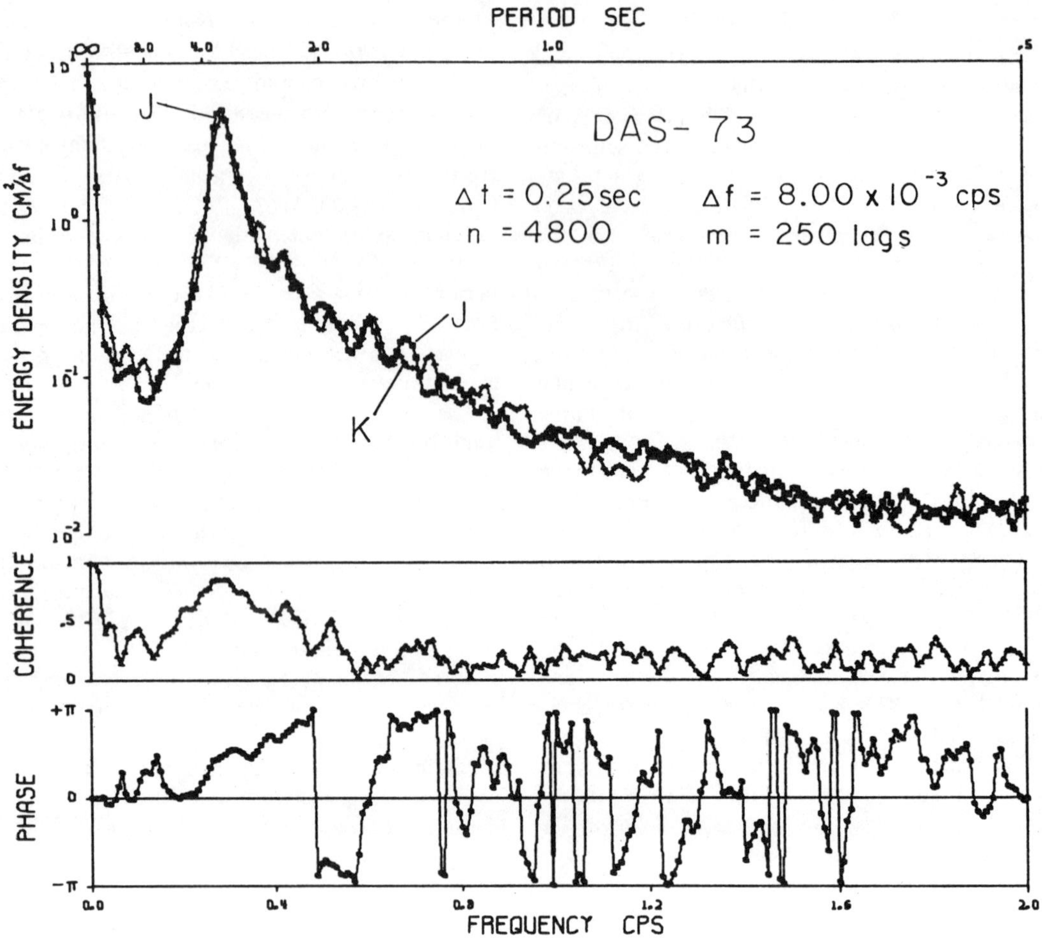

Fig. 9. Wave spectra and cross-spectral analysis for two members of an array of digital wave staffs placed near the surf zone at El Moreno Beach, Mexico. Run DAS 73.

modes n are possible, as there is no longshore restriction on the wavelength. Second, it is no longer necessary for the edge waves to be standing waves; they may equally well be progressive.

If the incoming wave and a progressive edge wave are of the same period, they interact to produce a steady variation in the wave height in much the same way as an incoming wave and a standing edge wave. If an incoming wave crest and an edge wave crest coincide at the break point at a given time, one period later the next wave crest and the edge wave crest one wavelength behind the first will meet at the same point. The interaction can therefore be steady.

The agreement between the spacing of rip currents and the modal wavelength suggests that an edge wave interaction is indeed taking place. The value of $\omega^2 x_b/(g \tan \beta)$ is also in good agreement with the suggested modal number. As there is little longshore variation in the bottom topography and the power spectra show no major peaks other than the fundamental, all the evidence suggests that edge waves of the input frequency are formed and do interact with the incoming waves to produce circulation patterns and rip currents. The generality of the result is extended by the fact that the waves did not approach the beach normally in these experiments.

It appears from field observations that the

spacing of rip currents on very flat beaches is at times related to the interaction of the edge wave with a beat of the incident waves, rather than with the incident waves themselves. For example, the spacing between rip currents at Silver Strand Beach, Coronado, California, may be very long and at times may exceed $2\pi g/\sigma^2$, where σ is the radian frequency of the energy peak of the incident waves.

The emphasis in the study has been on the flow of water in the nearshore region. It is clear, however, that the combined flow associated with the incoming waves, the edge waves, and the nearshore circulation may rearrange the beach sediment to produce a regular longshore pattern of beach cusps. The slow onshore flow between the rip currents will carry the sediment, put into motion by the wave action, onshore and deposit it in the surf zone. The seaward flowing rip current will tend to erode a channel for itself. The results should be a system of beach cusps, which on steep beaches have the same longshore wavelength as the nearshore circulation and the predominant edge wave.

Where the beach slope is small and the surf zone is wide, however, cusps that form on the inner beach face are usually small as compared with the spacing between rip currents. In this case, it seems likely that the formation of cusps is in response to an interaction of the reformed wave, associated with the bore of the breaking wave, and edge waves, traveling within the surf zone, which have a smaller modal number n than that of the predominant flow pattern. These flow patterns may make no significant contribution to the over-all circulation. Small beach cusps that may be of this type are often observed on the beach at La Jolla, California.

Acknowledgments. This study was supported by the Coastal Engineering Research Center, U. S. Army, and the Office of Naval Research, under contract with the University of California.

References

Bowen, A. J., Rip currents, Ph.D. thesis, University of California at San Diego, 115 pp., 1967.

Eckart, C., Surface waves in water of variable depth, *Univ. Calif., Scripps Inst. Oceanogr. Wave Rep. 100, S10 Ref 51-12,* 99 pp., 1951.

Galvin, C. J., Resonant edge waves on laboratory beaches (abstract), *Trans. Amer. Geophys. Union, 46*(1), 112, 1965.

Harris, T. F. W., Field and model studies of the nearshore circulation, Ph.D. thesis, University of Natal, Durban, South Africa, 188 pp., 1967.

Koontz, W. A., and D. L. Inman, A multipurpose data acquisition system for field and laboratory instrumentation of the nearshore environment, *Coastal Eng. Res. Center, Tech Memo 21, U.S. Army Corps Engrs.,* 38 pp., 1967.

Longuet-Higgins, M. S., and D. W. Parkin, Sea waves and beach cusps, *Geogr. J., 128,* 194–201, 1962.

Longuet-Higgins, M. S., and R. W. Stewart, Radiation stress in water waves: a physical discussion with applications, *Deep-Sea Res., 11,* 529–563, 1964.

McKenzie, R., Rip current systems, *J. Geol., 66,* 103–113, 1958.

Shepard, F. P., and D. L. Inman, Nearshore circulation related to bottom topography and wave refraction, *Trans, Amer. Geophys. Union, 31*(4), 555–565, 1950.

Shepard, F. P., and D. L. Inman, Nearshore circulation, *Proc. 1st Conf. Coastal Eng.,* pp. 50–59, University of California, Council on Wave Research, 1951.

Ursell, F., Edge waves on a sloping beach, *Proc. Roy. Soc., London, A, 214,* 79–97, 1952.

12

Reprinted from *Jour. Geophys. Research* 75(33):6778-6789 (1970)

Longshore Currents Generated by Obliquely Incident Sea Waves, 1

M. S. LONGUET-HIGGINS[1]

Oregon State University, Corvallis 97331

By using known results on the radiation stress associated with gravity waves, the total lateral thrust exerted by incoming waves on the beach and in the nearshore zone is rigorously shown to equal $(E_0/4) \sin 2\theta_0$ per unit distance parallel to the coastline, where E_0 denotes the energy density of the waves in deep water and θ_0 denotes the waves' angle of incidence. The local stress exerted on the surf zone in steady conditions is shown to be given by $(D/c) \sin \theta$ per unit area, where D is the local rate of energy dissipation and c is the phase velocity. These relations are independent of the manner of the energy dissipation, but, because breaker height is related to local depth in shallow water, it is argued that ordinarily most of the dissipation is due to wave breaking, not to bottom friction. Under these conditions the local mean longshore stress in the surf zone will be given by $(5/4)\rho u_{max}^2 s \sin \theta$, where ρ is the density, u_{max} is the maximum orbital velocity in the waves, s is the local beach slope, and θ is the angle of incidence. It is further shown that, if the friction coefficient C on the bottom is assumed constant and if horizontal mixing is neglected, the mean longshore component of velocity is given by $(5\pi/8)(s/C)\, u_{max} \sin \theta$. This value is proportional to the longshore component of the orbital velocity. When the horizontal mixing is taken into account, the longshore currents observed in field observations and laboratory experiments are consistent with a friction coefficient of about 0.010.

1. INTRODUCTION

It is well known [*Wiegel*, 1963; *Inman and Bagnold*, 1963] that when sea waves or swell approach a straight coastline at an oblique angle (Figure 1) a mean current tends to be set up parallel to the coastline. Such longshore currents and the associated longshore transport of sand or other sedimentary material are of prime importance for both the coastal engineer and the submarine geologist.

Many hypotheses, of a very rough kind, have been advanced to account for this phenomenon. However, a recent review of the subject by *Galvin* [1967] arrives at the justifiable conclusion that, 'A proven prediction of longshore current velocity is not available, and reliable data on longshore currents are lacking over a significant range of possible flows.'

It has often been suggested [e.g., *Putnam et al.*, [1949] that the magnitude of the longshore current is related in some way to the energy or the momentum of the incoming waves. Of these two approaches, that employing momentum is the more promising since momentum is conserved, whereas energy can be dissipated by breaking and other processes not immediately associated with sediment transport.

It goes without saying that any momentum theory must be correctly formulated. The estimate of the momentum made by *Putnam et al.* [1949] has been already criticized on theoretical grounds by *Galvin* [1967]. Moreover, *Inman and Quinn* [1952] showed that, in order to make the theory fit the observations, the friction coefficient C would have to be assumed to vary over a wide range of 3½ orders of magnitude. A version of the theory of Putnam et al. modified by *Galvin and Eagleson* [1965] also requires a large variation in C.

The aim of this paper is to introduce a more satisfactory estimate of the momentum of the incoming waves, which is based on the concept of the radiation stress as developed by *Longuet-Higgins and Stewart* [1960, 1961, 1962, 1963, 1964]. This estimate of the excess transfer of momentum due to the waves has already proved remarkably successful in the prediction of several wave phenomena, particularly the setup, or change in mean level of the sea sur-

[1] Now at National Institute of Oceanography, Wormley, Godalming, England, and Department of Applied Mathematics and Theoretical Physics, Silver Street, Cambridge, England.

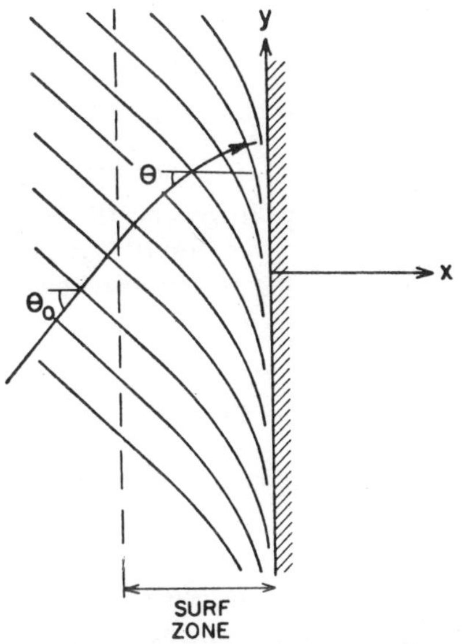

Fig. 1. Definition diagram for waves approaching a straight shoreline at an oblique angle.

face in the breaker zone [*Longuet-Higgins and Stewart*, 1963, 1964; *Bowen*, 1967].

In the present paper it is pointed out, first, that there exists a simple and precise relationship between the total longshore thrust exerted by the incoming waves on the one hand and their direction and amplitude in deep water on the other (see equation 10). This result can be derived either from the concept of the radiation stress mentioned earlier, or by a direct evaluation of the momentum flux due to the waves.

Next it is shown that the local longshore stress due to the waves is very simply related to the local rate of dissipation of wave energy, regardless of whether the dissipation is due to wave breaking or to bottom friction. Hence, using the known relation of breaker height to local depth in the surf zone, one can estimate accurately the local longshore stress due to the waves (section 4).

When the local longshore wave stress is known, it is possible to write an equation of motion for the longshore current that involves in general both the bottom friction and the horizontal mixing by turbulent eddies. If the horizontal mixing is negligible, the momentum balance gives an exceedingly simple expression for the longshore current $\langle v \rangle$. The addition of horizontal mixing generally reduces the current, although not drastically.

A comparison with the available data (section 7) shows that even without the assumption of mixing there is already an order- of-magnitude agreement between the observed and the theoretical current if one takes an a priori estimate of the friction coefficient (about 0.010) based on experiments with flow in rough pipes [*Prandtl*, 1952]. The comparison indicates also that horizontal mixing is significant, though not dominant, in most circumstances.

Note added in processing. Since this paper was prepared, a somewhat similar approach to the theory of longshore currents has been published by *Bowen* [1969]. Besides containing new results, the present treatment differs both in the derivation of equation 34 (since Bowen takes θ to be constant during differentiation) and in the assumed form of the bottom friction. For further comparisons see the companion paper.

2. Waves Approaching Coastline

Imagine a straight coastline, as in Figure 1, in which the local still water depth h is some function of the coordinate x normal to the shoreline and is independent of the distance y along the shore. The shoreline itself is at $x = 0$. A train of two-dimensional waves of amplitude a is advancing from deep water toward the coast, the local direction of propagation being inclined at an angle of incidence θ to the normal, as shown.

Both θ and a will vary with the distance $|x|$ from the shoreline. If σ denotes the frequency of the waves and k denotes the local wave number, Snell's law, which expresses the constancy of the wave number in the direction parallel to the shoreline, can be written as

$$k \sin \theta = \text{constant} \tag{1}$$

or equivalently

$$(\sin \theta)/c = \text{constant} \tag{2}$$

where $c = \sigma/k$ denotes the local phase velocity. If the bottom slope is gradual, so that the proportional change in depth over one wavelength is small, it is reasonable to assume that σ and k are related to the local depth $h(x)$ by the Stokes relation for waves of small amplitude:

$$\sigma^2 = gk \tanh kh \tag{3}$$

The phase velocity c is then given by

$$c = \sigma/k = [(g/k) \tanh kh]^{1/2} \tag{4}$$

and the group velocity, or velocity of energy propagation, is given by

$$c_g = \frac{d\sigma}{dk} = \frac{\sigma}{2k}\left(1 + \frac{2kh}{\sinh 2kh}\right) \tag{5}$$

With the local energy density per unit horizontal area being given by

$$E = \tfrac{1}{2}\rho g a^2 \tag{6}$$

correct to second order, the flux of energy toward the coast, per unit distance parallel to the shoreline, is given by

$$F_x = E c_g \cos \theta \tag{7}$$

If the waves are losing no energy by breaking, bottom friction, or otherwise, we have

$$F_x = \text{constant} \tag{8}$$

independently of x, from which one can deduce the law of variation of the wave amplitude a with distance offshore [*Burnside*, 1915; *Longuet-Higgins*, 1956]. Inside the breaker zone, however, some energy will be lost, and hence a will diminish toward the shoreline and become zero at or near $x = 0$. If D denotes the rate of dissipation of wave energy, either by breaking or friction, we have identically

$$\partial F_x/\partial x = -D \tag{9}$$

3. Radiation Stresses

So much is well accepted. We propose now to calculate the force exerted on the nearshore region by the incoming waves, by using the notion of the radiation stress, as introduced by *Longuet-Higgins and Stewart* [1960, 1961, 1962, 1963, 1964].

It can be shown [*Longuet-Higgins and Stewart,* 1960] that the presence of a wave train of amplitude a in water of depth h increases the flux of momentum parallel to the direction of propagation across any plane normal to that direction by an amount

$$S_{11} = E\left(\frac{1}{2} + \frac{2kh}{\sinh 2kh}\right) \tag{10}$$

Similarly the flux of momentum normal to the direction of wave propagation across a plane parallel to the direction of propagation is increased by an amount

$$S_{22} = E(kh/\sinh 2kh) \tag{11}$$

where $E = \frac{1}{2}\rho g a^2$. In general the momentum flux tensor, referred to coordinates (ξ_1, ξ_2) parallel and perpendicular to the direction of wave propagation, is given by

$$S_{ij} = \begin{bmatrix} E\left(\frac{1}{2} + \frac{2kh}{\sinh 2kh}\right) & 0 \\ 0 & E\frac{kh}{\sinh 2kh} \end{bmatrix} \tag{12}$$

the off-diagonal elements being zero.

Now let us calculate the flux of y momentum parallel to the shoreline across a plane $x =$ constant, parallel to the shoreline. Since the axes (x, y) are inclined at an angle θ to the principal axes (ξ_1, ξ_2) of the waves, we have

$$\begin{aligned} S_{xy} &= \sum_{i,j} S_{ij} \frac{\partial x}{\partial \xi_i}\frac{\partial y}{\partial \xi_j} \\ &= S_{11} \sin\theta \cos\theta + S_{22} \cos\theta(-\sin\theta) \\ &= E\left(\frac{1}{2} + \frac{kh}{\sinh 2kh}\right) \cos\theta \sin\theta \\ &= E(c_g/c) \cos\theta \sin\theta \end{aligned} \tag{13}$$

By (7) this relation can be written as

$$S_{xy} = F_x(\sin\theta)/c \tag{14}$$

or, if we make use of Snell's law in the form of (2), we then have

$$S_{xy} = F_x(\sin\theta_0)/c_0 \tag{15}$$

where θ_0 and c_0 refer to the (constant) values of θ and c in deep water.

This very simple and exact relation states that the flux of y momentum across the plane $x =$ constant is proportional, by a fixed, known constant, to the energy flux across the same plane.

Because of the simplicity and fundamental importance of relation 15 we give here an alternative proof.

The flux of y momentum across any vertical plane $x =$ constant is simply equal to ρuv, where u and v are the components of velocity in the x and y directions. On integrating this with respect to the vertical coordinate z we find

$$S_{xy} = \left\langle \int_{-h}^{\zeta} \rho uv \, dz \right\rangle \quad (16)$$

The angle brackets denote the mean value with respect to time. Now for waves traveling at an angle θ to the x axis we have

$$u = u_1 \cos \theta \qquad v = u_1 \sin \theta \quad (17)$$

where u_1 denotes the horizontal component of the orbital velocity in the direction of wave propagation. Also in (16) the upper limit of integration can be replaced by the mean value $z = 0$, since the difference $\int_0^\zeta \rho uv \, dz$ is only of the third order at most in the wave amplitude. (The mean value is actually of fourth order.) We have then, correct to second order,

$$S_{xy} = \int_{-h}^{0} \rho \langle u_1^2 \rangle \, dz \cos \theta \sin \theta \quad (18)$$

Now the flux of energy in the direction of wave propagation is given by

$$F = \left\langle \int_{-h}^{\zeta} [p + \tfrac{1}{2}\rho(u_1^2 + v_1^2)]u_1 \right\rangle dz \quad (19)$$

So to the same order of approximation

$$F = \int_{-h}^{0} \langle p u_1 \rangle \, dz \quad (20)$$

From the linearized equation of horizontal momentum, however, we have

$$\frac{\partial u_1}{\partial t} = -\frac{1}{\rho}\frac{\partial p}{\partial \xi_1} = \frac{1}{\rho c}\frac{\partial p}{\partial t} \quad (21)$$

since in progressive wave motion $\partial/\partial t \sim c \, \partial/\partial \xi_1$. Then on integration with respect to time we have

$$u_1 = (p/\rho c) + \text{constant} \quad (22)$$

On substituting in (18) and noting that for irrotational waves $\langle u_1 \rangle = 0$ correct to second order we obtain from (18) and (20)

$$S_{xy} = (1/c)F \cos \theta \sin \theta \quad (23)$$

The energy flux F_x being equal to $F \cos \theta$, we obtain (14) and hence (15) as before.

From (15) we can at once calculate the total longshore thrust of the waves, as follows.

Outside the breaker line (or the line at which energy losses become significant) we have

$$F_x = \text{constant} = E_0(\tfrac{1}{2}c_0) \cos \theta_0 \quad (24)$$

c_0 being the phase velocity in deep water, where the group velocity $c_g = \tfrac{1}{2}c_0$, and E_0 being the energy density in deep water. Therefore from (15)

$$(S_{xy})_\infty = \tfrac{1}{2}E_0 \cos \theta_0 \sin \theta_0 \quad (25)$$

On the other hand, at the shoreline $x = \delta > 0$ (just beyond the reach of the waves) we have

$$F_x = 0 \qquad S_{xy} = 0 \qquad x = \delta \quad (26)$$

Therefore, by considering the balance of momentum of the water between the breaker line and the shoreline, we see that the total external force G_y parallel to the shoreline acting on the water and sediment inside the breaker zone is given by

$$(S_{xy})_\infty + G_y = 0 \quad (27)$$

In the absence of wind or other surface stresses the only external force must come from bottom friction. Hence the total lateral littoral force exerted by the waves on the bottom is given by $H_y = -G_y$, that is to say

$$H_y = \tfrac{1}{4}E_0 \sin 2\theta_0 \quad (28)$$

It is interesting that the force is a maximum, for a given wave amplitude at infinity, when $\sin 2\theta_0 = 1$ or $\theta_0 = 45°$.

4. Local Wave Stress

Inside the breaker zone F_x gradually diminishes toward the shoreline. A consideration of the momentum balance between two planes $x = x_1$ and $x_1 + dx$ parallel to the shoreline and separated by a distance dx shows at once that the net stress τ_y per unit area exerted by the waves on the water in the surf zone is given by

$$\tau_y = -\partial S_{xy}/\partial x \quad (29)$$

and by (15) this equation becomes

$$\tau_y = -\frac{\partial F_x}{\partial x}\left(\frac{\sin \theta_0}{c_0}\right) = D\left(\frac{\sin \theta_0}{c_0}\right) \quad (30)$$

where D denotes the local rate of energy dissipation. In other words, the local stress exerted by the waves is directly proportional to the local rate of dissipation of wave energy. Outside the breaker zone the mean bottom stress vanishes.

In some situations the loss of wave energy can be attributed to bottom friction (due mainly to the orbital velocity of the waves). However, the observation by *Munk* [1949] that in the surf zone the breaker height is proportional to the mean depth suggests that under normal circumstances most of the loss of wave energy is due to wave breaking, not to bottom friction.

It is found that the rule

$$a = \alpha h \tag{31}$$

where α is a constant between 0.3 and 0.6 is in agreement both with direct observations (see Table 1 below) and with laboratory measurements of wave setup [*Longuet-Higgins and Stewart*, 1963, 1964; *Bowen*, 1967] the approximate linear shallow-water theory is used. On the basis of this theory we have from section 2, when $kh \ll 1$,

$$c = (gh)^{1/2} = c_g \tag{32}$$

If it is assumed that in the breaker zone θ is small enough that cos θ can be approximated by unity, we have from (7), (31), and (32)

$$F_x = \tfrac{1}{2}\rho g a^2 c_g = \tfrac{1}{2}\alpha^2 \rho g^{3/2} h^{5/2} \tag{33}$$

and so from (30)

$$\tau_y = -\tfrac{5}{4}\alpha^2 \rho (gh)^{3/2} \frac{dh}{dx}\frac{\sin\theta}{c}$$

$$= \tfrac{5}{4}\alpha^2 \rho g h (s \sin\theta) \tag{34}$$

where $s = -dh/dx$ denotes the local bottom slope.

Some values of α as determined by various authors are shown in Table 1. Though the later determinations of α tend to be higher than the earlier ones, no determination departs by more than 50% from the theoretical value of 0.41 calculated by Davies and Long for the solitary wave.

Using (31) and the linear shallow-water theory, we can also express (34) in terms of the maximum horizontal orbital velocity given by

$$u_{\max} = (a\sigma)/(kh) = \alpha(\sigma/k) = \alpha(gh)^{1/2} \tag{35}$$

Then we have simply

$$\tau_y = \tfrac{5}{4}\rho u_{\max}^2 (s \sin\theta) \tag{36}$$

where s denotes the bottom slope and θ denotes the local angle of incidence.

We note that in this simple relation there are no adjustable parameters.

Beyond the breaker line, i.e. where the energy dissipation is negligibly small, D vanishes, and so by (31)

$$\tau_y = 0 \tag{37}$$

TABLE 1. Observed and Theoretical Values of α

Investigator	s	α	$\langle\alpha\rangle$
Observed Values			
Putnam et al. [1949]	0.066	0.37	0.35
	0.098	0.36	
	0.100	0.33	
	0.139	0.32	
	0.143	0.37	
	0.144	0.32	
	0.241	0.35	
	0.260	0.36	
Iverson [1952]	0.020	0.41	0.44
	0.033	0.38	
	0.050	0.42	
	0.100	0.52	
Larras [1952]	0.010	0.34	0.39
	0.020	0.37	
	0.091	0.43	
Ippen and Kulin [1955]	0.023	0.60	0.60
Eagleson [1956]	0.067	0.56	0.56
Galvin and Eagleson [1965]	0.104	0.59	0.59
Bowen [1968]	0.082	0.45–0.62	0.56
Values Determined from Solitary Wave Theory			
McCowan [1894]	0.000	0.39	
Davies [1952]	0.000	0.41	
Long [1956]	0.000	0.406	

5. Bottom Friction

The tangential stress **B** excited by the water on the bottom will be assumed to be given adequately by a relation of the form

$$\mathbf{B} = C\rho\,|\mathbf{u}|\,\mathbf{u} \tag{38}$$

where **u** is the instantaneous velocity vector near the bottom and C is a constant coefficient.

If there were no longshore velocity, and if the amplitude of the motion were small and the bottom impermeable, the horizontal orbital velocity would be expected to be to-and-fro in the same straight line, making an angle θ with the normal to the shoreline (see Figure 2*a*).

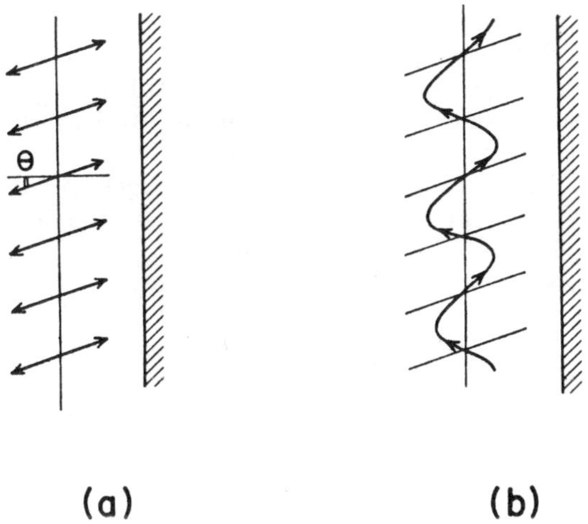

Fig. 2. Schematic representation of particle orbits (*a*) with zero mean littoral velocity and (*b*) with positive littoral velocity $\langle v \rangle$.

The frictional stress **B** given by

$$\mathbf{B} = C\rho \, |\mathbf{u}_{orb}| \, \mathbf{u}_{orb} \qquad (39)$$

would then vanish in the mean (according to linear theory).

Now suppose that a small component of velocity $\langle v \rangle$ in the longshore direction is added to the orbital velocity (Figure 2*b*). When θ is small, this component of velocity is almost perpendicular to the orbital velocity. Therefore the magnitude of the velocity $\mathbf{u} = \mathbf{u}_{orb} + (0, \langle v \rangle)$ is unchanged, to first order, but the direction of the bottom stress is changed by a small angle $\langle v \rangle / |\mathbf{u}_{orb}|$ approximately. Hence there is an additional stress in the y direction given by

$$B_y = C\rho \, |\mathbf{u}_{orb}|^2 \, (\langle v \rangle / |\mathbf{u}_{orb}|) = C\rho \, |\mathbf{u}_{orb}| \, \langle v \rangle \qquad (40)$$

Physically, when the orbital velocity is onshore, the direction of the bottom stress is inclined more toward the positive y direction (if $\langle v \rangle$ is positive); when the orbital velocity is offshore, the bottom stress, now almost in the opposite direction, is again more toward the positive y direction. Taking mean values in (40), we have the relation

$$\langle B_y \rangle = C\rho \langle |\mathbf{u}_{orb}| \rangle \langle v \rangle \qquad (41)$$

Assuming $\mathbf{u}_{orb}$ to be sinusoidal, we have

$$\langle |\mathbf{u}_{orb}| \rangle = (2/\pi) u_{max} \qquad (42)$$

and hence

$$\langle B_y \rangle = (2/\pi) C \rho u_{max} \, \langle v \rangle \qquad (43)$$

As a guide to the appropriate value of the friction coefficient we consider first the values for a rough horizontal plate in uniform flow, as given for example by *Prandtl* [1952] and based on Nikuradse's experiments with roughened pipes. For convenience we reproduce *Prandtl's* [1952, p. 195] diagram as Figure 3 below. The friction coefficient appears to depend on just two parameters. The first is the Reynolds number

$$Re = Ul/\nu \qquad (44)$$

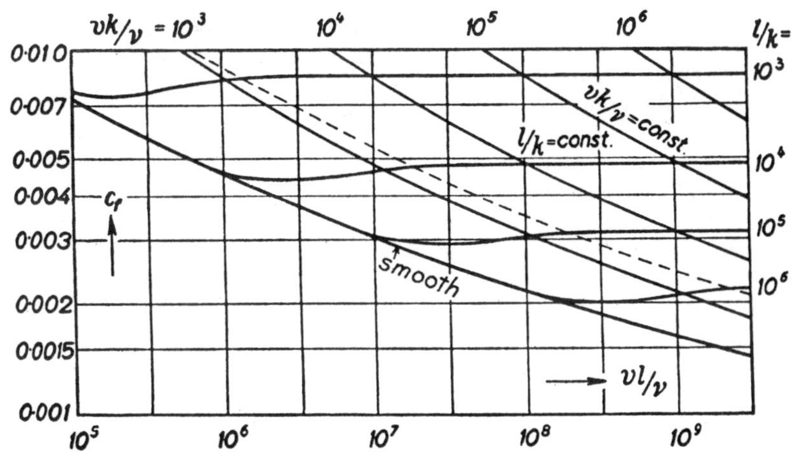

Fig. 3. Values of the friction coefficient C for flow over rough plates, as deduced from the experiments of Nikuradse [from *Prandtl*, 1952]. (Figure reprinted by permission of Haffner Co.)

where U denotes the horizontal velocity, l denotes the length of the plate, and ν is the kinematic viscosity. The second parameter is the ratio (l/K), where K denotes a typical scale for a roughness element. Here we can take as an appropriate value of U the horizontal component of the orbital velocity, u_{max}, and for l the horizontal excursion of a water particle from its mean position, that is $l = u_{max}/\sigma$. Thus we have

$$Re = u_{max}^2/\nu\sigma = \alpha^2 gh/\nu\sigma \qquad (45)$$

As typical values for field data we can take

$$\alpha = 0.4 \qquad g = 10\text{m/sec}^2 \qquad h = 1 \text{ meter} \qquad \sigma = 1 \text{ rad/sec} \qquad (46)$$

corresponding to 6-sec waves 0.8 meter high. With the approximate value $\nu = 1.3 \times 10^{-6}$ m^2/sec and with a sand grain diameter of 1 mm we obtain

$$Re = 1.3 \times 10^6 \qquad l/K = 1.3 \times 10^3 \qquad (47)$$

and so from Figure 3 $C_f \doteqdot 0.007$. On the other hand, for laboratory data more typical values are

$$h = 0.1\text{meter} \qquad \sigma = 5\text{rad/sec} \qquad (48)$$

With the same values of α, g, and ν, this leads to

$$Re = 2.5 \times 10^4 \qquad l/K = 20 \qquad (49)$$

if the roughness scale K is the same. In that case Figure 3 suggests that C_f is somewhat larger, about 0.010.

Bretschneider [1954] has found that the observed damping of swell which is propagated over a smooth, level, impermeable sea bed is consistent with a friction coefficient lying between 0.034 and 0.097. These values appear to agree well with Prandtl's values. On the other hand, Bretschneider also found that the spectral limitation of wave growth under the action of wind suggested higher values of C, between 0.01 and 0.02. These coefficients may include other significant effects such as bottom percolation. R. E. Mayer (personal communication) has found, however, that the theory of run-up of surf on beaches [*Shen and Meyer*, 1963; *Freeman and Le Méhauté*, 1964] can be made to agree fairly well with the model experiments of *Miller* [1968] over a hard sloping concrete bottom by assuming that C lies between 0.01 and 0.02. These values cannot be the result of bottom percolation, but might be attributable in part to turbulence arising from the breaking of the waves as they run up the slope.

Taken together, the above data suggest that it is not unreasonable to expect a friction coefficient C of the order of 0.01.

6. Equations for Longshore Current

To estimate the longshore current $\langle v \rangle$, let us assume first that the mean current is steady and two-dimensional, being independent of the time t and of the longshore coordinate y. Then the equation of motion in the longshore direction can be written as

$$0 = \tau_y + \frac{\partial}{\partial x}\left(N\frac{\partial\langle v\rangle}{\partial x}\right) - \langle B_y\rangle \qquad (50)$$

where in the surf zone τ_y and $\langle B_y \rangle$ are given by (37) and (43), respectively. The second term represents the exchange of momentum due to horizontal turbulent eddies, with eddy coefficient N.

In this equation the magnitude of N is unknown. Suppose first that the exchange of momentum by turbulence is negligible in comparison with that due to the waves; then in general the second term on the right of (50) can be neglected in comparison with the first. There remains a balance between the first and third terms:

$$\langle B_y \rangle = \tau_y \qquad (51)$$

Substituting from equations (36) and (43), we have in the breaker zone

$$\frac{2}{\pi} C\rho u_{max} \langle v \rangle = \tfrac{5}{4}\rho u_{max}^2 (s \sin \theta) \qquad (52)$$

and hence

$$\langle v \rangle = (5\pi/8C)u_{max} (s \sin \theta) \qquad (53)$$

This very simple relation implies that for constant values of C and s the longshore current is simply proportional to $u_{max} \sin \theta$, or to the longshore component of the orbital velocity.

The proportionality of $\langle v \rangle$ and u_{max} has been inferred on quite different grounds by P. Komar (personal communication, 1969).

Using (36) the relation between u_{max} and the local phase velocity c, we can also write (52) in the form

$$\langle v \rangle = \frac{5\pi}{8}\frac{\alpha}{C}\, ghs\left(\frac{\sin\theta}{c}\right) \qquad (53')$$

where $c = (gh)^{1/2}$. Now by Snell's law the last factor is a constant. Equation (53′) then states that in a given wave situation, if both C and s are constant, the longshore velocity $\langle v \rangle$ is simply proportional to the local depth h.

If we assume that the shallow-water theory is valid as far out as the breaker line where the depth h is equal to h_B, the mean longshore current, in the absence of horizontal mixing, can be written as

$$\langle v \rangle = \left(\frac{h}{h_B}\right) \times \begin{cases} v_0 & h < h_B \\ 0 & h > h_B \end{cases} \qquad (54)$$

where

$$v_0 = \frac{5\pi}{8}\frac{\alpha}{C}\,(gh_B)^{1/2}(s \sin\theta_B) \qquad (55)$$

This relation is shown in Figure 4 by the dashed line (corresponding to $\gamma = 0$). The total longshore flux in the surf zone is given by

$$Q = \int_{h=h_B}^{h=0} h\langle v \rangle\, dx$$

$$= \int_0^{h_B} h(h/h_B)v_0\, dh/s$$

$$= \tfrac{1}{3}h_B^2 v_0/s = \tfrac{1}{3}h_B\,|x_B|\,v_0 \qquad (56)$$

We have so far neglected the horizontal mixing entirely. In this idealized model there is a sharp discontinuity in the velocity profile at the breaker line. The presence of any horizontal

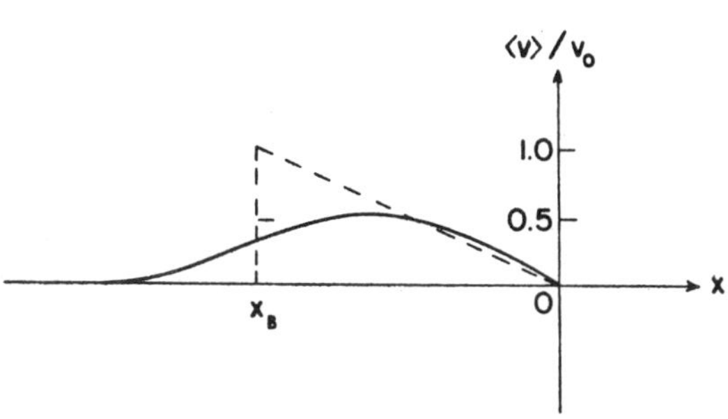

Fig. 4. Schematic representation of the longshore velocity profile as a function of distance offshore. Broken line denotes values without horizontal mixing; full line, with horizontal mixing.

TABLE 2. Theoretical Values of β

$\gamma = L/\lvert x_B \rvert$	β
0.00	0.500
0.25	0.386
0.50	0.290
0.75	0.218
1.00	0.167

mixing, as well as any variability in wave height and position of breaker line, will tend to smooth out the discontinuity at the breaker line and produce a smoother velocity profile; this shifts the maximum velocity closer to shore, as in Figure 4.

A very rough estimate of the effect of mixing on the velocity v_B at the breaker line can be obtained by taking the average of the momentum $h\langle v \rangle$ over a distance L on either side of the breaker line, where L represents a mixing length. When L is small in comparison with the width $|x_B|$ of the surf zone, the velocity v_B is equal to the mean value of the velocities on the two sides of the discontinuity in Figure 4. Hence $v_B = \frac{1}{2} v_0$. More generally if we take

$$L = \gamma\,|x_B| \qquad 0 < \gamma < 1 \qquad (57)$$

we obtain for constant bottom gradient

$$v_B = \beta v_0 \qquad (58)$$

where

$$\beta = \frac{1-\gamma}{2} + \frac{\gamma^2}{6} \qquad (59)$$

As y increases from 0 to 1, β decreases from $\frac{1}{2}$ to $\frac{1}{6}$. Then we can write

$$v_B = \frac{5\pi}{8}\frac{\alpha\beta}{C}\,(gh_B)^{1/2}(s \sin\theta_B) \qquad (60)$$

where θ_B denotes the angle of incidence at the breaker line and β is a constant between 0.5 and about 0.167. The dependence of β on γ is shown in Table 2.

7. Comparison with Observation

Because of the dependence of the longshore velocity on the distance from the shoreline it is particularly important to define precisely the position of the point of observation relative to the shoreline and breaker line.

TABLE 3. Laboratory Data by Brebner and Kamphuis from *Galvin and Nelson* [1965, p. 12]

H_B, feet	θ_B, deg	s	v_B, ft/sec	$(gH_B)^{1/2}/v_B$	$s \sin \theta_B$	r
0.092	7.0	0.10	0.44	3.92	0.012	0.048
0.097	7.5	0.10	0.47	3.77	0.013	0.049
0.110	9.0	0.10	0.67	2.81	0.016	0.045
0.118	10.0	0.10	0.82	2.38	0.017	0.041
0.118	7.5	0.10	0.49	3.97	0.013	0.052
0.138	8.0	0.10	0.67	3.15	0.014	0.044
0.153	10.0	0.10	0.83	2.68	0.017	0.047
0.159	12.0	0.10	0.99	2.29	0.021	0.047
0.157	9.0	0.10	0.63	3.57	0.016	0.056
0.159	9.5	0.10	0.80	2.83	0.016	0.047
0.200	12.0	0.10	0.96	2.65	0.021	0.055
0.203	13.0	0.10	1.07	2.39	0.022	0.054
0.177	9.0	0.10	0.63	3.79	0.016	0.059
0.220	11.0	0.10	0.88	3.02	0.019	0.058
0.228	12.5	0.10	1.04	2.60	0.022	0.056
0.231	14.0	0.10	1.16	2.35	0.024	0.057
0.092	10.0	0.10	0.60	2.87	0.017	0.050
0.112	11.0	0.10	0.81	2.35	0.019	0.045
0.110	13.0	0.10	0.84	2.24	0.022	0.050
0.118	15.0	0.10	0.91	2.14	0.026	0.055
0.118	11.0	0.10	0.83	2.35	0.019	0.145
0.133	12.5	0.10	0.97	2.14	0.022	0.046
0.153	15.0	0.10	1.04	2.14	0.026	0.055
0.159	17.0	0.10	1.14	1.99	0.029	0.058
0.170	13.0	0.10	0.94	2.49	0.022	0.056
0.158	14.0	0.10	1.12	2.01	0.024	0.049
0.200	17.0	0.10	1.25	2.03	0.029	0.059
0.194	18.0	0.10	1.32	1.89	0.031	0.058
0.184	13.0	0.10	1.07	2.28	0.022	0.051
0.204	16.0	0.10	1.25	2.05	0.027	0.056
0.231	18.0	0.10	1.29	2.12	0.031	0.065
0.234	21.0	0.10	1.32	2.08	0.036	0.074
0.085	12.0	0.10	0.70	2.36	0.021	0.049
0.097	14.0	0.10	0.83	2.13	0.024	0.052
0.110	17.0	0.10	0.88	2.14	0.029	0.062
0.112	18.0	0.10	1.05	1.81	0.031	0.056
0.118	14.0	0.10	0.91	2.14	0.024	0.052
0.133	16.0	0.10	0.96	2.16	0.027	0.059
0.141	18.0	0.10	1.10	1.94	0.031	0.060
0.147	21.0	0.10	1.22	1.79	0.036	0.064
0.151	17.0	0.10	1.08	2.04	0.029	0.060
0.153	18.0	0.10	1.18	1.88	0.031	0.059
0.176	22.0	0.10	1.36	2.75	0.037	0.066
0.187	24.0	0.10	1.53	1.60	0.041	0.065
0.177	17.0	0.10	1.21	1.97	0.029	0.057

The profiles of velocity versus offshore distance measured by *Galvin and Eagleson* [1965] show a maximum velocity about halfway between the mean shoreline (not the still water level) and the breaker line, as one would expect from section 6 if horizontal mixing were important. In the above instance, however, the flow was being accelerated downstream from a side wall, so that the compensating inflow would also contribute to the redistribution of longshore momentum and could have an effect similar to the presence of a large horizontal eddy viscosity.

A useful summary of the available field and laboratory data has been compiled by *Galvin and Nelson* [1967]; these data have been critically discussed by *Galvin* [1967]. It seems that the most commonly observed parameters of the wave field are the breaker height

$$H_B = 2\alpha h_B \tag{61}$$

and the angle of incidence θ_B at the breaker line, though in some instances these quantities must be deduced from the wave height and angle of incidence as measured in deep water. *Galvin and Eagleson* [1965] have shown that there is considerable uncertainty in the measurement of H_B and θ_B (especially θ_B) even under laboratory conditions.

Now substituting for h_B in (60) we have for the longshore velocity v_B at the breaker line

$$v_B = \frac{5\pi}{8\sqrt{2}} \frac{\sqrt{\alpha}\,\beta}{C} (gH_B)^{1/2}(s \sin \theta_B) \tag{62}$$

In other words, if we write

$$\frac{(gH_B)^{1/2}}{v_B} (s \sin \theta_B) = r \tag{63}$$

a dimensionless ratio, we have

$$C = 1.39\sqrt{\alpha}\,\beta r \tag{64}$$

With little uncertainty we can take α to be the mean value of the entries in the last column of Table 1, namely $\alpha = 0.42$; then (64) further simplifies to

$$C = 0.90\beta r \tag{65}$$

For each entry in the data compiled by *Galvin and Nelson* [1967] we have computed the quantity r as given by (63). The results of these computations for a typical page of laboratory data [*Brebner and Kamphuis*, 1963] are shown in Table 3 and for the field data of *Inman and Quinn* [1952] in Table 4. Despite the great range in the values of the breaker height H_B it will be seen that the computed value of r remains remarkably consistent. There is somewhat more scatter in the field data than

TABLE 4. Field Data by Inman and Quinn from *Galvin and Nelson* [1965, p. 17]

H_B, feet	θ_B, deg	s	v_B, ft/sec	$(gH_B)^{1/2}/v_B$	$s \sin \theta_B$	r
2.8	6.5	0.027	0.38	25.0	0.0030	0.076
3.1*	1.5*	0.027	0.04	25.0	0.0007	0.018*
3.7	4.0	0.027	0.22	49.6	0.0019	0.093
3.6*	0. *	0.027	0.04	269.0	0.0000	0.000*
4.9	5.0	0.027	0.84	14.9	0.0024	0.035
3.8	5.0	0.027	0.21	52.6	0.0024	0.124
3.4*	0. *	0.027	0.55	19.0	0.0000	0.000*
2.6*	0. *	0.035	0.04	22.9	0.0000	0.000*
3.0*	1.0*	0.035	0.01	98.2	0.0006	0.600*
2.7*	0. *	0.035	0.15	62.1	0.0000	0.000*
3.5*	0. *	0.035	0.09	117.9	0.0000	0.000*
4.9*	0. *	0.035	0.21	59.8	0.0000	0.000*
2.9*	0. *	0.035	0.50	19.3	0.0000	0.000*
4.6*	0. *	0.035	0.88	13.8	0.0000	0.000*
3.7*	0. *	0.028	0.20	54.5	0.0000	0.000*
5.1	6.0	0.027	0.29	44.1	0.0028	0.124
4.7	7.0	0.027	0.53	23.2	0.0033	0.076
4.5	4.0	0.027	0.70	17.2	0.0019	0.032
4.8	4.0	0.027	1.19	10.4	0.0019	0.020
4.2	4.5	0.027	0.40	29.1	0.0021	0.062
2.0	4.0	0.027	0.36	22.3	0.0019	0.042
1.7	7.0	0.027	0.23	31.2	0.0033	0.103
2.9	5.0	0.027	0.56	17.2	0.0023	0.041
1.6	5.0	0.027	0.11	65.2	0.0023	0.153
6.2	5.0	0.014	0.54	26.1	0.0012	0.032
3.1	7.0	0.014	0.62	16.1	0.0017	0.028
4.5	3.0	0.014	0.49	24.6	0.0007	0.018
3.5	4.0	0.014	0.17	62.4	0.0010	0.061
2.7	3.5	0.014	0.13	71.7	0.0009	0.061
4.7	7.0	0.014	1.37	9.0	0.0017	0.015
2.6*	2.0*	0.014	0.04	228.6	0.0005	0.116*
2.0	4.0	0.014	0.11	72.9	0.0010	0.071
1.8	2.5	0.014	0.06	126.8	0.0006	0.077

* Values for which θ_B is reckoned to be 2° or less.

TABLE 5. Summary of Observations: Mean Values

Investigators	Type of Beach	$\langle s \rangle$	$\langle H_B \rangle$	$\langle \theta_B \rangle$	N	$\langle r \rangle$
Putnam et al. [1949]*	Bonded sand	0.133	0.28	14.4	14	0.121
	Metal or smooth cement	0.172	0.23	36.8	14	0.134
	Gravel, 1/4 inch in diam.	0.123	0.22	22.0	9	0.322
Saville [1950]*	Concrete or 0.3 mm sand	0.100	0.14	6.5	7	0.087
Brebner and Kamphuis		0.100	0.15	13.9	45	0.054
[1963]	Roughened concrete	0.100	0.14	21.2	48	0.068
		0.100	0.16	14.6	48	0.035
Galvin and Eagleson [1965]	Smooth concrete	0.109	0.16	11.8	18	0.044
Putnam et al. [1949]	Oceanside	0.021	6.42	11.1	18	0.020
Inman and Quinn [1951]	Torrey Pines and Pacific Beach	0.022	3.58	4.9	21	0.064
Galvin and Savage [1966]	Nags Head	0.027	3.75	15.4	4	0.035

* Data rejected by *Galvin* [1967].

in the laboratory data, as is to be expected, especially considering the difficulty in measuring the angle of incidence θ_B. If we omit from consideration all observations (marked with an asterisk) for which θ_B is reckoned to be 2° or less, the mean value of the entries in the last column is $\langle r \rangle = 0.054$ for the laboratory measurements and $\langle r \rangle = 0.064$ for the field data.

A summary of such mean values is given in Table 5, for all the data compiled by *Galvin and Nelson* [1967] with the exception of the field observations of *Moore and Scholl* [1961], which contained a large proportion of zero or negative values of v_B and were thought to be influenced by disturbances other than wave action. In the laboratory measurements of *Saville* [1950] and of *Galvin and Eagleson* [1965] the entries corresponding to angles θ_B less than 6° have also been discarded on the grounds of unreliability.

On quite different grounds *Galvin* [1967] has rejected all the early laboratory measurements of *Putnam et al.* [1949] since they were found not to be reproducible under almost the same conditions either by *Brebner and Kamphuis* [1963] or by *Galvin and Eagleson* [1965]. It is possible that Putnam et al. employed a different definition of breaker height than Brebner and Kamphuis or Galvin and Eagleson. Galvin also suggests that less weight should be attached to the observations of *Saville* [1950], since he did not actually measure H_B and θ_B; these entries in the table are estimated from H_0 and θ_0.

Retaining then only the most reliable measurements in Table 5 (namely those not rejected by *Galvin* [1967]), we find for the field observations $\langle r \rangle = 0.040$ and for the laboratory data $\langle r \rangle = 0.050$.

According to (65), these values of r correspond to mean values of the friction coefficient C given by

$$C = 0.036\beta \qquad C = 0.045\beta \qquad (66)$$

where β, as we have seen, is between 0.50 and 0.167, depending on the horizontal mixing.

Assuming a friction coefficient C of about 0.010, we see that both field observations and laboratory data are consistent with a mean value of β equal to about 0.2. This suggests that horizontal mixing played some part, though not a dominant one, in the distribution of the longshore current.

A more precise estimate of the effects of horizontal mixing are given in the accompanying paper.

8. Conclusions

By the use of the concept of radiation stress and the small-amplitude theory of water waves, we have shown that the total longshore thrust exerted by the waves on the water and sea bed inside the surf zone is very simply related to the energy density and direction of propagation of the waves in deep water (equation 28). This relation is quite different from that given by previous authors, and it would be interesting to test it directly by experiment.

The local wave stress τ_y is also simply related to the local rate of energy dissipation, and, it would be interesting to test this relation also.

The comparisons so far made between theory and observation suggest that the rational prediction of longshore currents may be practically possible. There is no need, as some authors have suggested, to fall back on empirical correlations.

Acknowledgments. My interest in this topic was stimulated by a seminar given in Corvallis by Mr. Paul Komar. I am indebted to him and to my colleague, Dr. Victor Neal for references to recent literature. Professor Inman, Dr. Elliott, and Dr. Quinn kindly commented on a first draft of this paper.

The present research was supported under NSF grant GA-1452.

References

Bowen, A. J., Rip currents, Ph.D., thesis, Scripps Institution of Oceanography, University of California, San Diego, 115 pp., 1967.

Bowen, A. J., The generation of longshore currents on a plane beach, *J. Mar. Res., 27,* 206–215, 1969.

Brebner, A., and J. W. Kamphuis, Model tests on relationship between deep-water wave characteristics and longshore currents, *Queens Univ. Civil Eng. Res. Rep., 31,* 1–25, 1963.

Bretschneider, C. L., Field investigation of wave energy loss of shallow water ocean waves, *U.S. Army, Beach Erosion Board, Tech. Mem. 46,* 1954.

Burnside, W., On the modification of a train of waves as it advances into shallow water, *Proc. London Math. Soc., 14,* 131–133, 1915.

Davies, T. V., Symmetrical, finite amplitude gravity waves, in *Gravity Waves,* chap. 9, *NBS Circ. 521,* 55–60, 1952.

Eagleson, P. S., Properties of shoaling waves by theory and experiment, *Trans. AGU, 37,* 565–572, 1956.

Freeman, J. C., and B. Le Méhauté, Wave breakers on a beach and surges on a dry bed, *J. Hydr. Div. Amer. Soc. Civil Eng., 90,* 187–216, 1964.

Galvin, C. J., Longshore current velocity: A review of theory and data, *Rev. Geophys., 5,* 287–304, 1967.

Galvin, C. J., and P. S. Eagleson, Experimental study of longshore currents on a plane beach, *U.S. Army Coast. Eng. Res. Center, Tech. Mem. 10,* 1–80, 1965.

Galvin, C. J., and R. A. Nelson, Compilation on longshore current data, *U.S. Army Coast. Eng. Res. Center, Misc Pap. 2-67,* 1–19, 1967.

Inman, D. L., and R. A. Bagnold, Beach and nearshore processes, 2, Littoral processes, in *The Sea,* edited by M. N. Hill, pp. 529–553, Interscience, New York, 1963.

Inman, D. L., and W. H. Quinn, Currents in the surf zone, *Proc. 2nd Conf. on Coastal Eng.,* Council on Wave Research, University of California Berkeley, 24–36, 1952.

Ippen, A. T., and G. Kulin, Shoaling and breaking characteristics of the solitary wave, *MIT Hydrodynamics Lab. Rep. 15,* 1955.

Iverson, H. W., Laboratory study of breakers, in *Gravity Waves,* chap. 3, *NBS Circ. 521,* 9–32, 1952.

Larras, J., Experimental research on the breaking of waves, *Ann. Ponts Chaussees, 122,* 525–542, 1952.

Long, R. R., Solitary waves in one- and two-fluid systems, *Tellus, 8,* 460–471, 1956.

Longuet-Higgins, M. S., The refraction of sea waves in shallow water, *J. Fluid Mech., 1,* 163–176, 1956.

Longuet-Higgins, M. S., and R. W. Stewart, Changes in the form of short gravity waves on long waves and tidal currents, *J. Fluid Mech., 8,* 565–583, 1960.

Longuet-Higgins, M. S., and R. W. Stewart, The changes in amplitude of short gravity waves on steady non-uniform currents, *J. Fluid Mech., 10,* 529–549, 1961.

Longuet-Higgins, M. S., and R. W. Stewart, Radiation stress and mass transport in gravity waves, *J. Fluid Mech., 13,* 481–504, 1962.

Longuet-Higgins, M. S., and R. W. Stewart, A note on wave set-up, *J. Mar. Res., 21,* 4–10, 1963.

Longuet-Higgins, M. S., and R. W. Stewart, Radiation stresses in water waves; a physical discussion, with applications, *Deep-Sea Res., 11,* 529–562, 1964.

McCowan, J., On the highest wave of permanent type, *Phil. Mag., 39,* 351–359, 1894.

Miller, R. L., Experimental determination of run-up of undular and fully developed bores, *J. Geophys. Res., 73,* 4497–4510, 1968.

Moore, G. W., and D. W. Scholl, Coastal sedimentation in northwestern Alaska, *U.S. Geol. Surv. Rep. TE 1-779,* 43–65, 1961.

Munk, W. H., The solitary wave and its application to surf problems, *Ann. N.Y. Acad. Sci., 51,* 376–424, 1949.

Prandtl, L., *Essentials of Fluid Dynamics,* 452 pp., Haffner, New York, 1952.

Putnam, J. A., W. H. Munk, and M. A. Traylor, The prediction of longshore currents, *Trans. AGU, 30,* 337–345, 1949.

Saville, T., Model study of sand transport along an infinitely long straight beach, *Trans. AGU, 31,* 555–565, 1950.

Shen, M. C., and R. E. Meyer, Climb of a bore on a beach, 3, Run-up, *J. Fluid Mech., 16,* 113–125, 1963.

Wiegel, R. L., *Oceanographical Engineering,* Prentice-Hall, 532 pp. Englewood Cliffs, N. J., 1963.

13

Reprinted from *Jour. Geophysical Research* 77(18):3232–3247 (1972)

Field Observation of Nearshore Circulation and Meandering Currents

Choule J. Sonu

Coastal Studies Institute, Louisiana State University, Baton Rouge, Louisiana 70803

Field observations were made of wave-induced nearshore circulations and meandering longshore currents on an undulatory surf-zone bed, under the action of uniform incident waves. Circulations were associated with normal-wave incidence; meandering currents were associated with oblique-wave incidence. The transport in the observed circulations generally agreed with Bowen's (1967) linear theory based on the concept of radiation stress (Longuet-Higgins and Stewart, 1962, 1964), provided that a friction coefficient $C = 0.014$ was assumed. The longshore current near the shore line moved from a shoal to a depression as predicted, but this movement was also directed from an area of high waves to one of low waves, which is different from the case of a circulation driven by nonuniform breaker heights on the bar. Spilling breakers over a shoal underwent greater energy dissipation than plunging breakers in the rip current. Observed streamlines were narrow in the outflow and broad in the inflow, a characteristic that was probably associated with a nonlinear mechanism arising from a steep depression in the rip channel, as previously explained by Arthur (1962). These circulations were pulsational, unlike the circulations of a steady-state solution. Occasional strong outflows at beat frequencies caused water to escape from the circulation. For a given surf-zone undulation, breaking over the inner bar was essential to the formation of a circulation, and the intensity of breaking, controlled by tide, corresponded with a proportionally stronger circulation. Thus, circulations were generally stronger during low tide than during high tide. Low rip-current velocities at high tide fluctuated with incoming swells, whereas high velocities at low tide tended to fluctuate at surf beat frequencies. In proportion to increasing rip velocities, the rip pulsation tended toward lower intervals. Mean surface slopes caused by wave set-up and set-down agreed with trajectories of neutral-density balls released in the circulation. Meandering currents associated with oblique-wave incidence could be explained as a combined effect of circulation cells and parallel longshore flows.

Qualitative descriptions of wave-induced nearshore circulation appeared early in the literature [*Evans,* 1939; *Shepard et al.,* 1941]; however, quantitative studies have been scarce. This study reports actual measurements of velocity fields in the nearshore zone.

Driving mechanisms for nearshore circulation depend on wave characteristics and bottom topography. In the circulation observed by *Shepard and Inman* [1950] off Scripps Beach, California, longshore currents adjacent to the shore diverged from areas of wave convergence and flowed seaward as rips at areas of wave divergence. These longshore wave-height variations were caused by refraction over submarine canyons outside the bar [*Munk and Traylor,* 1947].

Studies on coasts with smooth offshore topography [*McKenzie,* 1958; *Sonu et al.,* 1966] revealed, however, little evidence of correlation between circulation and incident wave-height variation. McKenzie, for example, concluded that rip currents 'do not, as might be expected, seek that part of the beach with least energy concentration but tend to move seaward in the vicinity of greatest wave activity.' In these observations, a rather distinct correlation between circulation and bottom topography was recognized. The surf-zone topography was typically undulatory; it contained alternate shoals and depressions along the shore. Generally, shoreward currents occurred over the shoals and seaward currents occurred over the depressions, and thus the longshore currents moved from shoals to depressions near the shore line.

A theoretical study by *Bowen* [1967, 1969*b*] indicated that a driving force for circulation was provided by a characteristic distribution of radiation stress in the nearshore zone [*Longuet-Higgins and Stewart,* 1962, 1964]. *Bowen and Inman* [1969] suggested the possibility of periodic longshore variations of breaker height on a plane beach as a result of stationary coupling between incident surface waves and edge waves of equal periods.

The present study deals with a coast with smooth offshore topography and surf-zone undulations. The site was located at Seagrove, Florida, where an uninterrupted coastline extends about 60 km in the E-W direction and faces the Gulf of Mexico on the south. The 30- to 50-meter wide surf zone was shallow; the depth over the inner bar was 1 meter or less. Outside the inner bar the bottom dropped steeply to about 5 meters, and then rose to about 4 meters at the outer bar, approximately 200 meters from the shore line. The waves were generally small (local wind waves, up to 50 cm high and 4 sec in period; swells, ~20 cm high and up to 10 sec in period). Because of the absence of distinct bottom diversification outside the inner bar, the breaker height at the inner bar was essentially uniform.

Repeated visits to the study area were made in 1967 and 1968. It was possible to distinguish two major current patterns, depending on wave incidence angles (Figure 1). A pattern associated with normal-wave incidence was characterized by a series of discrete circulation cells with more or less closed streamlines. These cells were of comparable size but varied from time to time between 30 and 60 meters in the longshore diameter. Adjacent cells rotated in opposite directions so that the rip currents, which developed at the confluence of the seaward currents of these cells, were spaced at every two cells. An oblique-wave incidence induced a meandering pattern consisting of chains of oblique seaward and shoreward currents. The oblique type of current system, rarely reported in the literature, was quite common in the study area. Both patterns are investigated in this study.

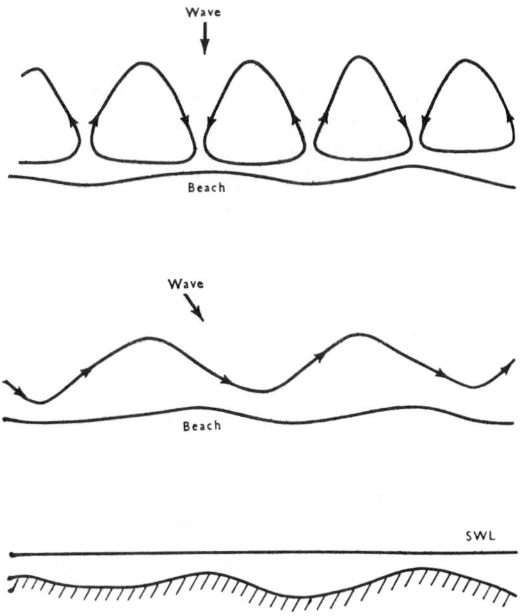

Fig. 1. Dependence of current patterns on wave-incidence angles and surf-zone topography. (Top) Circulation under normal-wave incidence. (Middle) Meander under oblique-wave incidence. (Bottom) Longshore surf-zone profile.

MEASUREMENT TECHNIQUES

Spatial velocity distribution was determined with water-filled polyethylene balls 20 cm in diameter that were carefully adjusted to a neutral density so as to minimize the surfboarding effect and bottom friction. Comparison with dye releases indicated that the ball movement, as it drifted freely in the water column, represented mean depth velocity.

A bidirectional current meter was used to obtain time-series records of velocities. The meter (Marine Advisers, Model B-9) consisted of a three-bladed impeller suspended in a duct, the rotation of which triggered paired reed switches on the duct through magnetic coupling. The output signals gave speeds and directions in the form of pulse density and polarity, respectively. The threshold sensitivity of the meter was rated at 5 cm/sec. The off-axis response was flat within ±30° about the main axis, falling rapidly toward zero beyond this limit. A similar device has been used to measure particle motions in ocean waves [*Shonting*, 1967].

To delineate a synoptic current pattern, the movement of dye trajectories was photographed by an air-borne radio-controlled camera (Nikon F35) suspended at a constant altitude of 120 meters from a helium-filled tethered balloon directly above the surf zone [*Sonu*, 1969]. A radio transmitter on the beach triggered the camera, which was housed in a water-tight capsule.

Waves were monitored with parallel resistance wires. A low-pass hydraulic filter was designed to determine local mean water levels in the presence of waves. The device was a clear plastic tube (250 cm long, 12 cm in diameter) with six 1-mm-diameter orifices at the closed bottom and a 1-cm air hole at the top. The frequency response of this device was <5% against exterior wave heights up to 40 cm and frequencies of >0.2 Hz; this response sufficiently suppressed most of the local waves. The tube was fastened to a survey staff and carried manually. The height of the water column in the tube was read directly from the survey staff, whereas the bottom elevation was determined simultaneously from an alidade on the beach. The difference between the two readings gave the local mean water level relative to the alidade elevation. The probable error from this procedure was held to ±3 cm.

SPATIAL VELOCITY DISTRIBUTIONS

Circulations associated with normal-wave incidence. The velocities in the nearshore currents were unsteady. Velocity fluctuations, both in speed and direction, occurred in a wide range of frequencies between those of incident waves and surf beats. To obtain a mean velocity field, each station was occupied for approximately 2 min. A neutral-density ball was released three times in succession at each of 39 stations, and its movement in a radius of 3.3 meters was timed. The three velocity vectors obtained were averaged to find local mean velocity. Following *Arthur* [1962], a transport stream function was defined such that

$$hu = -\partial\psi/\partial y \qquad hv = \partial\psi/\partial x$$

where h is the water depth and u and v are the velocity components in the x and y directions taken normal and parallel to the shore line, respectively. The lines, where ψ is constant (Figure 2), define transport streamlines. Streamline separation in the figure is 0.4 m³/sec.

Figure 2 reveals typical characteristics of a circulation system associated with normal-wave incidence. Two coupled cells with opposite rotations constituted a unit circulation system, occupying a longshore wavelength of approximately 60 meters. In each of these cells, of approximately equal size, the streamline distribution was skewed in such a way that streamlines were broad in the shoreward currents ($y \simeq \pm 30$ meters) and narrow in the rip current ($y \simeq 0$). The water converged into the rip along gently curved streamlines, and, although a feeder channel existed close to the shore line, the

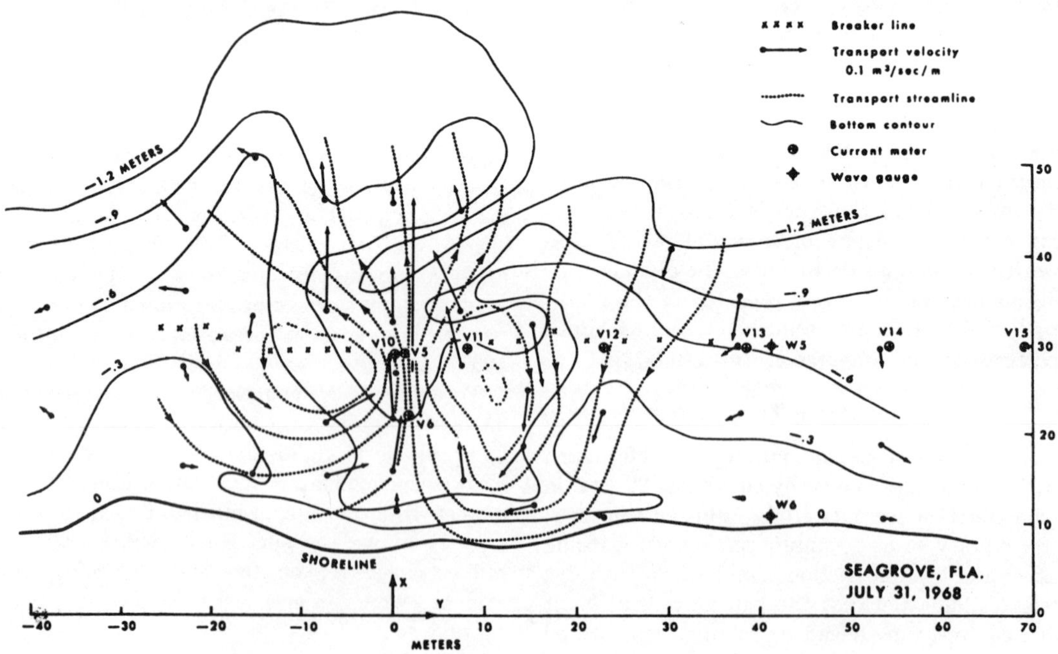

Fig. 2. Distribution of transport velocities and streamlines in the circulation cell. Streamline separation is 0.4 m³/sec, $H_b = 0.395$ meter, $T_b = 5.0$ sec, and $\alpha_b = 0$.

streamlines failed to define a distinct feeder current. Generally a feeder current was more readily distinguished during the time of small wave activity. In a high surf, proportionately more water was fed into the rip directly from its sides than through the feeder channel. A similar observation was reported by *McKenzie* [1958].

Whereas the shoreward currents were of the order of 20 to 30 cm/sec, the outflow velocities in the rip attained as much as 2 m/sec. Between these two areas velocities increased progressively, as indicated by gradual narrowing of streamlines toward the rip current. The so-called neck of the rip current represented the locus of maximum streamline concentration. When circulation velocities were low, as during a light surf or a high tide, the rip current was broad and failed to delineate a neck. A distinct neck appeared in a generally narrow and fast rip current associated with high surf and was located slightly shoreward of the breaker line, as shown in Figure 2.

Outside the breaker line, the streamlines broadened considerably, which made it difficult to measure velocities with a neutral-density ball. Dye trajectories generally indicated that at approximately twice the surf-zone width from the shore line the rip velocities were effectively masked by ambient coastal currents. It was evident that only a certain portion of the water leaving with the rip returned to the surf zone to complete a closed circulation. The water in the outstretched rip head usually drifted away and was lost to the circulation. The outstretching of the rip occurred intermittently at frequencies corresponding to pulsations of rip velocities of the order of 10^{-2} sec^{-1}. The rip's maximum reach generally remained inside the outer bar.

Figure 2 reveals a good correlation between the streamline distribution and the surf-zone topography. The contours indicate shoals at $y \simeq \pm 30$ meters and a depression at $y \simeq 0$. Shoreward currents in the circulation cell typically occurred over the shoals, and the rip current, over the depression. For several days prior to the measurements shown in Figure 2 the bottom undulation in the surf zone was less pronounced, and the corresponding circulation velocities were low. However, the same correlation persisted (Figure 5). In all these instances the breakers, though essentially uniform in height over the inner bar, displayed a characteristic variation in type. Waves entering the shoal broke by spilling, and tended to maintain the broken crests in the surf zone. Waves entering the rip broke by plunging in a narrow strip near the bar, and traveled the remainder of the surf zone with relatively unbroken crests. As a consequence, wave heights at the shore line were varied, i.e., low at the shoal (less than 20 cm) and high near the depression (about 30 cm). This feature indicated that the currents near the shore line moved from areas of low waves to those of high waves. This situation is in contrast to nearshore circulation associated with nonuniform incident-wave heights as reported off Scripps Beach [*Shepard and Inman*, 1950], where the opposite situation was observed.

Development of the concept of radiation stress and its application to a surf zone [*Longuet-Higgins and Stewart*, 1962, 1964] presented a possibility for explaining physical mechanisms of wave-induced near shore circulation. With this concept, *Bowen* [1967] successfully obtained a forcing function for circulation. The radiation stress is defined as the excess flux of momentum due to the presence of waves and is expressed by a tensor

$$S = \begin{pmatrix} S_{rr} & 0 \\ 0 & S_{ss} \end{pmatrix} = E \begin{pmatrix} \dfrac{2kh}{\sinh 2kh} + \dfrac{1}{2} & 0 \\ 0 & \dfrac{kh}{\sinh 2kh} \end{pmatrix} \tag{1}$$

in which r and s are parallel and normal to the wave direction. When the wave incidence is normal to the shore line, $r = x$ and $s = y$; hence $S_{rr} = S_{xx}$ and $S_{ss} = S_{yy}$. Longuet-Higgins and Stewart demonstrated that on a sloping bed $\partial S_{xx}/\partial x$ accounts for a set-down at the breaker line and a set-up at the shore line if the wave energy is dissipated continuously in the surf zone, as by breaking. If incident-wave heights vary along the shore (either by offshore refraction or longshore variation in depths in the surf zone), there will be longshore variations of set-down and set-up, resulting in longshore pressure gradients. Bowen demonstrated that inside the surf zone this longshore pressure gradient acts in the same manner as the longshore thrust caused by $\partial S_{yy}/\partial y$ to produce a net forcing of circulation, whereas outside the breaker line they are opposed and exactly balanced. This mechanism could give rise to a circulation where (1) incident waves have variable heights parallel to the shore line and the surf zone is a plane bed, and (2) incident waves are uniform at the bar (or breaker line) and the surf-zone bed has undulations. In case (1), longshore currents near the shore line move from high-wave areas to low-wave areas, as on Scripps Beach; in case (2), they move from shallow to deep areas, as in our observation.

A computation was carried out to compare our observation with Bowen's theory. Following Bowen, the undulatory surf-zone bed was approximated by

$$h = x \tan \beta(1 + \mu \cos \lambda y) \qquad (2)$$

and the bottom outside the breaker line, by a plane slope

$$h = x \tan \beta \qquad (3)$$

in which h is the still-water depth, β is bottom slope, μ is a parameter relating to the amplitude of undulation $\lambda = 2\pi/L$, L is the wavelength of undulation, and x and y are taken normal and parallel to the shore line. In our observation $L \simeq 60$ meters, $\mu = 0.4$, $\tan \beta = 0.03$, and the breaker position from the mean sea-level shore line was 23 meters.

Bowen defined a linearized friction coefficient C_f such that

$$\tau = [\rho d] \frac{C_f u}{d} = C\rho u^2 \qquad (4)$$

in which τ is the shear stress, ρ is the water density, d is the height of the water column including wave set-up or set-down, u is the horizontal velocity component, and C is the nondimensional friction coefficient according to conventional definition. The coefficient C_f is then

$$C_f = Cu \text{ (cm/sec)} \qquad (5)$$

and must be determined from observation. By considering only bottom friction to balance the driving force, the discharge in a single circulation cell for various values of C_f was computed, as shown in Figure 3. The value for C_f, corresponding to our observed discharge per unit cell 1.4 m³/sec, was 0.348 cm/sec. From (5),

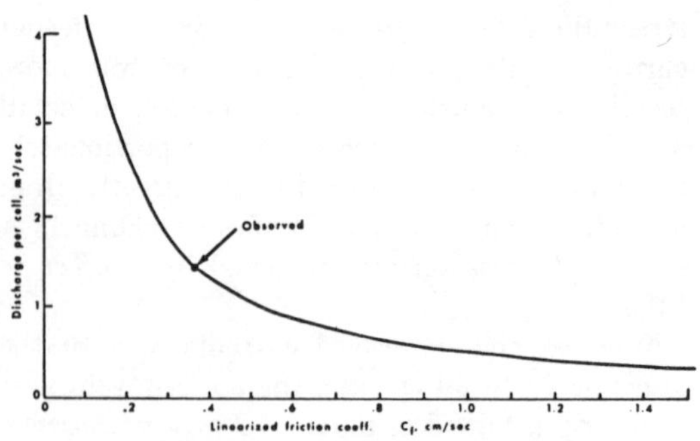

Fig. 3. Dependence of discharge in a circulation cell on friction coefficient.

assuming a typical velocity of circulation u = 25 cm/sec, we get $C = 0.014$. In general, C is is of the order of 0.01 in the surf zone [*Longuet-Higgins*, 1970*a*]. Our friction coefficient was slightly greater than this value; it probably included the effect of eddy viscosity and friction.

The transport streamlines for $C_f = 0.348$ cm/sec computed by Bowen's linear solution are shown in Figure 4. The computation correctly predicted an onshore current over the shoal ($y \simeq \pm(15 - 30$ meters)) and an outflow over the depression ($y \simeq (0 - \pm 15$ meters)). The nearshore currents moved from shoals to the depression. The streamlines were narrow nearshore and broad offshore. The computation also predicted the maximum streamline concentration in the rip to occur somewhat shoreward of the breaker line. The computed streamlines were equally spaced in the areas of shoreward and seaward currents, which shows a discrepancy from the observation. Bowen attributed this discrepancy to the linearity of the solution. The satisfactory convergence of the nonlinear solution by means of successive over-relaxation procedures required, however, small values for the Reynolds number, namely,

$$Re = L_s U_s/\mu_e < 0.2$$

In the field observation $Re \gg 0.2$; hence a nonlinear solution is not available by a numerical method. Previously, *Arthur* [1962], who considered only the nonlinear terms and neglected bottom friction, demonstrated that a potential vorticity [*Rossby*, 1940] must be conserved along a streamline. As a consequence, the streamlines would tend to crowd where a current moves from shallow to deep water, and to broaden where a current moves from deep to shallow water.

Our observation, along with Bowen's theoretical explanation, clearly demonstrated that for uniform incident waves the surf-zone undulation constitutes a factor of major importance to circulation. Our observation also showed that the breaking of incident waves over the bar, rather than their heights, was the controlling factor on the velocities in the circulation. This is consistent with the concept of radiation stress, since the driving force in the surf zone, which is proportional to $\partial S_{xx}/\partial x$ and $\partial S_{yy}/\partial y$, requires continuous reduction in wave height as a result of breaking. Consequently, for a given undulation of the surf zone the circulation velocities accelerated distinctly as soon as waves began to break over the bar. On the other hand, since tidal fluctuation controlled wave breaking, it was repeatedly observed that the circulation velocities would change considerably under constant wave input, depending on the phase of the tide in the course of a day. During summer, when the local waves were dominated by a diurnal sea- and land-breeze system, the displayed intensity of circulation combined effects of solar (sea and land breeze) and lunar (diurnal tide) cycles. Because the solar-cycle period was exactly 24 hours and the lunar-cycle period nearly 25 hours, the time of optimum coupling for these two independent cycles to produce effective wave breaking (hence a strong circulation) was repeated, with a progressive lag of nearly an hour, every day.

Wave set-ups in the surf zone. An attempt was made to confirm the extent of wave set-ups caused by radiation stress in the field condition. The hydraulic filter device already described was positioned at a total of 64 stations in the nearshore zone, and neutral-density balls and dye were released simultaneously. Figure 5 shows the result (water levels are expressed in terms of the deviation from the overall mean level in the survey area). These measurements were conducted 4 days before the circulation measurement of Figure 2, when the bottom undulation had not been fully developed. Nevertheless, there was a distinct rip channel near $y = 0$, which was flanked on either side by a broad-crested shoal. The left shoal had a slight depression on its crest ($y = -35$ meters), but it was not deep enough to behave as a rip channel, as indicated by the shoreward ball movement through it. All the balls were released in pairs, but because the paired balls described essentially identical trajectories, only one is shown for each run. As expected from the weak surf-zone undulation, currents were generally slow, being about 10 cm/sec in the onshore flow and about 60 cm/sec in the rip.

The probable error of ±3 cm in the water-level determination may render the results sig-

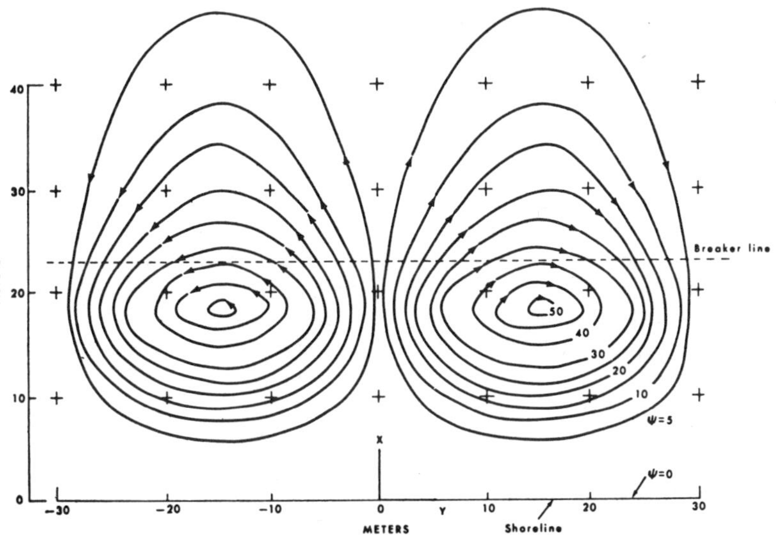

Fig. 4. Computed transport streamlines of a circulation system for normal-wave incidence and undulatory surf zone: $H_b = 0.395$ meter, $T_b = 5.0$ sec, $\alpha_b = 0$, $\tan\beta = 0.03$.

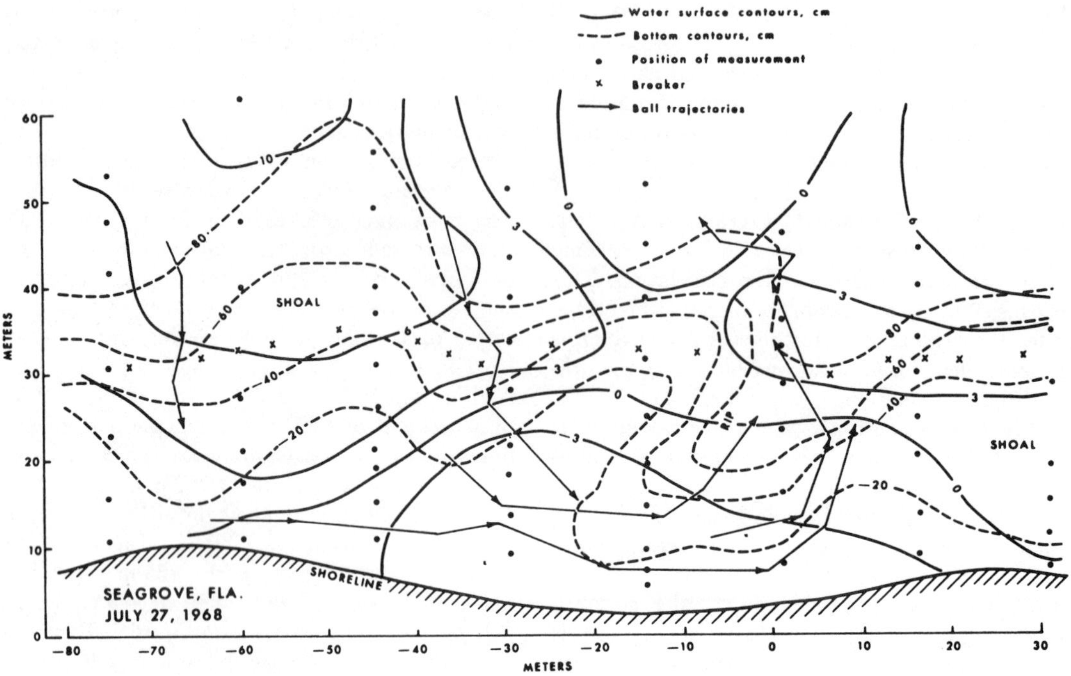

Fig. 5. Water surface contours in a circulation system. Water levels shown are mean positive (z upward) deviations from the overall mean level: $H_b = 0.35$ meter, $T_b = 5.1$ sec, $\alpha_b = 10°$ from left of normal.

nificant only in gross features. However, the zonations of set-up and set-down patterns were definitely indicated. A high-water level was found on the shoals and a low-water level in the rip channel, as expected, but these features extended beyond the breaker line. A large set-up, as much as 10 cm outside the breaker line off the left shoal ($y = -60$ meters), is not readily explained. A similar but somewhat smaller set-up (6 cm) also existed beyond the breaker line off the right shoal ($y = 20$ meters), suggesting that in spite of the probable survey error of ± 3 cm this phenomenon was real. It could be that an interaction between the circulation and incident waves produced modification of the radiation stress field to cause redistribution of set-ups.

The absence of set-down along the breaker line was not surprising because the known amount (only about 4% of the breaker height in laboratory tests) was too small to be distinguished by our technique. On the shoal inside the breaker line the water level continued to fall toward a final breaker line near the shore line, contrary to laboratory observations, in which there was a wave set-up from the breaker line toward the shore. However, a set-up requires continuous wave dissipation, as by breaking. In our observation, the waves broke only in a narrow zone inside the breaker line near $x = 30$ meters and traveled the remaining part of the surf zone with unbroken crests, which should have resulted in a set-down toward the shore.

The ball trajectories followed the directions of the surface gradient, moving shoreward over the shoal and converging on the rip channel. Reflecting the presence of an upward surface gradient from the rip to the shoal outside the breaker line, the balls escaping outward with the rip failed to recirculate.

It is interesting that there was an order of magnitude agreement between the field and laboratory set-ups [*Saville*, 1961; *Bowen*, 1967; *Bowen et al.*, 1968]. The maximum water-level differences in the study area (i.e., set-up plus set-down) amounted to about 12 cm, or about 33% of the breaker height at the bar (35 cm), representing approximately the same relative magnitude as discovered in laboratory tests.

Currents associated with oblique-wave inci-

dence. A meandering current pattern associated with oblique-wave incidence was observed by means of balloon-borne photography, as shown in Figures 6*a*–6*c*. Photographs were made at 1-min intervals, but only three samples are presented (lapse times $t = 6$, 8, and 12 min after dye injection). They reveal a wave-incidence angle of about 10°. The bottom in the surf zone contained longshore sequences of shoals and depressions, the effect of which was revealed in the rhythmic shore lines with a wavelength of about 100 meters. Initially, a potassium permanganate solution was injected in a straight line about 100 meters long parallel to the shore line in the surf zone.

Time-lapse positions of dye patches indicated three distinct current subsystems: (1) meandering current inside the surf zone, (2) longshore current parallel to the shore line outside the breaker line, and (3) eddy motion in the lee of a rip current. In these photographs, currents are labeled as follows: on the shoals, A; in the rip, B; outside the breaker line, C; and in the lee of the rip, D. Subscripts 1, 2, and 3 are numbered in the down-wave direction.

1. *Meandering current.* The meandering current can be traced by following B_0, A_0, B_1, and A_1, in that order. The patch at A_0 (Figure 6*a*) was formed by a mass branching out of the rip head B_0, which subsequently moved diagonally shoreward across the shoal toward the downstream rip B_1. The rip B_1 issued diagonally seaward, starting at the embayment of the shore line. There was another branching at the rip head B_1 (Figures 6*b* and 6*c*), followed by yet another diagonal shoreward current at A_1, which was bound for a third rip, farther downstream (not seen in the photos).

2. *Longshore currents.* Dye separating from the rip head traveled parallel to the shore line outside the breaker line (C_0 and C_1, Figure 6*a*). This dye patch tended to shed part of its mass into the surf zone as it passed across the front of the shoal (Figures 6*b* and 6*c*). Little seaward diffusion from this patch took place, however. Although this parallel current outside the breaker line was continuous, the contribution from the rip into this current was sporadic since only during outbursts of rip velocities did a large amount of water from inside the surf zone penetrate the breaker line. As a result, this parallel current outside the breaker was delineated by discrete patches of dye issuing from the rip at intervals of rip pulsation. In the example shown here (see C_0 and C_0', produced by two successive outbursts at rip B_0), the period of pulsation was about 2 min and the speed of the parallel current was about 0.3 m/sec.

3. *Closed-eddy motion.* A small eddy motion was detected in the lee of the rip (D_1),

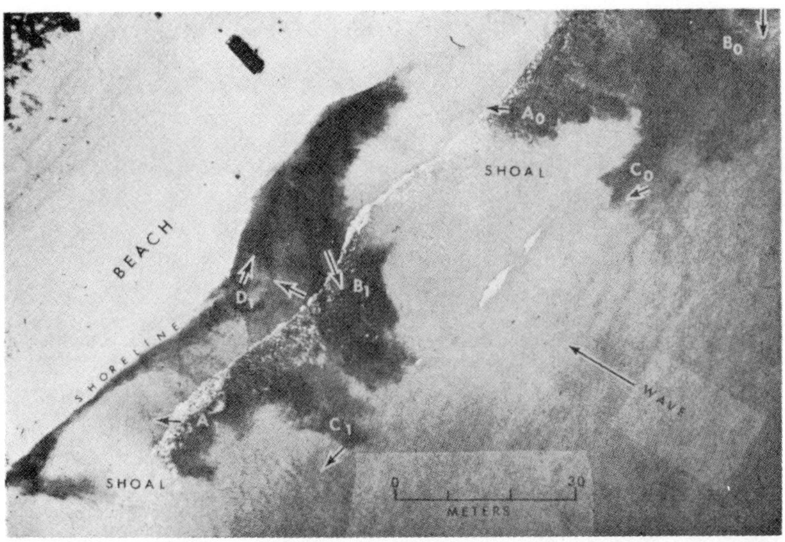

Fig. 6*a*. Meandering current pattern photographed from balloon-borne camera. Lapse time, $T = 6$ min.

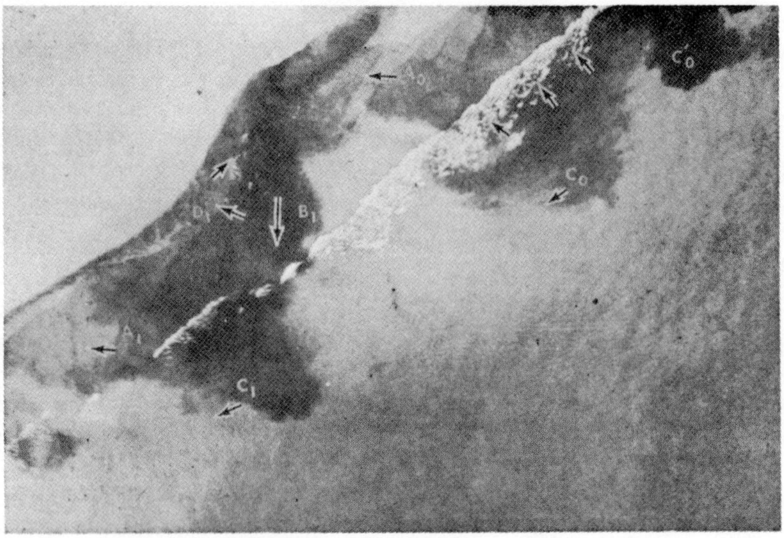

Fig. 6b. Meandering current. Lapse time, $T \doteq 8$ min.

which recycled shoreward part of the water moving outward with the rip. No similar eddy motion was present on the upwave side of the rip.

Although an exact mathematical solution for a meandering current system requires further study, a plausible explanation appears possible by considering a superposition of stream functions for parallel and circulation currents. Consider a wave train with an incidence angle α_b. By tensor transformation we get

$$\begin{aligned} S_{xx} &= S_{rr} \cos^2 \alpha_b + S_{ss} \sin^2 \alpha_b \\ S_{yy} &= S_{rr} \sin^2 \alpha_b + S_{ss} \cos^2 \alpha_b \end{aligned} \qquad (6)$$

Since $S_{rr} = (3/16)\rho g H^2$ and $S_{ss} = (1/16)\rho g H^2$ [*Longuet-Higgins and Stewart,* 1962, 1964], these are rewritten to

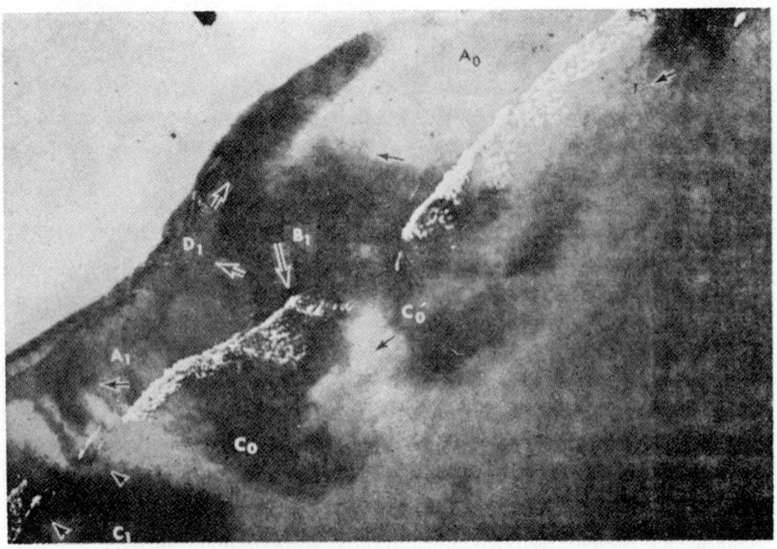

Fig. 6c. Meandering current. Lapse time, $T \doteq 12$ min.

$$S_{xx} = (\tfrac{3}{16})\rho g[H(1 - \tfrac{2}{3}\sin^2\alpha_b)^{1/2}]^2 \leq S_{rr}$$
$$S_{yy} = (\tfrac{1}{16})\rho g[H(1 - \tfrac{2}{3}\cos^2\alpha_b)^{1/2}]^2 \leq S_{ss} \quad (7)$$

The diagonal elements of the radiation stress tensor, which account for a driving force for circulation, i.e., S_{xx} and S_{yy}, now behave as if the wave height were reduced by factors $(1 - 2/3 \sin^2 \alpha_b)^{1/2}$ and $(1 - 2/3 \cos^2 \alpha_b)^{1/2}$, respectively. We will denote the transport stream function for this circulation by ψ_1. Because of the oblique-wave incidence, the flux of momentum across a plane parallel to the shore line is not zero; hence,

$$S_{yx} = -(S_{rr} - S_{ss}) \sin \alpha_b \cdot \cos \alpha_b$$
$$= -\tfrac{1}{8}\rho g H^2 \sin \alpha_b \cdot \cos \alpha_b \neq 0 \quad (8)$$
$$\alpha_b \neq 0$$

and a longshore current is driven. *Bowen* [1969*a*], *Longuet-Higgins* [1970*a, b*], and *Iwata* [1970] independently derived formulas for longshore-current velocity distribution on a plane beach. The longshore current will be distorted by an undulatory surf-zone bed; however, to be able to employ one of these formulas, we will assume a plane bed for the longshore current (see equation 3). Bowen's formula was used to compute the longshore velocity for the same wave properties as in Figure 4, i.e., $H_b = 0.39$ meters, $T = 5$ sec, and $\tan \beta = 0.03$. The transport stream function for the parallel longshore current was obtained by

$$\psi_2 = -\int_0^x (vd)\, dx$$

The resultant current system was given by

$$\psi = \psi_1 + \psi_2$$

assuming no interaction between circulation cells and parallel currents.

The result of the computation is shown in Figure 7, which reveals a meandering current pattern featuring an oblique shoreward current on the shoal ($y \simeq \pm(15 - 30$ meters)) and an oblique seaward current on the depression ($y \simeq (0 - \pm 15$ meters)). It also predicts a small recirculation eddy in the lee of the rip, as was also observed. The current speed drops markedly outside the breaker line, and a tendency for curved streamlines is retained to some extent.

In the field two wave trains that consisted of swells of normal incidence and local wind waves of oblique incidence usually coexisted. Under this condition, circulation (ψ_1) was reinforced by swells and the tendency for meander was greater than when there was only a single oblique wave train.

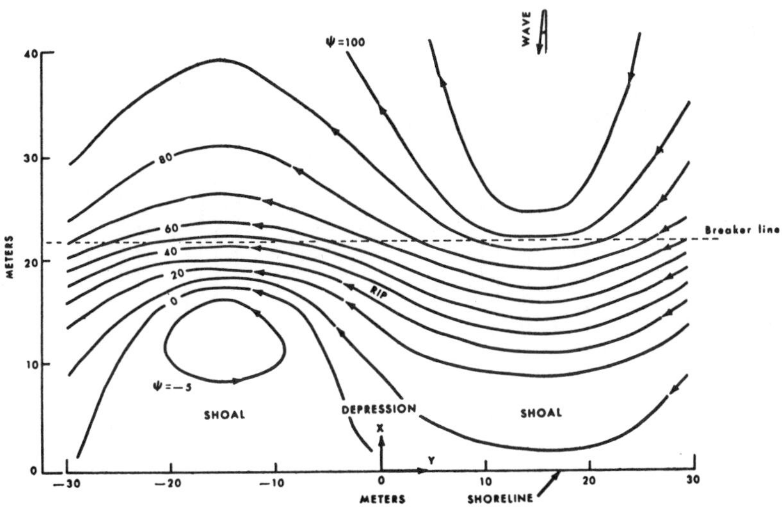

Fig. 7. Computed transport streamlines of a meander pattern for oblique-wave incidence (5°) and undulatory surf zone.

VELOCITIES IN THE RIP CURRENT

Figure 8 shows an example of rip-current velocities obtained with a bidirectional current meter; the data were digitized at 2-sec intervals. The high-frequency fluctuations corresponded to incident-wave frequencies, whereas the banding of high-frequency fluctuations represented pulsations at beat frequencies. It is important to note that the rip is not steady, as a steady-state solution suggests [*Bowen*, 1967]. As already described in relation to Figure 6, the rip outflow in most cases undergoes periodic eruptions at beat frequencies.

Figure 9 shows an example of velocity profiles in the rip current in which each plot represents an average of 2-min recordings. All profiles showed relatively high velocities near the bed. This was especially pronounced nearer the shore; there was an indication that the locus of peak velocity in the profiles shifted upward with distance seaward. The measurement outside the breaker line in the upper layer was unsatisfactory because the meter was exposed to the air in the wave troughs. However, the velocities in the lower half of the water depth were essentially nil, which suggests that outflow outside the breaker line probably concentrated near the surface. *Shepard et al.* [1941] quoted a statement by a lifeguard that 'where the neck is very narrow the flow underneath is stronger than at the surface and where the rip is spread over considerable area the flow decreases with depth.' *Shepard and Inman* [1950] also reported a concentrated surface outflow near the rip head. The problem is one of interaction between waves and opposing currents, the mechanism of which has had limited investigation [*Longuet-Higgins and Stewart*, 1960; *Hamada*, 1961].

Intensities of velocity fluctuation associated with rip currents can be shown to be substantially different from turbulence intensities normally observed in an open-channel flow. Figure 10 shows the vertical distribution of turbulence intensity, defined by $(\langle u'^2\rangle^{1/2})/U$, in which u' is a deviation from the mean velocity at a given elevation and U is the depth mean velocity. For the purpose of comparison, turbulence intensities observed in a laboratory open-channel flow over a rough bed are also shown [*McQuivey and Richardson*, 1969]. The intensities of velocity fluctuation in the rip current are found to be an order of magnitude larger than those in open-channel flow. In addition, whereas they decrease exponentially upward in open-channel flow, they tend to increase upward in a rip current. The discrepancy reflects the difference in

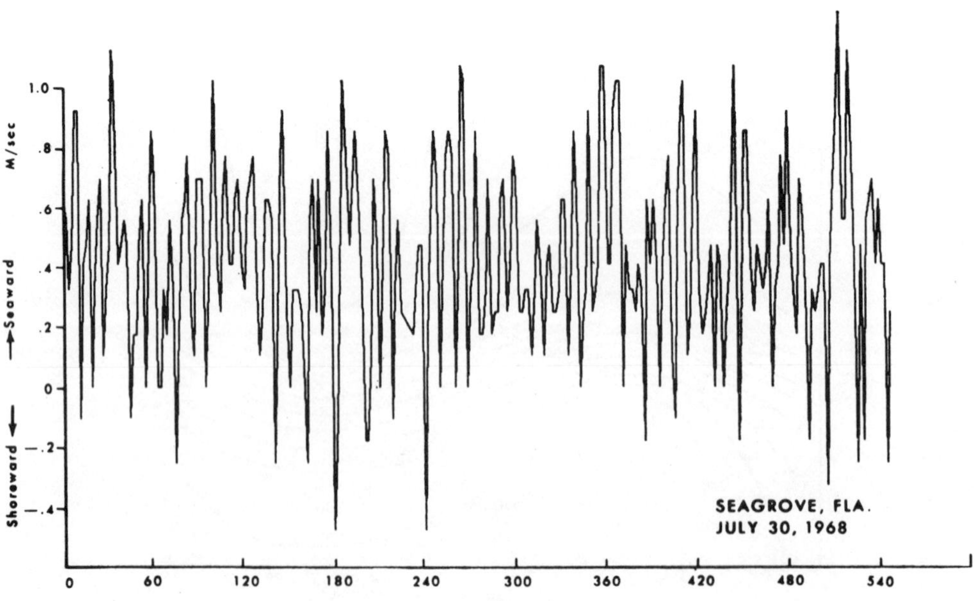

Fig. 8. Sample time history of rip-current velocities. Data from sensor V-6.

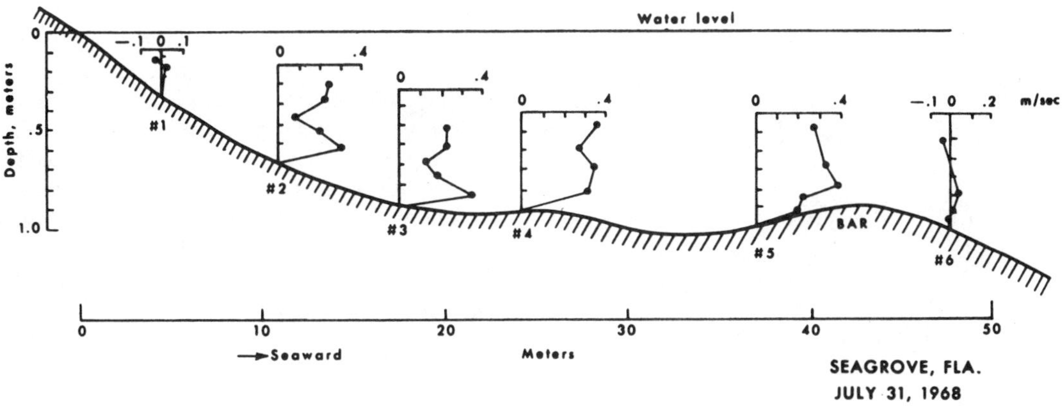

Fig. 9. Velocity profiles in a rip current.

the origin of turbulence in the two cases; whereas the turbulence in the open-channel flow originates at the rough bed, diffusing upward, the velocity fluctuations in the rip current are imposed by orbital motion as a result of surface waves.

To investigate cross-spectral relationships between rip-current velocities and wave characteristics, two sets of measurements at both high and low tides were taken. In the high-tide measurement, the wave and current sensors were placed on the breaker line over the bar (W-5 and V-5, respectively; see Figure 2). In the low-tide measurement, the wave sensor (W-6) was moved near the shore line, directly shoreward across a shoal from W-5, whereas the current sensor (V-6) was moved to the neck of the rip. In both measurements the longshore separation between wave and current sensors was 40 meters, the former being located on a shoal.

The wave power spectrum at the breaker line (Figure 11, top) revealed three significant peaks of power density for wind waves ($\simeq$0.2 sec^{-1}), swells ($\simeq$0.1 sec^{-1}), and beat oscillations ($\simeq$0.02 sec^{-1}). Significant breaker heights and periods, with a combination of both swells and wind waves, were 0.395 meters and 5.0 sec in this measurement (made simultaneously with current measurement of Figure 2). At the shore line, however, the power density at wind-wave and swell frequencies attenuated as a result of breaking in the surf zone. This effect was most pronounced across a shoal and particularly at low tide, as shown by the bottom graph of Figure 11, in which the power spectrum retains a significant density peak only at beat frequencies. The corresponding significant breaker heights and periods on the bar were 0.42 meters and 7.5 sec, hence larger than the top graph of Figure 11. These latter measurements were taken simultaneously with the rip-current record shown in Figure 8. The field observation showed that at all phases of tide more wave energy reached the shore line across a depression, i.e., against a rip current, than across a shoal.

The spectrum of high-tide rip currents (Figure 12, V-5; compare with Figure 11, top) displayed peak densities at wind-wave, swell, and beat frequencies, and lesser peaks were caused by secondary oscillations. The only significant coupling between currents and waves was recognized at the swell frequencies, as shown by a coherence level of about 0.8 over a relatively

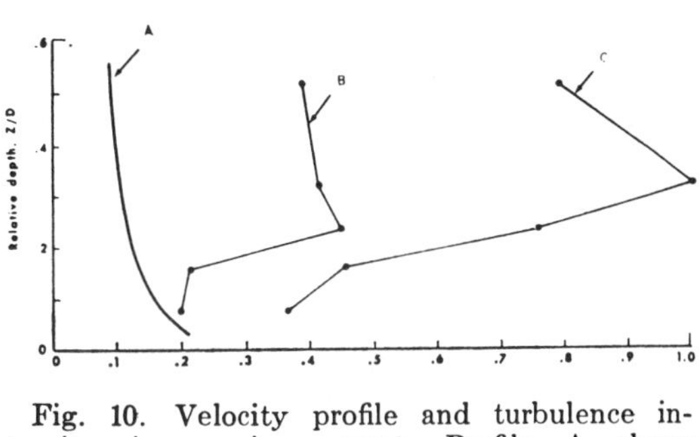

Fig. 10. Velocity profile and turbulence intensity in a rip current. Profile A shows $(\langle u'^2\rangle)^{1/2}/U$, the unidirectional open channel flow. Profile B shows rip current (cm/sec). Profile C shows rip current $(\langle u'^2\rangle)^{1/2}/U$.

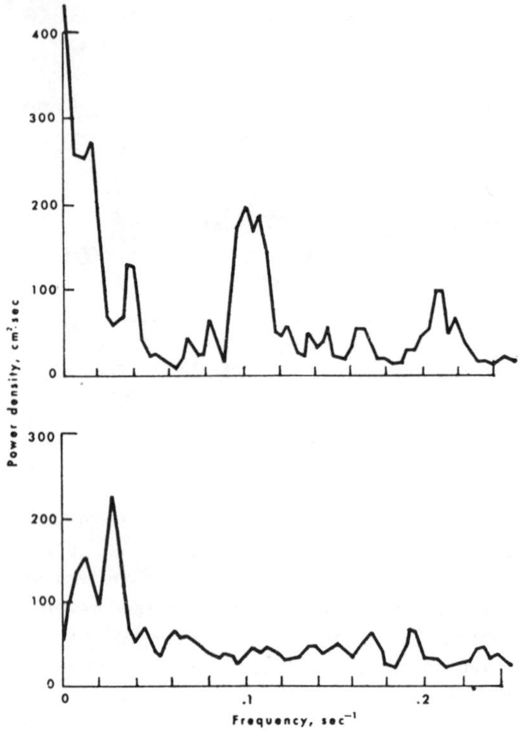

Fig. 11. Power spectra of incident waves. (Top) Breaker at bar at high tide, July 31, 1968. Data from sensor W-5, where $\Delta t = 2$ sec, $n = 1000$, and $m = 80$ lags. (Bottom) Breaker nearshore at low tide, July 30, 1968. Data from sensor W-6.

broad frequency band around 0.1 sec^{-1}. The corresponding phase lag was about $\pm\pi$, meaning that the seaward (positive) eruption of rip velocities coincided with the troughs of swells. (A high power density near zero frequency represents the presence of a trend in the velocity record, not the effect of the mean rip velocity, which was subtracted from the data prior to spectral analysis.)

The spectrum of low-tide rip currents (Figure 13, V-6; compare with Figure 11, bottom) was characterized by a greater overall power density (hence a higher level of turbulence), a prominent peak at swell frequencies ($\simeq$0.1 sec^{-1}), and a somewhat smaller, yet significant, peak at best frequencies ($\simeq$0.04 sec^{-1}). Wind waves were effectively dissipated as a result of intense breaking in the rip current near the bar and failed to influence the current at the neck. Coherence at swell frequencies was low, but this partly reflected the fact that the current and wave sensors were offset relative to the shore line. A significant feature in the low-tide rip-current spectrum was the high level of coupling at beat frequencies, as shown by a coherence of close to 0.8 for frequencies $\leq$0.04 sec^{-1}. This value agreed with the observation that as the rip-current velocities increased with increasing breaker activity over the bar during a falling tide the periodic pulsation of velocities in the rip current tended to occur at progressively longer intervals. Phase lag at these beat frequencies was about $-\pi/2$, indicating that the current advanced the beat oscillations by one-quarter cycle. This may be explained by assuming the reflection of surf beats from the beach and the consequent formation of standing waves. In a linear-wave theory, the horizontal particle velocity in a standing wave precedes the surface oscillation by $\pi/2$ at all positions in the wave

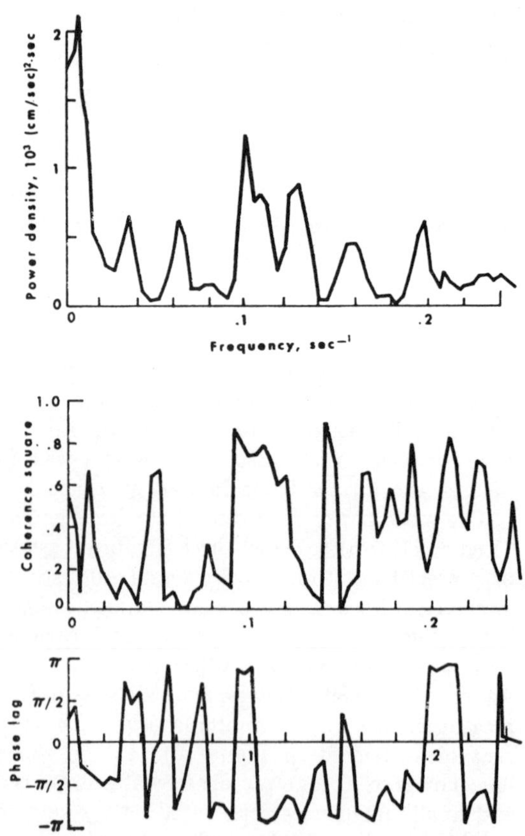

Fig. 12. Spectral properties of rip velocities at high tide. Rip current at high tide, July 31, 1968. Data from sensor V-5, where $\Delta t = 2$ sec, $n = 1000$, and $m = 80$ lags.

train. Although the current and wave sensors were offset in this measurement, their separation was essentially negligible as compared to the beat wavelength. Interestingly, the phase lag at beat frequencies in the high-tide measurement (Figure 12) was also nearly $-\pi/2$.

DISCUSSION

Our field observations in 1967 and 1968 demonstrated that undulation of the surf-zone bed was an essential condition to the formation of two major patterns of wave-induced nearshore currents: circulation and meander. Three additional field sessions in the same general area from 1969 through 1971 further confirmed this point. *Bowen* [1967, 1969*b*] theoretically showed that variable breaker heights away from the shore line could induce circulations on a plane beach. Unfortunately, such a situation is almost impossible to observe on the Florida gulf coasts, which are consistently characterized by various degrees of surf-zone undulation and little longshore variation in breaker height. A salient effect of the surf-zone relief was particularly noticed in the shape of the rip outflow. As evident from a comparison between Figures 2 and 5, a jetlike rip current with a typical prominent neck and an expanding head was invariably associated with a channel considerably deeper than the neighboring shoals. Where the surf-zone relief was weak, as in Figure 5, the outflow had indistinct boundary delineation and little lateral expansion seaward of the breaker line. These observations bear out the nonlinear mechanism of streamline concentration in a deepening channel, as explained by *Arthur* [1962].

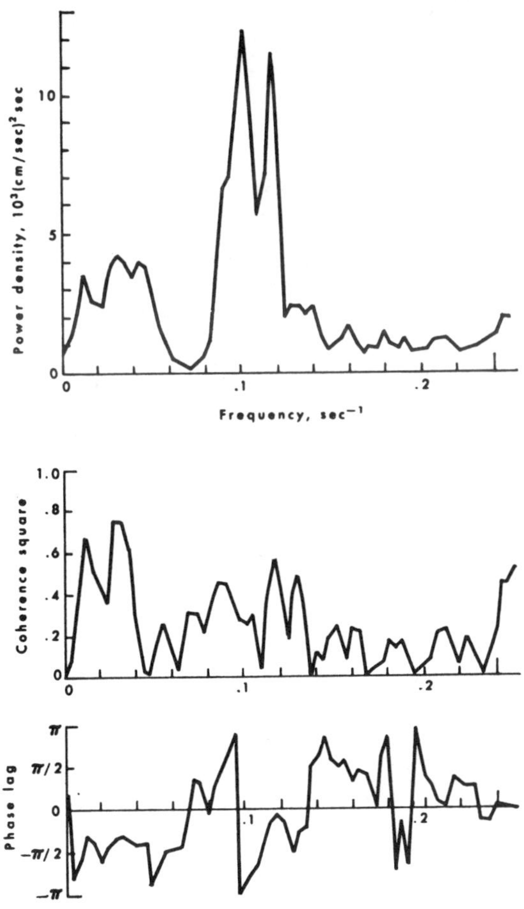

Fig. 13. Spectral properties of rip velocities at low tide. Rip current at low tide July 30, 1968. Data from sensor V-6, where $\Delta t = 2$ sec, $n = 1000$, and $m = 80$ lags.

Another salient aspect of the current systems observed in this area was that the current patterns, either circulatory or meandering, were stationary because they were arrested by the existing surf-zone topography. Between July 27 and July 30 in this study there was a brief storm during which waves with a maximum height of nearly 1 meter and a period of 5 sec arrived at 10° to the shore line, accompanied by strong longshore currents up to 50 cm/sec. During this storm the initially irregular topography shown in Figure 5 evolved into a series of periodic undulations with a wavelength of about 60 meters, as shown in Figure 2. The current patterns associated with this topography contained rip currents at precisely the same spacings as the undulation wavelength, and their directions were oriented either normal or oblique to the shore line, depending upon the wave angle.

Development of surf-zone undulations is essentially an unsolved problem. Alternative explanations have appeared in the literature. *Sonu* [1968] reasoned that beach material which had been deposited in the surf zone during the height of a storm could be rearranged into a train of sand-wave-type formations in response to longshore currents during the period of wave subsidence. Successive daily surveys of surf-zone topography illustrating this process have been presented elsewhere (C. J. Sonu, unpublished data, 1972). *Clos-Arceduc* [1962, 1964] suggested that on a beach bounded by promontories a seiche trapped between the promontories would induce stationary periodic nodes and antinodes, and the consequent redeposition of bottom material would take the form of crescentic bars. This approach was recently further exploited by *Bowen and Inman* [1971] on the basis of consideration of drift velocities in a standing edge wave. However, their experiment showed that crescentic bars are formed away from the shore, and therefore it is not clear whether this mechanism is related to surf-zone undulations. *Bowen and Inman* [1969] suggested also that a circulation developing on an initially plane bottom as a result of coupling between an incident surface wave and an edge wave could cause redistribution of sediment to form undulatory surf-zone topography. It can be shown, however, by using the dispersion relation of *Ursell* [1952] and *Eckart* [1965], that typical circulation spacings on the order of 60 meters or more can only be explained from incident-wave periods far in excess of those of ordinary surface waves. Since such long waves usually fail to break away from the shore, they are unable to produce a driving force for circulation in the surf zone, as explained by *Bowen* [1969*b*]. In addition, there is evidence (C. J. Sonu, unpublished data, 1972) that the surf-zone undulations migrate with longshore currents as they are formed, a process that cannot be explained by a stationary coupling mechanism between incident and edge waves. Further clarification of this subject apparently needs more field studies combining waves, currents, and topography.

Acknowledgments. We are indebted to the authorities at Elgin AFB, Florida, who kindly supplied all the helium used for our balloon and offered a site and valuable facilities within the Air Proving Ground, Santa Rosa Island, Florida, for our experiments, through the intervention of Mr. Marshal Cartledge. The author is grateful to Mr. Rodney Fredericks, Mr. Rodolfo Cruz, and Dr. L. D. Wright, who at the time of the study was a graduate assistant, for providing essential contributions to the execution of field operations.

This study was supported by the Office of Naval Research, Geography Programs, under contract N00014-69-A-0211-0003, project NR 388 002, with the Coastal Studies Institute, Louisiana State University, Baton Rouge, Louisiana.

REFERENCES

Arthur, R. S., A note on the dynamics of rip currents, *J. Geophys. Res.*, *67*(7), 2777–2779, 1962.

Bowen, A. J., Rip currents, Ph.D. dissertation, 115 pp., Univ. of Calif. at San Diego, La Jolla, 1967.

Bowen, A. J., The generation of longshore currents on a plane beach, *J. Mar. Res.*, *27*(2), 206–215, 1969*a*.

Bowen, A. J., Rip currents, 1, Theoretical investigations, *J. Geophys. Res.*, *74*(23), 5467–5478, 1969*b*.

Bowen, A. J., and D. L. Inman, Rip currents, 2, Laboratory and field observations, *J. Geophys. Res.*, *74*(23), 5479–5490, 1969.

Bowen, A. J., and D. L. Inman, Edge waves and crescentic bars, *J. Geophys. Res.*, *76*(36), 8662–8671, 1971.

Bowen, A. J., D. L. Inman, and V. P. Simmons, Wave set-down and set-up, *J. Geophys. Res.*, *73*(8), 2569–2577, 1968.

Clos-Arceduc, A., Etude sur les vues aeriennes, des alluvions littorales d'allure periodique, cordons littoraux et festons, *Soc. Fr. Photogramm. Bull.*, *4*, 13–22, 1962.

Clos-Arceduc, A., La photographie aerienne et l'etude des depots prelittoraux, *Etud. Photo-Interpretation 1*, 53 pp., Inst. Geogr. Nat., Paris, 1964.

Eckart, C., Surface waves on water of variable depth, *Wave Rep. 100, Ref. 51–12*, 99 pp., Scripps Inst. of Oceanogr., Univ. of Calif., 1965.

Evans, O. F., Mass transportation of sediments on subaqueous terraces, *J. Geol.*, *47*, 325–334, 1939.

Hamada, T., A computation of wave characteristics in an opposing current (in Japanese), *Proc. 8th Japan Coastal Eng. Conf., Japan Soc. Civil Eng.*, 25–29, 1961.

Iwata, N., A note on the wave set-up, longshore currents, and undertows, *J. Oceanogr. Soc. Japan*, *26*(4), 233–236, 1970.

Longuet-Higgins, M. S., Longshore currents generated by obliquely incident sea waves, 1, *J. Geophys. Res.*, *75*(33), 6778–6789, 1970*a*.

Longuet-Higgins, M. S., Longshore currents generated by obliquely incident sea waves, 2, *J. Geophys. Res.*, *75*(33), 6790–6801, 1970*b*.

Longuet-Higgins, M. S., and R. W. Stewart, The changes in amplitude of short gravity waves on steady non-uniform currents, *J. Fluid Mech.*, *10*(4), 529–549, 1960.

Longuet-Higgins, M. S., and R. W. Stewart, Radiation stress and mass transport in gravity waves with application to surf beats, *J. Fluid Mech.*, *13*, 481–504, 1962.

Longuet-Higgins, M. S., and R. W. Stewart, Radiation stress in water waves, A physical discussion with applications, *Deep Sea Res.*, *11*(4), 529–563, 1964.

McKenzie, R., Rip current system, *J. Geol.*, *66*(2), 103–113, 1958.

McQuivey, R. S., and E. V. Richardson, Some turbulence measurements in open-channel flow, *Proc. Amer. Soc. Civil Eng.*, *95*(HY1), 209–223, 1969.

Munk, W. H., and M. A. Traylor, Refraction of ocean waves: A process linking underwater topography to beach erosion, *J. Geol., 55,* 1–26, 1947.

Rossby, C. G., Planetary flow patterns in the atmosphere, *Quart. J. Roy. Meteorol. Soc., 66* (suppl.), 72, 1940.

Saville, T., Experimental determination of wave set-up, *Proc. 2nd Tech. Conf. Hurricanes,* 242–252, 1961.

Shepard, F. P., and D. L. Inman, Nearshore water circulation related to bottom topography and wave refraction, *Eos Trans. AGU, 31*(2), 196–212, 1950.

Shepard, F. P., K. O. Emery, and E. C. LaFond, Rip currents: A process of geological importance, *J. Geol., 49*(4), 337–369, 1941.

Shonting, D. H., Observations of particle motions in ocean waves, *TM 377,* 1 and 2, Naval Underwater Weapons Res. and Eng. Sta., Newport, R.I., 1967.

Sonu, C. J., Collective movement of sediment in littoral environment, *Proc. 11th Conf. Coastal Eng.,* 373–399, 1968.

Sonu, C. J., Tethered balloon for study of coastal dynamics, *Proc. Symp. Earth Observ. Balloons,* 91–102, 1969.

Sonu, C. J., J. M. McCloy, and D. S. McArthur, Longshore currents and nearshore topographies, *Proc. 10th Conf. Coastal Eng.,* 525–549, 1966.

Ursell, F., Edge waves on a sloping beach, *Proc. Roy. Soc., Ser. A, 214,* 79–97, 1952.

14

Reprinted from *Jour. Geophys. Research* 75(30):5914–5927 (1970)

Longshore Sand Transport on Beaches

PAUL D. KOMAR AND DOUGLAS L. INMAN

Scripps Institution of Oceanography, University of California, La Jolla 92037

Simultaneous field measurements of wave and current parameters in the surf zone and the resulting longshore transport of sand have been made on two beaches under a variety of conditions. The direction and flux of wave energy was measured from an array of digital wave sensors placed in and near the surf zone. Quantitative measurements of the longshore sand transport rate were obtained from the time history of the center of gravity of sand tracer. The measurements have been used to test two models for the prediction of the longshore transport rate of sand. The first model gives the immersed weight longshore transport rate of sand, I_l, as proportional to the longshore component of wave energy flux (power), $I_l = K(ECn)_b \sin \alpha_b \cos \alpha_b$, where E is the energy density, Cn is the wave group velocity, and α_b is the breaker angle. The second model assumes that the waves provide the power to move and support the sand and that the superimposed longshore current $\langle v_l \rangle$ provides a longshore component that results in the longshore transport of sand according to the relationship $I_l = K' (ECn)_b \cos \alpha_b \langle v_l \rangle / u_m$, where u_m is the magnitude of the maximum horizontal component of orbital velocity near the bottom under the breaking wave, assumed to be proportional to the rate of energy dissipation by friction on the beach bed. The measurements show that both models successfully predict the sand transport rate, with values of the dimensionless coefficients $K = 0.77$ and $K' = 0.28$. The coherence of the models implies that they are interrelated, their common solution giving the relation as $\langle v_l \rangle = K'' u_m \sin \alpha_b$, where K'' is a dimensionless constant equal to 2.7. This relation can be obtained directly by equating the longshore current and the longshore component of the momentum flux (radiation stress) of the breaking waves. Thus, the coherence of the models appears to be based on the generation of the longshore currents by the longshore radiation stress. The models will not be equivalent if $\langle v_l \rangle$ owes its origin to some other generating mechanism such as tides or winds.

It has been recognized for many years that, when waves break at an angle to the shoreline, they cause a transport of sand along the beach. This longshore transport of sand is of great interest to the coastal engineer who is to build structures such as jetties and breakwaters that act as dams to the longshore movement of sand. It is also of interest to marine geologists in understanding the formation of spits and the transport of sand along the coasts of the world.

Grant [1943] has suggested that the longshore current combined with the agitating action of the breaking waves provides the driving force for moving sand along the beach. A report of the *Scripps Institution of Oceanography* [1947] suggests that the work performed by waves might be a useful parameter for predicting the longshore transport rate of sand by wave action. This concept has been applied by the Los Angeles District of the U.S. Army Corps of Engineers to estimate the predominant direction of sand transport from hindcast wave data [*Eaton,* 1951]. Since then, numerous laboratory and field investigations of the relation between the sand transport rate and wave and current parameters have been conducted. The field experiments of *Watts* [1953], *Caldwell* [1956], and *Ingle* [1966] all indicate that the longshore transport of sand is proportional to the longshore component of the energy flux of the waves measured at the break point. However, none of these experiments were made with sufficient resolution to permit an accurate assessment of both the sand transport rate and the energy flux of the waves.

Thus, a major limitation of our understanding of the relation between the longshore transport of sand and wave and current parameters has been the almost total lack of reliable field measurements relating the two. The development of radioactive tagging of sand grains and the more readily applied techniques of fluorescent tagging of sand together with the procedures of *Crickmore and Lean* [1962] made the accurate assessment of sand transport rates possible. To reliably measure the energy flux of the waves one must install wave sensor arrays in and near the surf zone and compute wave spectra and cross spectra [*Koontz and Inman*, 1967]. Field measurements using these newer tools began in 1966. In a preliminary report *Inman et al.* [1969] verified that the longshore transport rate of sand is directly proportional to the longshore component of wave power for fully developed transport conditions and showed that the factor of proportionality between the transport rate and the energy flux of the waves was larger than previously thought. Our study also shows that transport conditions in the laboratory are not fully developed and that therefore laboratory data cannot be linearly extrapolated to predict field conditions. This paper presents the results of the completed study; it is also in part a summary of the Ph.D. thesis of *Komar* [1969], which may be consulted for greater detail.

Mechanics of Transport

Two concepts or models have been proposed for the longshore transport of sand. The first is principally intuitive and attempts to correlate the sand transport rate to the longshore component of the wave energy flux per unit length of beach

$$P_l = (ECn)_b \sin \alpha_b \cos \alpha_b \tag{1}$$

where E is the wave energy density, Cn is the wave group velocity, and α_b is the breaker angle. The subscript b denotes the wave breaker position. Energy flux has the units of power, and for that reason P_l has generally been termed 'the longshore component of wave power.' However, this term is somewhat misleading because in physics power is a scalar quantity that lacks direction.

Initial attempts to correlate sand transport to wave parameters focused on demonstrating empirical relationships between P_l and the longshore volume transport rate of sand S_l [*Watts*, 1953; *Caldwell*, 1956; *Savage*, 1959]. Such purely empirical correlations summarize the available data but have the disadvantage of being dimensionally incorrect.

Inman and Bagnold [1963] point out that the transport rate of sediment should be expressed as an immersed weight transport rate given by

$$I_l = (\rho_s - \rho) g a' S_l \tag{2}$$

where ρ_s and ρ are respectively the sand and water densities and a' is the correction factor for pore space and can be taken as 0.6. It is advantageous to express the sand transport as an immersed weight rather than a volume because doing so takes into consideration the density of the sediment grains and permits I_l and P_l, which have the same units, to be related by

$$I_l = KP_l = K(ECn)_b \sin \alpha_b \cos \alpha_b \tag{3}$$

where K is a dimensionless coefficient or proportionality.

This relationship was basically intuitive because no real consideration of the actual mechanics of sand transport went into its formulation. However, *Komar* [1970] shows that equation 3 can be derived directly under the assumption that the longshore transport rate of sand is equal to the longshore component of the sediment being carried forward under the bore from the breaking wave as the wave advances obliquely toward the shoreline.

Inman and Frautschy [1966] calculated the immersed weight transport rates I_l for the available field and laboratory data. A straight line with a slope of unity on a log-log plot of I_l versus P_l fitted the data then available, giving a value of K in equation 3 of 0.25.

An alternate, seemingly independent relation between the sand transport rate and the wave and current parameters was proposed by *Bagnold* [1963, p. 518] and applied to the littoral zone by *Inman and Bagnold* [1963, p. 545]. This model is based on the supposition that the orbital motion of the waves supports and moves sediment back and forth but produces no net transport even though wave energy is expended. Since the back and forth motion of the sand is supported by the wave motion, any net current superimposed on this motion will produce a net transport of sand. If some portion of the energy flux is dissipated in placing the sand in motion, the mean stress applied to the beach face is proportional to

$(ECn)_b \cos \alpha_b/u_0$, where $(ECn)_b \cos \alpha_b$ is the energy flux per unit length of beach and u_0 is the mean frictional velocity relative to the bed within the surf zone. We assume that u_0 is proportional to u_m, the maximum horizontal component of the orbital velocity near the bottom just before the wave breaks. If $\langle v_l \rangle$ is the average longshore current in the surf zone, the resulting immersed weight sand transport rate past a section of beach becomes

$$I_l = K'(ECn)_b \cos \alpha_b \langle v_l \rangle / u_m \qquad (4)$$

where K' is a dimensionless factor of proportionality. The origin of the current $\langle v_l \rangle$ is left unspecified so that the model permits a longshore transport of sand to result from a wave train arriving parallel to the shoreline ($\alpha_b = 0$) in the presence of a longshore current whose generation is not dependent on an oblique wave approach. For example, equation 4 would presumably predict the transport of sand resulting from tidal or wind-driven longshore currents in the presence of wave action.

Equations 3 and 4 provide two seemingly independent relationships for the prediction of the longshore sand transport rate from wave and current parameters. It is apparent that both models can be correct only when they are equivalent, and it will be shown that this equivalence occurs when the longshore current is due only to the longshore component of the breaking waves.

Bowen [1969a] suggested that longshore currents are generated from the longshore component of the radiation stress, where the radiation stress is defined by *Longuet-Higgins and Stewart* [1964] as the excess flow of momentum due to the presence of the waves. Following this suggestion, *Inman et al.* [1969] and *Longuet-Higgins* [1970] have independently obtained the relationship

$$\langle v_l \rangle = c_1 u_m \frac{\tan \beta}{c_f} \sin \alpha_b \cos \alpha_b \qquad (5)$$

for the longshore current, where $\tan \beta$ is the beach slope, assumed to be constant, c_f is the bottom frictional drag coefficient, u_m is the maximum horizontal component of the orbital velocity of the waves, and c_1 is a dimensionless coefficient of proportionality that *Longuet-Higgins* [1970] gives as $5\pi/8$. Substitution of this expression of $\langle v_l \rangle$ in equation 4 gives a relation that is similar to (3) but is also dependent on the additional factor ($\tan \beta \cos \alpha_b / c_f$). However, considerations by *Komar* [1969] suggest that

$$(\tan \beta \cos \alpha_b)/c_f = \text{constant} \qquad (6)$$

so that variation in the beach slope $\tan \beta$ does not produce a change in the longshore transport rate of sand. This possible dependence on $\tan \beta$ will be reexamined with the data obtained in this study.

Methods and Procedures

Longshore transport of sand. We obtained quantitative measurements of the longshore transport rate of sand by using sand tracers, i.e., native sand colored with a thin coating of fluorescent dye. The sand tracer was introduced onto the beach face as a single large injection of between 45 and 90 kg (100 to 200 pounds). The time history of the position of the center of gravity of the tracer as it moved along the beach was utilized to deduce the rate of advection of the sand on the beach. The movement of the sand was determined by collecting up to 200 volume samples from a grid that extended up to 450 meters along the shore and 60 meters offshore. The sample locations were closest together near the injection site and become progressively farther apart with distance from the injection line. Sampling was generally conducted between 2 and 4 hours after injection. All samples were collected with a volume sampler that penetrates the beach face to a depth of 5 cm and obtains a volume of about 450 cm^3 of sand. A sample of this size permits detection of tracer concentrations as low as one tracer grain in 10^7 to 10^8 untagged grains.

After the samples were brought back to the laboratory, washed, and dried, the number of tracer grains was counted by scanning the sample with an ultraviolet light. Tracer concentrations were expressed as the number of tracer grains per kilogram of sample. The values of tracer concentration were contoured as shown in Figure 1. Other distributions used in the study can be found in *Komar* [1969]. Following the procedure developed by *Crickmore and Lean* [1962] for rivers, we calculated the mean distance of transport from the tracer distribution by taking the first moment of the concentration in the longshore direction about the injection line. Dividing this mean longshore transport distance by the elapsed time between injection and sampling then yields a mean speed for the longshore sand transport $\langle V_l \rangle$ that is applicable to the total thickness of sand moved.

Meaningful results are obtained by this procedure only when the volume samples that the concentration diagrams are based on extend well into the beach face to the depth of burial of the sand tracer. *Ingle* [1966] used sticky cards that sampled only the top layer of sand on the beach face. Such sampling yields results that are not representative of the entire distribution of tracer and therefore are not suitable for quantitative purposes. In our study the grid sampling on the average accounted for over 80% of the total tracer introduced onto the beach.

The thickness of the layer of motion on the beach face as well as the layer's advection rate are required to compute the longshore transport rate. The thickness of the layer of motion was obtained in two ways: (1) We always assumed that the thickness of motion was equal to the depth of burial of the tracer in cores of the beach face. This requires that sufficient time for burial has elapsed, but burial appears to be quick, being complete within a half hour after tracer introduction. (2) Following *King*'s [1951] approach (which can be used only at locations where there is a sizable tidal range) we dug a series of small holes into the beach face and filled them with natural beach sand dyed with India ink. Near each we placed a stake used both to relocate the position and to determine changes in the level of the beach face. During the subsequent low tide we excavated the dyed sand sites and took the depth to which the dyed sand had been removed as a measure of the thickness of movement.

It is significant that the results of both approaches for determining the thickness of the layer of motion generally agree quite well. In spite of this, the thickness of movement is still the most uncertain of all the parameters measured. In this connection it should be recognized that the depth of burial of tracer used in this study as the thickness of the layer of motion is not an instantaneous thickness; instead it is an average that includes instantaneous motions (which may be only a few grain layers thick) as well as the effects of scour and fill produced by local adjustments in beach profile. However, because each experiment is based on a finite amount of tracer that is dispersed during a finite length of time, the depth of burial of tracer is the measure that is consistent with a transport velocity based on the new center of gravity of the tracer distribution and on the elapsed time between injection and sampling. The depth of

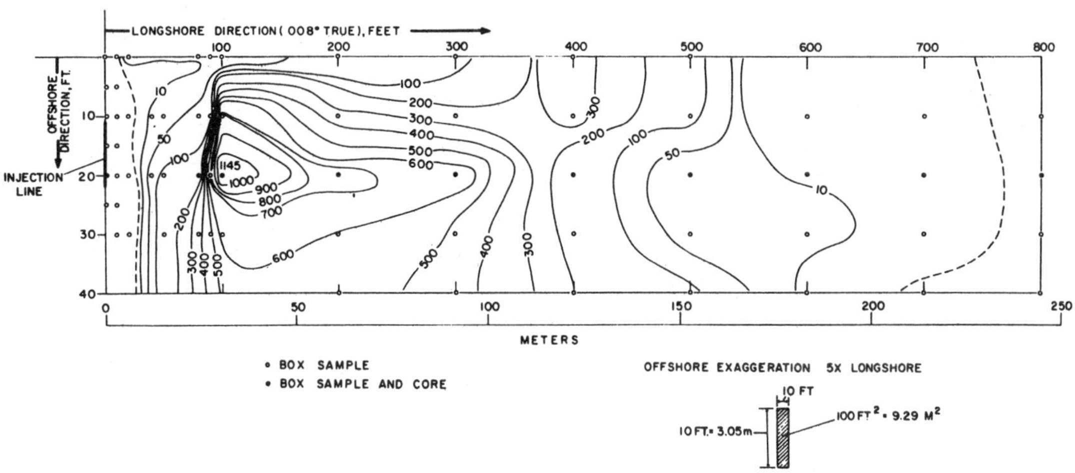

Fig. 1. Example of distribution of sand tracer along El Moreno Beach on May 5, 1966. Concentrations are in tracer grains per kilogram. Injection was 45.9 kg of El Moreno sand treated with fluorescent dye. The center of gravity of the new distribution is 75 meters from the injection line, 190 minutes after injection.

burial properly considers the total budget of tracer, and experience shows that it gives data that are coherent. This measure would lead to erroneous results if the measured thickness included layers of untagged sand, a situation which was not observed in these experiments.

If b denotes the measured thickness of movement of sand on the beach face, the bulk volume sand transport rate along the beach is given by

$$S_l = bX_b\langle V_l\rangle \tag{7}$$

where X_b is either the width of the beach face or the width of the surf zone. The immersed weight sand transport rate is then given by equation 2.

Wave and current measurements. We measured the direction and flux of wave energy from an array of digital wave sensors placed in the nearshore zone (Figure 2). The array has both onshore-offshore and longshore components with spacings between staffs of about one-quarter of the prevailing wavelength [*Koontz and Inman*, 1967].

The data output from each sensor is recorded on magnetic tape by a multi-purpose data acquisition system (DAS) and is later processed by a computer to obtain the wave spectra and cross-spectral analysis for various pairs of sensors. The wave spectrum is obtained by Fourier analysis of the time series of water level fluctuations at each sensor by using a rapid procedure modified from *Cooley and Tukey* [1965]. The individual estimates of energy density are printed out numerically and then plotted on a graph as illustrated in Figure 3. Such a graph permits the rapid identification of the wave trains arriving at the beach, a determination of their peak frequencies, and comparisons of their energies. The sum of the energy densities under the spectral peak of the incident wave train yields the mean-square elevation of the water surface $\langle\eta^2\rangle$. The wave energy per unit area of water surface is then

$$E = \rho g\langle\eta^2\rangle = \tfrac{1}{8}\rho g H_{\mathrm{rms}}^{\;2} \tag{8}$$

where H_{rms} is the root-mean-square wave height. The flux of energy per unit length of wave crest is then $P = ECn$, where the group velocity Cn is calculated from the peak period T of the wave train.

The direction of energy flux α was determined from the phase difference between pairs of sensors parallel to the shoreline and in, or just seaward of, the breakpoint of the waves. At beach locations where the wave breaker heights were low, wave breaker angles were also measured directly by hand-held sextant. (Where the breaking waves are too high and the water too deep, the breaker angle is calculated by correcting the angle measured at the sensors to that at the break point by using Snell's law.)

We measured longshore currents by following the center of a dye patch injected into the surf zone. At times we used floats of near neutral buoyancy to measure the longshore current. Because longshore currents vary in time and in space [*Inman and Quinn*, 1951], representative values of $\langle v_l\rangle$ can be obtained only from current measurements made along the complete length of beach, from the injection point of sand tracer to the maximum down coast transport position and across the surf zone.

Direct measurement of the orbital velocity near the bottom is extremely difficult and was not attempted in this study. Instead we calculated this velocity from the data obtained from

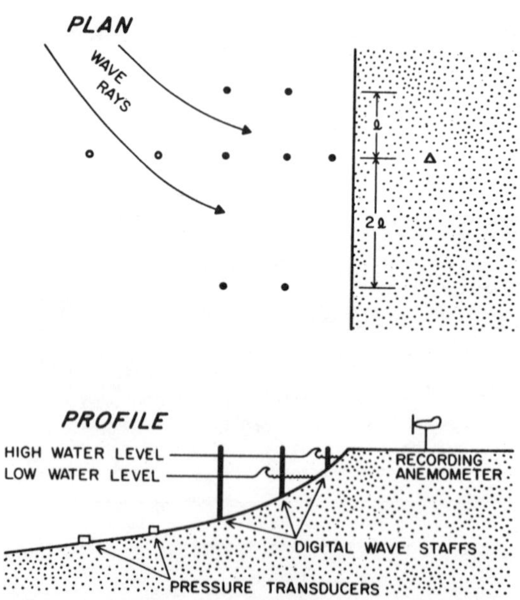

Fig. 2. Schematic diagram of sensor array used for measuring direction and flux of wave energy [after *Koontz and Inman*, 1967, Figure 16]. Spectra and cross-spectral analysis shown in Figure 3 are from a longshore pair of such an array.

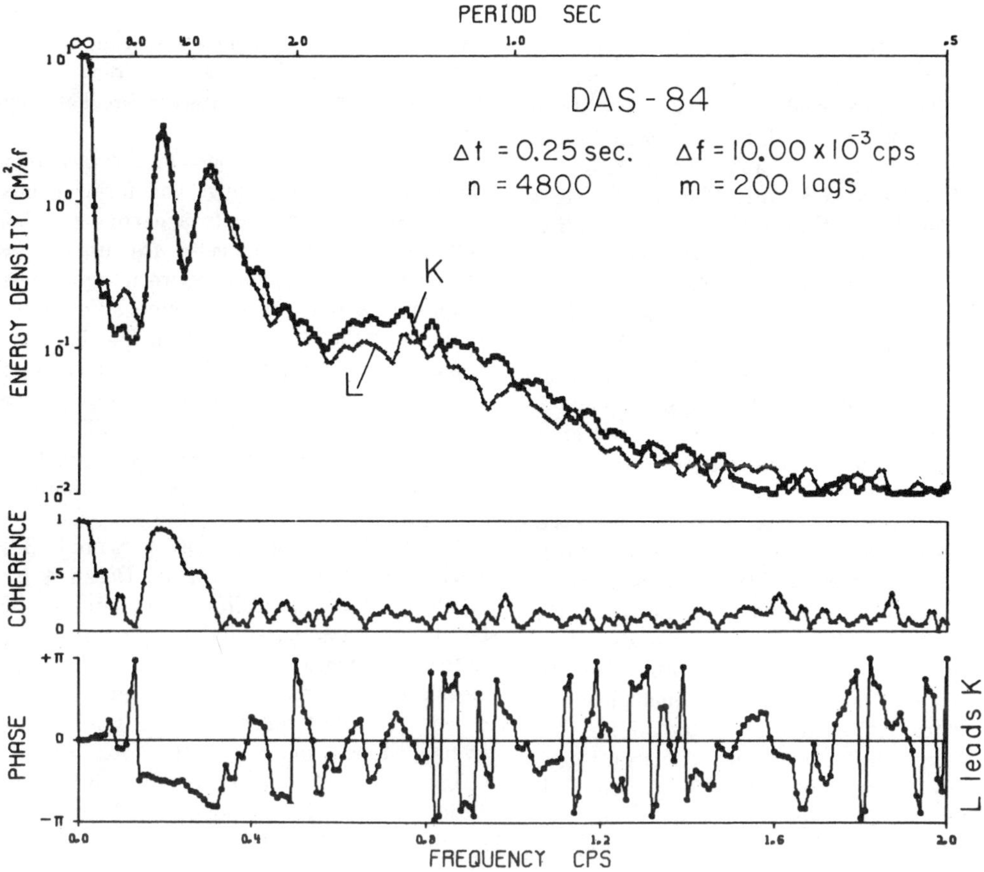

Fig. 3. Wave spectra and cross-spectral analysis for two members of an array of digital wave staffs. Staffs were parallel to the shoreline and just outside of the break point of the waves.

the energy spectrum. Whenever more than a single prominent wave train arrives at the beach, a simple calculation of a single value for u_m is not possible. On such occasions we calculated an orbital velocity that was representative of the total wave energy. Assume that H_b represents the height of the breaking wave of an imaginary wave train that combines the total energy fluxes from several superimposed wave trains according to

$$\frac{1}{8}\rho g H_b^2 C_b = (ECn)_{b1} + (ECn)_{b2} \quad (9)$$

where C_b is given by linearized wave theory as

$$C_b = (gh_b)^{1/2} = (gH_b/\gamma_b)^{\frac{1}{2}} \quad (10)$$

and $\gamma_b = H_b/h_b$ and is assumed to equal 0.78. The maximum horizontal component of the orbital velocity for this wave is

$$u_m = \frac{1}{2}\gamma_b C_b \quad (11)$$

Measurements of the wave and current parameters were obtained and combined over 20-min time periods. Therefore, each 20-min data collection provides a measure of the power quantities required in the sand transport models. The quantities show some variation during any 4-hour study of sand transport. Often these variations are progressive, especially where locally generated waves are important. The sand transport results from the time average of these changing wave conditions. Therefore, the energy flux of the waves were averaged to obtain a quantitative representation of the entire period of time.

Localities of Study

We conducted field measurements at two natural beaches under a variety of wave conditions. The two localities differ markedly in beach configuration and in the magnitude of the waves causing the sand transport.

El Moreno Beach. El Moreno Beach is located on the northwest shore of the Gulf of California in Baja California, Mexico, approximately 15 miles north of San Felipe (Figure 4). The beach is composed of a coarse-grained (median diameter, 600 μ) quartz-feldspar sand. The coarse sand is confined to a steep beach face (average slope, $\tan \beta = 0.138$), some 30 meters wide, and is abruptly terminated at its base by a broad low-tide terrace composed of fine-grained muddy sand.

The tidal range is extreme, reaching a maximum diurnal range of about 8.6 meters during spring tide. This large tidal range permits the sand tracer to be injected onto the beach face at low tide in a shallow trench and also allows the grid sampling to be performed on the exposed beach face at the subsequent low tide.

The area landward of El Moreno Beach is arid desert. The principal wave energy is generated by the sea breeze that results from the pressure gradient between the warm air over the land and the cooler air over the sea. The

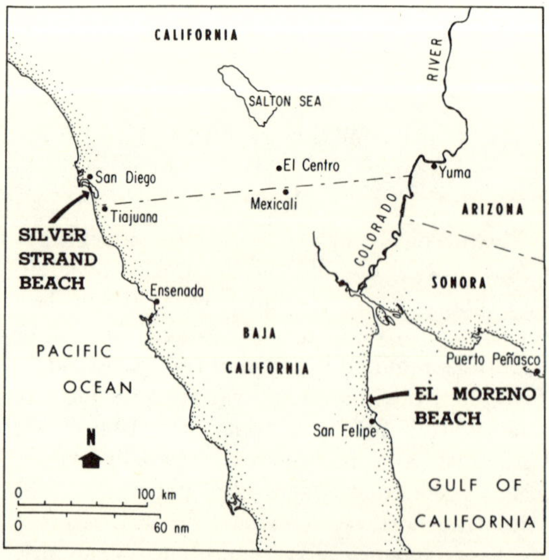

Fig. 4. Locations of El Moreno Beach and Silver Strand Beach, where simultaneous measurements of the longshore transport of sand and the direction and flux of wave energy were made.

waves generated by these local onshore winds reach heights up to 1 meter. During the study the sea breeze generally was not directly onshore and so generated waves with breaker angles that ranged up to 16°.

Because of the steepness of the beach face, the breaking waves plunge and form an intensive swash zone with only a narrow surf zone. Nearshore circulation cells are usually well developed with spacing between rip currents commonly about 10 to 20 meters. We observed that, when the waves approach the shore at an angle, the circulation cells move slowly in the longshore direction. The speeds of the longshore currents range up to 75 cm/sec. Tidal and wind induced currents are negligible in comparison with wave-induced longshore currents.

Silver Strand Beach. Silver Strand Beach Park is located on the Pacific Ocean 4 miles south of Coronado, California (Figure 4). This beach contrasts markedly in configuration and in magnitude of wave energy with the El Moreno Beach. It is composed of a fine-grained (median diameter, 175 μ) quartz-feldspar sand, and the average slope is much lower ($\tan \beta = 0.034$).

The principal wave energy at Silver Strand Beach is derived from long-period swell waves generated by distant storms and is frequently an order of magnitude greater than at El Moreno. During the study the breaker angles were generally small, however, and reached a maximum of only 8°. Therefore, even though the total energy flux is much larger at Silver Strand than at El Moreno Beach, the longshore component of the energy flux is nearly the same. When the wave energy and angle of approach are simultaneously large, the resulting longshore component can be an order of magnitude higher than at El Moreno Beach. This difference occurred once during the study.

Because the waves break in fairly deep water, the measurements of the sand transport rates were confined to the surf and to the inner breaker zones. In calculating the immersed weight transport rate I_l we assumed that the measured sand advection rate applied throughout the entire surf zone. The tracer diagrams appear to indicate that the transport rate increases toward the breaker zone so that one might expect that the values of I_l are somewhat low [see *Komar*, 1969]. Because of this and

TABLE 1. Data Relating Longshore Sand Transport to Wave Characteristics

Date	Tide	Sand Advection Rate $\langle V_l \rangle$, cm/sec	b, cm	Volume Transport Rate S_l, cm³/sec	Immersed Weight Transport I_l, dynes/sec	Longshore Energy Flux P_l, ergs/cm sec	$(ECn)_b \cos \alpha_b \frac{\langle v_l \rangle}{u_m}$, ergs/cm sec
				El Moreno Beach			
May 4, 1966	Low	0.50	8.5	4490	45.1×10^5	43×10^5	158×10^5
May 5, 1966	Low	0.65	10.5	8380	84.4×10^5	104×10^5	254×10^5
Oct. 11, 1966	Low	0.14	10.0	2560	25.7×10^5	30×10^5	72×10^5
Oct. 13, 1966	Low	0.065	8.2	981	9.9×10^5	15×10^5	29×10^5
May 22, 1967	High	0.076	9.0	1450	14.6×10^5	20×10^5	77×10^5
May 22, 1967	Low	0.22	9.0	4260	42.8×10^5	38×10^5	116×10^5
May 23, 1967	High	0.038	7.5	604	6.1×10^5	6.0×10^5	40×10^5
May 23, 1967	Low	0.055	7.5	881	8.9×10^5	18×10^5	27×10^5
Jan. 28, 1968	Low	0.032	7.5	292	2.9×10^5	6.0×10^5	13×10^5
May 11, 1968	Low	0.14	6.0	2060	20.8×10^5	18×10^5	32×10^5
				Silver Strand Beach			
Nov. 14, 1967	High	0.099	2.0	1270	12.8×10^5	15×10^5	206×10^5
Nov. 22, 1967	High	0.55	6.0	30100	$302. \times 10^5$	380×10^5	1890×10^5
Sept. 4, 1968	High	0.15	5.0	4680	47.1×10^5	91×10^5	134×10^5
Sept. 5, 1968	High	0.12	5.5	3760	37.9×10^5	41×10^5	129×10^5

the difficulty of measuring the thickness of movement at Silver Strand Beach, the measurements of I_l at that location have a higher degree of uncertainty. Rip current spacings typically range between 100 and 500 meters and, as at El Moreno Beach, migrate slowly along shore whenever waves approach the shore at an angle. Individual longshore current measurements ranged up to 76 cm/sec.

Discussion of Results

Longshore sand transport rate. Table 1 contains the measured values of the sand advection rate $\langle V_l \rangle$ and the thickness of movement b, along with the calculated values of the volume transport rate S_l and immersed weight transport rate I_l, and also includes the corresponding time averages of the energy flux of the waves.

The corresponding measurements of the immersed weight sand transport rate I_l and the longshore component of the wave energy flux P_l are plotted in Figure 5 to test the relationship given in equation 3. It is apparent that a straight line of unit slope with a coefficient $K = 0.77$ fits the data obtained in this study very well, so that equation 3 becomes

$$I_l = 0.77 P_l \qquad (12)$$

In view of the uncertainties in the sand transport measurements at Silver Strand Beach, the data agree surprisingly well with the best-fit straight line. Additional measurements probably would reveal more variation than shown.

The good agreement between the data from the two locations, whose beach slopes differ by a factor of 4, indicates that the coefficient K in the relationship between I_l and P_l is independent of the beach slope. The results also indicate that the relationship is independent of the sand size within the range of about 175 to 600 μ included in the study. In suspended load transport the size of the sediment grains is an important variable because of the dependence of the settling velocity on size. The results therefore suggest that the suspended load transport of sand in the surf zone is less important than bed load transport.

The field and laboratory results from the previous studies are also included in Figure 5. It is apparent that the measurements of the longshore energy flux obtained in this study are about one-half the value from previous field studies for a given transport rate. This discrepancy appears to result from the erroneous use of the significant wave height in computing wave energy for the earlier field observations as suggested by *Inman and Frautschy* [1966]. *Longuet-Higgins* [1952], by assuming that a wave spectrum consists of a narrow band of fre-

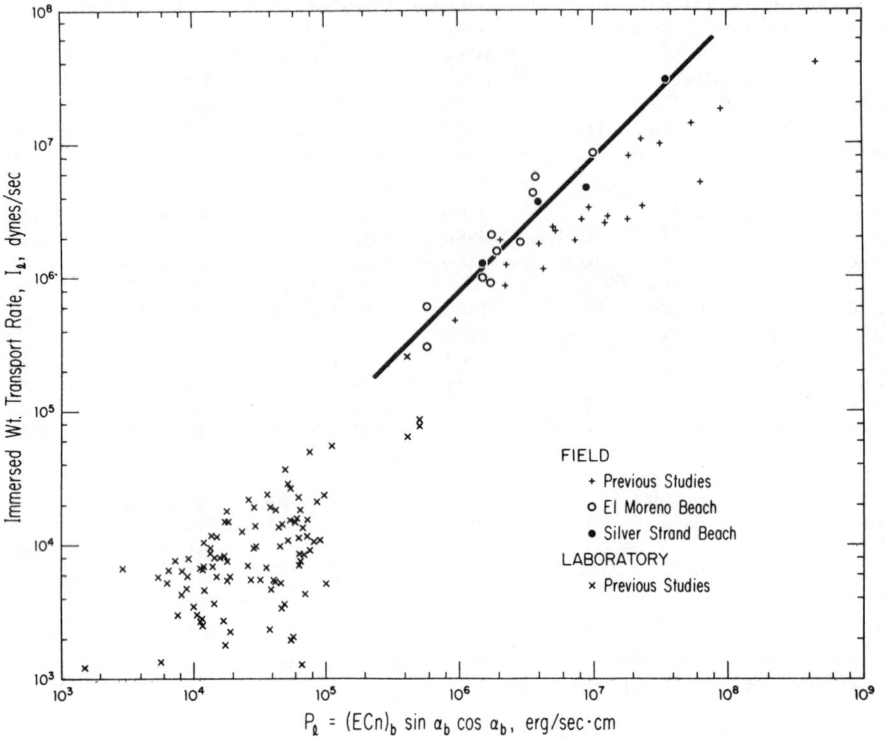

Fig. 5. Relationship between the immersed weight longshore sand transport rate and the longshore component of wave energy flux per unit length of beach. The new data are listed in Table 1 and the various sources for the 'previous studies' are given in *Inman et al.* [1969].

quencies and that the wave energy comes from a large number of different sources whose phase is random, deduced that the ratio of the significant wave height to the rms wave height $H_{1/3}/H_{rms}$ should have a value of 1.418. If this is true, the wave energy calculated from the significant wave height would be a factor of 2 higher than the energy correctly calculated from H_{rms}. This would then account for the difference between the energy flux calculations of this study and those of the previous studies.

To examine this, sets of wave data have been analyzed both by the significant wave height approach and by spectra analysis. The data acquisition system provides a 20-min analog record as well as a digital record of the waves; hence such a comparison can readily be made. The results of several such comparisons [*Komar*, 1969] indicate that the analysis of *Longuet-Higgins* [1952] generally applies to the field data of this study and thus explains the discrepancy between the results of this and previous field studies.

Figure 5 also shows that the laboratory measurements fall well below the extension of the straight line established by the field data. We believe that the reason for this is the lack of fully developed transport conditions in these laboratory studies. In the transport of sand by rivers the immersed weight sand transport rate is proportional to the available power with a constant proportionality factor only when fully developed flow is achieved [*Bagnold*, 1966]. Under lower flow regimes that are not fully developed the efficiency of utilization of the available power to transport sand is less and the proportionality factor is lower and is no longer constant. A similar effect can be expected to occur in the transport of sand on beaches so that the transport mechanism is less efficient for the laboratory studies than it is in the field. This difference in efficiency must

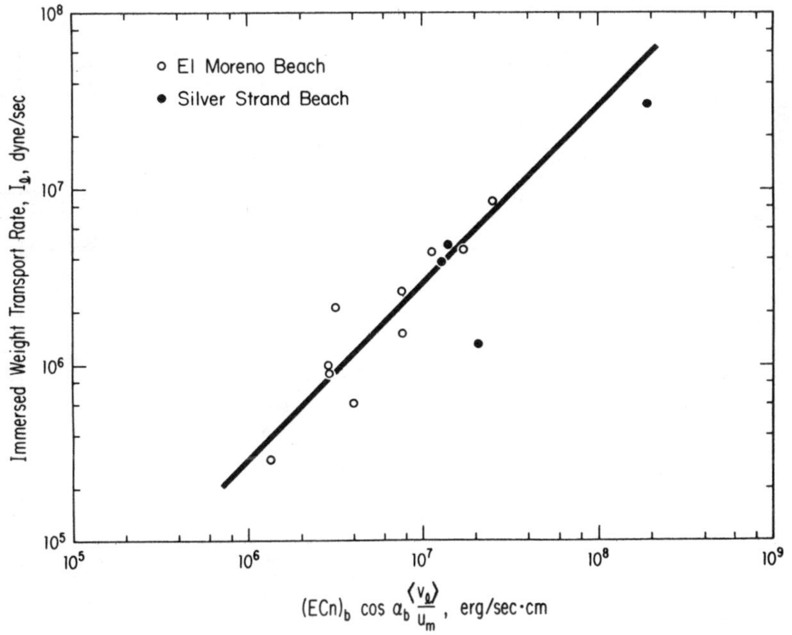

Fig. 6. Relationship between the immersed weight longshore sand transport rate and the energy flux terms of equation 4. Plotted data are listed in Table 1.

be considered in any attempts to extrapolate laboratory measurements to field conditions.

The straight line of Figure 5 and therefore equation 12 represent fully developed transport conditions and predict the maximum transport possible for a given wave situation. Plotted positions of less efficient transport conditions can be expected to fall below this line, which forms an upper limit. This is seen to be the situation for the laboratory data included in Figure 5.

The corresponding measurements of the sand transport rate I_l and the quantity $(ECn)_b \cos \alpha_b \langle v_l \rangle / u_m$ of equation 4, listed in Table 1, are plotted in Figure 6. Although the scatter of the data is somewhat greater than for the previous model, it is apparent that the data follow a trend that yields the relationship

$$I_l = 0.28(ECn)_b \cos \alpha_b \frac{\langle v \rangle_l}{u_m} \qquad (13)$$

as the best fit. Therefore, even though the two models are conceptually different, the data

TABLE 2. Longshore Currents

DAS	$\langle v_l \rangle$, cm/sec	u_m, cm/sec	α_b, deg	$u_m \tan \beta \sin \alpha_b \cos \alpha_b$, cm/sec	$u_m \sin \alpha_b$, cm/sec
55	50.0	77.7	10	1.84	13.4
59	61.0	85.6	13.5	2.68	20.0
60	54.0	88.7	14.5	2.96	22.2
61	46.4	88.3	13	2.67	19.9
62	49.4	86.8	14	2.82	21.0
63	47.0	86.3	15	2.98	22.4
73	45.7	68.9	16	2.52	19.0
76	40.0	63.8	16	2.33	17.6
79	14.1	64.4	8.8	1.34	9.8
84	15.2	66.8	8	1.27	9.2
92	27.8	68.4	7	1.14	8.3
94	48.8	77.7	11.5	2.10	15.5
95	36.0	81.5	13	2.46	18.3
96	48.8	81.9	11	2.12	15.6
97	30.5	85.7	11	2.22	16.4
1	7.6	67.7	3	0.49	3.5
3	6.9	65.5	3	0.47	3.4
173	8.1	73.0	2.8	0.48	3.6
174	15.2	74.8	2.8	0.49	3.7
175	20.1	72.7	7	1.21	8.9
176	22.3	76.1	7	1.27	9.3
198*	14.6	96.7	6	0.30	10.1

* Values for Silver Strand Beach. All other values are for El Moreno Beach.

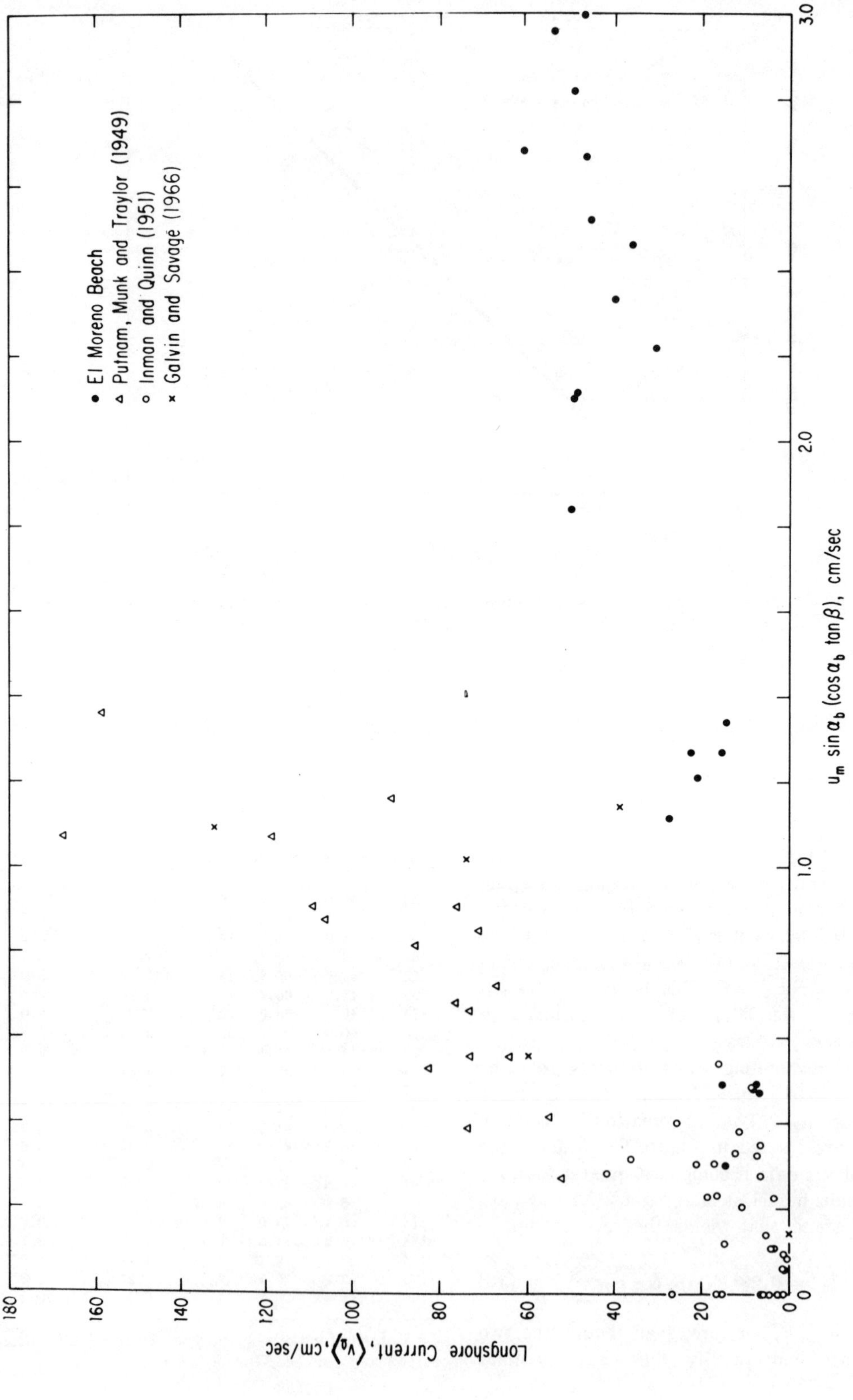

Fig. 7. Relationship between the average longshore current and the quantity determined from a consideration of the longshore component of radiation stress (equation 5). The data plotted for El Moreno Beach are listed in Table 2 (compare with Figure 8).

obtained in this study confirm both. It appears that, when the longshore current and the longshore sand transport are due to an oblique wave approach, the two models are equivalent. A simple simultaneous solution of equations 3 and 4 yields the relationship

$$\langle v_l \rangle = (K/K')u_m \sin \alpha_b \qquad (14)$$

This result is similar to equation 5 for the generation of the longshore current by the radiation stress, except for the presence of the factor $\tan \beta \cos \alpha_b/c_f$. *Komar* [1969] indicates that the factor $\tan \beta \cos \alpha_b/c_f$ appears to be nearly constant.

To examine this point further, the parameters $u_m (\tan \beta \cos \alpha_b) \sin \alpha_b$ and $u_m \sin \alpha_b$ have been computed for the runs in which only a single wave train predominated. The results are given in Table 2 along with the measured values of $\langle v_l \rangle$ and are plotted in Figures 7 and 8. Also included in the plots are the field measurements of *Putnam et al.* [1949], *Inman and Quinn* [1951], and *Galvin and Savage* [1966] as tabulated in *Galvin and Nelson* [1967]. Compari-

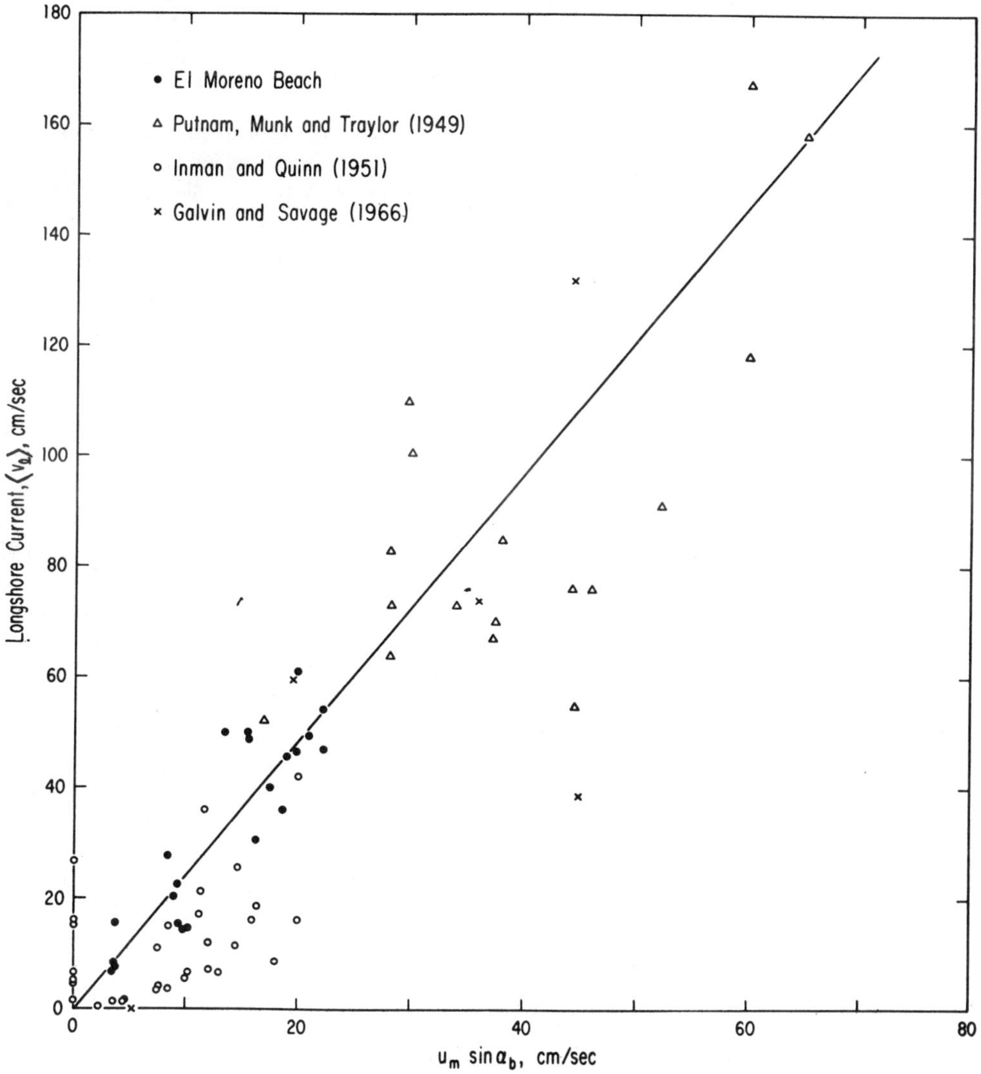

Fig. 8. A replot of the data shown in Figure 7, under the assumption that the ratio $\tan \beta \cos \alpha_b/c_f$ in equation 5 is constant as indicated by *Komar* [1969].

son of the two figures confirms the validity of equation 14 and indicates that the ratio $\tan \beta \cos \alpha_b / c_f$ is nearly constant. This explains the apparent lack of dependence on beach slope of both proportionality coefficients K and K'.

The relationship between the sand transport and the nearshore circulation of water requires further comment. *Bowen* [1967, 1969*b*] and *Bowen and Inman* [1969] have shown that interactions between an incoming wave train and edge waves generated in the nearshore region control the spacing between rip currents and result in circulation cells like those shown in Figure 9*A*. Under an oblique wave attack the circulation cells have been observed to move slowly along the shore; the motion is made apparent by the migration of the rip currents. Because of this migration there will be no net longshore transport of sand resulting from the symmetrical diverging currrents in the cell. At any one position on the beach the transport will be first in one direction and then in the opposite direction as the cells migrate and the cell current over the position reverses. In our measurements of the average longshore current over the entire cell $\langle v_l \rangle$ the currents of the circulation cell of Figure 9*A* average out owing to the symmetry. The residual average in the longshore current is the current generated by the longshore component of the radiation stress. It will be nearly constant in direction and magnitude so that its effect will be to produce a net longshore drift of sand. This current illustrated in Figure 9*B*, when added vectorally to the circulation system of Figure 9*A*, results in the nearshore circulation pattern shown in Figure 9*C*. It is this system which is observed in the field whenever waves approach obliquely to the shoreline.

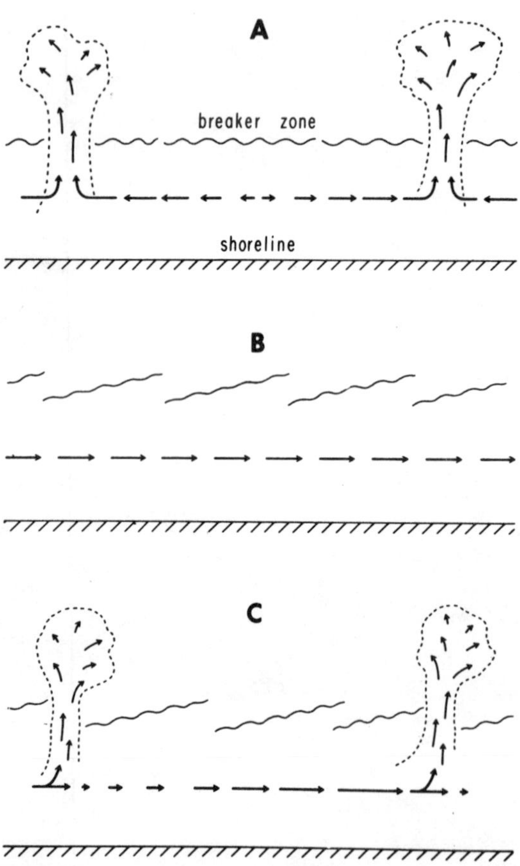

Fig. 9. (*A*) The nearshore circulation resulting from the interaction of the incident waves with edge waves. (*B*) The longshore water drift due to the longshore component of the wave momentum flux. (*C*) The vector sum of the currents shown in parts *A* and *B* to give the observed currents.

Summary

The data collected by this and previous studies clearly show that the immersed weight longshore transport rate of sand is directly proportional to the longshore component of the wave energy flux arriving at the beach as predicted by the first model (equations 3 and 12).

The second model, proposed by *Inman and Bagnold* [1963] whereby the wave motion provides the stress to support the sand and a superimposed longshore current causes a net transport, also successfully relates the sand transport rate to the wave and current parameters (equations 4 and 13).

From the equivalence of the two models it is deduced that the average longshore current resulting from the oblique approach of waves is proportional to $u_m \sin \alpha_b$. This differs from the relationship obtained from the generation of the longshore current by the longshore component of the radiation stress by only the factor $\tan \beta \cos \alpha_b / c_f$, which appears to be nearly constant.

The two models are not equivalent if a component of the longshore current is generated by some other mechanism such as tides or winds. In this situation the second model (equation 4) appears to be more flexible and would presumably predict the longshore transport of sand, providing that the longshore current is known.

Acknowledgments. We are indebted to Professors R. S. Arthur and M. S. Longuet-Higgins, whose advice has been most helpful in conducting this study, and to C. E. Nordstrom, W. A. Koontz, E. A. Murray, S. W. Roberts, and R. J. Mann, who aided in the collection, collocation, and analysis of the field data.

This study was supported by the Coastal Engineering Research Center, U.S. Army Corps of Engineers, and by the Office of Naval Research under contract with the University of California. Komar was supported for two years by a National Science Foundation Fellowship.

References

Bagnold, R. A., Mechanics of marine sedimentation, in *The Sea,* vol. 3, pp. 507–528, edited by M. N. Hill, Interscience, New York, 1963.

Bagnold, R. A., An approach to the sediment transport problem from general physics, *U.S. Geol. Surv. Prof. Pap. 422-1,* 37 pp., 1966.

Bowen, A. J., Rip currents, Ph.D. thesis, University of California, San Diego, 1967.

Bowen, A. J., The generation of longshore currents on a plane beach, *J. Mar. Res., 37,* 206–215. 1969*a*.

Bowen, A. J., Rip currents, 1, Theoretical investigations, *J. Geophys. Res., 74,* 5467–5478, 1969*b*.

Bowen, A. J., and D. L. Inman, Rip currents, 2, Laboratory and field observations, *J. Geophys. Res., 74,* 5479–5490, 1969.

Caldwell, J. M., Wave action and sand movement near Aneheim Bay, California, Beach Erosion Board, *U.S. Army Corps Eng. Tech. Memo. 68,* 21 pp., 1956.

Cooley, J. W., and J. W. Tukey, An algorithm for the machine calculation of complex Fourier series, *Math. Comput., 19,* 297–301, 1965.

Crickmore, M. J., and G. H. Lean, The measurement of sand transport by means of radioactive tracers, *Proc. Royal Soc. London, A, 266,* 402–421, 1962.

Eaton, R. O., Littoral processes on sandy coasts, *Proc. First Conf. Coastal Eng.,* pp. 140–154, Council on Wave Research, University of California, 1951.

Galvin, C. J., Jr., and R. A. Nelson, Compilation of longshore current data, *U.S. Army Coastal Eng. Res. Center Misc. Pap. 2-67,* 19 pp., 1967.

Galvin, C. J., Jr., and R. P. Savage, Longshore currents at Nags Head, North Carolina, *U.S. Army Coastal Eng. Res. Center Bull., 11,* 11–29, 1966.

Grant, U. S., Waves as a sand transporting agent, *Amer. J. Sci., 241,* 117–123, 1943.

Ingle, J. C., Jr., The m[illegible] *Developments in Sedi*[illegible] Elsevier, New York, 1[illegible]

Inman, D. L., and R. A[illegible] esses, in *The Sea,* vol.[illegible] M. N. Hill, Interscienc[illegible]

Inman, D. L., and J. D. [illegible] esses and the developme[illegible] *Engineering,* pp. 511–5[illegible] Civil Engineers, New Y[illegible]

Inman, D. L., P. D. Kom[illegible] J. Bowen, Longshore transport of sand, *Proc. Eleventh Conf. Coastal Eng., vol. 1,* pp. 298–306, American Society of Civil Engineers, New York, 1969.

Inman, D. L., and W. H. Quinn, Currents in the surf zone, *Proc. Second Conf. Coastal Eng.,* pp. 24–36, Council on Wave Research, University of California, 1951.

Inman, D. L., R. J. Tait, P. D. Komar, and C. E. Nordstrom, Dispersion of water and sediment in the surf zone, *Scripps Inst. Oceanogr. SIO Ref. 69-10,* University of California, San Diego, 1969.

King, C. A. M., Depth of disturbance of sand on sea beaches by waves, *J. Sediment. Petrol., 21,* 131–140, 1951.

Komar, P. D., The longshore transport of sand on beaches, Ph.D. thesis, University of California, San Diego, 1969.

Komar, P. D., The mechanics of sand transport on beaches, *J. Geophys. Res., 75,* in press, 1970.

Koontz, W. A., and D. L. Inman, A multipurpose data acquisition system for instrumentation of the nearshore environment, *U.S. Army Coastal Eng. Res. Center Tech. Memo. 21,* 38 pp., 1967.

Longuet-Higgins, M. S., On the statistical distribution of the height of sea waves, *J. Mar. Res., 11,* 245–266, 1952.

Longuet-Higgins, M. S., Longshore currents generated by obliquely incident sea waves, 1 and 2, *J. Geophys. Res., 75*(33), 1970.

Longuet-Higgins, M. S., and R. W. Stewart, Radiation stresses in water waves; a physical discussion with applications, *Deep-Sea Res., 11* 529–562, 1964.

Putnam, J. A., W. H. Munk, and M. A. Traylor, The prediction of longshore currents, *Trans. AGU, 30,* 337–345, 1969.

Savage, R. P., Laboratory study of the effect of groins on the rate of littoral transport, Beach Erosion Board, *U.S. Army Corps Eng. Tech. Memo. 114,* 55 pp., 1959.

Scripps Institution of Oceanography, A statistical study of wave conditions at five locations along the California coast, *Univ. Calif. Wave Rep. 68,* 34 pp., San Diego, 1947.

Watts, G. M., Study of sand movement at South Lake Worth Inlet, Florida, *Beach Erosion Board, U.S. Army Corps Eng. Tech. Memo. 42,* 24 pp., 1953.

Part III

BEACH PROCESSES

Editors' Comments on Papers 15 Through 20

The study of the transport of sediment is difficult in general, and particularly so in a shoaling wave environment. Thus it is unreasonable to expect the development of beach processes to proceed more rapidly than the development of generalized models for suspended and bed-load sediment transport. Even the more indirect subjects of beach profiles and the sorting and distribution of beach sediments depend on the understanding of the basic hydrodynamics for their eventual theoretical solutions. Having acknowledged these limitations, we can more fully appreciate the contributions presented here, because most are examples of how we have learned from observation rather than from mathematical derivation.

In contrast to the selections for Part II, these papers on beach processes do not trace the development of a single thread or element of their broader subject. Indeed, while they have been organized chronologically, they otherwise appear to range from Bagnold's classic wave tank experiments to Strahler's beach surveys with little interrelationship. The development of our understanding of beach processes follows a similar irregular course. While we have included papers published only through the late 1960s, the more recent publications have not yet provided a unified model for the beach.

We have had to omit from this section a number of excellent papers that deal with the design of coastal engineering structures for beach protection and stabilization. The Corps of Engineers' "Shore Protection Manual," cited in the references for the Introduction, is an invaluable source for these papers.

It is most appropriate to begin this section on beach processes with a contribution by R. A. Bagnold. No collection on this subject would be complete without one of his expositions on the experimental analysis of sediment transport, either by air or water waves.

This paper, which describes Bagnold's efforts in using a wave tank to model the interaction of the waves and beach, presents a thorough discussion of the basic physics of the phenomenon, as well as an excellent discussion of the application of a hydraulic model for the analysis of these processes. Reading this paper reminds one of the care and thoroughness with which Bagnold undertook his experiments. The reader is treated to a detailed discussion of the experimental results for a large number of conditions and flow fields. The concepts and analysis presented are frequently cited in the more recent literature, and they provide but a single example of the extraordinary influence Bagnold has had on these subjects. Because of space limitations we have omitted the author's last section about ripples, as well as three photographic plates.

Our second selection is also by one of the major researchers in this field. The contribution of Hans Albert Einstein to the understanding and quantification of sediment transport is monumental. During Einstein's lifetime he developed some of the most important concepts in the understanding of this subject. Although Einstein and his colleagues at the University of California, Berkeley, have produced many papers appropriate for this Benchmark collection, we have chosen a brief discussion of the nature and complexity of this sediment transport problem.

In Paper 16 Einstein presents a concise review of the interaction of a river flow and its transported sediment. Then he explains how the wave environment modifies this interaction. His discussion of a plan of research for additional understanding of this complex subject is as valid today as it was in 1949. It is to Einstein's credit that his own and his students' subsequent contributions have been a major factor in reaching our present level of understanding.

It is of course necessary to collect and analyze field data in order to expand Einstein's model for these processes. The next paper is a classic description of such a field study. Willard Bascom's *Waves and Beaches* is a popular general science text, and his papers in *Scientific American* have been invaluable in bridging the literature gap between the general public and coastal researchers. In its survey of the Pacific coastline, *Waves and Beaches* recounts the collection of the data presented here in Paper 17. Interestingly, this is yet another contribution motivated by the need for an increase in the understanding of coastal processes during World War II. Perhaps equally noteworthy is M. P. O'Brien's influence on the direction of these studies.

Bascom presents an analysis of the relationship between sand size and beach-face slope. This analysis remains one of the classic data sets of this relationship, indeed, this paper is frequently cited in the more recent literature.

Our next selection is also a solid member of the literature dealing with field studies. This contribution by Robert Miller and John Zeigler (Paper 18) is a benchmark of the 1950s. The authors investigated relationships between material transport within the surf zone and the resulting sediment distribution patterns. The roots of their research can easily be traced through several of our earlier selections, back to Cornaglia's paper (Paper 1).

In this study, Miller and Zeigler, both geologists, test their hypotheses in the field. The measurements are simple—so simple, in fact, that they could easily be repeated without a major investment. The results are presented in an interesting manner with a strong intuitive base, and the graphics provide an excellent visual summary of the complex relationships between sediment distributions and inshore processes.

The authors' model of sediment sorting and its comparison with field data provides a useful framework for viewing the sorting and distributions from the step, across the bar, to the zone of shoaling waves. Considering the complexities of these processes, the relative agreement of their model with the field data suggests that Miller and Zeigler have made an important contribution to the development of our understanding of beach processes.

Of course, as we learned from Bagnold, the laboratory can also be a valuable place to test theories of beach processes. Maurice Schwartz has contributed a number of papers concerning the interaction of sea level fluctuations and shoreline erosion. We have selected a discussion of the use of a laboratory stream table to test the ideas of Per Bruun. This paper (Paper 20) is an excellent example of how relatively simple experiments can be applied to the analysis of rather complex systems. The author is careful to point out the experimental limitations of the project with regard to the dimensions and operation of the hydraulic model.

Along with a second paper by Schwartz, "The Bruun Theory of Sea-Level Rise As a Cause of Shore Erosion," *Journal of Geology,* 1967, this paper provides a comprehensive discussion of the interaction of local sea level and beach volume changes. Numerous other articles in the literature (including Bruun's contributions that are listed in Schwartz's references) might rightly belong with this paper in a benchmark collection. We selected this one, however, for its concise presentation and clear explanation of how a simple model can be used in coastal processes research. This latter reason is clearly founded on Bagnold's work.

The final paper in this collection is, appropriately, a detailed field investigation of an equilibrium beach. Arthur Strahler's study of the New Jersey coast is in a class with the best tradition of the other investigations included in this Benchmark volume. His experimental design and discussion of data clearly reflect the influence of Krumbein and others who preceded him in these studies. And of course, more recent investigators have used Strahler's techniques to develop our understanding further.

The scope of this paper includes both an analysis of the seasonal variation of the beach profile as well as its variation within the tidal period. This latter discussion is an interesting extension of the early observations of Thompson and Thompson, included in Part I of this volume. Strahler also presents a discussion of the sorting of beach material and the mechanisms of deposition and scour. The work of Miller and Zeigler (Paper 18) is further confirmed by Strahler's field studies.

In general, this paper provides an excellent model of a well-designed field study of beach processes, including the experimental procedures as well as the careful examination and interpretation of the data.

REFERENCES

Dolan, Robert, and John Ferm (1966). "Swash Processes and Beach Characteristics," *Prof. Geographer* **18**(4):210–213.

Harrison, W., and W. C. Krumbein (1964). "Interaction of the Beach-Ocean-Atmospheric System at Virginia Beach, Virginia," CERC Tech. Memo 7.

Hicks, S. D., and J. E. Crosby (1973). "Trends and Variability of Yearly Mean Sea Level, 1893-1972," NOAA Tech. Memo no. 12.

Ingle, J. C. (1966). *The Movement of Beach Sand.* Developments Sediment. 5, Amsterdam: Elsevier.

Inman, Douglas L., and Jean Filloux (1960). "Beach Cycles Related to Tide and Local Wind Wave Regime," *Jour. Geology* **68**:225-231.

Kalkanis, G. (1964). "Transportation of Bed Material due to Wave Action," CERC Tech. Memo 2.

King, C. A. M. (1951). "Depth of Disturbance of Sand on Sea Beaches by Waves," *Jour. Sed. Petrology* **21**(3):131-140.
Krumbein, W. C. (1953). "Statistical Designs for Sampling Beach Sand," *Am. Geophys. Union Trans.* **34**(6):857-868.
Bruun, Per F. (1962). "Sea-Level Rise as a Cause of Shore Erosion," *Am. Soc. Civil Engineers Proc., Jour. Waterways and Harbors Div.* **88**:117-130.
Sonu, Choule J., and Johannes L. Van Beek (1971). "Systematic Beach Changes on the Outer Banks, North Carolina," *Jour. Geology* **79**(4):416-425.
Tanner, William F. (1958). "The Equilibrium Beach," *Am. Geophys Union Trans.* **39**(5):889-891.
Thornton, E. B. (1969). "A Field Investigation of Sand Transport in the Surf Zone," *Proc.* 11th Conf. Coastal Eng., London, 1968, ASCE, 335-351.
Watts, G. M. (1953). "A Study of Sand Movement at South Lake Worth Inlet, Florida," Beach Erosion Board Tech. Memo 42.

15

Reprinted from pp. 27–49 of *Inst. Civil Engineers Jour. Paper No. 5237:* 27–53 (1940)

"Beach Formation by Waves; Some Model-Experiments in a Wave Tank."

By MAJOR RALPH ALGER BAGNOLD, M.A.

(*Ordered by the Council to be published with written discussion.*)[1]

TABLE OF CONTENTS.

THE USE OF MODELS FOR RESEARCH ON BEACH FORMATION.

THE shaping of loose bed-material into beaches by the action of water waves has been studied in nature by geomorphologists and others, but, owing to the difficulties involved, only scanty information is available about what happens beneath the water-surface, and about the factors which control the shape of the mature beach profile. These difficulties are obvious. For instance, the lower portion of the beach cannot be seen during a storm; natural waves are necessarily uncontrolled, and are seldom constant in character for any length of time; and the whole process of beach formation is complicated by tidal variations, both in the mean water-level and in the direction and intensity of coastwise currents.

An investigation with models, on the other hand, is hindered by none of these difficulties; the movement of the water and of the material can be readily watched and measured over all portions of the beach, under the action of any required type of wave, repeated at will. Moreover the

[1] Correspondence on this Paper can be accepted until the 15th March, 1941, and will be published in the Institution Journal for October 1941.—SEC. INST. C.E.

influence of the particle size on the beach profile can be demonstrated directly by a change of beach material while the wave characteristics are kept constant.

It must be remembered, however, that model-experiments are subject to the following limitations:

(*a*) In a narrow wave-tank the movement, both of the water and of the beach material, is constrained to two dimensions, the horizontal movement being forward and backward only. But on an open beach there may be important coastwise components as well. These may be of two kinds: (i) a steady transverse current, which, although usually far too weak itself to set any bed-material in motion, adds a continuous coastwise drift to the oscillatory bed-movement due to the waves. It is this drift which ultimately controls at any given spot the supply or removal of beach-building material. Although it cannot be reproduced in the wave tank, its effect on the availability of beach-building material can nevertheless be imitated by the direct addition of material to, or by its removal from, the tank during the progress of an experiment. (ii) Even though there is no coastwise drift, and though the direction of wave-advance is at right-angles to the beach, the movement of the surge over the beach may deviate from this direction and take the form of a sideways sweep. Such a deviation is particularly noticeable on gently-sloping sand beaches; it cannot be imitated in the tank. Its effect is discussed later in this Paper.

(*b*) Before the results of model-research can be applied with any certainty to full-scale phenomena, a sufficiently sound picture of the physics must be obtained, so that reliable model rules can be formulated. It is clear that research in this case must be largely empirical, since no mathematical theory at present underlies the physics either of the breaking wave or of the carriage of solids over a bed by a steady stream, let alone their carriage by an oscillating stream with powerful accelerations. Hence model rules, that is to say, the definition of the conditions governing similarity of beach profile for differing linear dimensions, can be but approximate and can only emerge as the research proceeds.

It may be useful to make an initial assumption regarding the similarity of beach profiles, in order to determine the scale of the model to a first approximation, and later to examine the general validity of this assumption in the light of the experimental results. It will be assumed, therefore, that beach profiles should be similar if the ratio:

What about ν, g?

$$R = \frac{\text{amplitude of the wave}: h}{\text{diameter of the beach particles}: d}$$

is constant.

THE SCOPE OF THE EXPERIMENTS.

Experiments by the Author on the shock-pressures exerted against a vertical sea-wall by breaking waves have been described in a previous

$\nu = L^2T^{-1}$
$g = LT^{-2}$

$L = (\nu^2/g)^{1/3} = (10^{-4}/10^3)^{1/3} = (0.1)^{1/3}/100 = 0.005$ cm

Paper[1]. Full details of the wave-tank used (Fig. 1, Plate 1) were given therein. The tank was 110 centimetres high and 53 centimetres wide in cross-section, and was provided with glass sides. A 10-centimetre grid was fixed to one side for easy measurement, as shown in Figs. 3, 4, 5, *et seq.* (Plate 1). The left-hand edge of the glass coincided with a vertical concrete wall. The waves, which were created by a hydraulically-operated paddle, advanced from right to left, and were made to break by the distortion imposed upon them by the presence of a sloping beach. This beach consisted of a rigid steel plate abutting on to the wall and inclined at an arbitrary angle.

A further series of experiments, upon which this Paper is based, were planned with the object of discovering what the ultimate beach profile would be if the steel plate were to be replaced by loose shingle or sand, and the waves themselves were allowed to fashion their own beach. These experiments were in progress at the outbreak of war, and were interrupted by the departure of the Author on military service abroad. The results, although incomplete, may be worth recording.

An idea of the range of model-scales upon which the experiments were conducted may be obtained from the following figures. Assuming 3 metres to be the amplitude of an average real storm-wave, and a range of possible sizes of real bed-material between 8 centimetres and 0·05 centimetre, the values of R (as defined on p. 28), which are found in nature, range from 37 (shingle) to 6,000 (sand). In the model the waves varied in amplitude between 30 centimetres and 5 centimetres, and the size of the material varied between 0·7 centimetres and 0·05 centimetres. The values of R_m in the model therefore covered a range between 7·5 and 600. It will thus be seen that the whole range of natural shingle and pebble conditions was more than covered by the model, but that the model failed to imitate the extreme natural case of large waves and fine sand. It was intended to investigate this case later by the use of fine pit-sand of 0·007 centimetre diameter, which would have given values of R up to 4,300.

TERMINOLOGY.

Figs. 2 (*a*) and *2* (*c*), p. 30, show the profile of a typical shingle beach as thrown up in the wave tank by waves of the kind indicated. Though the following terms have special reference to such a beach, they can be applied with but slight modification to beaches formed under different conditions.

The Wave-Amplitude (h). The total amplitude, as measured by the difference of level between the crest and the trough of the wave " out at sea ", bears no known relation to the dimensions of the distorted wave near the beach. Moreover, owing to the limited length of the experimental tank and to the fact that the wave is never symmetrical about any

[1] Major R. A. Bagnold, " Interim Report on Wave-Pressure Research. Journal Inst. C.E., vol. 12, p. 202 (June 1939).

horizontal plane, the " mean level " is difficult to define and is clearly not the same as the " still-water level " (H_0). The wave amplitude has therefore been defined, for the purposes of this Paper, as the difference between the crest level " out at sea " and the level of the lowest exposed beach-line immediately in front of the advancing wave. As thus defined, the amplitude is independent of wave-length, wave-shape, and still-water height. Both the levels involved are easily measured. The level of the

Figs. 2.

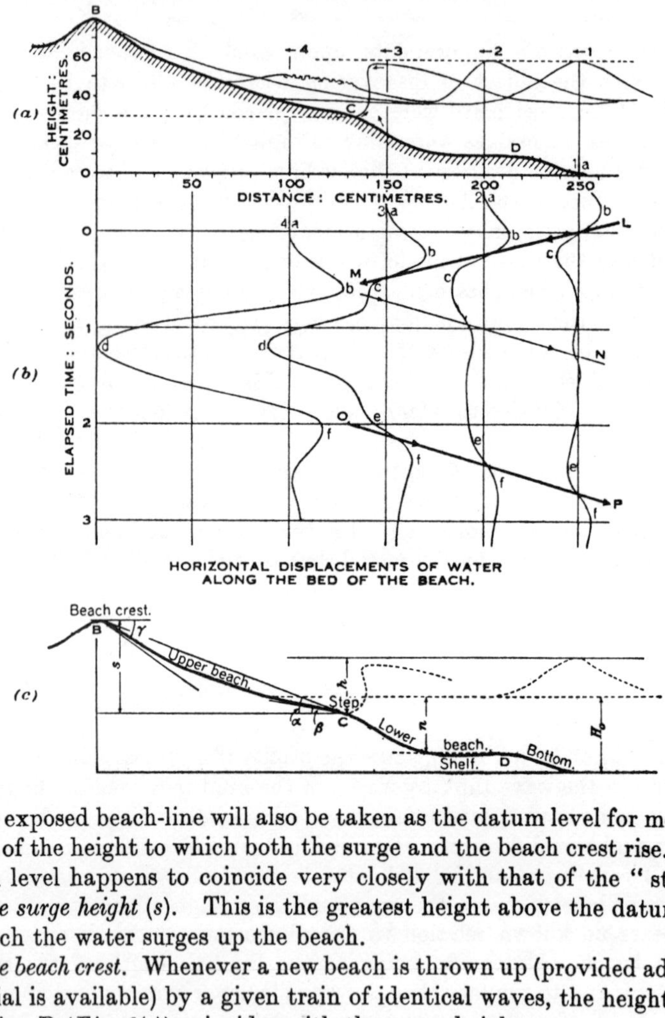

lowest exposed beach-line will also be taken as the datum level for measurements of the height to which both the surge and the beach crest rise. This datum level happens to coincide very closely with that of the " step."

The surge height (s). This is the greatest height above the datum level to which the water surges up the beach.

The beach crest. Whenever a new beach is thrown up (provided adequate material is available) by a given train of identical waves, the height of the crest line B (*Fig. 2(c)*) coincides with the surge height s.

The beach. This consists of two concave zones BC and CD, which will be called the " upper beach " and the " lower beach " respectively. They are separated by the convex " step " C. The centre of the step is found to coincide with the datum level of the lowest limit of the uncovered beach at the moment when the wave is in position 3 (*Fig. 2* (*a*)). The bottom portion of the lower beach is often horizontal, and may even terminate in an actual rise at D. This horizontal or nearly horizontal portion will be called the " shelf."

The shelf depth (n). This is the mean depth of the shelf below mean water-level during wave-action. For simplicity this mean water-level will be taken as approximately the same as the still-water level H_0.

The bottom. This refers to all the submerged surface to seaward of the limit D of beach movement.

The beach angle (α). This is the angle made by the line BC with the horizontal.

The upper-step angle (β). This is the slope of the flattest portion of the beach surface immediately above the step.

The crest angle (γ). This is the angle of the steepest beach slope immediately below the beach crest.

THE MOVEMENT OF THE WATER (CONSIDERING THE LOWER AND UPPER BEACHES SEPARATELY).

The velocity c of advance of a surface-water wave depends in general on the wave-length λ and on the depth H_0 of the water, according to the equation :

$$c^2 = \frac{g\lambda}{2\pi} \tanh \frac{2\pi H_0}{\lambda}.$$

If the wave advances over a very gently rising bottom, till the depth becomes less than a quarter of the wave-length, the effect of the wave-length on the wave-velocity disappears. In the above expression $\tanh \frac{2\pi H_0}{\lambda}$ now approaches $\frac{2\pi H_0}{\lambda}$, so that λ cancels out, leaving :

$$c^2 = gH_0, \text{ or } c = \sqrt{gH_0}.$$

Under such conditions the wave is called a " long wave." The actual wave-length decreases as the wave continues to advance, because the foremost waves are travelling more slowly in the shallower water and tend to be overtaken by those behind, which are still in deeper water. When the wave-length is about 7 times the amplitude the wave becomes unstable and breaks. If, however, the sea bottom is bordered by a comparatively steep shingle beach, the wave's velocity cannot slow down sufficiently fast to keep pace with the rapidly decreasing depth ; the equation, $c = \sqrt{gH_0}$, no longer holds good, and the fall in wave-velocity between the beginning of the beach and its extreme crest may be negligible.

The wave suffers one of two possible fates. (1) If its wave-length/amplitude ratio was previously approaching 7 the wave is broken by the sudden distortion caused by the slope of the beach. *Fig. 2* (*a*) shows succeeding stages of the distortion. In the final stage the wave flattens out after the break and surges up the beach. (2) If the wave-length/amplitude ratio was previously low the wave passes through the stage of maximum instability, when its forward face is raised to the steepest angle, and does not break. It rights itself by pushing out the foot of its forward face at a higher velocity than that at which its crest is moving. The wave thereafter flattens out into a beach surge, as in the former case. The main difference between the two cases is that in (1) the foremost water of the surge up the beach has come from the overturned crest, whereas in (2) it has come from the foot of the wave face. These two cases are but extremes of a complete range of possibilities corresponding to a total break, various degrees of partial break, and no break at all.

The experiments showed that on a shingle beach it makes no appreciable difference to the ultimate profile whether the wave breaks or not.

In *Fig. 2* (*b*) an attempt has been made to indicate, by means of displacement/time curves, the observed movement, along the bed, of water particles which were originally at rest in positions 1a, 2a, 3a, and 4a (all displacements along the inclined bed being represented as horizontal in the diagram). Comparative bed-water velocities are indicated, qualitatively only, by the relative flatness of the curves along the displacement axis, and the accelerations and retardations by the relative abruptness of their curvatures. The advance and recession of the wave proper in the water above is accompanied by a wave of movement along the bed, and the propagation of this forwards and backwards is indicated by the lines LM and OP. These lines are not continued to the ultimate beach crest, because the surge up and down the beach can hardly be considered as a travelling wave whose rate of propagation, as such, has any real meaning.

Before proceeding to discuss the experimental results in detail, the following important general features should be noted:

(i) The drag on the bed-surface, which presumably controls the movement of the bed-material, is known, from existing work on the carriage of solids by streams, to be some direct function of the velocity, the acceleration, and the degree of turbulence in the lowest water-layers.

(ii) Owing to inevitable losses due to bed friction and similar causes during the beach surge, to the breaking of the wave, and to the partial reflexion from the neighbourhood of the step (shown by the line MN in *Fig. 2* (*b*)), the energy of the main outward wave OP is considerably less than that of the approaching wave LM. Hence the outward velocities and accelerations of the water are also less.

(iii) As a result of (i) and (ii), it follows that, of those particles (shingle) which are so large that they are moved only during the most violent part of the cycle (the portions bc and ef of the displacement curves), a greater number are set in motion by the inward than by the outward waves. Such particles therefore creep beachward, and the bigger the particle the more pronounced is the creep. The biggest mobile particles can only be moved beachwards, and therefore collect highest up the beach, along the crest.

(iv) On all steeply sloping beaches such as those dealt with here, whether or not they are composed of movable material, there exists a certain critical contour-line, above which the advancing wave-front is most vertically inclined. On mobile beaches the material at this level becomes fashioned into the step; but if the beach surface is rigid, this level still coincides with a critical position in the history of the wave. At this position the shallow reverse current 4ab (*Fig. 2* (*b*)) is so delayed by the drag of the beach close under it that it collides at M with the advancing current 3bc. As a result, a partial node or *clapotis* occurs, and the bulk of the water at this node is forced upwards, as indicated by the small arrows in the top figure. At the bed, both currents are momentarily halted, and a small secondary wave MN is sent out into deeper water. On the beachward side the surge may similarly be regarded as a secondary wave, albeit of far greater energy, reflected forward from the node. On its return from the beach crest the surge collides at the same place with the now almost stationary water, and again forms a *clapotis*. It is at this collision, at O, that the main return wave OP is generated.

The above picture of the water motion gives some physical background to the conception of two separate beach zones, the lower beach and the upper beach. The profile of each appears to depend on a different set of factors and to obey different model-rules. It will be shown that, whilst the character of the lower beach seems to depend only on the value of the ratio R of wave amplitude to particle size, that of the upper beach depends on the absolute value of each of these quantities independently, the height of the upper beach being proportional to the wave amplitude h, and its angle α depending on the porosity of the material, which is itself a function of the grain diameter.

The Initial Beach Profile: Outline of the Main Series of Experiments.

The main series of experiments was made with rounded beach pebbles of diameters ranging from 0·5 centimetre to 0·9 centimetre, with a mean diameter of 0·7 centimetre. In subsequent experiments sufficient work was done with other material, of diameter ranging from 0·3 centimetre to 0·05 centimetre, to ascertain how far the results were modified by changes in the particle size.

The standard initial beach profile was taken as that left by wave action during a receding tide. To produce this initial beach, material was heaped steeply against the back wall, as shown in Fig. 3, Plate 1. The tank was then filled with water to a suitable level, and the wave-making apparatus was put into operation. When a beach of maximum height had been built up, the water in the tank was let out slowly, the wave action being maintained to give the effect of a falling tide. Apart from minor irregularities (due to certain resonant states being set up between the beach and the paddle) the resulting beach had a remarkably straight profile. Fig. 4, Plate 1, shows this low-tide beach for the 0·7-centimetre material, and Fig. 5, Plate 1, for the 0·3-centimetre material. In both cases the inclination of the beach was 19 degrees. (The grid in these and all subsequent photographs is divided into 10-centimetre squares.)

The tank was filled up again to a still-water level of 20 centimetres. Waves of very small amplitude were made to strike the beach. The amplitude was increased until the beach material began to move, and then it was kept constant, the wave action being continued till the resulting tiny beach became mature and stable. The wave amplitude was measured, and a photographic record was taken of the final beach profile.

With the water still at the same level, another series of waves of greater amplitude was run, until the new and larger beach became mature. The process was repeated with bigger and bigger waves up to the maximum amplitude possible for the particular water-depth used. The beach was then re-formed to the same initial profile, and the whole procedure was repeated in turn for a series of water depths ranging up to 50 centimetres. A selection of beach profiles for the 30-centimetre and 50-centimetre levels is shown in Figs. 6 and 7, Plate 1.

Formation of the Lower Beach.

For values of R below 45 (*large shingle*).

It will be seen from *Fig. 2* (*b*) that the initial acceleration at b, and the maximum value attained by the bed velocity of the water under the advancing wave, both increase as the wave passes over the rising bottom. At some point D (*Fig. 2* (*a*)) the violence of the bottom flow becomes sufficient to move a few of the most exposed pebbles forward up the beach. Farther on this motion is augmented by the addition of still more moving pebbles from the bed. The pebbles are carried to the neighbourhood of the step C, where their motion suffers a check (due to the momentary gathering-up of the water), before they are passed on to the upper beach. The returning wave ef causes a slight reversal of the pebble motion, but the net effect is a removal of pebbles from the whole zone between D and C. As the slope of the resulting shelf approaches the horizontal, fewer and fewer pebbles are moved forward by each wave, and the reverse motion ceases altogether. The final stages of the excavation are very slow indeed. Ultimately all pebble motion on the lower beach ceases.

That the limiting depth at which pebble motion ceases can be slightly greater on the beachward side of D than at D itself is, no doubt, due to the progressive development of turbulence in the water as it passes over a greater length of beach surface, and to the consequently greater drag which the same mean velocity can exert on individual pebbles.

The width of the shelf, according to experiments in which the initial beach inclination was varied, depends on the initial profile. The flatter

Fig. 9.

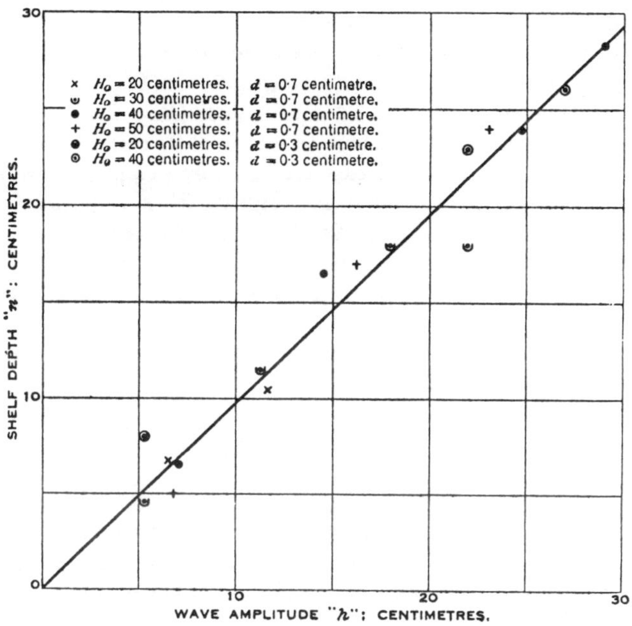

this is made, the wider is the shelf. In beaches formed by the breakdown of an initially very steep shingle bank, there is no shelf or lower beach at all, the " made ground " of the step falling directly to the bottom at the angle of repose. (Fig. 8, Plate 1.)

The shelf-depth n below the mean water-level H_0 was found to be independent of H_0 and also of the grain-size d (at all events between the limits $H_0 = 20$ to 50 centimetres, $d = 0{\cdot}3$ to $0{\cdot}7$ centimetre, and $R = 16$ to 90). Though the shape of the oncoming wave and the profile of the bottom seem to have some minor effect on it, n was found to be approximately equal to the wave-amplitude h over the whole of this range. The experimental evidence is shown graphically in *Fig. 9*. Further experiments with bigger grains are required to verify this relation for very small values of R.

For values of R above 45 (*Oscillation Ripples*).

That the depth to which the material is removed does not increase as the grain size is made smaller is a rather surprising result. The explanation appears to lie in the progressive changes which take place in the small-scale profile of the shelf surface and their effect on the water-flow near it. The flat shelf surface becomes unstable when R exceeds 45. Transverse ripples appear on it. These evidently exert an increasingly powerful drag on the movement of the lowest water-layers as the grain-size is made smaller and the grains are more mobile.

Figs. 10 and 11, Plate 1, show the effect of reducing the grain-size while keeping both H_0 and h nearly constant. In Fig. 10 the grain-size d was 0·7 centimetre, and h was 29 centimetres ($R = 41$). In Fig. 11, d was 0·3 centimetre and h was nearly the same at 27 centimetres ($R = 90$). It will be seen that in the latter case a single large ripple has appeared at E between the step C and the point D. Fig. 12, Plate 1, shows the further development of the ripple system as the value of R is increased. In this instance the grain-size was 0·05 centimetre, and h was 11 centimetres ($R = 220$). Here the concavity of the lower beach has entirely gone, and the beach has begun to assume the typical rippled appearance of a low-tide sand floor. The formation of these ripples and the movement of the water over them is dealt with on p. 50.

For large values of R.

As already mentioned, the experiments did not extend to cover those values of R corresponding in nature to the case of fine sand and large waves. Classical wave theory, however, rather indicates that at some high value of R a change should take place in the motion of the grains over the lower beach, for it has been shown that the advance of water-waves of large amplitude/wave-length ratio must be accompanied by a slow advance on the part of the upper layers of the water itself. This means that underneath there must be a corresponding drift of water outwards down the lower beach. The velocity of this drift is very feeble compared with the alternating water-velocities constituting the wave itself; but its effect on particles so small that they are kept in almost permanent suspension over the rippled surface may well be a net outward drift of material instead of the inward drift which certainly prevails for large particles. The material removed may be expected to come to rest in such deeper water that the bottom motion due to the waves is not sufficient to pick individual particles off the bed.

Although this outward drift has not been observed in the model-tank, evidence for it seems to be provided in nature by the very different effects of big waves on shingle and on sand beaches. For whilst, under the action of gentle seas, sand tends to drift shoreward like shingle, a storm (R very large) tends to carry the sand out to sea. The critical value of R at which the change from shingle to sand takes place is probably of the order of 2,000.

The Upper Beach : Behaviour when Beach Formation is Unimpeded.

Movement over the Beach.

The water above the step is violently accelerated, and surges up the beach. It drags with it not only material from the step and from the shoreward side of it, but also all the new material which may have been passed up from the lower beach. The velocity of the surge is subsequently checked by the action of gravity. It becomes zero instantaneously over the whole length of the surge at the moment when the foremost water has reached the maximum height s above the step-level. There then remains, however, an appreciable water-velocity perpendicular to the surface, for the water continues to flow inwards and downwards through the interstices between the grains. The mass of water taking part in the return surge is therefore considerably less than that in the up surge. There are, in addition, energy losses due to friction and turbulence which also reduce

Fig. 13.

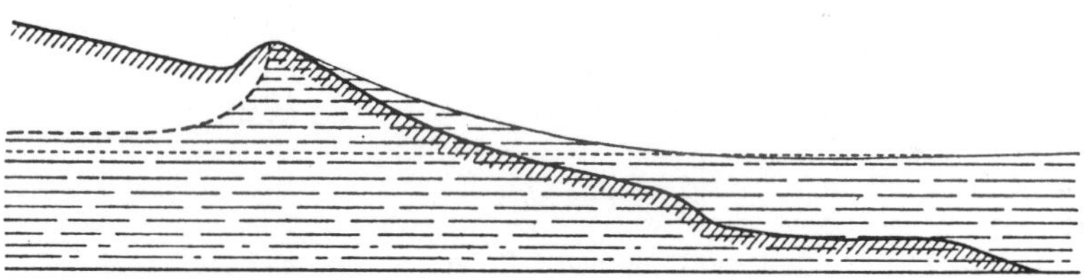

the velocity and the drag of the returning water. Hence, were it not for the inclination of the beach surface, which favours the downward movement of the material, more material would always be driven up the beach than was brought back again. Evidently once the beach is mature its surface assumes such an inclination and such a curvature that a state of equilibrium is everywhere set up, and just as much material returns as is driven up.

The water of the returning surge is violently checked as each element of it successively reaches, and piles up against, the stationary water above the step. The descending shingle comes to rest in a heap at the step by tumbling over its brink. The step therefore advances seawards by building out a steep slip-face inclined at the angle of repose.

The water which percolates into the shingle forms a water-table which, at the moment of maximum surge-height, has a cross-section as shown in *Fig. 13*, p. 37. The mean water-level within the shingle is maintained during the whole period of wave action at an appreciable height above the still-water level. Hence, except during the period of rising surge, there is an outflow of water through the submerged beach surface. This outflow is greatest during and immediately after the fall of the surge, when the shingle vertically beneath the crest is still nearly full, and the head of water is consequently high.

The Beach Angle α.

The beach angle was found to be entirely independent both of the wave-amplitude and of the water-depth over the whole of the experimental range. It appears to depend only on the absolute size of the grains composing the beach. The figures were as follows :—

Mean grain diameter : centimetre.	Beach angle : degrees.
0·7	22
0·3	19·5
0·05	14

The beach angle was remarkably constant for all the beaches thrown up with each kind of material, the variation being less than 4 per cent.

The Surge or Beach-Crest Height : *s*.

The final height of the mature built-up beach crest above the step was found to be proportional, for any given size of beach material, to the amplitude h of the "out-at-sea" wave above the same datum-level, so that :

$$s = bh$$

where b had the value 1·68 for the 0·7-centimetre material, and 1·78 for that of 0·3 centimetre diameter. Time did not permit of many readings being taken for the 0·05-centimetre material, but there is some evidence that for this b lies in the neighbourhood of 1·8. The experimental values of s in terms of h are shown graphically in *Fig. 14*. Once again the values are independent of the water depth H_0.

From these results, combined with those given under the heading "The Beach Angle", it seems possible to predict the maximum dimensions of any given storm beach from the amplitude of the waves and the size of the beach material. Conversely, under well-chosen conditions, the height of the highest storm beach known to have been formed during modern times should give useful information as to the maximum wave-amplitudes likely to be experienced on that particular stretch of coastline.

In cases where there was no newly built-up beach, for example where the value of h was too small for any grain movement to take place, or in the case of a breakdown beach such as that shown in Fig. 8, Plate 1, the surge still rises to the height given above.

Fig. 14.

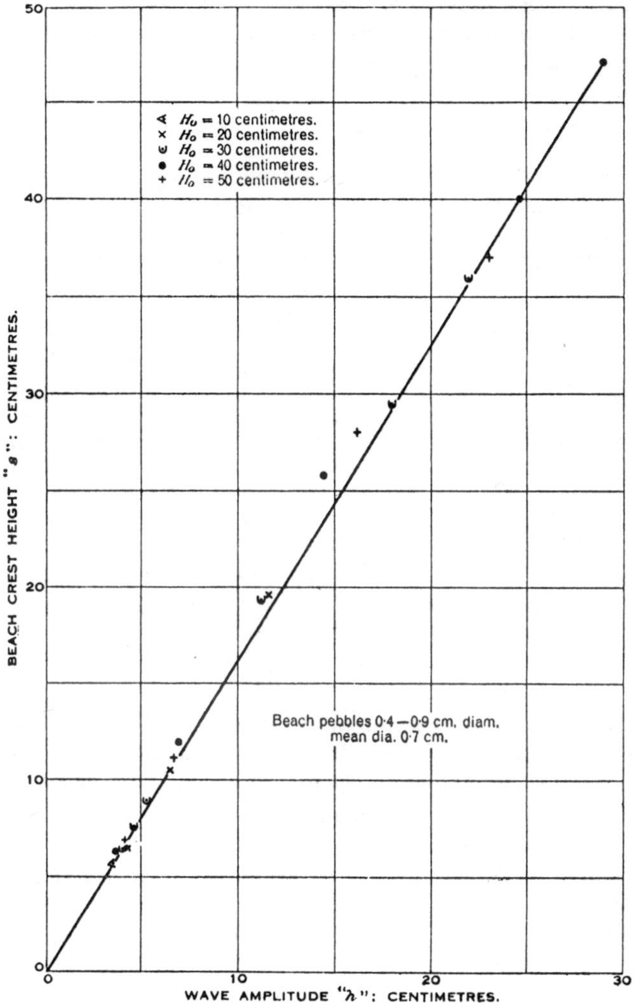

Lateral Stability of the Beach Contours.

In all the experiments where shingle was used, the model-beaches which were thrown up by the unimpeded action of the waves set themselves at right-angles to the line of the wave-advance. When more material was added on one side of the tank, this soon became distributed evenly over the width of the mobile portion of the beach. From this it appears that if waves strike a stretch of shingle beach obliquely, the beach-line will, provided that coastwise losses of material away from the farther boundary of the stretch are prevented, become re-oriented so that its contours run at right-angles to the line of advance of the waves. The conditions thus postulated are instanced in the accumulation of shingle in the corner cut off by a high groyne.

Continued supply of Material and Constancy of Profile.

A further experiment was made by adding a small but continuous supply of fresh material to the shelf after a mature beach had been thrown up. This was intended to imitate the accumulation of material by coast-wise drift. The effect was that the whole beach advanced seawards, but its profile remained entirely unchanged. The topmost surface, previously a narrow ridge, became a flat horizontal table as the beach top moved forward.

The Effect of the Presence of Impervious Material at the Crest, Lateral Instability, and the Breakdown of the Upper Beach.

As previously described, the original mass of beach material in the tank was heaped against a vertical concrete back-wall. In the course of the experiments, the wave-amplitude was gradually increased during the passage of a long train of waves until the water surging up the rising beach reached the neighbourhood of the exposed wall. That is to say, the vertical plane through the beach crest approached the plane of the wall.

It was found that the crest-line was now no longer stable and horizontal. Either one side of it or the other fell rapidly below the surge-level, and the surplus material accumulated on the opposite side. At the same time the surge-height up the wall on the denuded side rose from 1·68 h (for 0·7-centimetre material) to nearly 2 h, whilst that on the opposite side remained the same as before. A further slight increase in the wave-amplitude led to a breakdown of the whole beach. The crest-line became stable and horizontal once more, but at a considerably lower level, and was now wholly submerged by each wave. The surplus material went to extend the step to seaward (Fig. 15, Plate 1). The beach angle α made by the line joining the beach crest to the step had fallen from 22 degrees to 14 degrees.

The presence of the wall had caused a breakdown of the beach to occur before the free water of the surge had been allowed to reach such a height that it actually touched the wall above the shingle. Moreover, careful observation showed that the breakdown was not due to any slipping of the shingle down the wall, or to any internal collapse of the material. Clearly, therefore, it can only have been due to the effect of the wall in preventing the free percolation of water through the material. To test this, an impervious steel plate (the top of which had been roughened by sticking beach pebbles to it) was inserted just below the sloping beach surface (Fig. 16, Plate 1). If this plate had been absent the beach material would have piled up till the crest angle γ reached the angle of repose—33 degrees—and the overall beach angle α had reached 22 degrees (for the 0·7-centimetre shingle used). The plate was, in fact, found to act in just the same way as the vertical wall had done. Although, again, no internal slipping was detected along the plate-surface and no part of this surface was initially exposed to the free action of the surge, yet the thin layer of material above the plate, instead of increasing in thickness, at once began to be removed by the surge. At the same time lateral instability was again noticed. The plate became denuded at random on one side or the other, and the material began to pile up on the opposite side. When, however, the surge height was slightly raised so that the surge reached the exposed plate-surface on both sides, stability again obtained. The now submerged beach-top made an angle α with the step whose value was exactly the same as that with the vertical wall, that is to say, 14 degrees.

As already stated on p. 38 this 14-degree beach-angle was that attained by the mature unobstructed beach composed of the 0·05-centimetre material (Fig. 12, Plate 1). Here again, but this time owing to the relative fineness of the interstices between the grains, the porosity of the material was low.

It seems, therefore, from the experimental evidence, that if no water sank into the material the beach angle would remain approximately constant at 14 degrees for a wide range of grain-size and wave-amplitude (for all values of R up to 400 at least). It should be noted that this statement is subject to the limitation, imposed by the wave tank, that the motion of the water is in two dimensions only, lateral circulation being negligible. This proviso appears to be fulfilled in nature in the case of shingle beaches of high porosity. The effect on the beach angle of the lateral sweeping movement of the surge over a very flat sand beach will be discussed later (p. 45).

FACTORS GOVERNING THE UPPER-BEACH PROFILE. THE EFFECTS OF ENERGY LOSSES AND OF PERCOLATION (TWO-DIMENSIONAL WATER-MOVEMENT ONLY).

The following ideas as to the behaviour of the upper beach are suggested by the experimental facts.

(1) The surge may be regarded as a special case of a *clapotis*, or standing wave, in process of being reflected off an unyielding wall. The case of the vertical wall has been investigated by Sainflou[1], who found that the surge-

[1] G. Sainflou, "Essai sur les Digues Maritimes Verticales." *Annales des Ponts et Chaussées*, vol. 98-*ii*, 1928, p. 5.

heights should be equal to 2 h in the case of deep water, and slightly less for water of limited depth.

(2) If there are no losses of water energy by friction, internal or external, and no mass loss by percolation through the beach surface during the surge cycle, then the motion of the surge should be reversible, and as much material would always be carried back as was carried forward. Hence there would be no beachward drift of material and the beach angle would be zero. The surge would rise to the height calculated by Sainflou for a vertical wall.

(3) If there are such losses, then the mean drag force of the surge on the surface material is greater on the upstroke than on the downstroke. The excess force is just balanced by the component of gravity in the direction of the now inclined beach.

The surface angle of the beach at any given contour would, therefore, seem to depend on the ratio of the energy lost or dissipated by the surge over all the beach surface above that contour to the total energy which passed the contour on the upstroke. If this ratio is high, the relatively feeble downward drag of the returning surge cannot carry back a quantity of material equal to that previously brought up, unless the angle is steep and the assisting gravity component large.

The energy losses may be divided into two classes:

(*a*) A general dissipation of kinetic energy by friction against the bottom, and a dissipation of the eddy-energy of turbulence by internal fluid-friction;

(*b*) Potential energy carried away by the water which percolates into the beach pebbles and does not take part in the return surge.

In the case of fine beach material the porosity is small and the energy-loss (*b*) by percolation is negligible, so that the inclination of the surface at every level up the beach should be the result of (*a*) alone. Experiment (Fig. 12, Plate 1) appears to show that the inclination under these conditions is constant; the beach is flat and inclined at 14 degrees.

With material of larger grain-size the loss (*b*) by percolation is appreciable. Not only is the overall beach angle α increased, but since the proportional loss of energy by the disappearance of water through the surface is not constant from contour to contour, the surface is no longer plane. The proportional loss is greatest at the top, where the mass of the surge-water is nearly zero and where its potential energy is greatest. Here, in the case of pebbles, the entire flow of the water is into, instead of parallel to, the surface, so that all the pebbles which have been thrown up the crest must stay there. The top of the beach therefore piles up until the surface angle reaches the limiting value of 33 degrees. At this angle the pebbles are brought down by gravity alone.

Farther down the slope the proportional loss through the surface is smaller, and so the surface angle decreases. At a certain level the net inflow through the surface must be zero. This level is presumably marked by the contour at which the surface inclination is 14 degrees. Lower still there is, as has already been pointed out, an outflow of water; so that the loss of energy by percolation is negative. The outward pressure against the surface pebbles here assists the downward movement of the material, and the surface inclination is, in consequence, less than that of a non-porous beach. The inclination of the beach close to the step (the upper step angle) is usually about 12 degrees.

Considering the overall beach angle α made by the line joining the crest to the step, it seems possible to divide it into two component angles: the "friction angle," α_0, which remains approximately constant at 14 degrees for all values of R up to 400 and perhaps higher, and which is independent of the absolute size of the beach particles; and a "porosity increment," α_p, which appears to depend only on the absolute size of the material. The porosity increment varies between zero for fine sand and about 8 degrees for shingle.

(4) That the porosity increment depends only on the absolute size of the grains, and not on the wave-dimension, seems reasonable from the following known facts. The velocity of percolation through material of grain diameter d, where d is so small that the flow can be considered to be viscous, is given by:

$$v = \frac{Ag}{\nu} \cdot \frac{H}{L} d^2,$$

where H denotes the head of water, L the length of the path, and ν the kinematic viscosity of the fluid, which can be taken as 0·01 for water. Since the beach angle remains of the same order for a very large range of wave-amplitudes, the length L is approximately proportional to the crest height s, which is itself proportional to the wave-amplitude. Furthermore, the mean value of the head H is also nearly proportional to the wave-amplitude. Hence $\frac{H}{L}$ is approximately constant, and so for fine material with low porosity:

$$v \propto d^2.$$

For coarse material with high porosity (pebbles), on the other hand, the corresponding expression is:

$$v = B\sqrt{g \cdot \frac{H}{L} d},$$

whence:

$$v \propto \sqrt{d}$$

The porosity therefore increases very rapidly as the material changes from fine sand to pebbles, but varies little throughout the range of beach-shingle sizes.

That the friction angle α_0 is independent of the wave-amplitude can be verified experimentally. The plane beach of Fig. 12, Plate 1 (fine sand) was thrown up by waves of 11 centimetres amplitude ($R = 220$). Subsequently the surface was disturbed by a train of short choppy waves of only 1·5 centimetre amplitude ($R = 30$). Erosion took place, and the removed material collected farther down to form a step, as shown in *Fig. 17*. The surface of the excavation was, however, still plane, and it ran exactly parallel to the 14-degree plane of the original beach. (It is worth noting that, since R was less than 45, no ripples were formed on the lower beach, even though the material was fine sand.)

(5) The lateral instability of the beach crest when the surge top reaches the junction between a porous beach and an impervious layer (wall or plate) is easily explained by the percolation hypothesis. If the free-

Fig. 17.

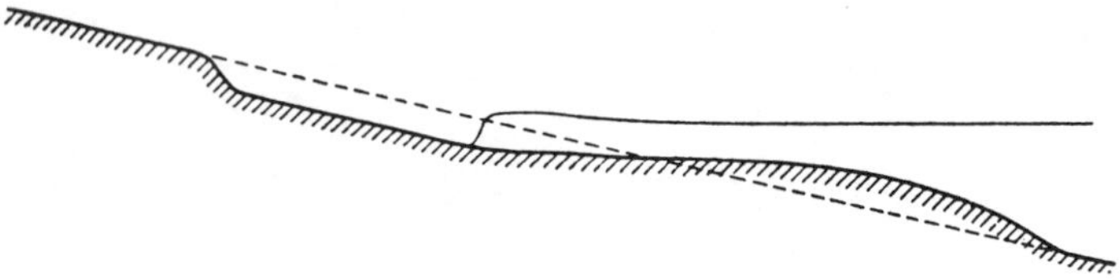

surface surge reaches the exposed layer before all of it has fallen through the intervening pebbles, a surface backwash is created where there was no backwash before. This backwash removes the uppermost pebbles and so lowers the crest. On the next surge the backwash is greater still, because the impervious layer is exposed and water has surged over it without any loss by percolation. More pebbles are removed, and so on, until the crest is submerged to such an extent that percolation through it is negligible, that is to say, the beach angle becomes that of an impervious beach—14 degrees.

Similarly, if by chance a group of pebbles lodges temporarily at the crest, it causes, locally, a greater separation between the top of the free surge and the impervious layer. Consequently there is increased percolation through the group of pebbles, the return surge is reduced in intensity, and so more pebbles are added to the existing group. The beach therefore builds up at this point.

(6) The effect of porosity on the behaviour of a shingle beach may have interesting applications to the problem of protecting shorelines from sea erosion. For instance, if it is required to induce the maximum possible accumulation of shingle, the presence of an impervious wall appears to be

undesirable. It would seem preferable to allow the shingle to pile up against an open palisade of vertical steel bars placed parallel to, and in front of, the sea wall which is to be protected.

THE EFFECT OF LATERAL WATER-MOVEMENT OVER SAND BEACHES.

This Paper has, so far, dealt with experimental beaches formed by surges which move in two dimensions only, that is to say, with those whose lateral velocity at right-angles to the direction of advance of the wave is negligible. In the case of large shingle of high porosity, observation on the seashore shows that when the direction of advance is perpendicular to the shore-line the waves do indeed retain this same direction throughout the rise and fall of the surge. Experimental shingle beaches thrown up within the narrow confines of the wave tank appear, for this reason, to imitate very closely the profiles of real shingle beaches in nature.

When, however, attempts are made to imitate in the tank the formation of sand beaches, the correspondence between the artificial and the natural beaches is not so close. The 14-degree beach thrown up in the tank (Fig. 12, Plate 1) is considerably steeper than most sand beaches found in nature. The reason for this discrepancy is suggested by two interesting observations, one (*a*), in the wave tank, and the other (*b*), in nature.

(*a*) Fig. 18, Plate 1, shows an apparently stable beach formed in 0·05 centimetre sand from an originally flat surface standing at 4 degrees by a series of waves of 6 centimetres amplitude which had so long a wave-length that they did not break. From the outset an instability was noticed. For though the beach contours were repeatedly straightened out by hand during the wave action, a low bank was, before long, always formed on one side or the other; this bank caused the surge to assume a circular sweeping motion across the tank instead of moving directly up and down as it had done on other beaches. Such a bank can be seen at the top of the beach in the figure. Measurement will show that its surface is inclined at the 14-degree angle. It lies on the camera side of the tank, and the lateral movement of the surge was away from it and across to, and then down, the other side of the tank, where there was no bank at all, and where the surface retained the general 5-degree inclination right up to the surge height.

(*b*) *Fig. 19* shows the ground plan of a typical sand beach near Mersa Matruh on the Egyptian coast west of Alexandria. Here an outer reef limits the wave-amplitude to about 60 centimetres, and the whole beach formation is simplified by the lack of tide in the Mediterranean and by a steady shoreward wind which blows at right-angles to the beach-line for long periods. It will be seen that the contour lines of the upper beach run in a regular succession of bays, which have an average width of about

12 metres, and are separated by promontories. The beach angle of these promontories approaches, but never exceeds, 14 degrees. The beach angle in the middle of each bay is considerably less, and may be as low as 3 degrees. The profile everywhere is a straight line, and all these straight lines run down to join a common horizontal and unindented step-line. Below the step the lower beach bears transverse ripples as in Fig. 12, Plate 1.

Waves with their straight crest-lines parallel to the general shore-line approach and break when they are immediately over the step. The surge, which piles up thickly against the steep face of each promontory, is divided by it into two diverging streams. The upper fringe of each stream, as it flows inwards towards the hollow of a bay, surrounds and heads off that

Fig. 19.

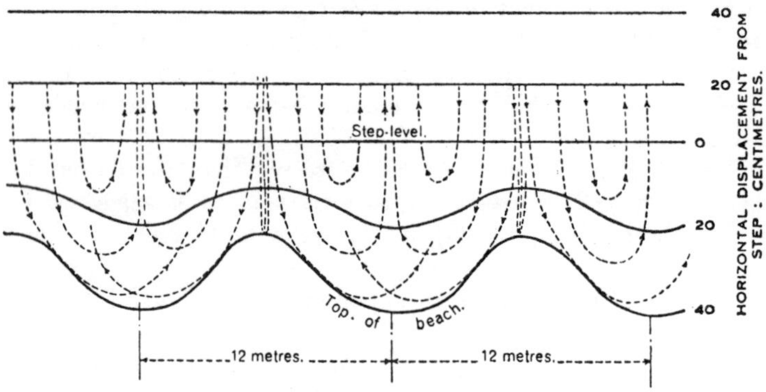

part of the surge which has flowed directly up the hollow. The two side streams from the promontories on either side meet in the centre and together form a return surge down the hollow of considerably greater intensity than the previous upward surge there. It should be noted also that owing to the introduction of a sideways flow there is no longer any dead period (except on the front of each promontory) when the water and the sand grains it carries come momentarily to rest. The grains are kept in continual movement and therefore do not get a chance to settle on the bed.

There can be little doubt that both of these effects, the incipient beginnings of which were previously noted in the wave-tank experiments (Fig. 18, Plate 1), are responsible for the fact that the angle of a natural sand beach is less than the experimental 14-degree friction angle, α_0.

On beaches exposed to less constant winds, waves, and water-levels, the rhythmic production of bays and promontories is not, of course, so pronounced as it is under the ideal conditions on the North African coast.

The Size-Grading of the Particles over a Pebble Beach.

When using shingle of mixed grain-size—0·5 to 0·9 centimetre—a very distinct sorting-out of the stones took place during the formation of a model-beach. From the lowest point D (Fig. 3, Plate 1) at which bed movement took place, the stones remaining on the surface of the mature shelf dwindled rapidly in size towards the step, until pebbles of the minimum diameter were found by themselves as a transverse belt immediately at the foot of the steep slip-face leading up to the step. The surface material composing this face was, on the contrary, of maximum diameter. From here to half-way up the upper beach the diameter fell to the general mean diameter of the whole mass. Above this again, the diameter increased with increasing steepness of the beach, till at the crest only stones of maximum size were found.

It was noticed, however, that when a fresh layer of new upper beach was thrown up from below and deposited, the selective grading was more apparent than real, being confined, except at the very top of the crest, to the surface only. The reason for this was clear: far more material was always set in motion forwards and backwards than was deposited, and this oscillating layer was a considerable number of grain diameters in thickness. During the motion the oscillating layer itself became graded vertically, because the small grains tended to fall to the bottom of the layer, leaving only the larger ones at the top. Whenever, therefore, a small permanent deposition took place from the layer, it tended to consist only of the small grains; the largest stones remained and accumulated on the surface of the oscillating layer as more material of the general beach composition arrived from below.

Small quantities of fine material originally mixed with the pebbles disappeared from the beach altogether. All the fine grains were washed downwards and outwards by the circulation of water through the body of the beach, and accumulated on the tank floor at the foot of the shingle beach.

On the whole, it may be said that the model-beach of shingle imitates, in a remarkable way, all the features of its full-scale counterpart.

The Effect of Variations in the Still-Water Level (tides) and of Wave-Amplitude, during Wave Action.

The effect of a slowly falling tide on an existing model-beach of large shingle ($R = 30$ to 70) has been described on pp. 33, 34 under the heading "The Initial Beach Profile." The material was left standing at an angle of 19 degrees whatever the wave-amplitude might be, and the beach surface was nearly plane. It was unfortunate that time did not permit experiments to be made with fine sand at higher values of R.

In the case of a steadily-rising tide, it is clear that at any given moment the beach has not reached the mature profile corresponding to the wave and water conditions at that moment, but that the profile approaches maturity as the rate of rise is made slower in relation to the frequency of the wave action. In this connexion it may be noted that the experiments brought out that the rate of bulk movement of beach material by waves of a given amplitude varies with the diameter of the grains. Each wave moves a far greater weight of coarse material than of fine material.

Since, for any given material, the beach form depends on the wave-amplitude, and since it does not approach maturity till many waves have acted on it, the profile of a natural beach which is formed by a succession of waves whose amplitude varies at random cannot be as clear cut as the profile of a model-beach formed in the wave tank.

Submerged Beaches at the Foot of a Vertical Wall.

All the foregoing refers to beaches of which the quantity of material was initially large. Experiments were made to see what happened to

Figs. 20.

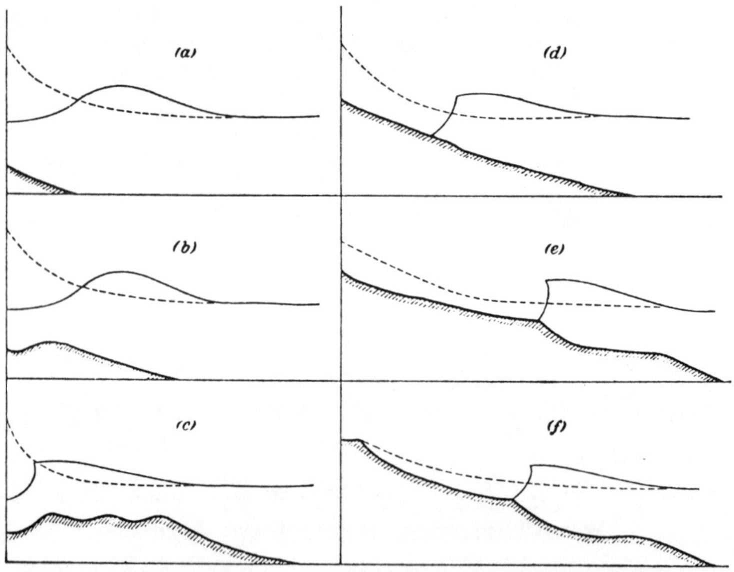

the beach profile when a small embryo beach at the foot of a wall was fed with a continuous supply of material, and was allowed to grow progressively bigger. The results showed that the growth of such a beach is discontinuous, and passes through at least two critical stages.

Figs. 20 show the successive stages which were noticed. Material

(0·7 centimetre) was fed in gradually at the foot of the beach, and was worked up it by the continued action of the waves. (The wave-amplitude remained the same throughout the experiment, but although in stages (*c*) to (*d*) (*Figs. 20*) the position of the break did not alter appreciably, there was no break at all in stages (*a*) and (*b*) because of the feeble distortion of the wave caused by the little embryo beach. In these early stages the wave merely surged vertically up the wall as a true *clapotis*.)

It will be seen that at stage (*b*) the beach had risen to such a height that the scour caused by the collapse of the *clapotis* down the wall had formed a trench in the beach immediately against the wall. As more material was added this scouring action at the wall prevented the beach there from rising any higher. The material therefore accumulated farther out. The growth of this accumulation presently caused the wave to break before reaching the wall. At this stage (*Fig. 20* (c)) a chain of two or three acceleration ripples appeared (Fig. 21, Plate 1). These travelled slowly outwards from the scour trench, which was consequently deepened and widened.

This ripple system appeared to be unstable, for it broke down without the addition of any more material. When the scour trench had grown to a maximum size, the accumulated material farther out was suddenly and rapidly moved forward to fill it up and obliterate it. The beach once more rose to a crest against the wall; but now the crest was exposed at low-water. The profile is shown in *Fig. 20* (*d*). This new crest rose steadily; but concurrently a fresh accumulation of material began farther out, so that the beach angle grew smaller. A second critical stage was reached when the beach angle attained the minimum value of 14 degrees (stage (*e*)). This stage appears to correspond with the condition described on p. 40 and in Fig. 15, Plate 1. Finally the addition of more material caused another sudden forward movement of the beach. The beach angle steepened to 21–22 degrees, and the shingle crest, now piled up to the maximum height of the surge, formed a porous barrier between the surge and the wall. After this final stage (*f*) the addition of more material merely had the effect, already described, of making the whole beach advance to seaward without change of profile. Fig. 22, Plate 1, shows a typical submerged beach formed with the fine 0·05-centimetre material.

[*Editors' Note:* Material, including Plate 1, has been omitted at this point.]

16

Reprinted from *Am. Geophys. Union Trans.* 29(5):653-655 (1948)

MOVEMENT OF BEACH SANDS BY WATER WAVES

Hans Albert Einstein

Abstract--The basic similarity of the sand movement on a beach and in river flow is discussed and from this similarity an approach is indicated to the solution of the beach problem.

In studying nature we encounter many phenomena which obviously move or change their shapes continuously such as the clouds in the sky and the water in the streams. Others appear to be stationary and to undergo no changes whatsoever. We have learned and are still learning, however, that even the most stable of things like the rocks constituting the Earth's surface undergo changes. They may be slow, even so slow that they become imperceptible to our eyes, but they still exist. One of the major tasks of all natural science is the study of such changes.

The beaches which line a great part of our coast are constantly undergoing changes. Those who live near the coast will observe that the width of beaches increase and decrease with the seasons. They will notice also that a single storm may alter the appearance of a beach entirely and that it may take a considerable length of time until the beach has regained its normal appearance. Conditions become critical, however, when beaches begin to regress steadily and finally disappear. It is then that a search is made for the reason of these changes for which usually no satisfactory explanation can be found locally. They cannot be explained by especially severe storms or by any other obvious cause. Only when we begin to study the normal sand movements on a beach systematically do we begin to realize that the reasons for any such changes may be found at a distant part of the beach or even in a stream supplying sediment to the beach. An interruption of the sediment flow there may cause the beach to become denuded.

Systematic studies show that the sand moves slowly along a beach. This littoral drift often assumes proportions which suggest the possibility of describing the beach as a flowing river rather than as the shore of a stationary body of water. Upon closer observation we find more and more basic similarity between a beach and a flowing river: (a) In both instances a body of water is moving over a large deposit of loose sediment particles; and (b) in both cases we may observe that sediment particles are moved along with the water, sometimes at a large rate, sometimes at a small rate, sometimes only along the bottom, and sometimes suspended over the entire water depth. The motion of sediment, as caused by river flow, has been studied systematically by geologists and engineers for many years. The movement of sediment on a beach did not attract as much interest as the river problem until very recently. Thus, we are confronted with the problem of planning the basic lines of approach to be followed in studying the motion of beach particles.

As stated above, the motion of sand along a beach resembles in many respects the motion of sand in a river. This latter problem has been solved in large part, and it is reasonable to expect that a solution of the beach problem might be found by application of these same principles. The sediment motion in rivers was first described by means of more or less empirical expressions. It was realized that the rates at which sediment is moved depend both on the strength of the flow and on the characteristics of the sediment particles involved. It was found that the strength of the flow can be expressed by the "tractive force," which is another expression for the unit shear exerted on the bed by the flow. The sediment of the bed may be described by a representative grain diameter and by its specific gravity. Flume experiments were used to determine the relationship between these variables. A great number of empirical formulas were proposed by different investigators. Each formula is applicable only under conditions similar to those for which the specific formula was derived. None can claim general applicability. There is very little hope that any one of these formulas may be applied to the beach problem. The greatest obstacle besides their small range of applicability is the fact that there is no continuous flow at the beach which would permit the determination of a representative shearing stress. Instead, the motion of the particles is generated by the oscillatory water movement due to wave action. The major part of this motion has its direction at right angles to the beach line. Correspondingly, the major part of the sediment motion is oscillatory and directed at right angles to the beach.

This much one can very easily observe on every beach. If the waves which create the surf approach with long crests parallel to the beach, the resulting water velocities are all directed at right angles to the beach and no long-shore movement of sediment can result. If the waves approach the beach at an angle and if the resulting oscillatory motions are symmetrical like the ocean waves in deep water, we should expect that the long shore-components of the velocities also are symmetrical. This means that we could not expect to find a resultant littoral drift to occur. In reality, however, all waves become asymmetric in the shallow water of the beach and finally break. Then velocities of the water particles must be described as an oscillation superimposed over a current at any point of the flow. If these current velocities are integrated over a beach cross section, it is found that the average current in a direction perpendicular to the beach becomes zero; while the components parallel to the shoreline give a resultant flow which is largely concentrated in the region between the breaker point and the shore. Must this current be introduced as a driving force into the bed-load equation or must it be combined with the oscillatory flow in order to predict the littoral sand movement? The existing empirical approach to the sediment transportation problem unfortunately cannot answer this question.

From this short analysis it becomes apparent that the beach problem is more complicated than the corresponding river problem. The littoral current which causes the littoral sand movement is a small secondary part of the flow while, in the river, sand motion is caused by the main flow itself. The pattern of the two motions is, therefore, different and we cannot expect them to be described by the same equations.

This seems to contradict our first statement that the two types of movement are basically similar, but it does not. Basic similarity between two physical phenomena does not necessarily imply that they are described by identical equations. It only means that the fundamental concepts which are used in the derivation of the equations are the same. For instance, it may mean that both occurrences follow the same basic differential equation which must be solved individually for each case. If we want to learn anything from the river problem that might be applicable to the beach problem, we must look for basic facts rather than final results. It is difficult if not impossible to define with a sentence what a "basic fact" is; however, an attempt will be made to enumerate some observations which have been made in the river problem and which may apply in the beach problem.

(1) It has been found that sediment particles can be suspended indefinitely in a turbulent flow. The vertical velocity components in the turbulent flow move sediment particles from lower layers with higher concentrations to higher layers with lower concentrations at a larger rate than in the opposite direction. An equilibrium condition is reached when the resulting upward movement of particles due to turbulence is just sufficient to counteract the normal settling of the sediment particles in water. This same description probably is applicable to the beach problem, but the application of the theory to practical beach problems is considerably more complicated than to stream problems unless the periodic changes of turbulence can be neglected.

(2) In the calculation of suspended load distributions it is possible to make a number of simplifying assumptions in the case of rivers. For instance, can the settling velocity of a particle, as determined in still water, be applied in turbulent flow? Furthermore, the turbulence in a river which embraces a very complicated pattern of constantly changing flow velocities in all directions can be sufficiently described for this problem by only an average value of the vertical velocity component. There is no reason why these shortcuts should not be equally valid for the beach problem.

(3) In a thin layer near the sediment bed the theory of suspension breaks down, and the particles there are not vertically supported by the flow turbulence in the case of the river flow, because at this point the scale of the turbulence becomes very small. The particles roll on the solid bed, instead. This zone has been called the "bed layer" and is about two grain diameters thick. The sediment which moves in this layer is usually called bed load. This same phenomenon may be expected to occur in the beach problem.

(4) In a river it has been found that a constant exchange of sediment occurs between the bed and the bed layer. The number of particles which are deposited every second in the bed depends on the rate of bed-load movement. The number of particles which are removed from the bed every second depends mainly on the flow pattern. The equation which expresses equality of the two rates is the bed-load equation for a stable bed. A similar equation also must exist for the beach problem. It is not clear, however, how the periodicity of the flow must be introduced into this equation, when the flow pattern is described.

(5) In deriving the bed-load equation for river flow it is sufficient to introduce the flow conditions in the immediate vicinity of the bed. This fact is probably correct also for the beach problem.

(6) The rate at which bed load is being deposited in the bed has been found to be proportional to the rate of bed-load movement and inversely proportional to the distance traveled by the bed-load particles between periods of deposition. This fact was observed for river flow and probably holds also for the beach problem.

(7) If the sediment bed of a river consists of various grain sizes, each grain size moves as if the entire bed consisted only of particles of this type. This basic independence with which the different fractions of a mixture move also can be used, with all probability, to describe the sediment movement on beaches.

Still more "basic facts" of this type probably could be enumerated. Together, they mark a definite line of attack which was successful in describing the river problem. The great difficulty in the case of the beach problem is our complete lack of knowledge concerning the flow details, especially the turbulence distribution. It seems that much basic hydraulic research is necessary for the beach problem before the sediment problem can be fully understood. Unfortunately, the turbulence studies conducted in wind tunnels which gave practically all the required information for the river problem are not directly applicable to the periodic flow on the beach.

Besides these basic studies, where the aim is to derive sediment transportation laws through analogy with the corresponding river problem, it will be necessary to make sediment measurements on natural and model beaches to check and support any such theoretical approach.

The stability of beaches is closely connected with the construction and maintenance of harbor installations. For this reason several public agencies have recently shown considerable interest in this problem. Geologists and oceanographers on the other hand have a large scientific interest in the solution of the sediment problem on beaches. It would be very desirable if these different efforts could be coordinated under a common program of systematic research.

Acknowledgment--This paper is an excerpt of the basic ideas given in a report entitled "Some thoughs on the organization of research on the movement of beach sands by water waves" for the Navy Wave Project at the University of California.

University of California,
Berkeley, California

17

Reprinted from *Am. Geophys. Union Trans.* 32(6):866-874 (1959)

THE RELATIONSHIP BETWEEN SAND SIZE AND BEACH-FACE SLOPE

Willard N. Bascom

Abstract--A series of diagrams is presented which illustrates the variability of sand size and beach-face slope and the relationship between the two as observed on Pacific Coast beaches of the United States. As a means of standardizing sampling procedures, a "reference point" (the part of the beach face subject to wave action at mid-tide level) was adopted, which permitted satisfactory comparisons of beach characteristics. Distribution of sand sizes along and across beaches is described, and slope changes with beach growth and erosion are illustrated. It is shown that beach-face slope is principally controlled by two factors: (1) the size of the sand and (2) the intensity of wave action.

Introduction--Under the sponsorship of the Office of Naval Research and the Beach Erosion Board, a field party of the University of California visited and repeatedly surveyed some 40 beaches on the Pacific Coast of the United States in the years 1944 to 1948 (see Fig. 1). These beaches included steep and flat beaches under winter and summer conditions and in high and low surf. Details of beach conditions from the dunes to beyond the breaker zone were surveyed and noted, sand samples were taken at many points, the effect of surf conditions on each beach was recorded, and an elaborate program of breaker measurements was conducted. A total of about 500 profiles (from dunes to minus 30 ft) was made and over 600 sand samples were taken in the course of these investigations. Details of these activities are described in the reports of the Institute of Engineering Research, University of California, Berkeley. Since these reports had a limited distribution, some of these data, screened to insure reliability, are used here in a somewhat different form. With its irregularities, the Pacific Coast of the United States is thousands of miles long and quite probably includes most of the possible ocean-beach conditions and configurations. The examples used in this paper represent its maximum variations, and it is believed that these graphs will be generally valid for use in comparing beaches throughout the world. The scope of this paper includes only sandy beaches in which sand is defined as ranging in size from 0.05 to 1.0 mm (U. S. Bureau of Soil Classification). Although there are a few pebble, gravel, and cobble beaches on the coast studied, the largest material found on extensive beaches subject to systematic sampling was 0.8-mm sand. It seems reasonable to believe that the curves can be extended smoothly upward when data on beaches composed of coarse material are available. The studies under which these data were obtained are still under way; this is not to be regarded as a final report but rather as a statement of progress to date.

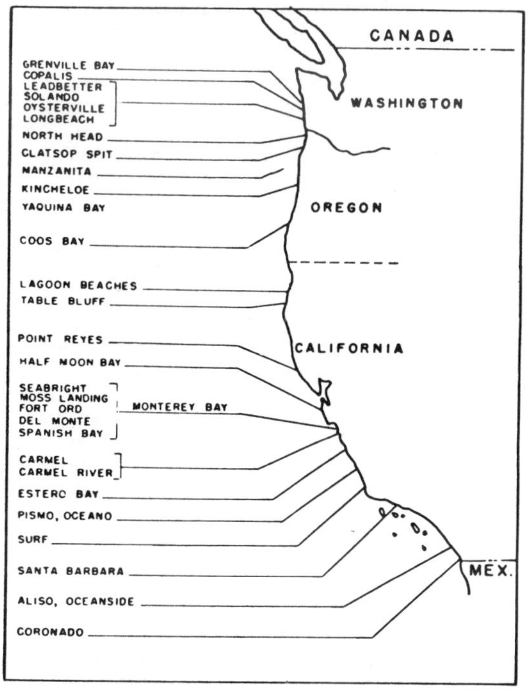

Fig. 1--Location of Pacific Coast beaches investigated for beach slope and sand characteristics

The importance of using standardized measurements can scarcely be overstated and other researchers are urged to consider the use of the "reference point" in future beach examinations so that data will be comparable. This paper is concerned with comparing western beaches of the United States on the basis of median sand size and beach-face slope and shows the probable limits of each. The beach face in this instance is synonomous with shore face or foreshore. The importance of surf action in changing slopes from a profile of deposition to one of erosion is shown, as are the variations in slope with exposure. Finally, Figure 2 gives the observed relation between slope and size.

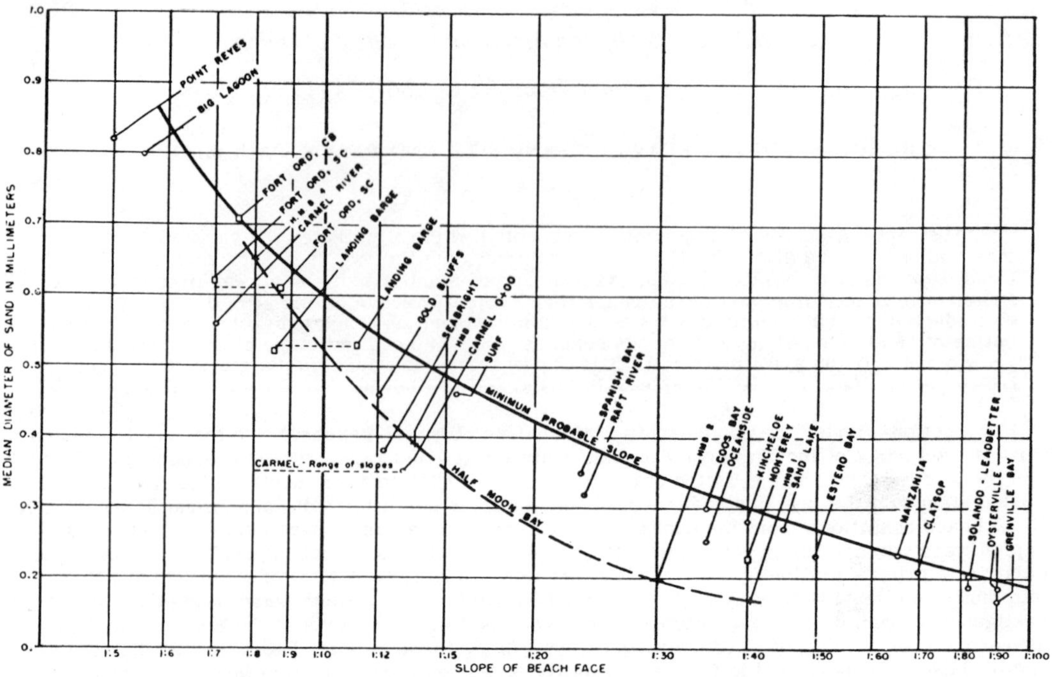

Fig. 2--Size - slope relationship at the reference point

Reference point--The necessity for standardizing beach-sampling and measurement procedures became increasingly evident as data were reduced from widely separated beaches between which it was desirable to make comparisons. The apparently simple job of determining the slope of a beach from a plotted profile raised perplexing questions of how and where to make the measurement. Sand samples from a single beach differed so much in median diameter (because of different sample locations) that first attempts at correlation were failures. Eventually, however, after the mass of data was plotted in various ways, patterns began to take shape. This work ultimately led to a decision to standardize on a specific location so that valid comparisons could be made across time and distance. The point selected was the part of the beach face subject to wave action at mid-tide elevation, now known as the "reference point." All slope measurements and sand samples referred to in this paper were taken at that point. The several illustrations discussed below develop the reasons for its selection and indicate the results that may be obtained by its use. Certain advantages are claimed for this technique: (1) A lone investigator with simple equipment can take measurements rapidly and easily. (2) The variables being sampled are those of the present time (the existing waves placed the sand sampled at the slope measured). (3) The inter-tidal beach-face slope seems to be a major criterion for the potential use of the beach. (4) Some previous investigators have used this location and their data are usable for comparison.

The reference zone may be fairly wide; it is deliberately described somewhat loosely as being the part of the beach subject to wave action at mid-tide stage since the technique would be highly impracticable if it were necessary to locate some exact elevation. Any reference point in this zone is presumed to have approximately the same characteristics. Since the backrush is only slightly below the average sea level, and the uprush may be considerably above, these measurements will be somewhat above the mid-tide level. This part of the beach seems to be the most stable, and consistent results can usually be obtained there. Mid-tide refers to a level half way between the previous high tide and the succeeding low. It is convenient to keep this dimension flexible because the intertidal zone on open ocean beaches may be above or below its predicted position; this is caused by piling up of water on a beach by wave action, wind drag on the sea surface, or changes in sea level from barometric pressure gradients. On steep beaches, the reference zone may be as narrow as 20 ft, while on flat beaches it may be over 100 ft wide.

Reference-point measurements and samples are most conveniently made at a low-tide stage. Slopes are measured perpendicular to the water line using simple instruments (a clinometer or Brunton compass and a driftwood board to average out minor irregularities). Sand samples

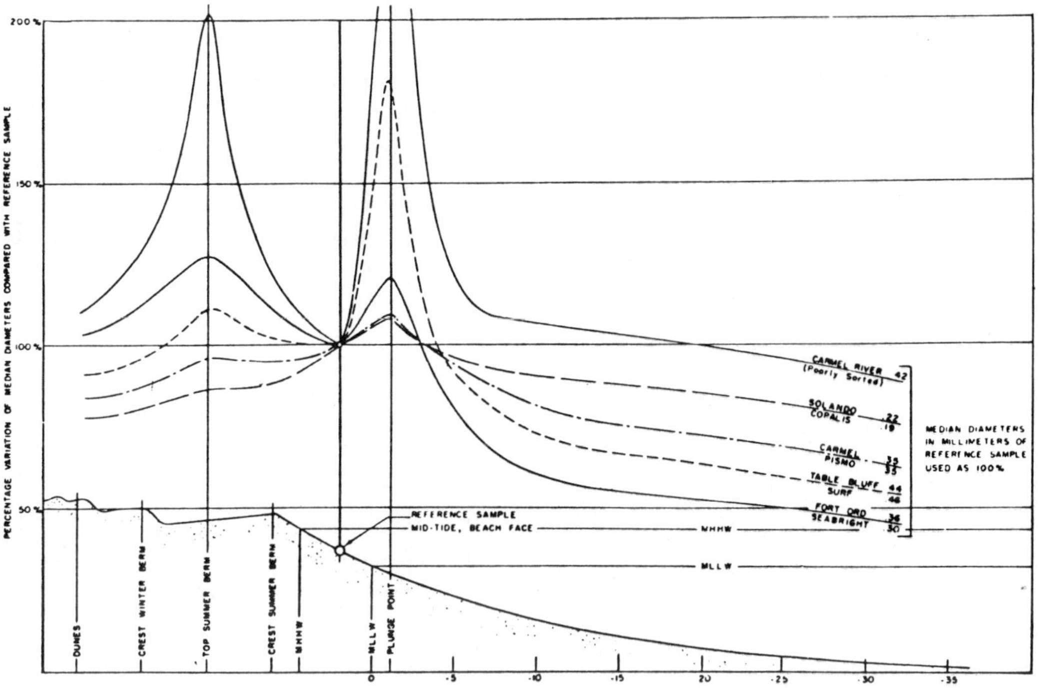

Fig. 3--Size distribution across a beach

should be taken at the same time. It is good practice to scrape away the upper one cm of sand, which often contains coarser material brought by recent swashes, and then take the sample from the top ten cm. A 1000-cc sample is ample, and no special sampling or quartering methods are necessary; quart cardboard cartons are satisfactory as sample containers. Slope, location, and other data can be written on the carton, thus insuring a correct combination of data. All samples mentioned here were subjected to the usual sieve analysis at the Engineering Materials Laboratory of the University of California.

Size distribution across a beach--The concept of a standardized reference point which will permit comparison of sands from widely separated beaches has been described. Figure 3 underlines the necessity for the selection of a specific point and the usefulness of the particular point chosen. From a glance at this figure, it is obvious that it makes a great deal of difference where a sand sample is taken. On a number of beaches sand samples were taken intermittently along the profile from the dunes to 30 ft of water (10 to 20 samples per profile); when plotted, these showed clearly the trends described here. This graphical presentation is a practical means of describing the general case; the stylized profile represents all beaches. A sand sample from the reference point is the zero or reference condition; other samples are plotted by their percentage relationship of their median diameters to the median diameter of the reference sample. At the time of sampling, detailed surveys were made and the sample points were accurately located on the profiles. In Figure 3 the curves have been rounded somewhat so that they represent average conditions and show the unmistakable trends which are characteristic of lateral sorting on ocean beaches. The large variations in the percentages between beaches are not regarded as significant by the author; these are probably due to sorting on the beach and to the size of the largest grains which exist there. The least variations from the reference median are at the beaches which have the smallest sand, and the greatest variations are found at the coarse sand beaches. The largest sand particles at any beach are found at the plunge point just seaward of the backrush, which, observation indicates, is the point of maximum turbulence. The next largest particles are found on top of the summer berm; churned up at the plunge point, the coarse material is held in suspension and carried by maximum uprushes up the beach face and across the crest of the berm from whence it cannot return. The finest sand on the beach is found in the dunes. Under water the sand becomes smaller with increasing depth.

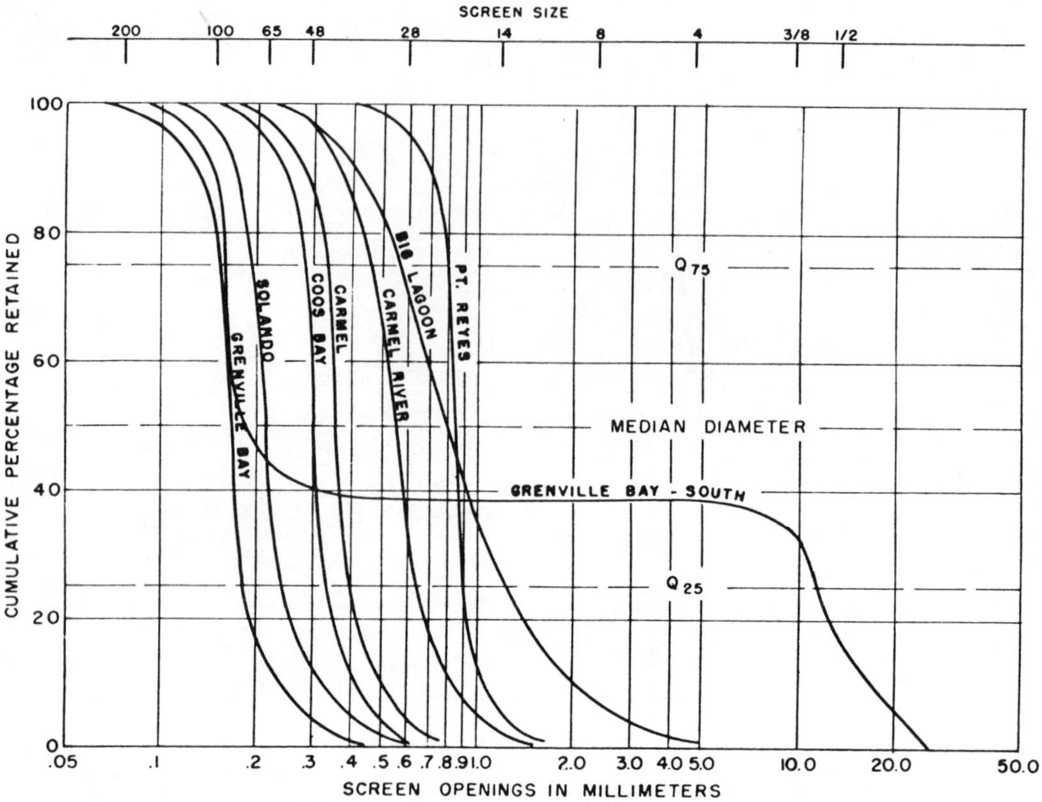

Fig. 4--Size variation of beach sands

Size variations in beach sands--Sands from reference points of Pacific Coast beaches have median diameters which range upward from 0.17 mm. Wave action, even in the most protected locations, seems to remove sand smaller than 0.17 mm from the beach face; the upper limit of 1.0 mm used in this discussion is fixed by definition. The largest sand size sampled was 0.8 mm. Figure 4 illustrates sorting curves which are typical of hundreds made during the investigation. The remarkable degree of sorting performed by wave action is evidenced by the steep slope of the curves and the good sorting coefficients (usually between 1.10 and 1.30). The sorting coefficient is defined as $\sqrt{Q_{25}/Q_{75}}$, where Q_{25} and Q_{75} are the grain sizes at the 25 and 75 pct finer points, respectively, on the size-analysis curve (1.0 = perfect sorting). The fact that sorting is generally very good makes the median diameter an acceptable standard.

Small amounts of both larger and smaller sand exist in each sample. It is believed that these grains are transient and were trapped by the sampling out of their equilibrium environment. Small grains, for example, may have been newly created from the destruction of larger ones and would migrate offshore when exposed to continued wave action. Wave variability also may account for some of the size variation. As might be expected, sorting is somewhat better among the fine sands than the coarse ones; this is probably because of the limits placed by the largest single grains in each case. Since fracturing of large grains can produce sand of any smaller size, the grains which are larger to begin with are likely to be associated with the greatest range of sizes.

The sand-size curve for Grenville Bay is of particular interest since it demonstrates the unusual phenomena of good sorting on two levels. Point Grenville, Washington, is a transition point on the coast; to its north, the beaches are composed of pebbles supplied by the Quinault River; to the south, the beaches are composed of typical Columbia River sand. At the mouth of the Columbia the median diameter is about 0.21 mm. Sand diameters become increasingly small to the north as the source becomes more distant until the finest possible material (0.17 mm) is found in Grenville Bay. Occasional storms, however, move some of the well-sorted pebbles from the beaches to the north around the point and into the bay; in these relatively quiet waters, both are able to exist together.

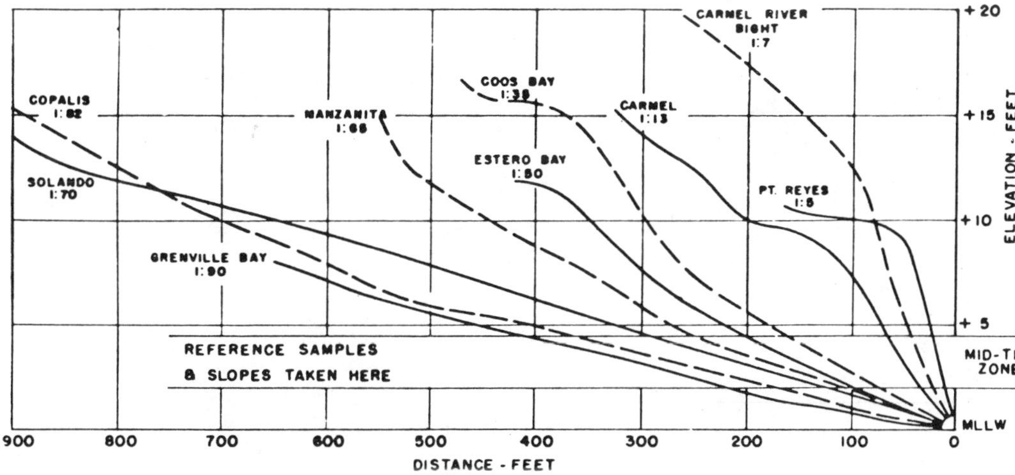

Fig. 5--Variation in sandy beach-face slope

Variation in sandy beach-face slopes--The beach profiles shown in Figure 5 are believed to show the limits of slopes possible for ocean beaches. It is emphasized that these slopes were measured at the reference point and do not include slopes formed by undercutting which can result in a vertical scarp, or slopes on the top of the berm which may have a reverse gradient. Beaches steeper than 1/4 require coarser material than that defined as sand; beaches flatter than 1/100 will not likely be found on exposed ocean shores. Each of the profiles shown is the average of several accurate surveys for the particular beach; surveys were made with transit and stadia and true elevations were carried from U. S. Coast and Geodetic Survey bench marks.

Slope changes with erosion and deposition--The principal difficulty in correlating sand size with beach-face slope is that the slope changes as the beach advances and retreats. Even though the median diameter of the sand remains the same, the beach becomes steeper with deposition and flatter with erosion. Model studies summarized by JOHNSON [1949] show that a relation exists between H_o/L_o values of waves and the characteristic profile of a beach. The conclusion reached was that H_o/L_o values of greater than 0.03 resulted in a "storm" profile (erosion) and that values of less than 0.03 caused an "ordinary" profile (deposition). It seems probable that such a relationship is the controlling factor and that the slope of the beach face responds readily to small changes in H/L values. If an equilibrium state ever exists, it is probably during periods when the waves have H/L values of about 0.03. This gives some clue as to how equilibrium standards may eventually be established; unfortunately H/L data do not exist for each of the points used on these graphs.

Figure 6 shows short-term changes on two beaches which were under close observation during periods of erosion and subsequent deposition. At Carmel, storm seas cut the beach back, flattening it in the reference zone and causing it to steepen near the crest of the berm. This erosion lasted for about one day; by the following day, wave conditions had changed to one favorable to deposition and the beach had steepened and widened. A week after the storm, the beach was wider than before and considerably steeper. Twenty profiles made of this station at weekly intervals throughout the summer of 1946 showed that there was no greater variation in beach-face slope than those caused by this storm.

The Santa Barbara example in Figure 6 shows the cumulative effects of two storms on a beach. The upper, original profile is typical of the summer (depositional) condition; two severe storms in quick succession cut the beach to much gentler slopes.

These bits of information indicate that a relation exists between the slope of the beach (for any given sand size) and the waves impinging upon it. Moreover, it appears that limits can be put on the amount of change (the range of slope) that can be expected for any beach.

The appearance of a beach after a storm emphasizes the need for a definite point at which to make measurements. For example, if a berm was formed during spring tides, and waves causing erosion occurred during the neap tides, a steep scarp may result. The beach will thus appear to be steeper, although it has actually become gentler in the reference zone.

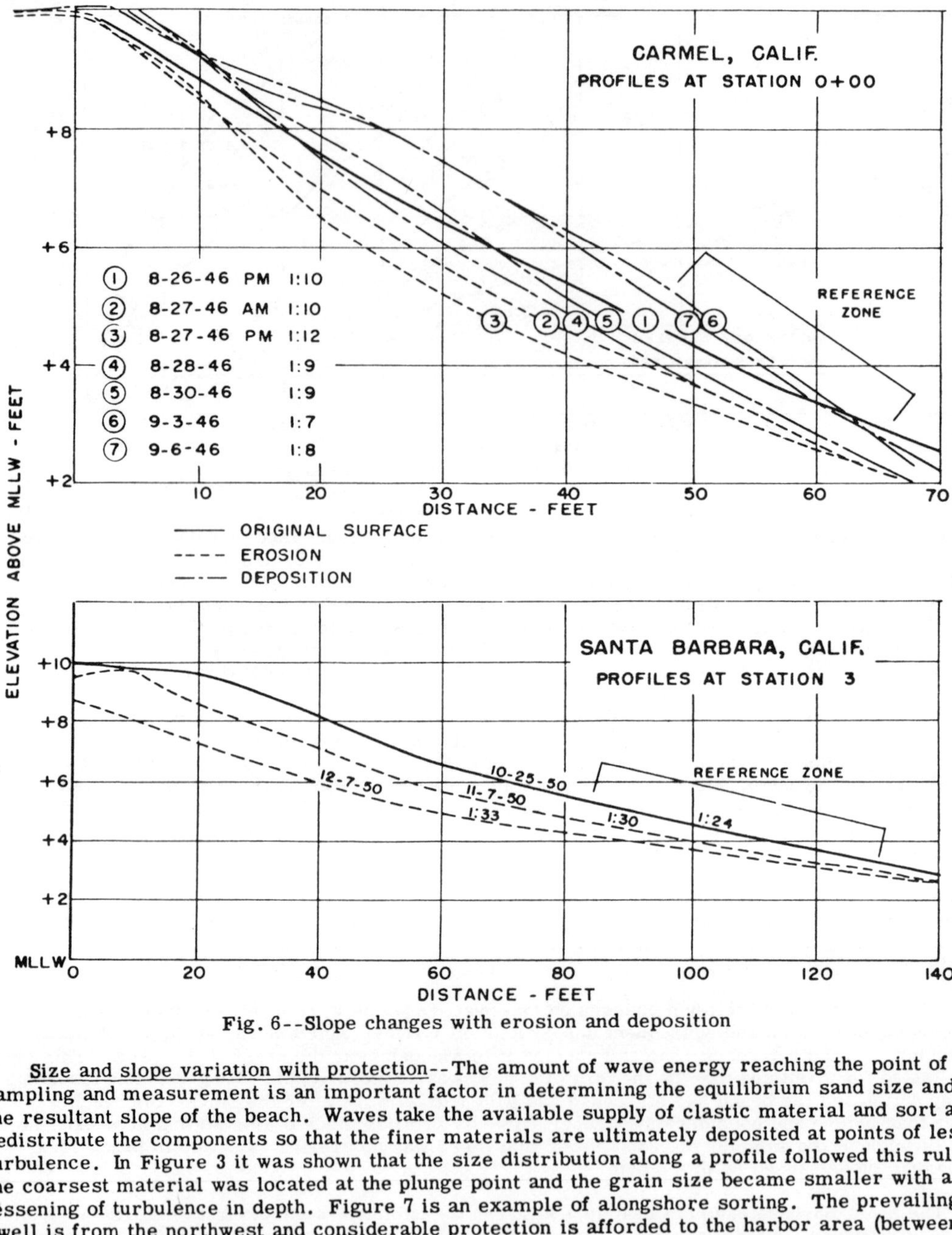

Fig. 6--Slope changes with erosion and deposition

Size and slope variation with protection--The amount of wave energy reaching the point of sampling and measurement is an important factor in determining the equilibrium sand size and the resultant slope of the beach. Waves take the available supply of clastic material and sort and redistribute the components so that the finer materials are ultimately deposited at points of lesser turbulence. In Figure 3 it was shown that the size distribution along a profile followed this rule; the coarsest material was located at the plunge point and the grain size became smaller with a lessening of turbulence in depth. Figure 7 is an example of alongshore sorting. The prevailing swell is from the northwest and considerable protection is afforded to the harbor area (between Profiles 1 and 2) by Pillar Point. South along the beach the protection gradually disappears until at Profile 4 the beach is exposed to nearly the full effect of the prevailing seas. The beach, which is continuous and unbroken, has reacted to fit itself to this environment. At Profile 1, where wave action is at a minimum, the beach is fine-grained (0.17 mm) and flat (1/41 slope). As the beach becomes more exposed, the slopes and sizes respond characteristically. Profile 4 has coarse sand (0.65 mm) and a steep beach face (1/8 slope). This alongshore sorting phenomenon is typical of a dozen or so similar locations, such as San Simeon, Shelter Cove, and Bodega Bay, California; and Cape Lookout, Oregon.

In the discussion of Figure 6, reference was made to changes along a profile caused by varia-

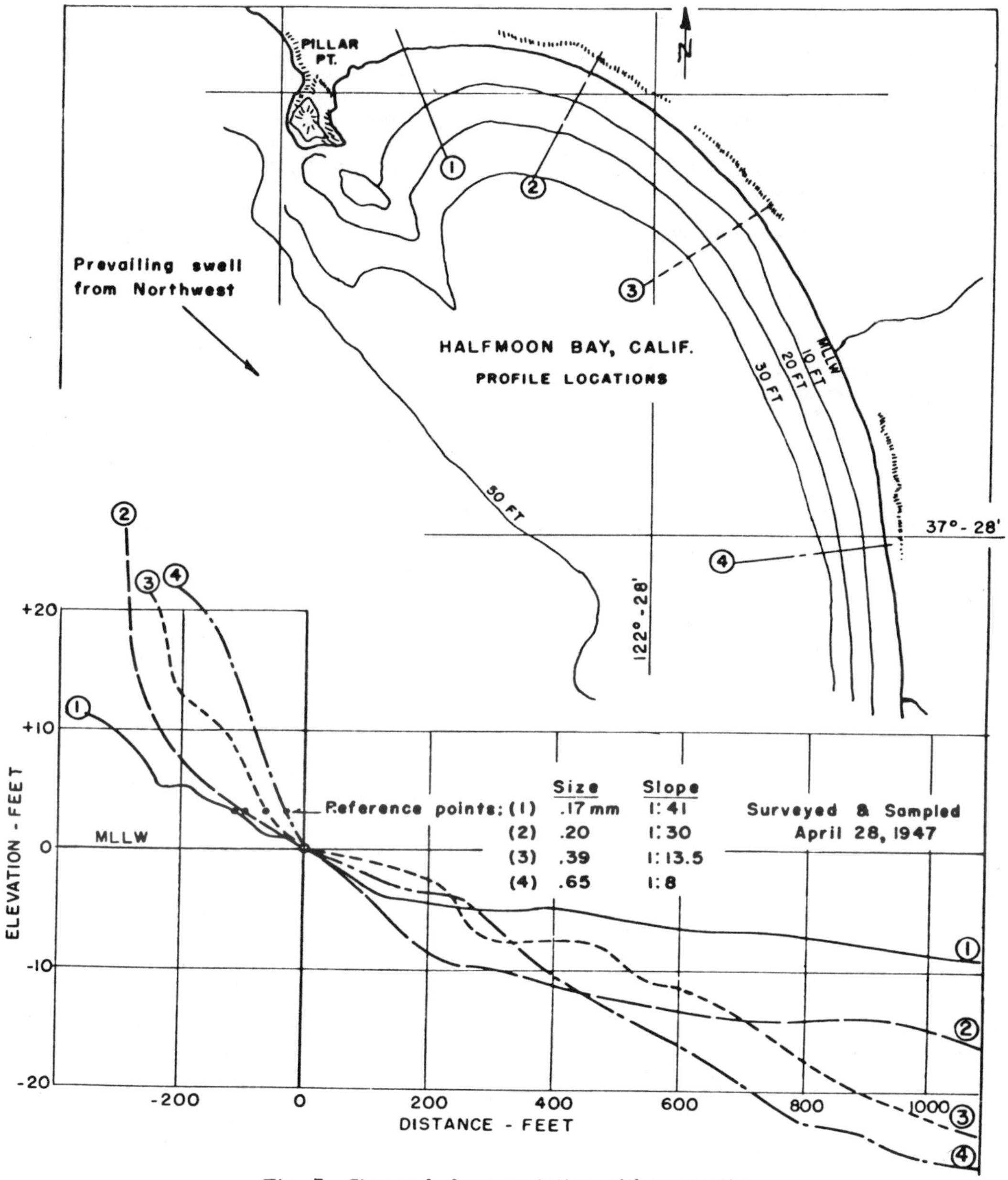

Fig. 7--Size and slope variation with protection

tions in wave action; storm seas cause the beach to flatten, quiet ones cause it to steepen. At first glance, Figure 7 may appear to show the opposite to be true by demonstrating the presence of steep beaches at points of maximum wave action. Sorting and distributing the sand on an extended beach apparently requires considerable time; it is probable that no static state is ever reached and that the sand shifts back and forth along the beach with changes in wave direction. On all of these beaches there is a preponderance of northwest swell (particularly the larger sand-moving waves). Therefore, we can regard each of the Half Moon Bay profiles as being an average of conditions along a half mile or so of beach throughout the year. In short, the long-term situation illustrated is one of approximate equilibrium.

When a storm arises, however, the energy of the surf zone increases and the sand on the beach face is too small to remain there in this turbulent water. Consequently, it is moved about until it reaches water quiet enough to settle to the bottom; this may be in deeper water or at any place of greater protection. This usually results in the formation of a bar (a storm feature); as the bar grows at the expense of the beach, it influences the waves and eventually causes them to break. As breaker energy is expended on the bar, the water on the beach face becomes less turbulent; this appears to permit the beach to reach another point of equilibrium which is in balance with the storm waves. It is of interest to note that when the sand supply is very large, the result is frequently the formation of three or more bars; where small, the beach is denuded and the back-beach is attacked.

The equilibrium beach face resulting from the storm, as well as that extant during erosion of the beach face, has a gentler slope than normal. It should be pointed out again that the median sand size remained the same throughout the cycle described.

Slope-size relationships at the beach face--A qualitative relationship between the slope of a beach and the size of its materials has long been observed, but a simple correlation between them has been precluded because of the many complexities involved. This paper suggests a standard sampling procedure and shows why it is necessary; it indicates the range of conditions that may be expected; and it describes the variations due to protection and to wave characteristics. There are other factors which undoubtedly have some effect, but in most cases they appear to be negligible; they include shape and density of grains, the stage of tide, and the total supply of sand available. Figure 2 is the first step in assembling this information in a compatible form. Previous work has pointed out this problem; it is the purpose of this graph and discussion to hold out hope for a solution. The limitations are readily admitted, but they emphasize the necessity for specific measurement techniques in the future which will supply the missing data.

The upper curve represents the minimum slope which an unprotected Pacific Coast beach is likely to reach in a storm. Exposed beaches such as Surf, Table Bluff, and Oysterville have slopes approaching this limit much of the time. Beaches such as Carmel, Oceanside, or Seabright, which are protected by their location in embayments or in the lee of offshore islands, can probably never reach it in any ordinary storm condition. The erosional power of waves reaching any point on a beach is a function of the refraction condition and the deep water H/L value. Since the slope is related to the value of H/L (Fig. 6) and to the refraction condition (Fig. 7), the curve of minimum slope represents a level of wave energy. Similar curves would result if each point were assigned a refraction coefficient for the "average Pacific storm" and lines of equal K values were drawn (over a large part of the year the direction is NW and T = 12 sec). The minimum slope curve would about equal the K = 1.0 curve. The distance of any point from the curve is thus seen to be an approximate measure of the intensity of the waves which were shaping the beach face at the time of sampling.

An admittedly more useful curve would be one for the equilibrium condition; at this writing there are insufficient data to make even an approximation; however, the investigations are still underway. Such a curve would determine the trend of a beach, since a point falling on the steep side would indicate building, and a point on the gentle (right-hand) side would indicate erosion.

The curve connecting the Half Moon Bay samples, details of which are shown in Figure 7, is particularly interesting. It is seen to substantiate this discussion by diverging from the curve of minimum slope as protection increases. A moderate northwest storm existed when these profiles and samples were taken, causing the storm profiles shown. If the storm had increased in intensity, the upper points (3 and 4) probably would have shifted to the right. However, if the storm had shifted its direction to the south, the upper points would have remained about the same and points 1 and 2 would have shifted to the right.

Other examples of this variation in slope with protection can be seen; Monterey, Estero Bay, and Manzanita, each with sand median diameters of about 0.23 mm, are distributed by their average protection. Manzanita is well exposed to northern Pacific seas, Monterey is well protected, and the exposure of Estero Bay is about midway between the two.

Conclusions--The following conclusions are justified by studies to this time.

(1) Sandy beach-face slopes (reference zone) on U. S. Pacific Coast beaches range from 1/4 to 1/100.

(2) Median diameter of beach material from the reference point on sandy beaches range upward from 0.17 mm.

(3) As beaches build and erode, the slope of the beach face at the reference point will change considerably, apparently in response to the H/L factor of the waves. An eroding beach will flatten; a building beach will steepen.

(4) Distribution of sand by size along a profile is predictable with reasonable accuracy if the median diameter of sand at the reference point is known. Large grains are found at points of maximum turbulence; sizes get smaller as turbulence decreases.

(5) The slope of the beach face is related to the median diameter of the sand and the amount of wave energy reaching that point. The amount of energy is a function of the refraction conditions; consequently, beaches that are protected are steeper for any given sand size than exposed beaches.

Acknowledgments--The author wishes to give a large measure of credit to his associates, M. P. O'Brien and J. D. Isaacs, who assisted in the planning of this work.

References

BASCOM, W. N., Field studies at Monterey (2 parts), HE-116-224 Inst. of Engr. Research, Univ. of Calif., Oct. 1946.

BASCOM, W. N., Beach and surf conditions on the Oregon and Washington coasts in 1946, HE-116-247 Inst. of Engr. Research, Univ. of Calif., April 1947.

ISAACS, J. D., Beach and surf conditions on the Oregon and Washington coasts in 1945, HE-116-229 Inst. of Engr. Research, Univ. of Calif., April 1946.

JOHNSON, J. W., Scale effects in hydraulic models involving wave motion, Trans. Amer. Geophys. Union, v. 30, pp. 517-525, 1949.

STUMP, R. S., and BASCOM, W. N., Surf observations at Table Bluff, California, HE-116-205 Inst. of Engr. Research, Univ. of Calif., May 16, 1947.

University of California,
Berkeley, California

18

Reprinted from *Jour. Geology* **66**(4):417–441 (1958)

A MODEL RELATING DYNAMICS AND SEDIMENT PATTERN IN EQUILIBRIUM IN THE REGION OF SHOALING WAVES, BREAKER ZONE, AND FORESHORE[1]

ROBERT L. MILLER AND JOHN M. ZEIGLER
University of Chicago and Woods Hole Oceanographic Institution

ABSTRACT

A theoretical model is presented for the expected patterns of sediment size and sorting in the region of shoaling wave, breaker, and beach foreshore. This model is based on theoretical considerations and published experimental results and is intended to hold for the above-mentioned dynamic zones in a state of equilibrium. Comparison with detailed observations in three natural situations and with published sediment size and sorting patterns appears to give good agreement with the model. The implication is that such generalized patterns may serve in the recognition of departures from equilibrium and change in general environment in the study of contemporary sediments and possibly in the recognition of dynamic environments in ancient rocks.

INTRODUCTION

The interpretation and prediction of contemporary sediment patterns on beach and near-shore areas have attracted the attention of a number of investigators. In the present study we focus attention on that portion which extends from the breaker zone to the top of the foreshore and seaward to a depth not exceeding 2 fathoms. Our purpose is to combine published theoretical and experimental work in such a way as to formulate or predict sediment patterns by using variables representing the water movements and the sediment size and degree of sorting. The predicted patterns are then compared with observed sediment patterns collected by us in various field locations and with published areal sediment patterns.

No attempt is made to cover thoroughly the extensive literature on this subject. Pertinent papers are referred to in the text.

We take the point of view that the beach and surf region can be subdivided into several zones which are characterized by dynamics peculiar to that zone, with resulting cut and fill leading to typical sorting and size patterns. The area of interest will be limited to a narrow band along the shore which has as its landward margin the well-defined upper limit of wave swash on the foreshore. The outer margin is seaward of the breaker zone at low tide. In all cases, the area limits fixed for sampling purposes at low tide remain the same, but the proportions of the zone change as the tide rises or falls, and the breaker line migrates in a shoreward or seaward direction.

The region in which the breakers form is surely an important one in sedimentary processes. Not only does it migrate with the tides from day to day, but it also migrates across the land with passage of time, as sea level changes. Clearly the great energy concentrated in this narrow band and its margins both seaward and landward is of primary importance in forming the final sediment pattern of all within its range. Although the band is narrow at any one time, its migrations with tide and long-term sea-level changes must leave its effect as the dominant process on an areally important proportion of the sedimentary record. The long-range purpose of this initial study is to ascertain to what extent and in how fine a detail certain sediment size and sorting patterns in ancient rocks can be recognized as typical of the breaker zone and its margins in contemporary sediments.

Figure 1 shows the dynamic zones which we feel characterize the surf region. The "step" as a topographic feature which ap-

[1] Woods Hole Oceanographic Institution Contribution No. 958. Manuscript received February 13, 1958.

pears when conditions are favorable is a distinct break in slope just under the average wave crest at the point of plunge. The zones will be discussed in some detail in the main body of the paper. In brief, Zone W is characterized by shoaling wave dynamics; Zone B, just at the breaking wave and including the step when present, is characterized by turbulence; and Zone S, the beach foreshore up to the limit of swash, is characterized by sheet flow, first up and then down the foreshore.

It is important to understand the distribution and application of the great energy concentrated in this narrow band. Those who study ancient environments by interpreting the sediment pattern of ancient rocks are well aware that the land-sea interface moved across a considerable part of inundated portions of continents as the sea encroached farther and farther inland. Sometimes the interface was marked by beaches which moved with the rise or fall of sea level, became buried and preserved (Thompson, 1937), or were destroyed. Workers with ancient rocks have immediate use for ideas or techniques which will aid in interpreting past environments and the effects of marine inundation. Certainly, one of the long-range purposes of this study is to ascertain to what extent and in how fine detail certain sediment size and sorting patterns in ancient rocks can be recognized as typical of modern environments—in this particular case the surf zone and beach foreshore.

FIELD METHODS

In those field studies made by us and referred to in this paper, the following properties were measured.

Median sediment diamter (mm.) . . . . . Md.

Degree of sorting[2] . . . . D.S. $= \dfrac{P_{80} - P_{20}}{P_{50}}$

Wave height . . . . H
Wave length L
Wave period . . . . P
Still-water depth . . . . S.W.D.
Still-water level . . . . S.W.L.
Average velocity of backwash on foreshore . . . . $\overline{V}_{max}$
Slope of foreshore . . . . S
Average depth of flow of backwash . . . t_d
Local topography of near-shore and beach foreshore

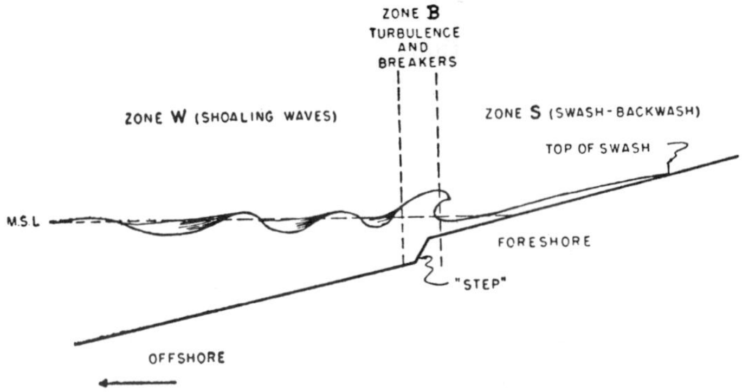

FIG. 1.—The dynamic zones

Other variables were also measured in the course of the study but are not directly used in the present study.

[2] The measure of sorting used in this paper differs from those customarily employed, in that the effect of size is removed from the measure of the "spread" of the frequency distribution by dividing by the median or 50th percentile. In an earlier paper (Miller and Olson, 1955) the coefficient of variation = standard deviation ÷ mean was applied for a similar purpose. The present measurement is simpler to compute and appears to serve equally well. The term "sorting" means different things to different people. To us, a handful of pebbles of roughly the same size class is as "well sorted" as a handful of sand grains in a single size class.

Clearly the standard deviation will faithfully reflect the size units and therefore will not be a useful measure for our purposes. For full discussion of various aspects of this problem see Doeglas, 1946; Inman, 1949.

Wave length was measured directly, using a tape to measure from crest to crest. The number of waves passing the tide gauge per minute was called *wave period*.

Wave height was recorded by observing the height of successive troughs and crests with respect to still-water level on a marked tide gauge. The tide gauge itself consisted of a plastic pipe forced into the bottom (fig. 2). A small hole in the pipe below water surface acted as a damping mechanism, permitting only small fluctuations of still-water level in the pipe. A Ping-Pong ball floating inside marked the still-water level.

All these measurements were made in a defined study area; therefore, there is a distribution-in-space aspect which is especially important in the sediment (and the cut-and-fill measurements) (see Inman and Rusnak, 1956). Time also is an important factor. It is obvious that all the afore-mentioned variables change with time; hence a number of repeated measurements were taken during each hour at Falmouth Beach and during each 2-hour period at Highland Light Beach over a full tidal cycle in each case (see fig. 2*a*).

The field procedure at Falmouth Beach will be described briefly. A portion of beach 75 feet long and 36 feet wide was selected. This area covered the zone of upwash, zone of breaking waves, and seaward of the breaking waves (fig. 1). Thirty-six sample stations were established within the area (fig. 3). The stations were marked by iron rods pushed into the sand. Sediment samples were taken in half-pint containers once each hour at each station for 11 consecutive hours.

A considerable problem was encountered in reducing the relatively large amount of data to a usable form. We chose to describe the sediment in terms of median and sorting values contoured over the study area (e.g.

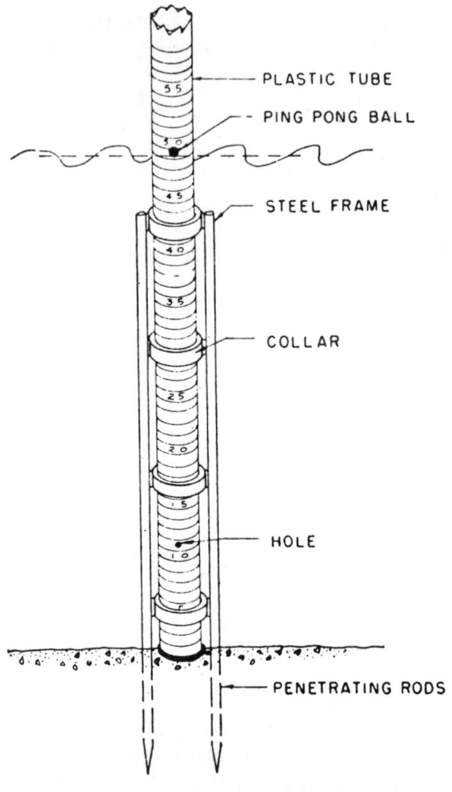

Fig. 2.—Still-water level gauge

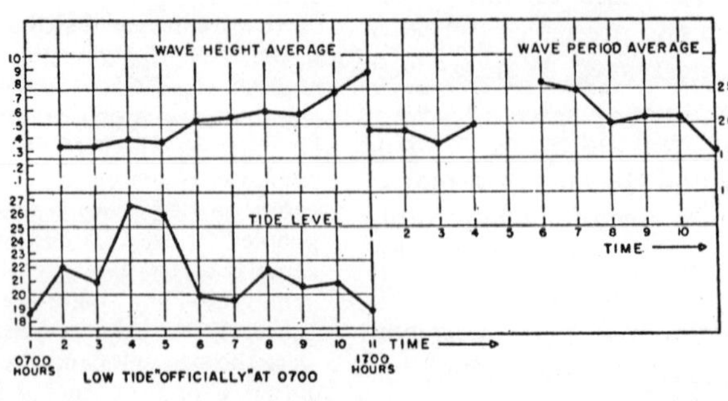

Fig. 2*a*—Falmouth study, 1956

fig. 9). The median and sorting values are not contoured directly; rather, a surface of best fit is passed through all of the sample points. This gives a pattern free of local irregularities in sediment size and degree of sorting. The technique for fitting surfaces to sedimentary data is described by Miller (1956); in this particular study the computations involved in fitting surfaces to our data were done by the high-speed computer AVIDAC, at Argonne Laboratories in Chicago, and by IBM.

An alternative to using trend maps for analysis of the areal size and sorting distribution is the application of the analysis of variance as described in Krumbein (1953) and Krumbein and Miller (1953).

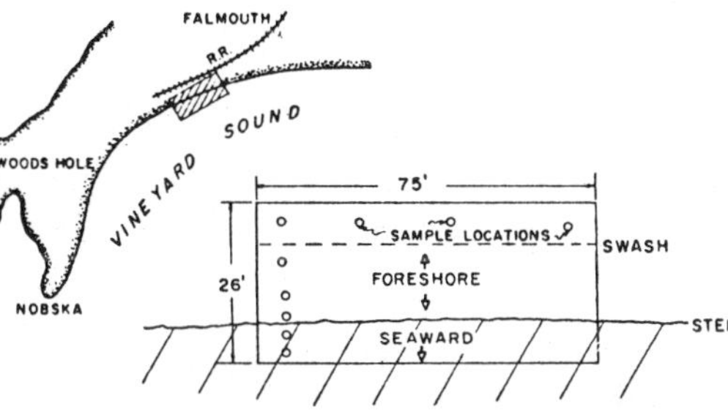

FIG. 3.—Location sketch of Falmouth Beach

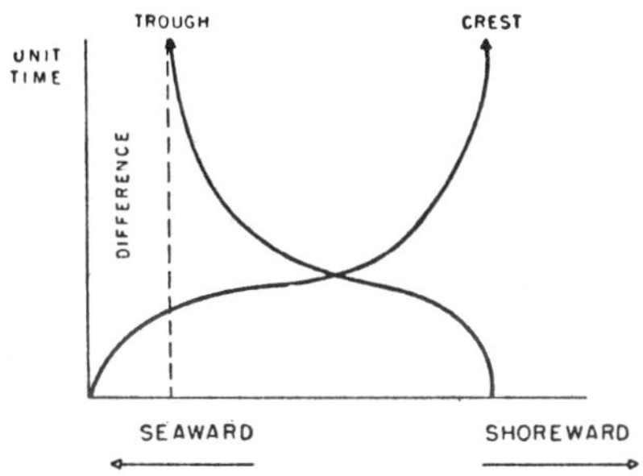

FIG. 4.—Directional movement of sediment particle under crest and trough in shoaling wave.

SHOALING WAVES (ZONE W)

A hypothesis relating the dimensions (length and internal-wave velocity) of shoaling waves to the movement of sediment was proposed by P. Cornaglia and described briefly in a speech by Münch-Peterson (1950). Cornaglia states in his proposal that, because crest velocity is greater than trough velocity in shallow water, the force moving a sediment particle will act to move the particle in the direction that the crests are moving, a distance proportional to the difference between crest and trough velocities. Figure 4 illustrates his theory. Cornaglia then states that the difference between trough and crest velocity becomes less as the water depth increases until there is not enough difference to move the particle a net distance in either direction. Meanwhile the influence of gravity becomes increasingly important in pulling sediment particles downslope—seaward. Somewhere, according to Cornaglia, all of these forces are equal, and each sediment particle has a "null point" where net movement along a line normal to the wave direction is zero—the particle simply moves back and forth. Seaward of this point the net movement is seaward, and landward of this point the movement is landward. If a lot of different sizes are present, only one size is at its "null point"; the larger sizes move downslope (seaward) and the finer sizes in the direction of the crests. It follows, therefore, that, for given slopes and given wave characteristics, each particle has a "null point" where net movement of that particle is zero.

Münch-Peterson (1950) holds that the above-described theory is not valid, in that the influence of gravity relative to wave pressure is too small to be considered important, no matter what the bottom slope is. Ippen and Eagleson (1955) point out that Cornaglia's hypothesis has not gained recognition in the United States. Scott (1954) describes equilibrium conditions for various grain sizes, as observed through an interesting series of wave-tank experiments.

Careful laboratory observations on single sedimentary particles in a wave tank by Ippen and Eagleson (1955), however, appear to verify at least a part of Cornaglia's hypothesis. For a fixed slope of 1:15, the existence of "null points" and zones of forward movement were recorded in this study, for several combinations of wave height and length ratios, with single particles varying in size and specific gravity. Seaward movement was also noted, but to a much less degree. They concluded that net sediment transport is dependent on the balance between transportation shoreward due to wave action and transportation seaward due to gravity. They further conclude that "a special 'null point' or zone of zero net sediment size on the given beach under a given incident wave . . . confirming the existence of the so-called Cornaglia 'neutral line.' For the plane fixed beach of 1:15 slope employed, and a range of incident wave steepness of $0.01 \leq H/L \leq 0.08$, the null point appears independent of the sediment diameter to roughness diameter ratio, D/K, and of boundary layer effects and may be given by the following relation:

$$\left(\frac{H}{d}\right)^2 \left(\frac{L}{H}\right) \left(\frac{C}{w}\right) = 11.6 .$$

"This equation holds, provided that fluid velocities are greater than that required for the incipient motion of the particle being considered. H, C, and L are *local* wave characteristics, and w is the fall velocity of the sediment particle under consideration. The constant 11.6 is valid only for a slope of 1:15 and must be presumed to vary with a function of the slope for the general case."

The present writers have applied the hypothesis to Zone W, the region of shoaling waves (see fig. 1). The equation given above was solved for depth for a given sediment particle size under the wave conditions observed at Falmouth beach (fig. 2*a*). Contour maps of the "neutral lines" were then made to scale on transparent paper so that they could be laid over the sediment size and sediment sorting, for comparisons. The following describes in detail the way in which the null point or neutral line maps were constructed:

1. Ippen and Eagleson's empirical equation,

$$\left(\frac{H}{d}\right)^2 \left(\frac{L}{H}\right) \left(\frac{C}{w}\right) = 11.6 ,$$

was rearranged into a suitable and dimensionally correct form:

$$\frac{HLC}{d^2}(0.0865) = w , \qquad (1W)$$

where

H = wave height,
L = wave length,
C = wave celerity,
d = still-water depth,
w = terminal fall velocity of the sediment particle.

2. Wave celerity varies as a function of L, H, and d. If the effect of H (wave height) is neglected, an equation due to Airy (1842) may be used, where

$$C^2 = \frac{gL}{2\pi} \tanh \frac{2\pi d}{L} . \qquad (2W)$$

3. H and L values characteristic of a given period of time were made available by averaging field observations taken over the time period. Thus for a time period, equation (1W) was solved for w, given (1) various values of d, (2) C values corresponding to d (eq. [2W]), and (3) fixed H and L.

4. The resulting terminal fall velocities were then transposed to grain diameters by the use of a nomogram devised by Rouse (1937).

Figure 5*a* shows the relationship between grain size and still-water depth for Cornaglia's null points. Curves were constructed for four separate hour periods during the 12-

hour tidal-cycle study at Falmouth Beach. From these curves still-water depths corresponding to the range of median grain diameters found at Falmouth were used to construct the contour maps of figure 6, *a–c*. Figure 5*b* shows curves for larger wave height and wave length.

The null-point maps are on the same scale as the maps of sediment size and sediment sorting. In this way, comparisons of several of the variables at corresponding times may be conveniently made. In figure 6, *a*, the null points contoured were for time T approximately at low tide. The observed median sediment size ranges for T_1 indicates a minimum median contour of 0.5 contour interval. We therefore computed null points for median sizes ranging from 0.5 to 4.0 mm. The dotted lines show where S.W.L. intersects the beach at about the middle of hour T_1. The small circles indicate the sampling points, and the double circles the position of the tide gauge. The observed median values are given at each

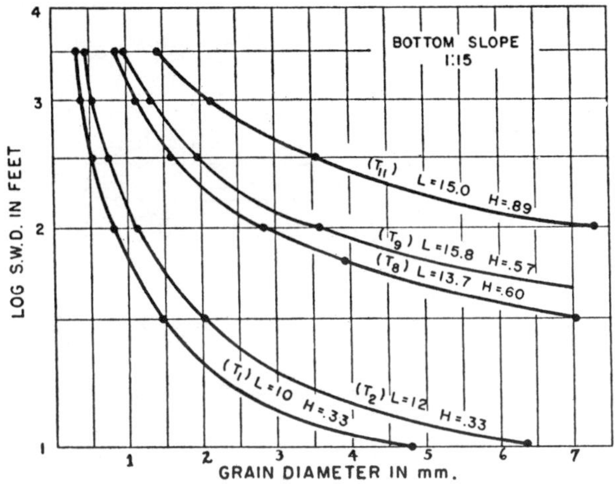

FIG. 5*a*.—Prediction in terms of SWD of position of null point for grain size with given H, L, and transforming wave celerity (for several tidal stages at Falmouth).

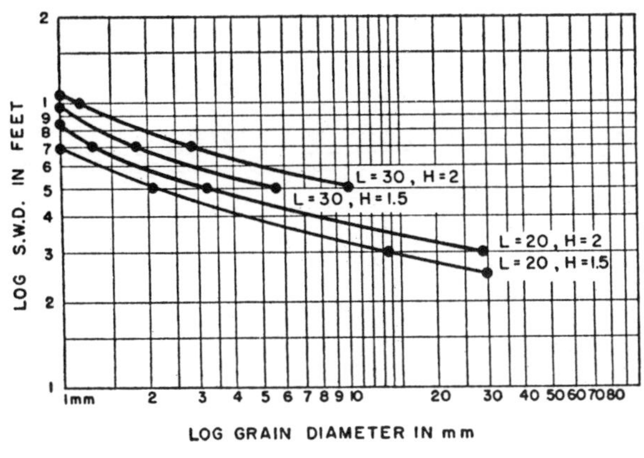

FIG. 5*b*—Prediction of null-point–SWD relation for an extended range of wave H, L. The legitimacy of extrapolation here is questionable. Note that, for wave height of 1.5 feet and wave length of 20 feet, small pebbles of 13.5 mm. in diameter begin net shoreward movement in SWD of just less than 3 feet.

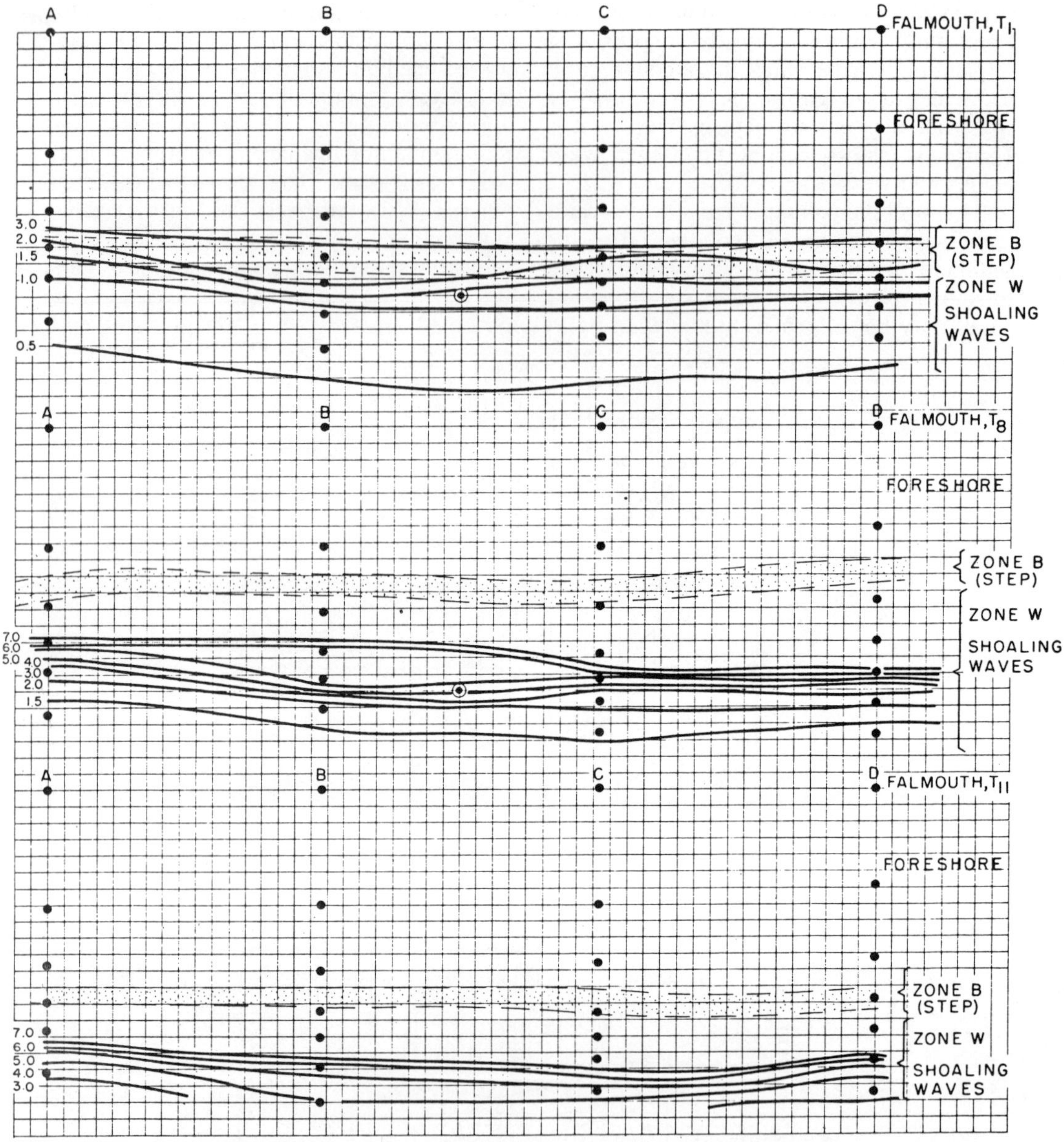

FIG. 6.—Computed position of null points (*neutral lines*) for 0.5–3.0-mm. grain size for three tidal stages at Falmouth in accord with the wave height, wave length, and SWD at those particular times.

of the sample points for comparison with the contours for null points.

Let us now examine the combined theories to see what we would expect if they are correct. We shall use two models: Model I, no seaward motion under gravity; Model II, both landward and seaward motion.

MODEL I, FIGURE 7

We may conveniently start with a hypothetical situation in which there is a wide range of sizes, well mixed so that there is an equal proportion of each size everywhere within Zone W. If wave action is then begun, the various sizes will be sorted in accord with their various null points. Let us suppose for illustration that there are five arbitrary size classes numbered 1 through 5. We shall hold beach slope and wave length and height constant. Then there will be five null points on a traverse normal to the beach, so spaced that the finest is seaward and the largest nearest land. As the waves begin to transport the material, everything finer than the value at the null point will move toward shore, accelerating until maximum velocity is reached just seaward of the plunge point. Just shoreward of the plunge point, the velocity of the sediment drops to zero. We refer here to net sediment velocity in a shoreward direction for convenience, even though the particles actually may move in a forward-backward path, with a net forward movement. Our expectation under this model is that the median sediment size will increase regularly toward the shore and, at a given null point, will be greater than, or equal to, that null-point value. The sorting will become better by regular degrees toward the shore.

Shoreward of the largest null point and where the net sediment velocity curve drops to zero, the accumulated mass of sediment in transport is abruptly deposited, to be once again put into motion in a different manner in Zone B. The sorting at this point should be poor, and in our illustrative model the median should exceed 2.0.

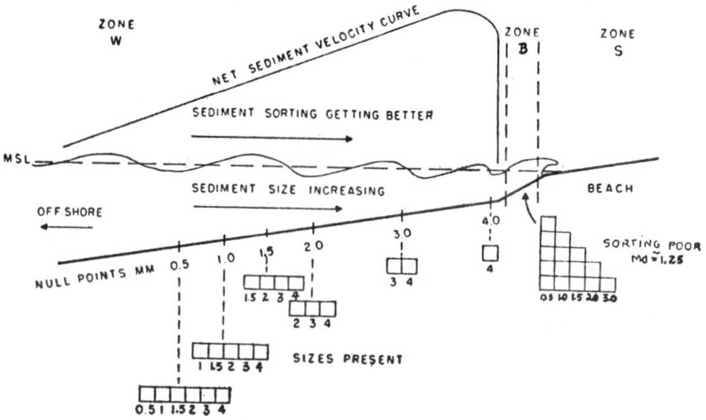

FIG. 7.—Model I

The expected accumulation of sediment medians at the boundary between Zones W and B is also shown in figure 7.

This model utilizes the region of oscillation for a given particle size with shoreward net movement on the land side but does not consider seaward movement on the seaward side of the null point.

MODEL II, FIGURE 8

Suppose we also take into consideration that on the seaward side of the null point all larger sediment particles will move offshore. Figure 8 then illustrates what our expectation would be for the hypothetical situation where the distribution of sediment sizes is initially homogeneous.

1. As before, sediment Md. size increases toward the shore, with the Md. size corresponding exactly to null-point values.

2. Sorting is uniformly excellent, with one size class at each null point.

3. The accumulation at Zone B drawn from seaward in Zone W will have the same appearance as in Model I with low Md. value and poor sorting.

4. There will be a seaward accumulation but less well defined, with an Md. value coarse, and poor sorting.

Before testing the agreement of the models with natural conditions at Falmouth Beach, we shall list a series of factors which should be kept in mind when figures 9–11 a e examined.

(1) The neutral line contour maps are based on an equation which is for a slope of 1:15. Falmouth Beach has a slope of 1:10 (This changes the constant in eq. [1W]).

(2) In addition, the equation is based on a perfectly plane beach. Local irregularities in the bottom slope established before the 12-hour study by markedly different wave conditions should not be expected to be completely erased.

(3) It is appropriate to consider the three zones defined in this paper as a "palimpsest" in which the patterns left by previous tides and changing wave conditions are only partially erased and show through the pattern put down at the time of study. This is particularly applicable to Zone W.

(4) The Falmouth Beach sampling grid was chosen sufficiently small on a traverse parallel to the beach so as to minimize the effect of longshore or littoral movement. In this way we effectively deal only with movements of wave and current orthogonal to the beach.

(5) No allowance for effect of wave period has been made.

We shall now turn to a comparison of the theoretical neutral lines with median size and sorting trend maps for several time stages during a single tidal cycle, as observed at Falmouth Beach.

Figure 9 facilitates a comparison of the expected position of the calculated neutral lines with observed size and sorting, for tidal period T_1 (low tide) in the form of a trend map (see fig. 6, *a*). We focus attention on Zone W (shoaling-wave zone).

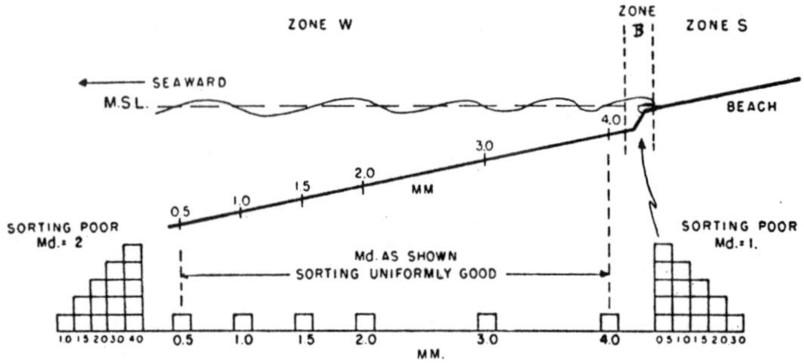

FIG. 8.—Model II

1. The trend-map median size contours are consistently higher in value on a traverse toward the beach than the calculated neutral-line contours. This fits either Model I or Model II.

2. There is a tendency for median size to increase toward the margin of Zone B (the breaker zone). This fits both Model I and Model II.

3. Sorting clearly improves shoreward in the trend map. This fits only Model I.

Figure 10, showing the trend map for tidal period T_8 (falling tide), appears to reverse the picture. In Zone W, median size decreases strongly toward the new position of Zone B. No clear-cut interpretation is possible for the sorting trend map.

However, at T_{11}, 1 hour before low tide,

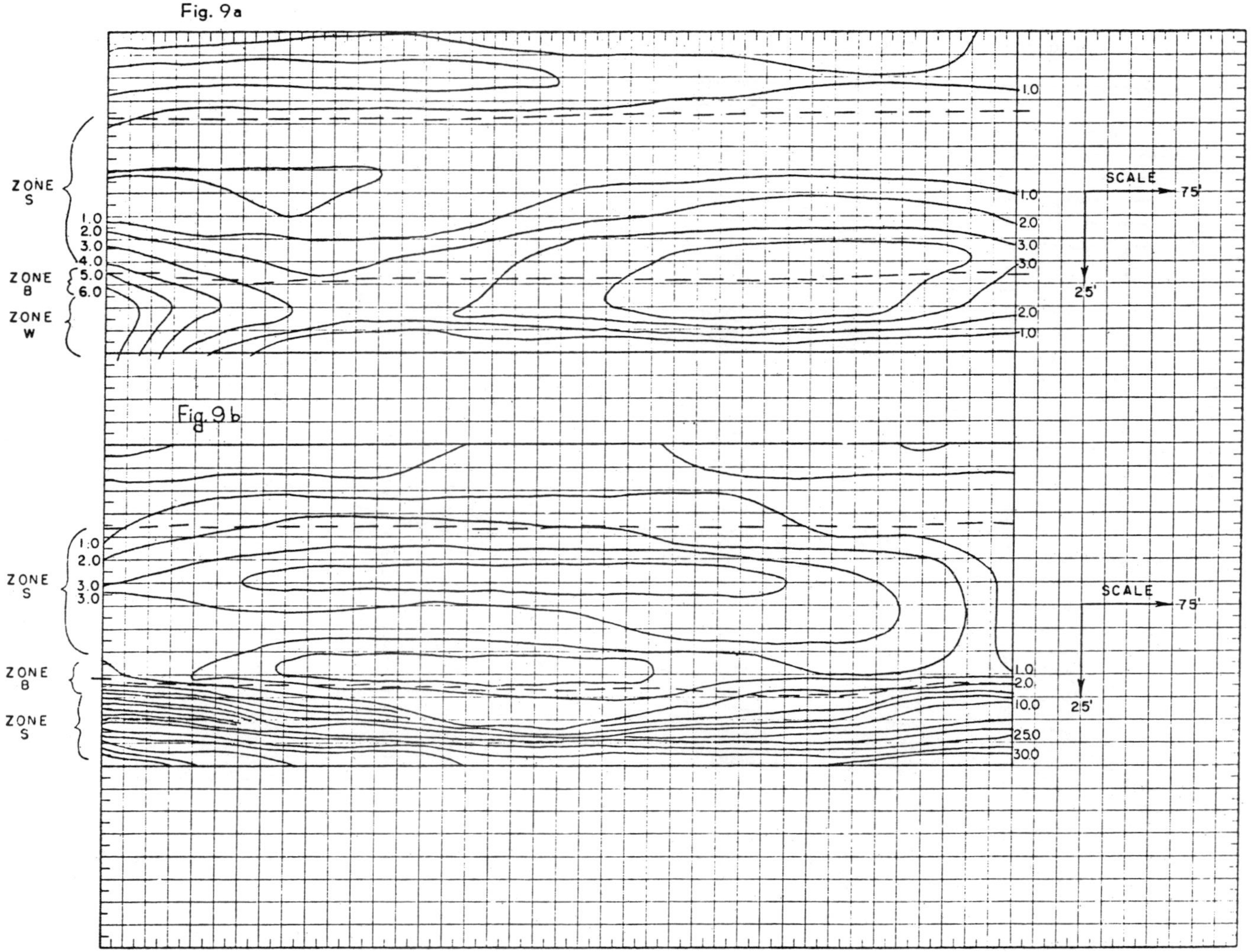

FIG. 9.—*a*, Median size trend map of T_1 Falmouth. Contour intervals = 1.0 mm. *b*, Sorting trend map of T_1 Falmouth. Contour interval = 1.0 mm.

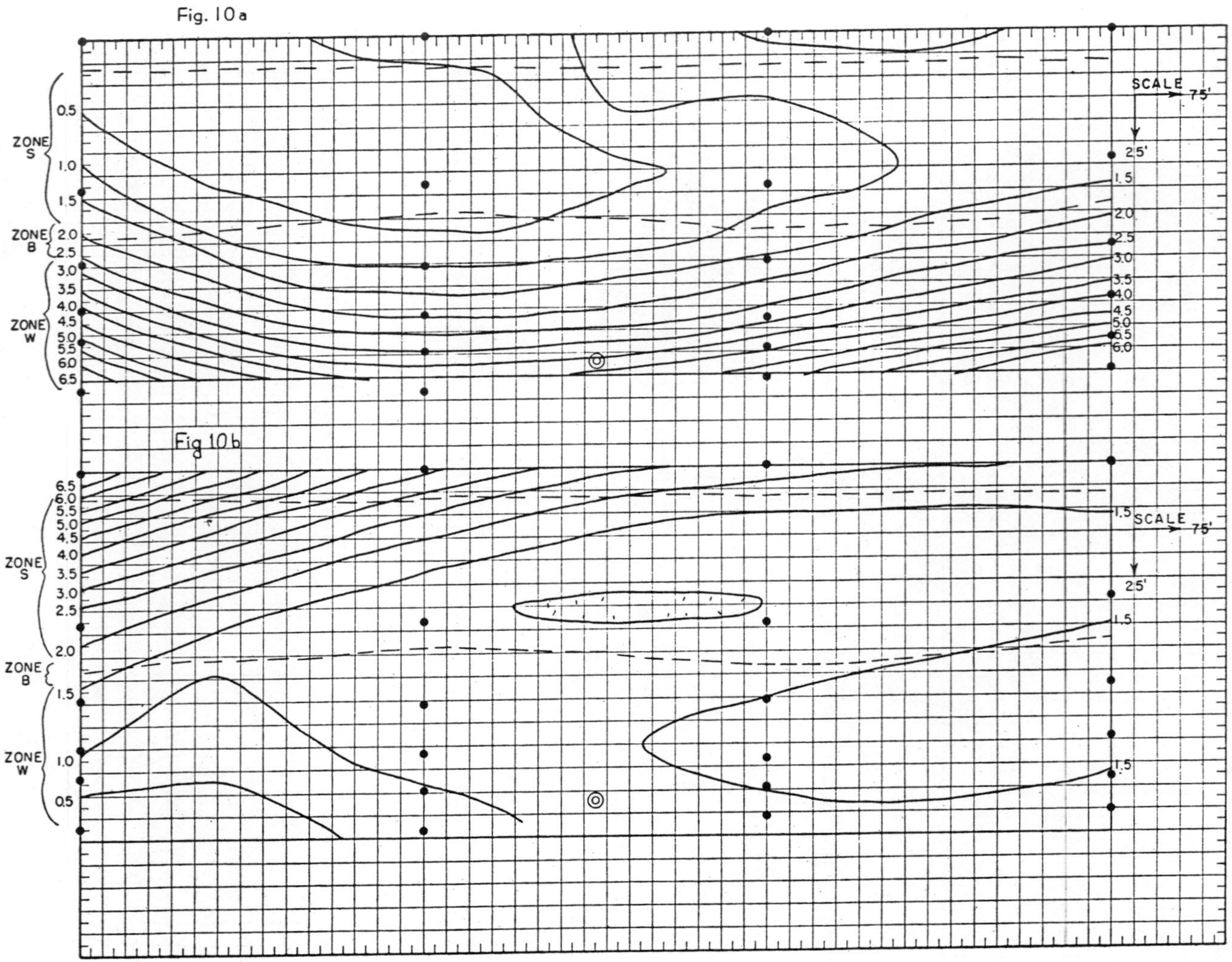

FIG. 10.—a, Median size trend map of T_8 Falmouth. Contour interval = 0.5 mm. b, Sorting trend map of T_8 Falmouth. Contour interval = 0.5 mm. b, Sorting trend map of T_8 Falmouth. Contour interval = 0.5 mm.

at the end of the tidal cycle studied, figure 11 appears to show that the pattern has reverted, at least in part, to that of tidal period T_1, the previous low tide. During this tidal cycle, wave height and length have increased, with the result that the neutral-line contours (fig. 6, *c*) are greater than the trend-map contours for median size. Whether or not a significant lag in the process is required to bring the expected into agreement with observed, we cannot say. Nevertheless, part of Zone W shows once again improvement in sorting toward Zone B, and similarly an increase, locally at least, in median size.

It seems striking to us that the combination of changing sea-state and tidal effect did not succeed in changing the basic pattern established at T_1 and returned to at T_{11}. There may be here a response to some underlying stability within this region. Clearly, repeated studies of this kind will be needed.

We now examine several other places. The trend map for Highland Beach on Cape Cod, taken at low tide, figure 12, shows the expected increase of median size toward the shore. Sorting also improves as expected, but a strong non-orthogonal component appears, so that the contours do not parallel the beach. It is possible that the near-shore littoral-current effect was not canceled out here, even though the length of traverse along the beach was only 75 feet.

The trend map for Indiana Dunes State Park beach on Lake Michigan shows the expected improvement of sorting in a shoreward direction, but the median-size trend map shows the expected increase in median size toward the shore in only one portion of Zone W (fig. 13).

Model I appears to be favored over Model II by the trend maps. We conclude, therefore, that the sand moved in accordance with a modified form of Cornaglia's theory and that, in the very shoal waters where our experiment was conducted, gravity played very little part in moving materials offshore. The above discussion is based on the initial assumption of a uniform distribution for the sediment. In the conclusions we shall take up the implications for Zone W when the initial distribution is symmetrical and also when it is polymodal. These two alternatives will give more realistic models. We felt that the uniform distribution would serve better as a first approximation.

A note on tidal effect.—The tidal effect is negligible in the Lake Michigan study. In the case of the Falmouth and Highland regions, we feel that low tide presents a pattern which is returned to in each successive low tide. Space does not permit the addition of eleven more trend maps. However, the following describes the pattern changes observed on an hourly basis through a full tidal cycle at Falmouth. The description is from a total of eleven trend maps completed for each of the hourly periods of the tidal cycle.

1. Low tide: as shown in figure 9. As the tide rises, the pattern established at low tide remains unchanged, but a new step is formed during the third hour, about 5 feet in a shoreward direction from the original step. Ripple marks are scattered through the interval between old and new steps.

2. High tide: the old step and the original Zone W pattern is now almost erased and does not show at all on the smoothed-surface trend map but can be seen clearly on contouring the original data. A series of incipient "steps" where tide and wave were long enough in a "steady state" are scattered from the original step to the high-tide breaker zone.

3. Falling tide: see, for example, figure 10. The pattern appears to be reversed. However, just at the next hour, the upper pattern of figure 8 is erased, and the original low-tide pattern begins to emerge until we reach the last period of the cycle (fig. 11).

At Highland, the trend maps were computed for every 4-hour interval in the tidal cycle. Once again, the low-tide pattern weakens as new steps are formed temporarily with the rising tide, and the low-tide pattern becomes dominant with the recurrence of low tide.

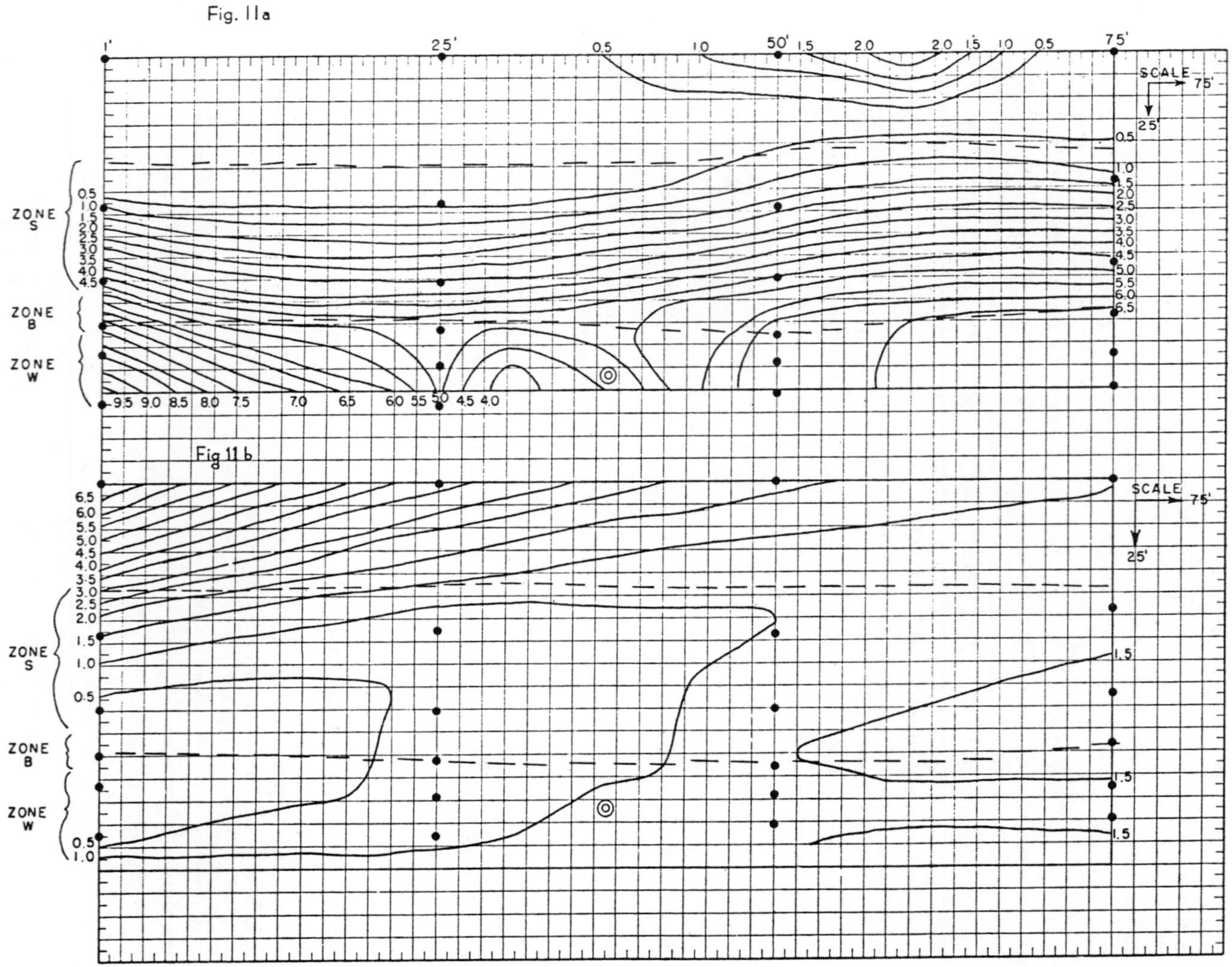

FIG. 11.—*a*, Median size trend map of T_{11} Falmouth. Contour interval = 0.5 mm. *b*, Sorting trend map of T_{11} Falmouth. Contour interval 0.5 mm.

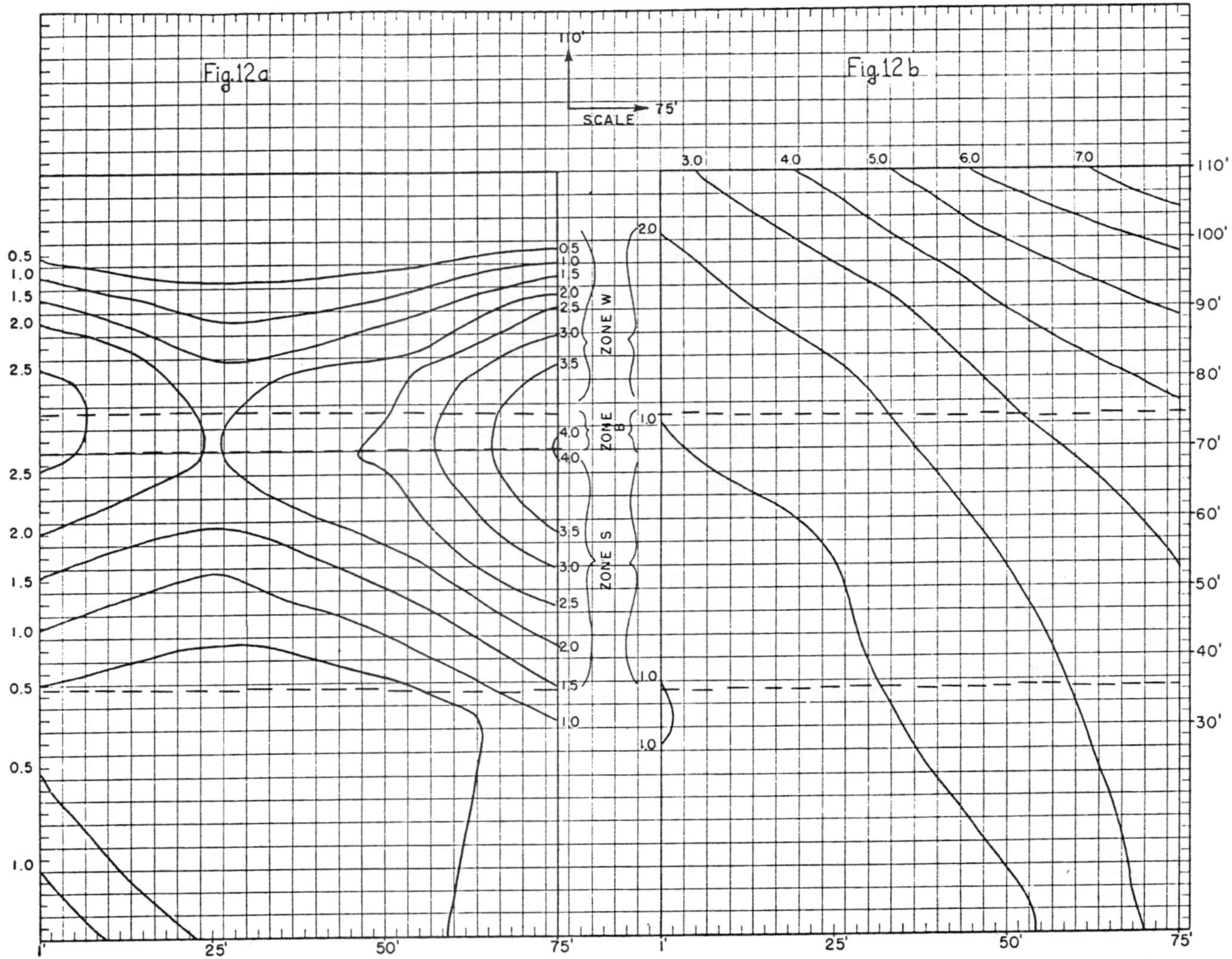

FIG. 12.—*a*, Median size trend map of Highland (low tide). Contour interval = 0.5 mm. *b*, Sorting trend map Highland (low tide). Contour interval = 1.0 mm.

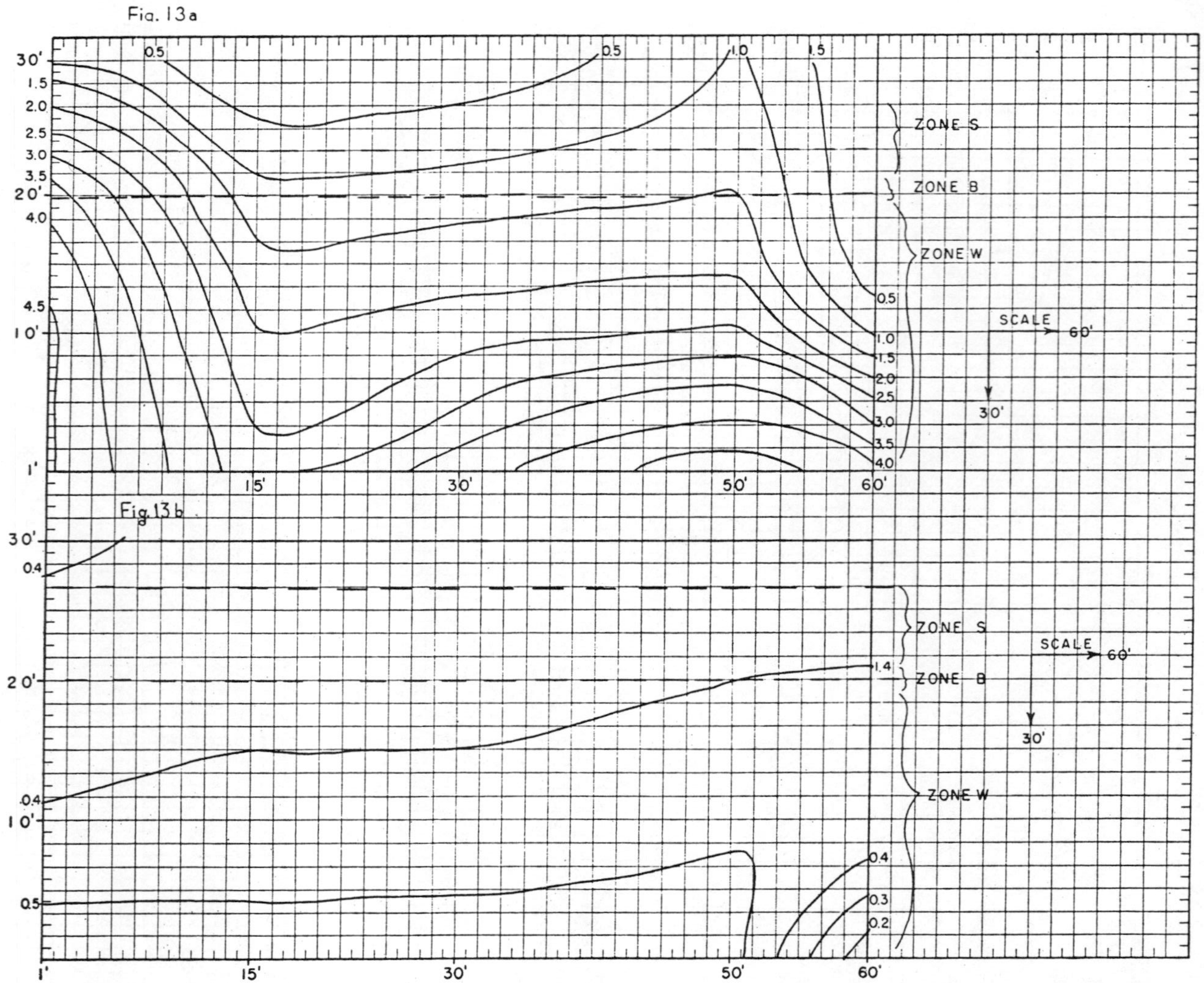

FIG. 13.—*a*, Sorting trend map of a Lake Michigan beach. Contour interval = 0.5 mm. *b*, Median size trend map of a Lake Michigan beach. Contour interval = 0.10 mm.

BREAKER ZONE (ZONE B)

Zone B is the region in which the forward net sediment velocity of Zone W drops to zero. It is where the breaking wave is met at its base by the returning backwash of a previous wave. The point of collision of the breaker and the swash, in our opinion is characterized by a turbulent region in which the net movement is vertically upward, so that a sorting action occurs. Although the returning swash does not always catch the base of the next wave just at the point of plunging, it does so a majority of the time. The flow characteristics of the backwash are a function of the slope of the foreshore, the size of wave creating the parent swash, and the size of sediment on the foreshore. Just as the incoming breaker deposits its load, it is met by the returning backwash, also carrying an accumulation of sediment. If the slope of the foreshore is sufficiently steep, a moderately well-sorted, relatively coarse "step" may be built in this turbulent zone.

The sediment which builds the "step" is derived in part from the load abruptly deposited by the incoming wave at this point. Thus the slope of the bottom in the near-offshore is also important (see discussion in the section on "Shoaling Waves" [Zone W]). The remainder of the sediment is carried down the foreshore slope by the backwash from the previous wave.

In the way described above, a grading or sorting action occurs in the vertical direction within the breaking wave over Zone B. The finer material does not settle out so rapidly as the coarse material, and the finest is probably carried forward onto the foreshore. We should, under these circumstances, expect the coarsest material to be concentrated at the step, whereas finer material should be found on both the seaward and the landward sides. In the specific terms of the sediment parameters dealt with in this paper, our expectation under the above mechanism is as follows:

1. Size. See figure 14.

2. Degree of sorting. This is difficult to specify. A sorting action occurs in all three zones. In Zone W, due to the action of shoaling waves on the bottom; in Zone S under the turbulent sheet flow downslope on the foreshore; and, finally, the above-described vertical movement in Zone B. Although less fine material is expected to remain in Zone B under the swirling turbulent motion, fine material may or may not be trapped with the coarser material as it settles out. This would greatly affect the numerical value of any measurement of degree of sorting.

Table 2 shows median size and degree of sorting for a traverse of samples collected across the breaker zone at a number of

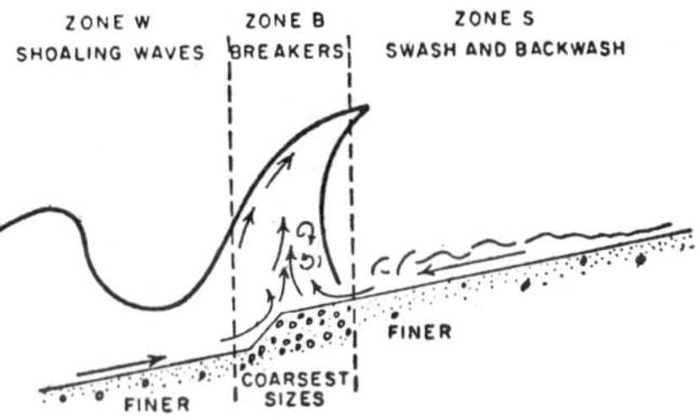

FIG. 14.—Mechanism for sediment pattern in the breaker zone (zone B). Arrows indicate net sediment movements at the moment the wave breaks.

localities and under varying surf conditions. The sample points were arranged as shown in figure 15, so that the first sample was taken just inside Zone W; the second sample was taken on the slope of the topographic step in Zone B; the third sample at the crest of the step in Zone B; and the fourth sample midway between the upper margin of swash on the foreshore and sample point 3.

Inspection of the table indicates that sediment samples taken at points 2 and 3 in the breaker zone have consistently larger medians than either sample point 1 in Zone W or sample point 4 in Zone S. The field evidence, within the limits of our data, thus supports the previous discussion regarding the expected size distribution across Zone B. This does not, of course, support the *uniqueness* of our hypothesis.

The data on degree of sorting seem to fall into two categories. The value of the sorting coefficients of the sample points in Zone B are either both higher or both lower than the sample in Zone W or Zone S. As suggested earlier, we are not, under the present relatively crude model, able to advance consistent expectations for degree of sorting in Zone B.

Further field data, bearing on the above discussion, were collected by sampling the wave itself. This was done in Zone B by using a series of cups fastened to a rod. Just as the wave is on the point of breaking, the rod is held vertically so that the cups sampled a vertical series of points through the wave (see fig. 16).

A single horizontal set of samples was caught by the same device at a later time as the wave plunged down on the foreshore (see fig. 16). The results as given in table 1 are not to be considered quantitatively accurate,[3] but should be of some interest.

Inspection of table 1 shows that the largest volume of sediment is found at the sample points nearest the bottom of the wave. The largest sizes are also found at the lower sample points, with one exception—in the vertical traverses. In two of the three vertical traverses, the sorting is poorest near the bottom. It is tempting to attribute the discrepancies to the failure of the sampling device. However, it is evident that further study is needed.

TABLE 1

Sample	Md. (Mm.)	Sorting	Weight (Gm.)	Location
Vertical traverse:				
1...........	0.38	1.39	46.19	Nauset, 1
2...........	.34	0.85	49.67	
3...........	.25	0.92	11.87	
1...........	.41	1.32	22.78	Nauset, 2
2...........	.35	0.77	7.87	
3...........	.30	0.90	1.07	
1...........	.26	0.79	15.88	Nauset, 3
2...........	.34	0.76	7.97	
3...........	.30	0.83	2.90	
*Horizontal traverse:**				
1...........	.34	0.71	23.48	
2...........	.33	0.79	74.68	
3...........	0.33	0.79	38.88	

* Average of three sets of samples.

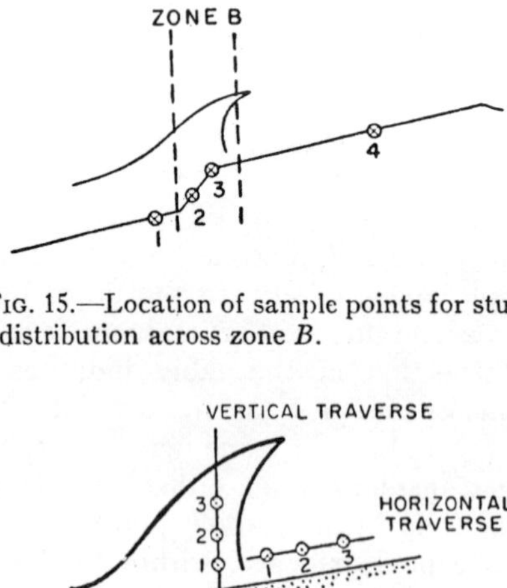

FIG. 15.—Location of sample points for study of size distribution across zone *B*.

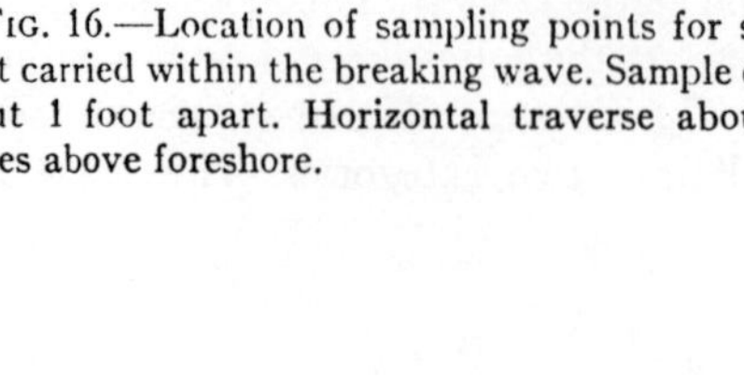

FIG. 16.—Location of sampling points for sediment carried within the breaking wave. Sample cups about 1 foot apart. Horizontal traverse about 6 inches above foreshore.

Some question arose as to whether the breaker itself deposited sediment on the foreshore. A suggestive bit of data is found in the averaged horizontal traverse of table 1. Note that median size and degrees of sorting are about the same in the several

[3] Effort was made to prevent the cup from disgorging part of its catch by downcurving the rear portion to create an eddying motion to the sediment-laden water. However, we do not, at this stage, know how much of the total catch was immediately lost.

sampling points. The weight column indicates that, at least for that particular sequence of waves, a measurable quantity of sediment was carried across the breaker zone and deposited on the foreshore.

The data of the vertical traverses within the limits of the small sample lend further credence to the hypothesis advanced earlier regarding the mechanism of water versus sediment within the breaker and the heaviest suspension to be lowest also.

In summary, this section presents a hypothesis for the mechanism in the breaker zone which results in the sediment pattern consistently observed at this place in nature. The field evidence presented in this section supports the hypothesis of net vertical movement in the base of the turbulent breaker, with a secondary sorting action on the material thrown into suspension. It is clear, however, that the hypothesis presented is qualitative and is only one possible explanation.

TABLE 2

Location	Median Size (Mm.) Sample Point 1	2	3	4	Degree of Sorting (Mm.) 1	2	3	4
Nobska Beach	0.33	0.88	0.44	0.29	4.12	4.35	1.61	0.93
Falmouth	4.60	4.90	5.20	1.09	0.83	0.37	0.31	1.04
Sippewissett	0.11	0.36	0.11	0.11	0.26	2.20	1.72	0.24
M. B. L. Beach	0.69	1.09	0.72	0.48	1.00	0.82	0.60	1.12
Old Silver Beach	0.24	0.72	0.70	0.68	3.38	0.56	0.60	0.68
Sandwich	1.84	3.60	9.60	5.40	1.93	2.75	1.19	0.96
Green Harbor	0.43	0.42	0.76	0.24	2.99	2.45	5.12	0.98
Poponesset	0.74	0.81	0.66	0.68	0.97	1.59	0.76	0.84

FORESHORE (ZONE S)

In this section we shall put together our field observation and field measurements of velocity of flow with published experimental observations, to form a model for the sediment pattern of the foreshore. This model will be used to express our expectation of sediment size and sediment sorting. The model is then applied to several beaches in which the following information is available: (1) slope of foreshore, (2) average depth of backwash, (3) average velocity at surface of backwash, and (4) an estimate of the velocity gradient from top of swash to the lower end of the backwash return.

We proceed to work out a sediment pattern for a given beach in terms of the model. This pattern is then compared with the sediment patterns actually found at that beach.

THE MODEL

The model consists of the following elements:

1. *An assumption as to the sorting or size pattern left just after a given upwash or swash has passed over the foreshore.*—Consider that a given wave breaks and carries up the foreshore a load of sediment in suspension as well as by rolling or saltation. The sediment load is derived partly from the lower part of the foreshore and partly from material carried in the breaker itself. As the velocity of the upwash approaches zero, the load is dropped along the way, presumably with the coarsest material first, and the finest carried upslope. Certainly the mechanism of erosion and subsequent deposition of the upwash is not so simple. A detailed discussion of the complexities of backwash from a qualitative viewpoint is given in Demarest (1947). In the present discussion however, we are not concerned with presenting a detailed mechanism for erosion and deposition during the upwash. We wish merely to establish a pattern of sediment size and sorting after the upwash has passed over the foreshore, an as arbitrary starting point. The subsequent backwash and its effect on this initial distribution will then be discussed in some detail, from the working part of the model. Field samples of the foreshore sediment are collected just after the backwash when the foreshore is

momentarily free of water. We shall therefore assume that, just prior to the backwash, the following sediment pattern is found: graded size, with the finest nearest the top of the foreshore. Graded sorting, with the best sorting near the top of the foreshore.

2. *Consideration of the velocity required to start a grain of a given size moving downslope under the backwash.*—A relation between turbulent flow and initiation of sediment movement on a rough bottom has been experimentally determined by Ippen and Verma (1953). In their paper, v_e is defined to be the "edge velocity" at the upper edge of a sediment particle resting on a rough bottom (see fig. 17) where conditions of fully turbulent, uniform, steady flow are present. Then, for v_e just large enough to initiate movement in a given size of particle, the following is empirically established:

$$\bar{V}_e = 0.12w\left[S^{1/2}\frac{d}{K_e}(S_s - 1)\right]^{-0.3}, \quad (1S)$$

where w = terminal fall velocity for given grain size, d = grain diameter, S_s = specific gravity of grain, S = slope of the energy gradient, and K_e = effective hydraulic roughness length. In actual computation we took S_s to be 2.67 for quartz, S to be equivalent to the average slope of the sediment-water interface on the foreshore, and K_e to be $2.5 \times K$ where K is the average grain diameter of the bottom. This last was based on inspection of the relation between K and K_e as given in table 1 of Ippen and Verma.

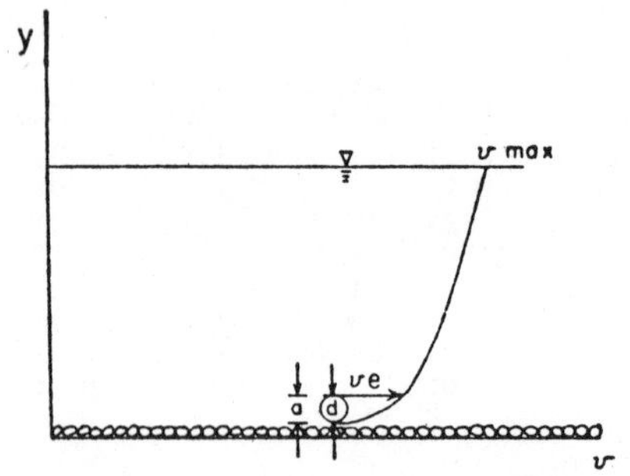

FIG. 17.—Definition sketch giving notation for velocity profile in the turbulent backwash. (Based in part on a figure in Ippen and Verma, 1953.)

3. *Velocity distribution on the foreshore.*—Velocity at the surface is measured on both backwash and upwash by timing a float, such as a Ping-Pong ball, over a measured distance. The traverses were normal to the breaker zone and so arranged that the upper end was well below the top of the swash and the lower end well above the breaker line. By timing the float over the measured distance for repeated trials, an average maximum (surface) velocity for upwash and for backwash was recorded.

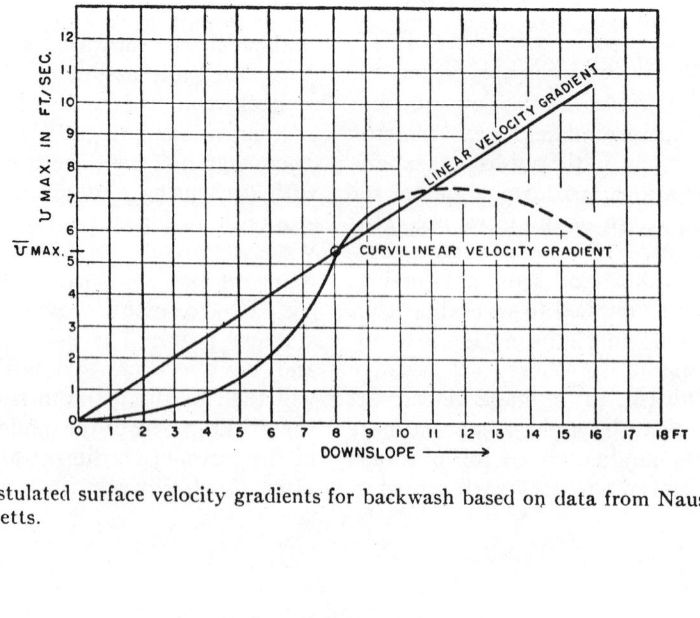

FIG. 18.—Postulated surface velocity gradients for backwash based on data from Nauset Beach, Cape Cod, Massachusetts.

We may, for simplicity, assume that the gradient for the surface velocity of the backwash is linear, starting at the top of the swash and proceeding to a maximum at the lower end of the traverse. Then the recorded average surface velocity forms the mid-point value of the traverse (see fig. 18). Since the velocity is usually at a maximum when a profile from sediment interface to free surface at the top is considered, the surface velocity will henceforth be referred to as the maximum velocity, v_{max}.

At Nauset Beach on Cape Cod, Massachusetts, the velocity was measured over a traverse, and an average was determined to be 5.32 ft/sec. Then, for zero velocity at the top of the traverse and 5.32 ft/sec at 8 feet downslope, we have the linear equation $v_{max} = 0.665x$. Field observations tend to indicate that a more realistic v_{max} gradient would not be linear but would take the general form shown in figure 18, where the dimensions are once again based on the Nauset Beach study.

The postulated drop in the velocity curve at the upper end of figure 18 is thought to be due to the accumulated sediment load entrained in the backwash, and subsequently the velocity is abruptly checked when backwash strikes the next oncoming wave or the "dead" water at the foot of the foreshore. It should be emphasized at this point that accurate measurements of flow characteristics on swash or backwash under field conditions is a very difficult problem. Adaptation of Pitot tube devices, for example, have thus far proved unsuccessful because of almost immediate clogging of the orifice with sand. The shallowness of the flow renders standard current meters inoperable, e.g., Price-Gurley or Ekman.

4. *Computation of theoretical contours of size and sorting distribution on the foreshore.*—For a given grain size we compute the edge velocity necessary to initiate movement in the direction of flow. This direction will be downslope under the backwash. We use the relation as given in equation (1S) of section 2 above. Then, for the resulting v_e, we compute the surface velocity v_{max}, by using a logarithmic equation due to Krey (1927) as cited in Rouse (1938):

$$V = V_{max}\left\{\frac{\log(1 + y/a)}{\log(1 + t_d/a)}\right\}, \qquad (2S)$$

where y = variable distance from the bottom, t_d = total depth of flow, and a = a very small distance between boundary and reference axis. The notation is illustrated in figure 17. By taking y equivalent to grain diameter and a equivalent to y, the equation takes the following convenient form:

$$V_{max} = \bar{V}_e\left\{\frac{\log(1 + t_d/a)}{\log 2}\right\}. \qquad (3S)$$

The V_{max} computed[4] for a given grain size is now translated into specific field conditions, either by referring to the straight-line first approximation for the velocity gradient or, graphically, by using the more realistic postulated curvilinear velocity gradient. For example, in the Nauset Beach study, previously referred to, the following data are available for a selected grain diameter:

Selected grain diameter	d = 1 mm.
	= 0.0033 feet
Average surface velocity of backwash	V_{max}, 5.34 ft/sec
Average grain size on foreshore	K = 1 mm.
	= 0.0033 feet
Specific gravity of quartz . . .	S_s = 2.67
Slope of foreshore	S = 0.10
Effective hydraulic roughness length	K_e = 2.5K
	= 0.0082
Terminal fall velocity for 1 mm.	w = 0.496 ft/sec
Length of traverse	16 feet
Depth of flow	t_d = 0.2 feet

Then $\bar{V}_e$ = 0.0992 ft/sec from equation (1S) and v_{max} = 0.587 ft/sec.

[4] We also tried the Prandtl form of the Von Karman velocity-deficiency equation for fully turbulent uniform steady flow, as cited in Bakmeteff (1936), where, in our present notation,

$$\frac{V_e - V_{max}}{\sqrt{gdS}} = \frac{1}{k}\left(\frac{d}{t_d}\right), \qquad k = 0.4 .$$

We found, however, that, within the range of our field data, the lower end of the velocity profile did not include our very small values for d, the distance from sediment interface to the top of the sediment particle. It thus was not possible to use the equation for extrapolation from V_{max} to V_e.

Referring to figure 18, we find that, for the linear velocity gradient, the 1-mm. grain size begins to move downslope at approximately 0.9 feet from the upper end of the traverse. If we use the postulated curvilinear velocity gradient, we find that the 1-mm. grain size begins to move downslope at approximately 3.2 feet from the top of the traverse. Table 3 presents the expected distance downslope that various grain sizes will begin to move under observed field conditions at Nauset and Highland Light Beaches, both on Cape Cod, Massachusetts.

We now return to the expected sediment pattern left by the upwash and construct an expected or theoretical sediment distribution on the foreshore for a given beach at a given time. We consider that the sediment distribution over the foreshore just after the upwash consists of graded size, with the finest near the top of the foreshore, and graded sorting, with the best sorted near the top of the foreshore. Under these circumstances, we may expect a zone near the upper end of the foreshore to show rapid increase in size on a traverse downslope and a rapid improvement in sorting. Continuing downslope, we expect a regular continued rise in median size, but poor sorting with irregular contour pattern. The above conclusions are arrived at by the following reasoning. Everything is put into motion under backwash within, let us say, the upper 10 per cent of the foreshore. However, as the backwash passes over successively lower portions of the foreshore, it is passing over an increasingly coarse bottom. Large particles put into motion higher in the traverse come to rest downslope as the velocity gradient passes its maximum. Still larger sizes in the lower foreshore are not appreciably moved at all. Finer sizes put into motion near the top of the foreshore are carried under the turbulent flow and trapped in an irregular manner between interstices or in eddies around larger particles; thus we expect a band near the top of the foreshore of graded small sizes, increasing rapidly downslopes followed by a regular coarsening of the sizes, to the edge of the breaker zone. We also expect good sorting, improving downslope, in the band of small sizes at the top of the foreshore. Below this band we expect poor sorting, irregularly distributed in contours over the lower 80–90 per cent of the foreshore within the sediment size and foreshore-slope ranges of this study (see fig. 19).

If the tide rises, the well-sorted band of finer material is continually erased or covered and appears higher on the beach, while the major active portion takes over a larger portion of the foreshore.

As the tide falls, the well-sorted band of finer material increases in width, reaching a maximum at low tide. This last is shown in the Highland Light trend maps (fig. 12). Comparison of the theoretical or expected

TABLE 3*

DISTANCE DOWNSLOPE AT WHICH GIVEN GRAIN SIZE BEGINS TO MOVE UNDER BACKWASH

Nauset		Highland Light	
Grain Size (Mm.)	Distance Downslope (Feet)	Grain Size (Mm.)	Distance Downslope (Feet)
0.5	2.7	0.5	1.5
1.0	3.2	1.0	2.0
2.0	3.6	2.0	2.5
3.0	3.6	3.0	2.6
Av. breaker height	1.5 feet	Av. breaker height	3.2 feet
Av. slope of foreshore	6° approx.	Av. slope of foreshore	13° approx.
Av. surface velocity for backwash	5.34 ft/sec	Av. surface velocity for backwash	6.13 ft/sec

* The velocity measurements are of the same order of magnitude as observed by Demarest (1947), who found average velocity for the backwash to be about 6 ft/sec for slopes from 9° to 13°.

sedimented pattern for size and sorting on the foreshore is now made with trend maps for various beaches based on sediment samples collected in the field. We shall first consider the Falmouth Beach tidal trend maps (figs. 9, 10, 11). The median size contours for tide stages T_1 and T_{11} increase downslope. This is in good agreement with our model. The sorting contours decrease downslope, as expected, for T_8 and T_{11} but are reversed for time T_8. The Highland Light trend maps show agreement for both size and sorting contours, with the expected contours. The Lake Michigan beach trend map (fig. 13) for sorting shows agreement with the model.

which includes the foreshore. Based on observations of a number of beaches on the Pacific Coast, he finds that there is a marked gradation in median size of sediment, from coarsest at the breaker line to successively finer up the slope of the foreshore up to, but not including, the crest of the berm. This is also in agreement with the observations of Evans (1939) on the foreshore of a Lake Michigan beach, near White Lake Harbor.

Krumbein (1938*a*) analyzed a short stretch of foreshore on Waverly Beach, Lake Michigan, by a grid of sampling points. Local irregularities tend partly to obscure the trend, but examination of a contour

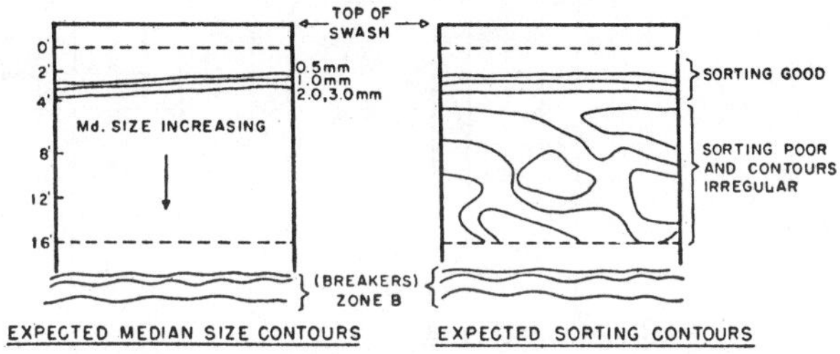

FIG. 19.—Plan view of foreshore, showing expected pattern of contours of median size and sorting of sediment.

In no case are contours concentrated toward the upper end of the foreshore as predicted by the model. Inspection of the trend maps and comparison, for example, with the calculated position of contours for Highland Light in table 3 indicated that agreement is not sensitive enough to warrant direct comparison of computed contour maps with the trend maps.

We feel, however, that the agreement between model and sediment patterns actually observed on various beach foreshores is sufficiently strong to suggest that some progress has been made toward relating foreshore sediment patterns in nature with the dynamics that created these patterns.

Published work appears to support the expectation of increase in size and down the foreshore toward the breaker zone.

Bascom (1951) describes the variation in median size of sediment on a traverse

map made by the present writers, based on the Md. size data given by Krumbein, is quite suggestive of a general trend toward larger Md. size toward the plunge line and finer sizes near the crest of the berm (see fig. 20). The writers have also constructed a size-free sorting map from Krumbein data using $\sigma_\phi/\mathrm{Md.}_\phi$ (see fig. 20).

In another study of areal variation by Krumbein (1938*b*) an irregularly spaced distribution of points was used in sampling about 1,000 feet along a beach at Little Sister Bay, Wisconsin. This beach is exceptionally coarse, ranging from pebble sizes to cobbles. Because of the large sizes relative to the local competency of waves to move the sediment, ice-shove patterns were apparently preserved in part through the summers. Our present study is definitely limited at the upper size ranges to particles which are small enough to be at least occasionally

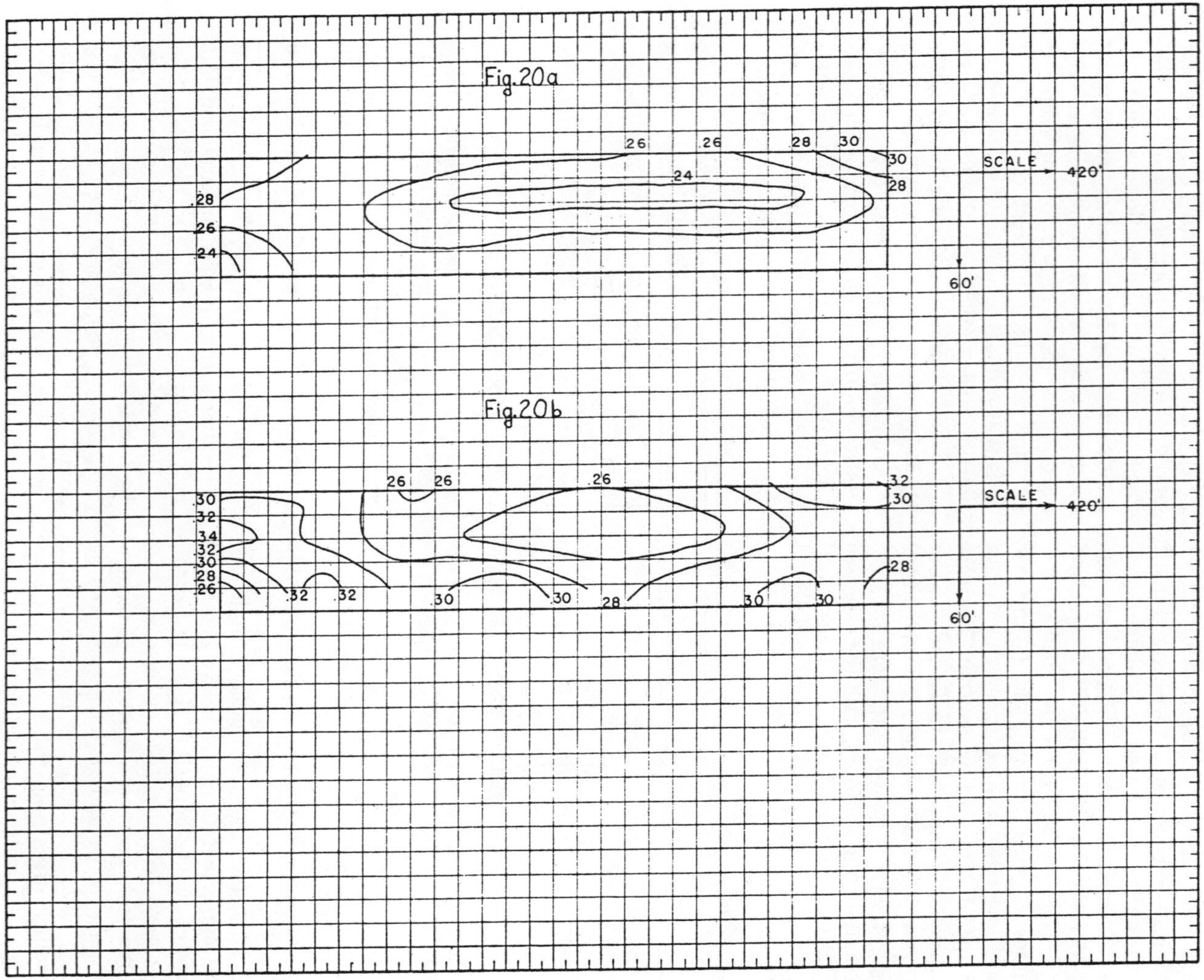

FIG. 20.—*a*, Map of median size of Lake Michigan, near Michigan City, Indiana (data from Krumbein, 1938*a*), Waverly Beach. Contour interval = 0.02 mm. *b*, Map of sorting σ_ϕ/M_ϕ at Lake Michigan (data from Krumbein, 1938*a*) near Michigan City, Indiana, Waverly Beach. Contour interval = 0.02 mm.

moved by local waves and swash. Although geometric mean size shows a tendency to increase toward the surf zone, in Krumbein's study no clear pattern emerges for sorting expressed as $\sigma_{\phi}/\mathrm{Md.}_{\phi}$. It is evident that further work is required to find the upper and lower size limits for the prediction equations and mechanisms discussed in the present paper.

SUMMARY

Essentially this paper represents an effort to establish a pattern of sediment size and sorting free of local fluctuation and representative of foreshore and near-shore in a state of equilibrium. The region is divided into three dynamic zones: Zone W that of shoaling waves landward of near-shore bars; Zone B, the breaker zone; and Zone S, the swash-backwash zone. Each zone is treated separately. Published hypotheses and experimental investigations, which included the properties we could also measure in the field, were combined by us into model form, representing a proposed mechanism and a resulting sediment pattern. The model was then compared with trend maps of several field locations in which very local fluctuations were removed. By substituting observed values for the parameters of the model, a detailed comparison was made in the case of Falmouth Beach and near-shore. Highland Beach and Lake Michigan Beach were compared with the model in a less detailed way. And, finally, Waverly Beach, representing unadjusted field data, was compared with the model for the foreshore zone.

In this way three levels of comparisons were made: first, where both wave and current data and a trend map were available; second, trend map only available; and, third, field data only. Agreement of model with the natural cases, although tempered by the information available, was still sufficiently good to warrant publication.

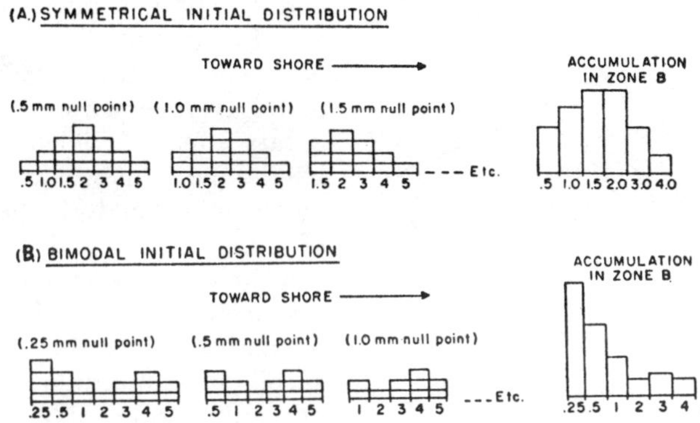

FIG. 21.—Expected displacement of sediment under Model I for a symmetrical initial distribution and bimodel initial distribution.

DISCUSSION

An immediate extension of the earlier discussion of Zone W is to consider the implications of two initial frequency distributions of sediment size which are closer to reality than that of the uniform distribution.

A. *Initial symmetrical frequency distribution, as an approximation to the normal distribution.*—A simplified frequency distribution using unit per cent by weight with a median of 2.0 mm. serves as a representative of the large class of approximately normal or mederately skewed distributions found in nature. As in the earlier discussion, assume that this initial distribution is distributed over the near-shore region under discussion and that wave motion subsequently begins and continues until a stable pattern is established. Figure 21, *a*, illustrates the initial distribution, the changes as successive null points are passed in a shoreward direction, and the nature of the distribution dropped in Zone B. We apply dynamic Model I (discounting seaward movement for the sizes under consideration). As in the simpler model using a uniform distribution, the same trends are evident. The median sizes increase in a shoreward direction, and the sorting improves in a shoreward direction. The most notable dif-

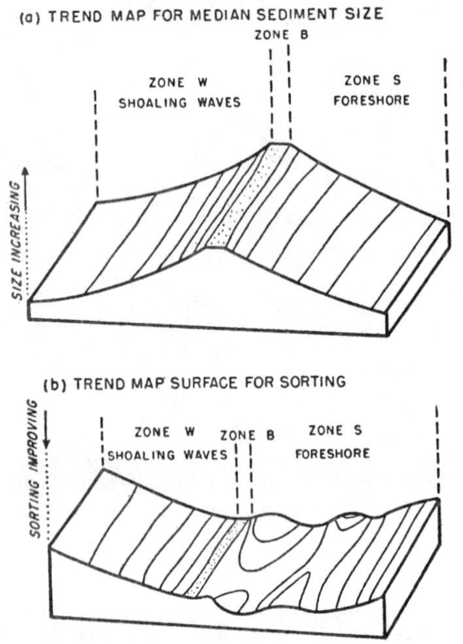

FIG. 22.—Postulated trend maps for nearshore breaker zone and foreshore in equilibrium with wave and current dynamics.

ferences seem to be (1) the rate at which the sorting and median sizes change in the shoreward direction (this alters the relative position of the individual contour lines but not the basic pattern) and (2) the nature of the accumulation finally dumped at the "step" in Zone B.

B. *Initial bimodal frequency distribution, representing the general form of certain glacial deposits.*—Figure 21, *b*, illustrates the inital distribution, the changes corresponding to successive null points in shoreward direction, and the form of the final accumulation dumped in Zone B. Interpretation is similar to that of the symmetrical distribution.

The underlying sediment pattern in the near-shore–foreshore region under consideration appears to us to be composed in part of the following factors: (1) the nature of the initial sediment distribution, including the range of sizes, as well as the relative amounts in the various size classes; (2) the availability and kind of sediment being brought in both from the seaward side of the shore line, e.g., by littoral drift, and from the landward side, e.g., by wind, storm, abnormally high tides, and slope wash (see, e.g., Inman and Chamberlin, 1955); (3) the nature of the local wave and current system in an average sense, discounting the temporary effect of violent storms, on the one hand, and abnormally quiescent periods, on the other; and (4) the nature of removal of sediment, e.g., along the shore. If the last three factors are considered to represent averaged conditions in a stable situation, we may consider this region to be an open system in equilibrium over a period of time. It should be possible to characterize such a situation in terms of an underlying pattern of median size and sorting of sediment from which extremely local variation has been removed.

Our investigations have led us to propose the following composite trend map as a model, figure 22, representing the near-shore–foreshore region in a state of equilibrium, as discussed above. Departures from this equilibrium state—e.g., a wasting or building beach—should be reflected in departures of the trend map from this composite model.

Further work along the lines of the present investigation appears to be worth doing. An extension in scope to include the phenomena of bars and an extension seaward beyond the region of shoaling waves are planned for the future. The relation of current and tide and waves in the larger sense, as discussed by Scruton (1955 and especially 1956) and by Van Andel (1954), must then be taken into account.

ACKNOWLEDGMENTS.—Our thanks for illuminating discussions are due to Dr. C. Carver and Dr. A. C. Redfield, of the Woods Hole Oceanographic Institution. We are indebted to C. Hayes, B. Gill, F. Fuglister, G. Hampsen, and J. Helwig for their aid on this project. Without their enthusiastic support this study could not have been reported. Finally, we are greatly indebted to S. Zawadski, of the Argonne Laboratories, A.E.C., for patient guidance and hours of work on the high-speed computer, AVIDAC. This work was carried out under ONR-Geography Branch (Contract No. Nonr-1254(00))(NR-388-018).

REFERENCES CITED

AIRY, G. B., 1842, On waves and tides, p. 289: London, Encyclopaeddia Metropolitan.

BAKMETEFF, B. Z., 1935, The mechanics of turbulent flow: Princeton, N.J., Princeton Univ. Press.

BASCOM, W. H., 1951, The relationship between sand size and beach face slope: Trans. Am. Geophys. Union. v. 32, no. 6. p. 866–874.

DEMAREST, D. F., 1947, Rhomboid ripple marks and beach slope: Jour. Sed. Petrology v. 17, no. 1, p. 18–22.

DOEGLAS, D. J., 1946, Interpretation of the results of mechanical analyses: Jour. Sed. Petrology, v. 16, p. 19–40.

EVANS, O. F., 1939, Sorting and transportation of material in the swash and backwash: Jour. Sed. Petrology, v. 9, no. 1, p. 28–31.

INMAN, D. L., 1949, Sorting of sediments in the light of fluid dynamics: Jour. Sed. Petrology, v. 19, no. 2, p. 51–70.

INMAN, D. L., and CHAMBERLIN, T. K., 1955, Particle size distribution in near-shore sediments: S.E.P.M. Special Bull. no. 3, p. 106–127.

INMAN, D. L., and RUSNAK, G., 1956, Changes in sand level on the beach and shelf at La Jolla, California: Beach Erosion Board, Tech. Mem. 82.

IPPEN, A. T., and EAGLESON, P., 1955, A study of sediment sorting by waves shoaling on a plane beach: M.I.T. Hydrodynamics Lab. Technol. Rept. no. 18.

IPPEN, A. T., and VERMA, R. P., 1953, The motion of discrete particles along the bed of a turbulent stream: Proc., Minn. Internat. Hydraulic Convention, Minneapolis, Minn.

KRUMBEIN, W. C., 1938*a*, Local areal variation of beach sands: Geol. Soc. America Bull., v. 49, p. 653–658.

——— 1938*b*, Beach environment in Little Sister Bay, Wisconsin: *ibid.*, p. 629–652.

——— 1953, Statistical designs for sampling beach sand: Trans. Am. Geophys. Union, v. 34, no. 6, p. 857–867.

KRUMBEIN, W. C., and MILLER, R. L., 1953, Design of experiments for statistical analysis of geological data: Jour. Geology, v. 61, no. 6, p. 510–522.

MILLER, R. L., 1956, Trend surfaces. 1. the relation of sediment size parameters to current-wave systems and physiography: Jour. Geology, v. 64, no. 5, p. 425–446.

MILLER, R. L., and OLSON, E. C., 1955, The statistical stability of quantitative properties as a fundamental criterion for the study of environments: Jour. Geology, v. 63, no. 4, p. 376–387.

MÜNCH-PETERSON, 1950, Münch Petersen's littoral drift formula: Bull. Beach Erosion Board, v. 4, no. 4, p. 1–31.

——— 1937, Monogram for the settling velocity of spheres, p. 57–64: Rept. Comm. Sed. 1926–37, Nat. Research Council. Div. Geology and Geography, Washington, D.C.

ROUSE, H., 1938, Fluid mechanics for hydraulic engineers: New York, McGraw-Hill Book Co.

SCOTT, T., 1954, Sand movement by waves: Beach Erosion Board Tech. Memo 48.

SCRUTON, P. C., 1955, Sediments of the eastern Mississippi delta: S.E.P.M. Spec. Pub. no. 3, p. 21–50.

——— 1956, Oceanography of Mississippi Delta sedimentary environments: Bull. A.A.P.G., v. 40, no. 12, p. 2864–2952.

THOMPSON, W. O., 1937, Original structures of beaches, bars, and dunes: Geol. Soc. America Bull., v 48, p. 723–752.

VAN ANDEL, T. J., and POSTMA, H., 1954, Recent sediments of the Gulf of Paria, v. I: Amsterdam, North Holland Pub. Co.

19

Reprinted from *Jour. Geology* 73(3):528–534 (1965)

LABORATORY STUDY OF SEA-LEVEL RISE AS A CAUSE OF SHORE EROSION[1]

MAURICE SCHWARTZ
Geology Department, Columbia University, New York, N.Y.

ABSTRACT

In contrast to classic concepts of the relationship of sea-level change to shore erosion, Per Bruun has proposed a new theory giving dimensional relationships to the process. To make a preliminary test of the validity of the theory, a small-scale model study was conducted, using constant beach and wave characteristics under variable conditions of water-level rise. Equilibrium was rapidly established with erosion and deposition taking place as predicted by Bruun.

INTRODUCTION

Purpose.—That the world mean sea level varies over longer than annual periods, and in the same sign, is now an accepted fact. Such eustatic changes in sea level have been studied extensively by Fairbridge (1960, 1961), who relates the recent world-wide rise over the first half of this century to a parallel climatic change (thus "glacio-eustasy"). Bruun, Morgan, and Purpura (1962), in considering this phenomenon, note a 5-mm.-per-year rise on the east coast of Florida in the decade from 1946 to 1956. Fairbridge and Krebs (1962) suggest this extraordinary rise on the East Coast may be related in part to a weakening of the Gulf Stream which under the Coriolis Force would develop a reduced "tilt," appearing on its west side as a rise of local mean sea level. Such a weakening of the Gulf Stream would likewise be climatically controlled, owing to the reduced equator-pole temperature gradient.

Exactly what occurs along the shore when there is a rise in mean sea level is not generally agreed upon. For many years geologists and coastal engineers have followed the school of thought expounded by D. W. Johnson (1919). This group visualized the eventual formation of an abrasion platform, 100 fathoms below sea level at its outer edge, gently sloping up shoreward for some indefinite distance to a point where the gradually dampened waves had lost their ability to erode the shore any further. Dietz and Menard (1951), followed by Guilcher (1958) and King (1959, 1963), have pointed out that in the light of current knowledge there must be a more limited and definitive process which limits marine abrasion to a depth on the order of 5 fathoms below sea level.

Recently, however, Bruun (1962) evolved a new theory of sea-level rise as a cause of shore erosion giving dimensions to a concrete concept. The theory assumes a profile of equilibrium, defined by Fenneman (1902), J. W. Johnson (1949), Saville (1950), and Bruun (1954*a*, *b*, 1955), as a beach profile which maintains its form more or less constantly through any secular fluctuations such as those exceeding 12 months. As the sea level rises, Bruun holds that material eroded from the upper beach is deposited on the nearshore bottom to a limiting depth between predominant nearshore and offshore material (fig. 1). Quantitative relationships in this exchange, when an equilibrium profile is developed, are as follows: (1) There is a shoreward displacement of the beach profile as the upper beach is eroded. (2) The material eroded from the upper beach is equal in volume to the material deposited on the nearshore bottom. (3) The rise of the nearshore bottom as a result of this deposition is equal to the rise in sea level, thus maintaining a constant water depth in that area.

The above delineates the erosion which will follow a rise in sea level. It takes no ac-

[1] Manuscript received April 17, 1964.

count of longshore drift, but along a straight coast without man-made obstructions, the amount of sediment removed in any one profile is likely to be matched by an equal volume of incoming material.

In order to test the validity of the theory, a small-scale laboratory model study was made of the aforementioned relationships. This preliminary report outlines the experimental procedure, data collected, and conclusions reached in the course of the study.

clamped to the side of the basin in two positions. One position was centered at the lower end of the unit, the other was 20 cm. from the lower end on the left side of the unit. (All references to left or right are taken as viewed from the lower, or drain-valve end of the stream table.)

The waves generated had a period of $\frac{1}{3}$ second $\pm$ 5 per cent, an amplitude of 8 $\pm$ 2 mm., and a length of 15 $\pm$ 1 cm. The wave length and period correspond fairly well

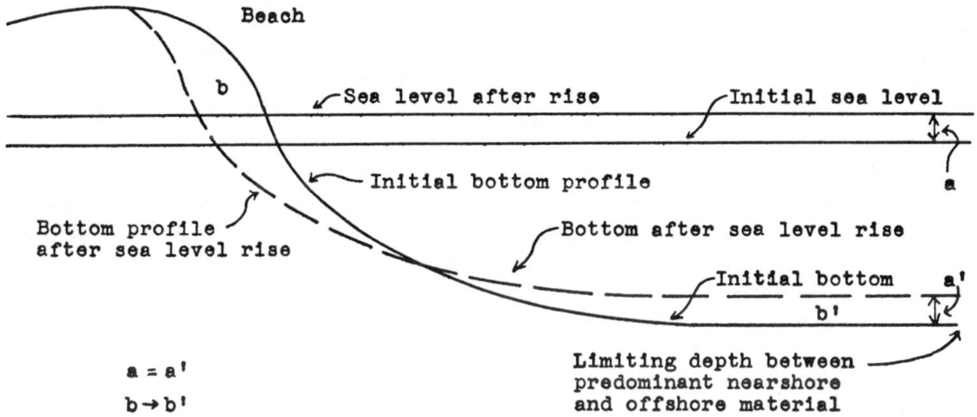

FIG. 1.—Shore erosion after a rise in sea level. (Source: Bruun)

EQUIPMENT

The equipment used was a stream table constructed by the author (Schwartz, 1962, 1963; Stong, 1963) for Columbia University. It has proved to be quite satisfactorily adaptable to the small-scale model study at hand.

Wave basin.—With the upper end of the stream table slightly elevated and a beach established across the width just below the hinge, the wave-basin region was 81.25 cm. wide and approximately 115 cm. long. Water depth ranged from 5 to 10 cm., depending on the gradient of the basin. The basin gradient was adjustable at 0, $2\frac{1}{2}$, and $4\frac{1}{2}°$, through the use of various blocks at each end.

Wave generator.—The wave generator used was the aquarium aerator pump described in the aforementioned stream-table papers. The wave generator could be

with Krumbein's formula, $L = 5.12\ T^2$; where L = length, and T = period. Slight variables above and below the normal period were taken to be inconsequential, as Watts (1954) has found that balanced variations in period at different intervals result in slopes essentially identical to those produced by a constant period wave.

Beach material.—The sand used in the experiment was washed and sorted Ottawa sand, of predominantly white silica grains. A mechanical analysis indicated the range of grain sizes to be: under $\frac{1}{8}$ mm., 6 per cent; $\frac{1}{8}$–$\frac{1}{4}$ mm., 73 per cent; $\frac{1}{4}$–$\frac{1}{2}$ mm., 20 per cent; and $\frac{1}{2}$–1 mm., 1 per cent. The sand responded quite naturally and in fine detail to wave action and water-level changes. The beach sand was mixed, prior to restoring the slope after each test, to counter any sorting which may have occurred.

PROCEDURE

Basic considerations.—The value of experiments as a means of arriving at a qualitative facsimile of a natural phenomenon has been expounded by Reynolds (1933), Hubbert (1937), Krumbein (1944), Bagnold (1946), Einstein (1948), J. W. Johnson (1949), Saville (1950), Bruun (1954*a*), and King (1959). It should be borne in mind that, while the results are not actual model reproductions of any larger-scale natural processes (Rector, 1954), scale models do simulate with reasonable accuracy the changes on natural beaches (King, 1959). The value of such a scale model therefore, is to reduce to a controlled experiment the principles which appear to be involved in a given situation.

Variable factors.—Since the development of the beach profile depends mainly on the characteristics of both waves and the shore material, and the relative elevation of the water level (Rector, 1954; Bruun, 1954*b*), it was deemed necessary in this experiment to maintain the first two as constants and change the latter under varying conditions. Based on this premise, each set of conditions tested was conducted at all three variations of the basin floor gradient.

The main variation was the direction of the wave attack. Though the wave period, amplitude, and length were always within the same range, placing the wave generator at the end of the basin produced waves that approached normal to the beach, while placing the wave generator at the side position produced waves that approached the beach at an angle. These latter waves were refracted so as to break almost parallel to the beach (LeMéhauté and Brebner, 1961; Strahler, 1960). It was found empirically that the best combination of waves and beach development was produced when the wave generator was at the side position. Therefore, for other than wave-approach determinations, this was the position used for all wave generation.

In recognition of the scale factor as discussed by Hubbert (1937), J. W. Johnson (1949), Rector (1954), and King (1959), a wetting agent was added to the water during one set of tests to reduce the surface tension. Triton CF-32, a non-ionic surfactant manufactured by the Rohm and Haas Company of Philadelphia, was used in a concentration on the order of 50 parts per million. In this, as in all the tests, the water was maintained at a temperature of 20 ± 1° C.

A water-level rise of 10 mm. was taken as a standard for most of the tests. As will be shown, this later proved to be a most fortunate choice, for irrelevant difficulties arose with changes which varied greatly from this. Tests involving a 15-mm. rise were conducted as a check on the 10-mm. data.

Taking the above variations into account, there were twelve possible tests from which data could be obtained. These are charted in table 1. It was felt that these were a sufficient number from which to collect empirical data. A similar approach was set forth by Watts and Dearduff (1954) in their tidal-action study.

Procedure.—At the start of the individual tests the desired conditions were set up, and a gentle beach slope formed. Identical slopes were not required in each instance, for according to Rector (1954), the initial beach slope has little or no effect on the final shape of the profile in the zones under consideration. With the basin filled to a satisfactory depth, as required by the basin gradient, the wave generator was set in motion. A half-hour of wave attack was found to be sufficient time for the development of an equilibrium profile. Careful observation failed to reveal any further significant changes in the profiles after this point. Rector (1954) has found that the same average equilibrium profiles are developed under the same conditions if the tests are run for a sufficient time.

When a profile of equilibrium was thus established, the wave generator was stopped. Both water depth and sand depth were then measured on the left, center, and right regions of the offshore edge of the developed shelf. In the early tests, direct measurement of the sand depth was attempted but proved to be liable to error because of the looseness of the submerged sand surface. Later the sand depth was determined by subtracting the water depth from the distance between the water surface and the floor of the basin.

After the initial beach profile measurements had been recorded, the water level was raised the desired amount and the wave generator set in motion for another half-hour. When the wave generator was discontinued, the same measurements were made again upon the newly developed profile. All data and pertinent observations were duly recorded for each test.

TABLE 1

EXPERIMENTAL COMBINATIONS

Possible No. of Tests	Slope (°)	Wave Orientation	Triton	Water-Level Rise (Mm.)
3..........	0, $2\frac{1}{2}$, $4\frac{1}{2}$	End	No	10
3..........	0, $2\frac{1}{2}$, $4\frac{1}{2}$	Side	No	10
3..........	0, $2\frac{1}{2}$, $4\frac{1}{2}$	Side	Yes	10
3..........	0,* $2\frac{1}{2}$,* $4\frac{1}{2}$	Side	No	15

* Tests not completed.

OBSERVATIONS

Errors.—Observation of the determinations involved indicates the possibility of errors which may occur. Due to the meniscus formed along the basin sides and on the immersed millimeter scale, measurement accuracy is limited to $\pm .5$ mm. This holds when measuring water-level rise, or water and sand depths. Then too, as the basin gradient departs from the level position, the bottom (from which the sand depth is computed) develops a slope of 1:33.5 at $2\frac{1}{2}°$ and 2:33.5 at $4\frac{1}{2}°$. The magnitude of the shoreward displacement of the shelf edge resulting from a 10-mm. water-level rise is at least 33.5 mm. Thus the basin rises from 1 to 2 mm., or more, between measurements in a sloped-basin test. In addition, after a measured water-level rise, there is some absorption of the water by the sand at the new upper horizon.

It is felt that in practice these errors might tend to cancel one another. They were borne in mind however, when drawing conclusions from the data collected.

Experimental limitations.—As mentioned previously, attempts were made at raising the water level to a greater and lesser degree than the basic 10-mm. rise. At a 5-mm. rise it was found that the minuteness of the changes, combined with the margin of error involved, produced results that could not be considered as significant for the theory under study. At a 20-mm. rise, the greater shoreward displacement of the newly developed shelf edge so increased the magnitude of the error inherent in the basin-bottom slope as to render the results unsatisfactory. Finally, at a 15-mm. rise, with the basin level, equilibrium profiles were developed which could be measured and recorded for analysis. The measurements on the left side of the beach proved to be unsatisfactory, however, due to the development of subaqueous dunes caused by malfunctioning of the wave generator. Sagging of the rubber panel induced a current rather than wave motion. The series of tests was then terminated at this point, with empirical data collected for ten combinations.

Profile translation.—In every test, the equilibrium profile developed subsequent to a water-level rise, showed a shoreward displacement relative to the equilibrium profile developed at the pre-rise water level. Neither the magnitude of the landward displacement nor its quantitative relationship to the water-level rise by which it was produced were measured in this experiment. For the objectives set forth at the beginning of this paper, it sufficed to find if in fact such a landward displacement actually occurred. The fact that it does corresponds with the findings of Watts and Dearduff (1954) and others.

Sediment loss.—In only one of the ten tests cited was there a loss of transported material over the edge of the first developed outer beach margin when developing a new profile with a higher water level. Determinations of any loss, or rather lack of it, were by visual observation. The offshore edge of the second profile could be observed to cease its development short of the corresponding portion of the first profile upon which it was built following a water-level rise. Measurement marks on the first shelf edge, which remained undisturbed by subsequent wave action after a water-level rise, aided in distinguishing the visible areas of the first profile. In test number 9, there was a slight loss of sediment over the first shelf edge in the development of the second profile. This occurred in the center region of the beach, but seems to have been compensated for, because measurements there indicated no greater deviation than was found in most other tests.

Surface tension.—Observations and measurements made in the series of tests involving Triton did not reveal any differences in profile developments, as compared to those tests conducted without a wetting agent. There did appear to be a tendency for the waves to spill or roll rather than break on the beach, but this did not produce any visible differences. It was concluded therefore, that in the concentration used, there was no physical effect upon beach changes produced by the lessening of the surface tension.

DATA

Presentation.—Though a separate work sheet was maintained for each test, a summary of the pertinent data collected is presented in table 2. Such constants as time lapse for profile development, water temperature, and wave characteristics are not repeated. The *mean change in water depth* and *mean increase in sand depth* for each test are given, since these are the criteria from which subsequent conclusions were drawn.

TABLE 2

SUMMARY OF DATA COLLECTED

Experiment No.	Slope (°)	Wave Orientation	Triton	Water-Level Rise (Mm.)	Mean Change in Water Depth (Mm.)	Mean Increase in Sand Depth (Mm.)
1.	$2\frac{1}{2}$	End	No	10	0.00	9.25
2.	$4\frac{1}{2}$	End	No	10	− .50	9.75
3.	0	End	No	10	.33	10.50
4.	0	Side	No	10	.50	8.00
5.	$2\frac{1}{2}$	Side	No	10	.00	10.00
6.	$4\frac{1}{2}$	Side	No	10	.83	8.33
7.	$4\frac{1}{2}$	Side	Yes	10	.17	8.17
8.	$2\frac{1}{2}$	Side	Yes	10	.00	8.00
9.	0	Side	Yes	10	.50	9.00
10.	0	Side	No	15	0.25	13.50

Empirical relationships.—Calculations indicate that, when taking the means of the measurements across the beach during each test, the grand means are: (1) change in water depth, 0.208 mm.; (2) rise in profile bottom, 9.00 mm. (for 10-mm. water-level rise). These are considered to be acceptable in consideration of the possible errors discussed previously.

In view of the collected data and computations, it is apparent that following a rise in water level: (1) The water depth on the shelf edge remains constant. (2) The shelf sediment rise equals the water-level rise. Based on these empirical relationships and the aforementioned observations, certain conclusions have been reached which are delineated in the following section.

SUMMARY AND CONCLUSIONS

The findings of this report are summarized by a review of the quantitative relationships previously set forth:

1. *There is a shoreward displacement of the entire beach profile as the upper beach is eroded.*—Observed in all of the tests conducted as a part of this experiment.

2. *The material eroded from the upper beach is equal in volume to the material deposited on the nearshore bottom.*—This was observed to be the case since, in all but test number 9, both were one and the same material.

3. *The rise of the nearshore bottom as a result of this deposition is equal to the rise in sea level, thus maintaining a constant water depth in that area.*—The measurements indicate that there is a one-to-one correspondence between the magnitude of the water-level rise and the rise of the nearshore bottom. These same measurements reveal the constancy of water depth in the same region.

Conclusions.—In view of the foregoing empirical affirmation, it is concluded that Per Bruun's theory of sea-level rise as a cause of shore erosion properly defines and gives dimensions to this much-misunderstood geologic process.

As a result of Bruun's concept, much of the thinking that has taken place along classic Johnsonian lines should be reviewed. Applying the relationships between sea-level rise and subsequent shore erosion put forth here, beach retrogression, past and present, can be considered in more realistic terms.

ACKNOWLEDGMENTS.—The writer wishes to thank the following persons who have assisted in this endeavor: Professor R. W. Fairbridge, for help in developing the research; Dr. Per Bruun, for useful advice and discussion; and Warren Yasso, for valuable counsel based on his beach studies with fluorescent markers. With the guidance of Professor R. W. Fairbridge of Columbia University, the writer is continuing this program with further laboratory studies and beach surveys.

REFERENCES CITED

BAGNOLD, R. A., 1946, Motion of waves in shallow water, interaction between waves and sand bottoms: Roy. Soc. London Proc., ser. A, v. 187, p. 1–18.

BRUUN, P., 1954*a*, Use of small-scale experiments with equilibrium profiles in studying actual problems and developing plans for coastal protection: Am. Geophys. Union Trans., v. 35, p. 445–452.

——— 1954*b*, Coast erosion and the development of beach profiles: Beach Erosion Board, Tech. Memo no. 44, 79 p.

——— 1955, Coastal development and coastal protection. Florida Eng. and Ind. Exp. Sta. Bull., Ser. no. 76, 30 p.

——— 1962, Sea level rise as a cause of shore erosion: Jour. Waterways and Harbors Div., Am. Soc. Civ. Engineers Proc., v. 88, p. 117–130, (also: Eng. Prog. Univ. Fla., Leaflet 152, 15 p.).

———, MORGAN, W. H., and PURPURA, J. A., 1962, Review of beach erosion and storm tide conditions in Florida 1961–1962: Florida Eng. and Ind. Expt. Sta., Rep. no. 13, 104 p.

DIETZ, R. S., and MENARD, H. W., 1951, Origin of abrupt change in slope at continental shelf margin: Am. Assoc. Petroleum Geologists Bull., v. 35, no. 9, p. 1994–2016.

EINSTEIN, H. A., 1948, Movement of beach sands by water waves: Am. Geophys. Union Trans., v. 29, p. 653–655.

FAIRBRIDGE, R. W., 1960, The changing level of the sea: Scientific American, v. 202, p. 70–79.

——— 1961, Eustatic changes in sea level, *in* Physics and chemistry of the earth: New York, Pergamon Press, p. 99–185.

——— and KREBS, O. A., JR., 1962, Sea level and the southern oscillation: Geophys. Jour. Roy. Astron. Soc., v. 6, no. 4, p. 532–545.

FENNEMAN, M. M., 1902, Development of the profile of equilibrium of the subaqueous shore terrace: Jour. Geology, v. 10, p. 1–32.

GUILCHER, A. 1958, Coastal and submarine morphology: London, Methuen & Co., 274 p.

HUBBERT, M. K., 1937, Theory of scale models as applied to the study of geologic structures: Geol. Soc. America Bull., v. 48, p. 1459–1519.

JOHNSON, D. W., 1919, Shore processes and shoreline development: New York, John Wiley & Sons, 583 p.

JOHNSON, J. W., 1949, Scale effects in hydraulic models involving wave motion: Am. Geophys. Union Trans., v. 30, p. 517–525.

KING, C. A. M., 1959, Beaches and coasts: London, Edward Arnold Ltd., 403 p.

——— 1963, An introduction to oceanography: New York, McGraw-Hill Book Co., 337 p.

KRUMBEIN, W. C., 1944, Shore currents and sand movement on a model beach: Beach Erosion Board Tech. Memo no. 7, 44 p.

LEMÉHAUTÉ, B., and BREBNER, A., 1961, An introduction to coastal morphology and littoral processes: Queen's University (Kingston, Ontario), Civ. Eng. Research Rept. no. 14, 46 p.

RECTOR, R. L., 1954, Laboratory study of equilibrium profiles of beaches: Beach Erosion Board Tech. Memo no. 41, 38 p.

REYNOLDS, K. C., 1933, Investigation of wave-action on sea walls by the use of models: Am. Geophys. Union Trans., v. 14, p. 512–516.

SAVILLE, T., 1950, Model study of sand transport along an indefinitely long, straight beach: Am. Geophys. Union Trans., v. 31, p. 555–565.

SCHWARTZ, M. L., 1962, Stream table construction and operation: Unpublished undergraduate research project, Columbia University, 37 p.

——— 1963, Stream table construction supplement I: unpublished senior seminar project, Columbia University, 44 p.

STONG, C. L., 1963, How to construct a stream table to simulate geological processes, in The amateur scientist: Scientific American, v. 208, p. 168–179.

STRAHLER, S. N., 1960, Physical geography: New York, John Wiley & Sons, 442 p.

WATTS, G. M., 1954, Laboratory study of effect of varying wave periods on beach profiles: Beach Erosion Board Tech. Memo no. 53, 19 p.

——— and DEARDUFF, R. F., 1954. Laboratory study of effect of tidal action on wave-formed beach profiles: *ibid.* Tech. Memo no. 52, 21 p.

20

Reprinted from *Jour. Geology* 74(3):247–268 (1966)

TIDAL CYCLE OF CHANGES IN AN EQUILIBRIUM BEACH, SANDY HOOK, NEW JERSEY[1]

ARTHUR N. STRAHLER
Columbia University

ABSTRACT

Repeated transverse profile surveys were made in the summers of 1959, 1960, and 1961 on the Atlantic beach of Sandy Hook, New Jersey. The beach experienced accretion (progradation) throughout early and middle summer at the same time that a summer berm was built. By late July progradation largely ceased, yielding an equilibrium profile closely fixed in position and slope.

Superimposed upon the late-summer steady state are rhythmic minor changes in beach profile attending the semidaily tidal cycle in which the swash-backwash zone and breaker zone shift up and down the foreshore. Elevation changes at selected points on the foreshore transverse profile were measured at half-hour intervals using engineer's level, level rod, and steel tape.

From accumulated level data, a model cycle was synthesized to represent elevation and form changes at various levels on the beach. As tide rises, a short phase of initial deposition of about 0.02 foot of medium sand takes place near the inner limit of swash. There follows a scour phase in which about 0.2 foot of sand is removed in a period of 1½–2 hours. Scour is followed by deposition, representing advance of a wedge of mixed sand and pebbles, terminating in the well-sorted pebble and gravel step formed under the breaker point. As step-phase deposition continues, median diameter increases, and sorting becomes poorer. Total step deposition ranges from 0.5 to 0.7 foot.

During falling tide, changes at a given survey point are repeated but in reverse order. Thus, under equilibrium conditions, the beach is restored to its original elevation, slope, and composition. About 0.2 foot of sand is completely removed and redeposited during each cycle at points in the mid-tide region of the zone of swash and backwash under characteristic summer wave conditions.

INTRODUCTION

The purpose of the field study described in this report was to determine the characteristic summer equilibrium profile form and composition of the Atlantic Ocean beach of Sandy Hook, New Jersey, and to determine in detail the minor cyclic fluctuations in profile and sediment composition imposed by the semidiurnal tidal cycle. To conduct the required instrumental field surveys, reference grids were laid out on the beach at two field stations to provide horizontal and vertical control for profile lines surveyed transverse to the shore line in the foreshore zone. Surveys carried out during the summers of 1959, 1960, and 1961 provided data on the characteristic form and evolution of the beach during the summer regime of low waves and persistent obliquity of wave approach from the southeast. A series of half-hourly profile surveys permitted documentation of the small changes of beach form accompanying the rise and fall of tide and led to the formulation of a schematic model for the sequences of erosion and deposition in a semidiurnal tidal cycle.

FIELD STATIONS

Two field stations were established along the eastern shore line of Sandy Hook on beaches exposed to waves of the open At-

[1] Manuscript received July 6, 1965; revised manuscript received October 20, 1965.

lantic Ocean. These stations are designated *Kingmill* and *Spermaceti* (fig. 1). The name "Kingmill" derives from a U.S. Coast and Geodetic Survey triangulation station which forms a convenient reference point nearby. The name "Spermaceti" derives from the old Spermaceti Cove Coast Guard Station lying a short distance to the north of the field station. Kingmill Station is situated 0.45 miles due north of Bench Mark L 17, elev. 9.970 feet (fig. 1). Spermaceti Station lies 0.2 miles due east of Bench Mark B 36, elev. 5.843 feet (fig. 1).

Kingmill Station, the more northerly of the two localities, includes a nearly straight stretch of beach bearing N. 24° W. It is relatively less exposed to the Atlantic fetch than Spermaceti Station and presents a greater obliquity with respect to waves approaching from the southeast. Figure 2, *A* is a portion of a map prepared in 1953 by the Corps of Engineers, U.S. Army, New York District, showing positions of the shore line for various years from 1836 to 1953. The total horizontal range at Kingmill Station in this period has been scarcely more than 100 yards, indicating a high degree of stability of the shore. The portion of beach used for observations described in this report lies in a slight re-entrant in the shore line. To the south this re-entrant culminates in a minor cuspate beach deposit from which an offshore-zone bar has been built northward by longshore drift past the field station. Kingmill Station is thus characterized by an offshore-zone bar lying about 80 yards seaward of the foot of the beach face and appearing above water during low waters of spring tides. A line of breakers usually forms over the offshore bar in the lower half of the tide cycle. The bar causes initial refraction of waves approaching from the open Atlantic Ocean, and, depending upon the stage of tide, the bar serves as a wave filter acting selectively against waves of large height or long period. Low waves of short period pass freely over the offshore-zone bar. Therefore, an equilibrium shore-line configuration is favored by relative uniformity of waves passing the offshore-zone bar.

Spermaceti Station, the more southerly of the two stations, includes a straight stretch of beach bearing N. 6° E. This azimuth represents a clockwise rotation of about 30° from the azimuth of the beach at Kingmill Station. The beach at Spermaceti Station is thus more directly exposed to waves approaching from the southeasterly quadrant, which is the direction of approach of dominant waves of the summer regime. Figure 2, *B* is a portion of the same source map as that from which figure 2, *A* was taken. Note that the shore line at Spermaceti Station has shifted through a total range of some 300 yards in the period 1836–1953. Progradation to a maximum position in 1926 was followed by 240 yards of retrogradation to 1953, bringing the shore to a line near its present position. Retrogradation is the consequence of a new shore-line equilibrium position caused by construction of a seawall, trending somewhat east of north, which ends about 700 yards south of the Spermaceti Station. The beach at Spermaceti Station showed no offshore bar during the period of the surveys. Most waves were observed to break only upon reaching the step at the foot of the beach face. Because of greater incident wave energy and the lack of an offshore bar, the breaker heights and range of swash were markedly greater at Spermaceti Station than at Kingmill Station, resulting in a higher berm and longer, higher beach face at the former locality.

At both Kingmill and Spermaceti field stations a square grid of markers spaced 100 feet apart was established upon the berm and bordering dune belt for purposes of accurate relocation of profile lines with respect to both horizontal and vertical control. The markers consisted of 5- and 6-foot galvanized steel pipes and fence posts. In 1960 a level line was run to Kingmill Station from BM L 17 (fig. 1) using the Kern NK-3 level and conforming to second-order level standards. Bench mark elevation is corrected to the 1929 North American Datum. The markers were then leveled with the Kern instrument. During Hurricane Donna, (September, 1960), the outer three rows of

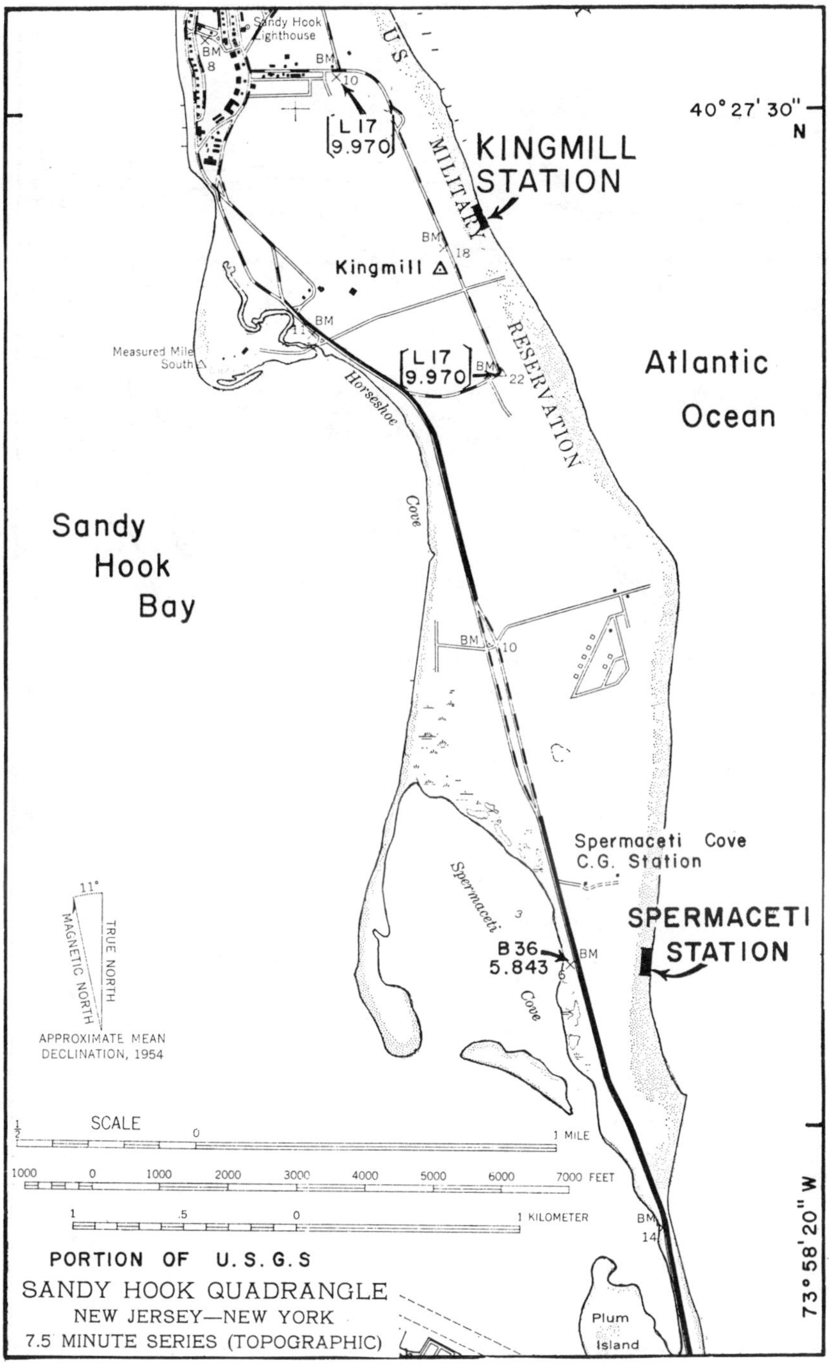

FIG. 1.—Location map, Sandy Hook, New Jersey

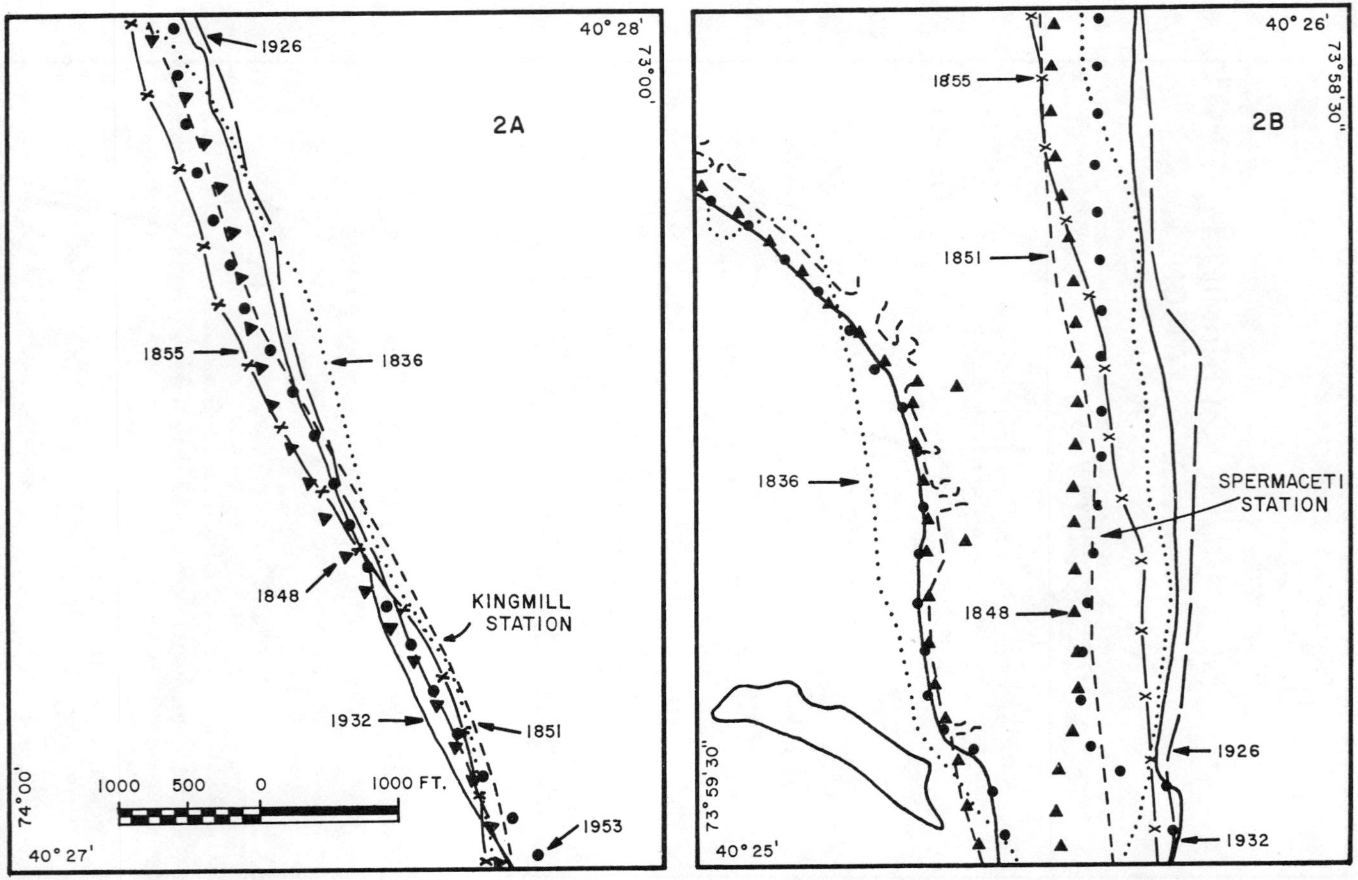

FIG. 2.—Sandy Hook shore line surveys, 1836–1953, at Kingmill and Spermaceti field stations. After Corps of Engineers, U.S. Army, New York District.

the grid were destroyed or buried by wave action. Only the first row of pipes, situated in the dune belt, survived the storm. It is believed that the impacts of huge logs and driftwood timbers were largely responsible for loss of markers during the storm.

In late August, 1960, a similar control grid was laid out at Spermaceti Station. A level line was run from U.S. Coast and Geodetic Survey BM 36 (fig. 1). On September 13, 1960, Hurricane Donna brought high water levels and storm surf to Sandy Hook. Only the markers of the innermost row, lying in the dune belt, survived the storm. Leveling of markers was done with the Kern NK-3 level and Kern level rod. Elevations shown in figures 4, 5, and 6 are those related to the bench marks bearing elevations corrected to the 1929 North American Datum.

Marker grids at both field stations were laid out with engineer's transit. Distances between points were taped to the nearest 0.01 foot after emplacement of the pipes or posts.

CHARACTERISTIC ELEMENTS OF THE FORESHORE ZONE

Characteristic elements of the beach in the foreshore zone are shown in figure 3, a composite profile patterned from surveyed profiles considered representative of low-wave conditions of mid and late summer along the Atlantic coast of Sandy Hook. This profile serves to illustrate the terms defined below.

The landward part of the beach consists of the winter berm, a broad platform with surface elevation about 8 or 9 feet above mean sea level of the 1929 North American Datum. Locally, the winter berm is a composite of several berms in steplike arrangement. The winter berm terminates on the seaward side by a steeply descending slope representing the upper region of swash and backwash of large winter waves during high waters. At about 6-feet elevation lies the summer berm, in places a nearly level platform, but elsewhere having a landward slope and rising seaward to a distinct ridge crest. The summer berm crest may be built up during high waters of spring and perigee tides, or it may be rapidly pared off by steep waves of summer tropical storms. Restoration after a storm is completed within a few days at most. Undercutting may occur just below the berm crest, locally producing a sand scarp several inches to 1 or 2 feet in height.

Extending seaward from the berm crest is the beach face, a remarkably uniform beach slope of medium grade sand, usually on about a 10–12 per cent grade, formed in the zone of swash and backwash and terminated at the base by the step. The step is an abrupt steepening of the slope associated with a gravel deposit. The total elevation range between summer berm crest and step at low-water springs is on the order of 8 or 9 feet. The range of semidaily tide, about 4–5 feet much of the time, causes the zone of swash and backwash to migrate upward and landward with rising tide, downward and seaward with falling tide. As the upper limit of swash rises higher on the beach, the breaker point and its associated step similarly migrate to higher points. As a rule, the elevation difference between swash limit and step increases as the tide rises because in the deepening water of the offshore zone the shoaling waves meet with less resistance and therefore break with greater energy, causing the swash to ride higher above still-water level than at low tide. In some cases, an offshore bar greatly reduces or even completely cuts off the waves at low water, so that the zone of swash and backwash is not defined.

The gravel step is commonly about 0.5 foot high and may be imagined to hold approximately to a subaqueous angle of repose on the seaward side, although continual agitation by breakers makes it difficult to observe this slope. Seaward of the step begins the offshore region, whose surface is under water at all times except for low waters of spring and perigee tides or where an offshore bar exists. In the profile shown in figure 3, the offshore profile reduces rapidly in steepness seaward of the step base, but in other locations and at other times a nearly

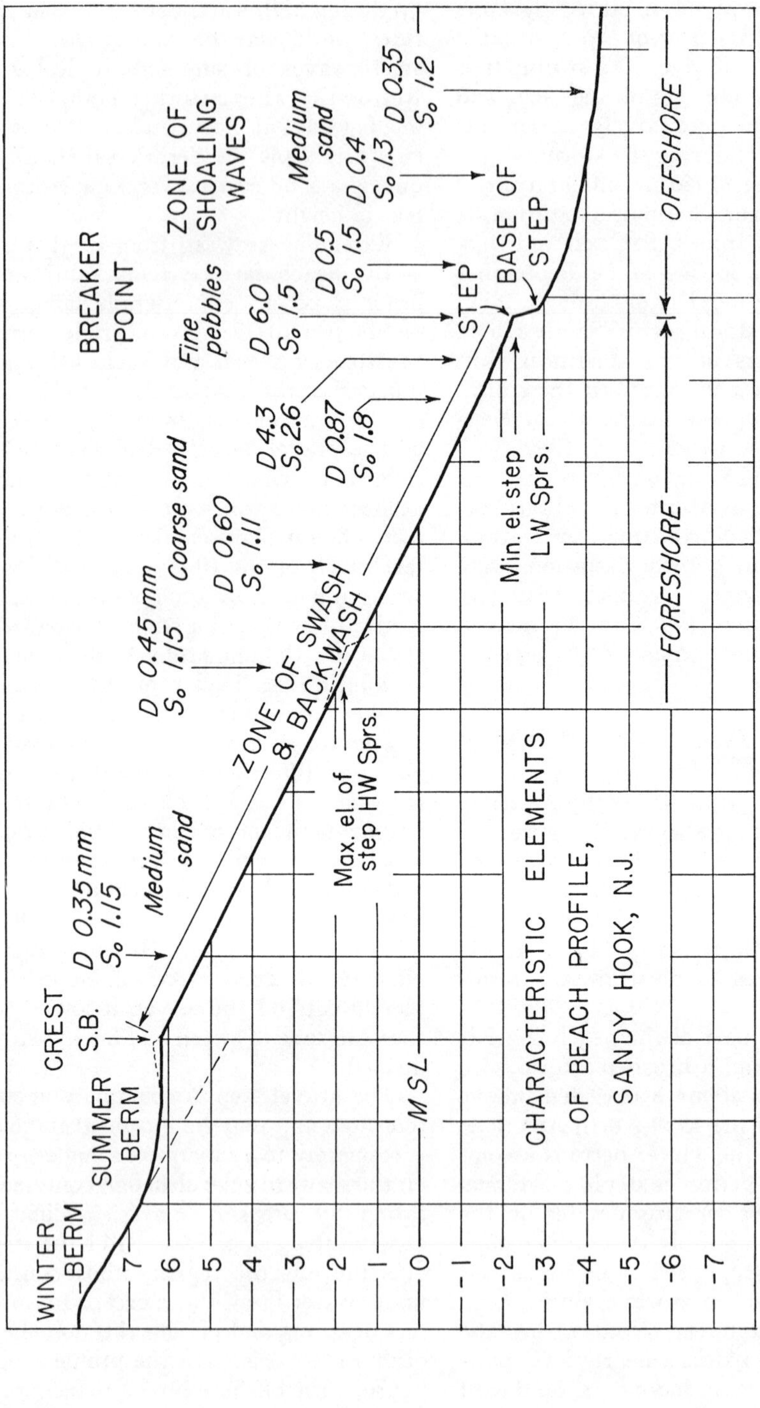

FIG. 3.—Characteristic elements of the beach profile, Sandy Hook, New Jersey

flat offshore platform or low offshore bar may lie immediately adjacent to the base of the step.

Sediment samples taken from the uppermost 2 cm. of the beach during the half-hourly elevation determinations and other profile surveys could be expected to show median diameters and Trask sorting coefficients about on the order of magnitude of the values given on the profile in figure 3. Median diameter is least at the top of the zone of swash and backwash, near the summer berm crest, but increases steadily down the beach face. As the step is approached, the diameter increases very rapidly and reaches a maximum in the range of 4–6 mm. on the step. Here the gravel is rather well sorted. Immediately seaward of the step base the surface is composed of sand of about 0.5 mm. median diameter. Diameter decreases and sorting improves in a seaward direction.

PROFILE DEVELOPMENT IN THE SUMMER REGIME

The summer changes in beach profile at Spermaceti Station were observed in 1961 by profiles surveyed along four profile lines at intervals of 2 weeks (fig. 4). Profiles were surveyed at low water and carried from the berm to the breaker zone. Sand surface elevations were obtained by use of the Kern NK-3 level and a level rod held at intervals of 5–10 feet on the profile and at additional key points as required. Elevations were read to the nearest 0.01 foot. Distances were taped from posts on the outer line of the reference grid. The first set of profiles was run on June 29, 1961, the fourth on September 7, 1961. All four are shown superimposed in figure 4 with a vertical exaggeration of ×20. Profile data were then plotted on a plan to scale and contours drawn to yield a map of the beach for a distance of about 100 yards (fig. 5). Along lines 1, 2, and 3, the beach face from berm crest to base was almost perfectly planar, with a slope of 14 per cent (8°), there being no appreciable difference in slope for the three lines. Berm crest elevation was not, however, initially the same but fell from line 1 to line 4 (elevation decreasing to south). The beach face at line 4, by contrast, initially had a sigmoidal profile and a lower maximum slope (10 per cent, $5\frac{1}{2}$°).

On June 29, 1961, the foot of the beach face in profile lines 1, 2, and 3 was marked by a steep scarp 0.5 foot high at the normal position of the gravel step (fig. 4). Adjacent to the scarp was a parallel trough acting at low water as a channel to conduct water supplied by ground water seepage and from overwash of waves across the adjacent bar. This excess water flowed north along the base of the beach face and escaped seaward at intervals of 50 yards or so by means of transverse channels cut across the bar. One such escape channel lies between lines 2 and 3 (fig. 5). Traced south, the trough ended 55 feet south of line 3 in an abrupt, steep-walled head which was being actively over-steepened and-extended by a number of springs of ground water emerging at −0.7 foot elevation. Emergence of small springs all along the base of the scarp was a striking feature of the beach profile. On line 1, the springs emerged at −0.4 foot, on line 2 at 0.0 foot. Combined flow from ground water effluence in the distance of about 100 feet was sufficient to create a stream 3 feet wide and 2 inches deep at the place where the water discharged through a gap in the bar. At the time the stream was observed, no water was being brought into the trough by swash over the bar, hence the observed flow was entirely from ground water effluence. It is apparent that spring sapping by ground water at times of maximum tide range is an important agent in steepening the base of the beach face at low water.[2] Perhaps this inference is supported by the observation that the beach face in profile 4, which had no trough and bar at the base, was lower in angle and had an upwardly concave lower

[2] On June 29, 1961, the tide range was abnormally great (6.6–7.0 feet) because full moon occurred the day previously and perigee the day following. The tide was essentially a combination of perigee and springs, with a maximum diurnal inequality seen in high waters.

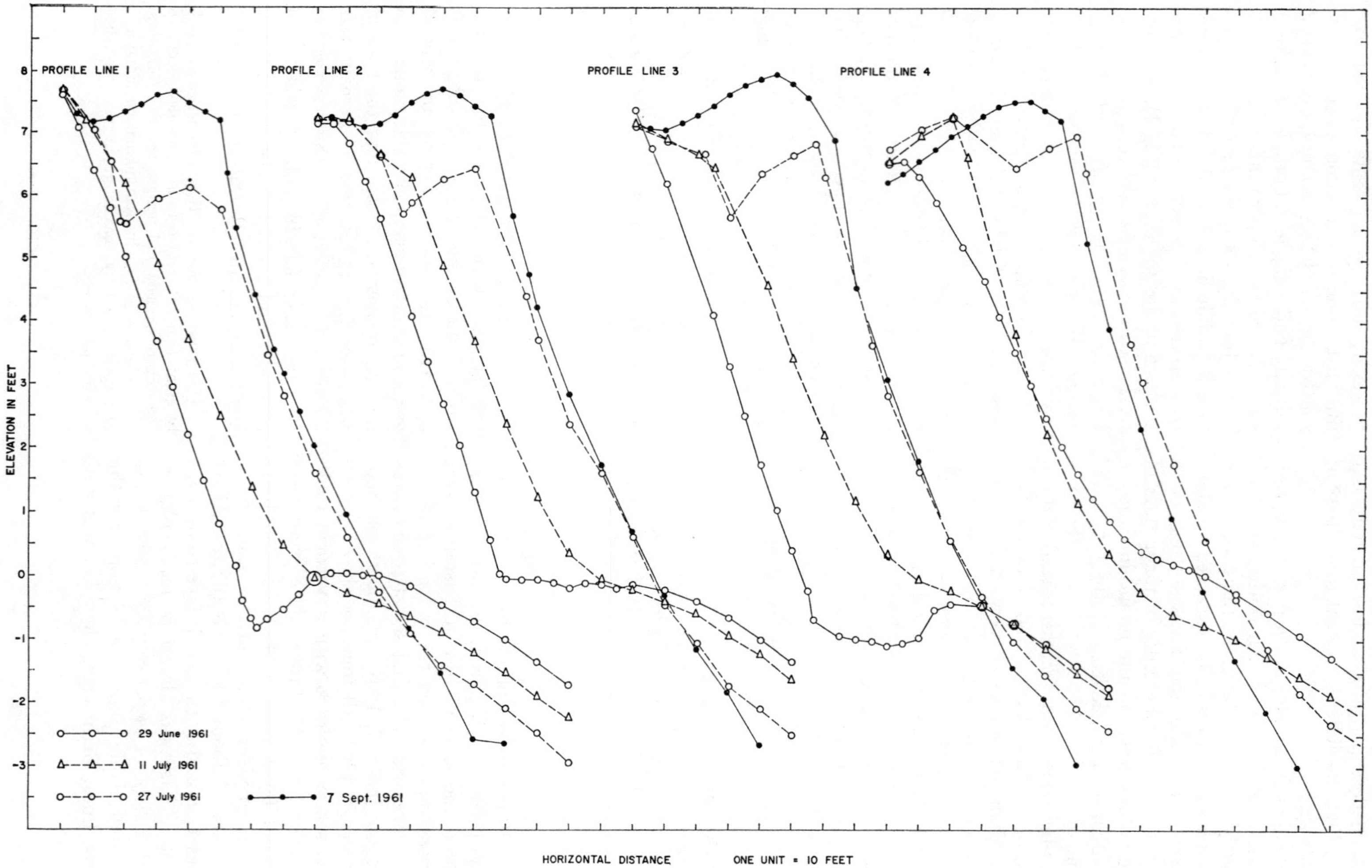

FIG. 4.—Transverse beach profiles at Spermaceti Station, 1961

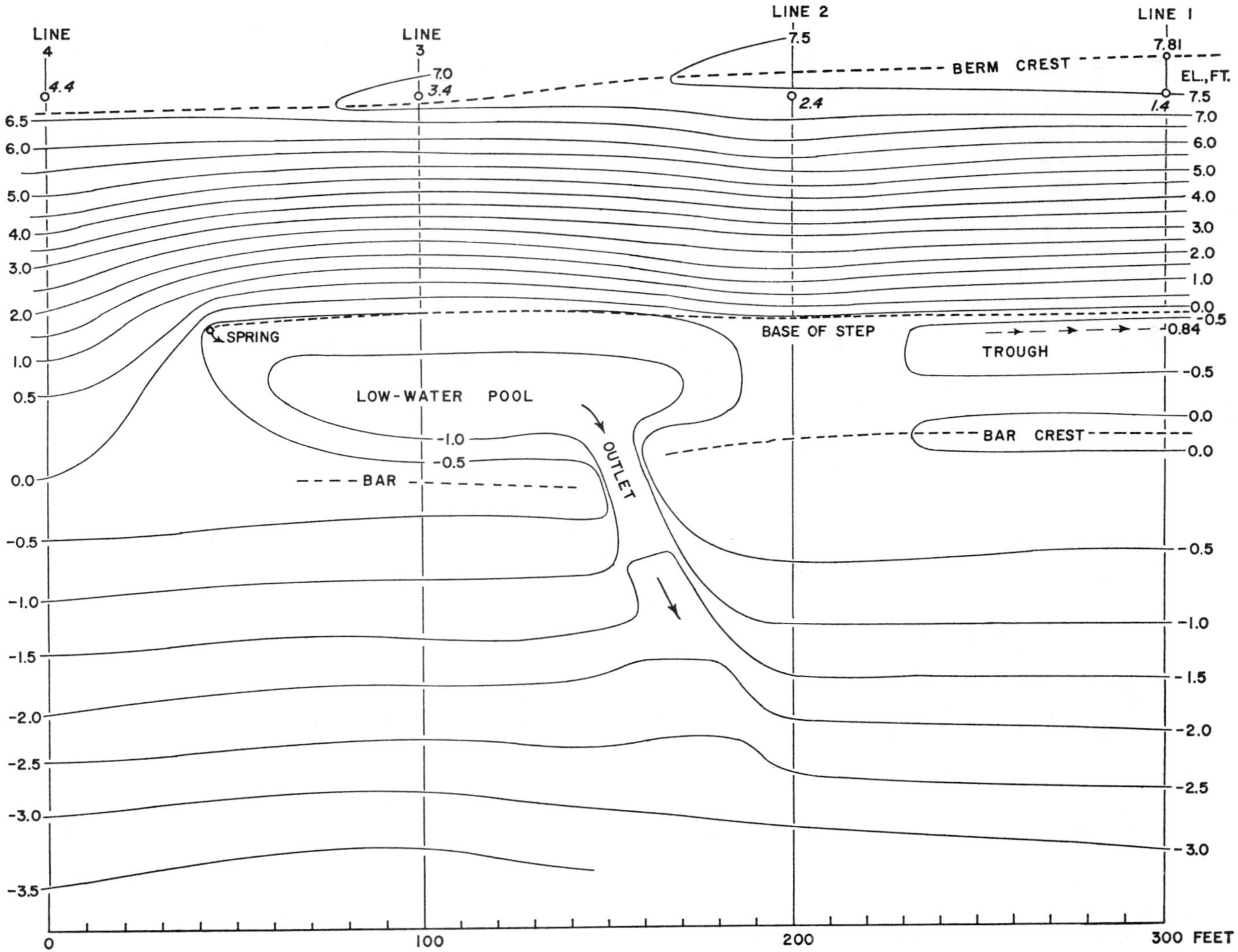

FIG. 5.—Contour topographic map of foreshore zone of beach, Spermaceti Station, June, 1961

profile, as one might expect of a slope not continually undercut at the base.

On June 29, 1961, the trough in profiles 1, 2, and 3 (fig. 4) was bounded on the seaward side by a low, broad-topped bar, rising to a crest 0.8 foot and 0.7 foot higher than the trough floor in profiles 1 and 3, respectively. Profile 2 showed no trough; here the bar took the form of a bench. Bar crest elevation declined southward from 0.0 foot on line 1 to −0.5 foot on line 3. Line 4 showed no trough or bar, merely a decrease in slope at a point equivalent to the bar in the other three profiles.

Seaward slope of the bar of 3.5 per cent (2°) was remarkably uniform for all four profiles over a horizontal distance of 80 feet from an elevation of −1.0 feet to −3.0 feet, the limit of the profiles. A striking feature of the beach profiles generally was, therefore, the contrast in steepness of slopes of the beach face and the bar face, combined with the straightness and uniformity of slope within each location.

As the summer progressed, the Spermaceti Station profiles showed important changes. Progradation of between 40 and 50 feet occurred on all profiles in the 4-week period from June 29 to July 27, causing the formation of a crested berm (fig. 4). The berm crest, at elevation of 6 feet on line 1 and 7 feet on line 4 thus showed a marked increase in elevation southward. The trough behind this berm crest was scoured to a maximum depth of 1.2 feet by escaping overwash. The axis of the trough held a channel 6 feet wide containing current ripples. Channel gradient was northward. At a point about 75 feet north of line 1 the channel turned seaward, forming a gap in the berm crest.

In the 2-week period between July 27 and September 7, the beach face held to about the same average position, but the berm was increased in height by 1.0–1.5 feet. Upward berm building is interpreted as the effect of a series of larger waves or stronger breakers acting at high water to bring the swash to a higher limit, enabling sediment to be thrust higher above sea level.

A noteworthy development of the 4-week period of survey was the removal of the bar and the seaward extension of the beach face to a level between −2.0 and −3.0 feet. At this depth, the step was encountered and with it breakers of sufficient energy to prevent further seaward continuation of the profile surveys. Thus the water had been deepened in the immediate offshore zone, and breaker energy increased. This change alone would account for the stronger swash and the up-building of the berm.

TIDAL CYCLE AND THE EQUILIBRIUM BEACH

The equilibrium profiles at Kingmill and Spermaceti stations, held for long periods during the summer, represent a general steady state in the action of waves and rate of transport of sediment. Superimposed upon this average steady state is a series of fluctuations controlled by the tidal cycle of rise and fall of water level and by changes in the height, steepness, and angle of incidence of the waves. The purpose of a series of half-hourly elevation surveys, described below, was to discover what minute but systematic changes in the beach profile might typically occur throughout the cycle of rising and falling tide. During the rising tide, the inner limit of swash moves steadily higher upon the beach; the breaker point, represented by a step of coarse gravel, also moves to successively higher points. To what extent are zones of deposition and scour associated with these shifting zones of wave action? What depth of reworking is indicated by the profile changes? Are there an accompanying change in grade and sorting of the surficial sand layer as the zones shift landward and seaward?

SURVEYING PROCEDURES

To observe and document in detail the many minor changes in beach profile occurring during the tidal cycle, a program of half-hourly elevation surveys of a given set of profile points was undertaken at Kingmill Station during July and August of 1960 and in July of 1961. At the beginning of a 4- to 6-hour survey period, the Kern NK-3 level instrument was set up on an established transverse beach profile line. Elevation of instrument was determined from permanent grid reference posts whose elevation had been previously determined with respect to U.S. Coast and Geodetic Survey bench marks of the 1929 Sea Level Datum. A steel tape was attached to the base of a specially constructed 15-foot level rod and stretched from a control point to points at successive 5-foot horizontal intervals seaward along the profile line. The tape was laid flat on the beach, and no attempt was made to correct for slope, the object being solely to reoccupy precisely the same rod station at successive half-hour intervals. Rod stations could thus be reoccupied within an error of perhaps no more than 0.01 foot in the horizontal direction. The level rod was read to the nearest 0.002 foot for the closer points, points lying shoreward of the swash limit serving as controls for frequent checking. Farther seaward, where swash and breaker activity was vigorous and the rod difficult to hold steady, rod readings were recorded to the nearest 0.01 foot. The rod was fitted with a flat baseboard, measuring about 4 × 6 inches and hinged to the rod base, so that the rod rested on the sand surface without compacting it appreciably. During the period of several seconds before the rod base was undermined by the backwash, it was possible for the instrument operator to take a rod reading.

HALF-HOURLY LEVEL DATA

When plotted on a greatly expanded vertical scale, level data revealed that a rather orderly series of profile changes takes place as the tide rises and falls. Despite various irregularities that may be attributed to surveying errors and to changes in the state of the wind and sea during the survey period, there emerged an unmistakable overall pattern of changes seen in figures 6*A*–6*F*.

These level data have been plotted to show elevation (ordinate) as a function of time (abscissa). Each series of points moving from left to right represents a point on the beach surface at a fixed horizontal distance from the control point. Series of points are arranged in order on the page according to increasing distance seaward at 5-foot distance intervals. Rise and fall in elevation of the points on any given series are plotted on the vertical scale according to the units shown on the ordinate of the graph. Obviously, then, when the points of a given series rise in elevation, deposition has occurred; where the points fall in elevation, scour (erosion) has occurred. Where the points lie on a horizontal line, there is no elevation change, a condition found landward of the inner limit of swash action.

Figure 6*C* serves as an example of changes accompanying rising tide. Refer, for example, to the points at the distance +40 feet. During the first two half-hour periods, the swash had not yet reached this point, hence there was no elevation change. At the fourth half-hour, the swash had just reached the survey point, and a small amount of deposition (about 0.02 foot) had taken place. By the fifth half-hour, a reversal had set in; removal of the thin deposited layer had taken place, and a strong scour had set in. By the seventh half-hour, scour of about 0.2 foot had occurred. Then deposition again set in and continued steadily through the tenth half-hour, when the survey period ended. Note that the phase of scour had traveled steadily up the beach and is clearly shown in each point series, but at successively later times at higher levels. The small initial deposition preceding scour is shown on several of the point series but is on an order of magnitude difficult to distinguish from surveying error. Figures 6*D*, and 6*F* also show periods of rising tide, with similar changes occurring.

Figure 6*B* shows changes with falling tide. In the point series for +45, +50, and +55 feet a rapid decline in elevation is seen, indicating scour during the first five to six half-hour periods. Scour was then followed by deposition, but to much less extent, and the changes were then terminated by the inner limit of swash passing beyond the limit of the survey point. In general, these changes are a reversed or mirror image of

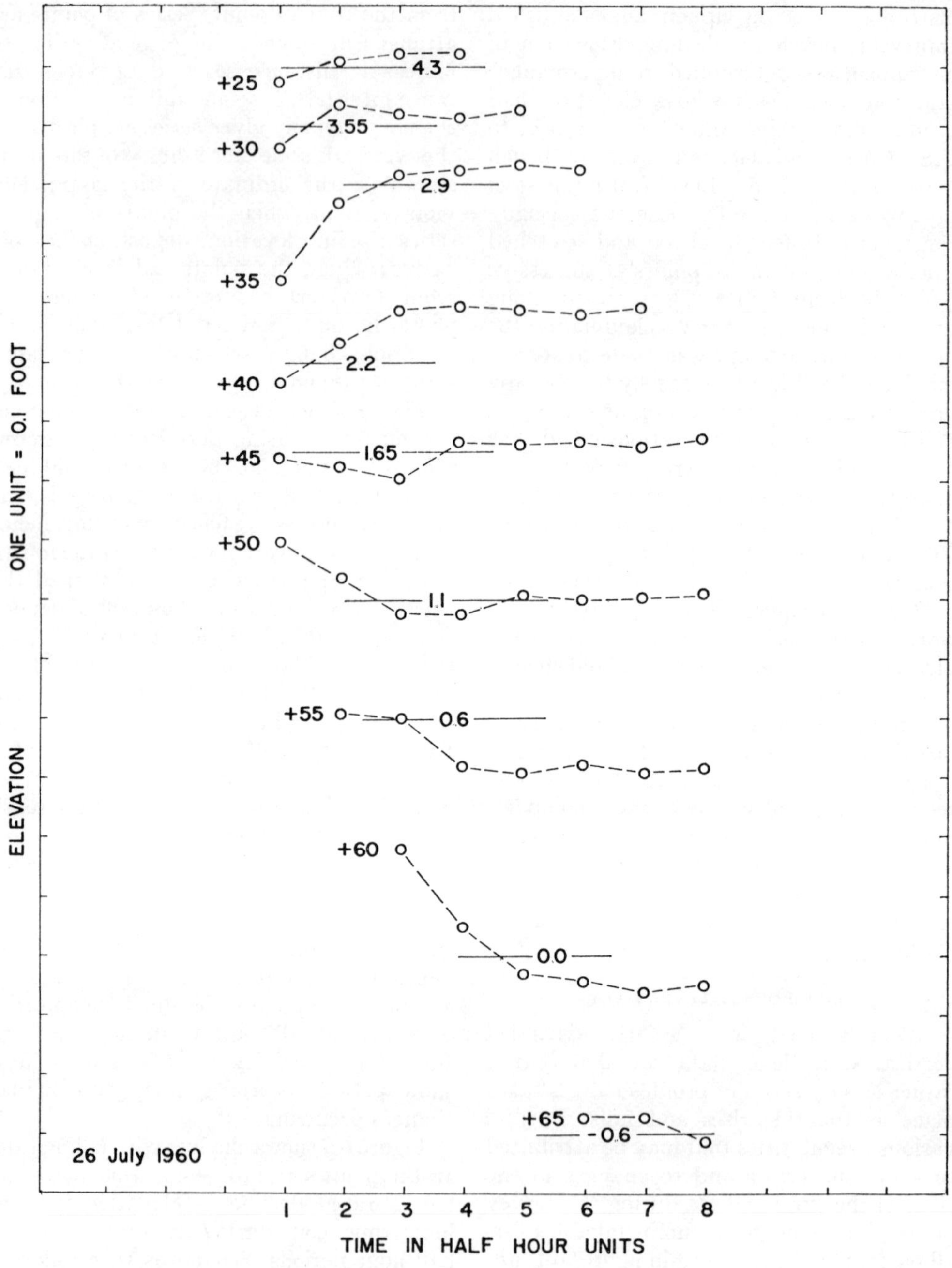

FIG. 6*A*.—Half-hourly changes in elevation, July 26, 1960, Kingmill Station

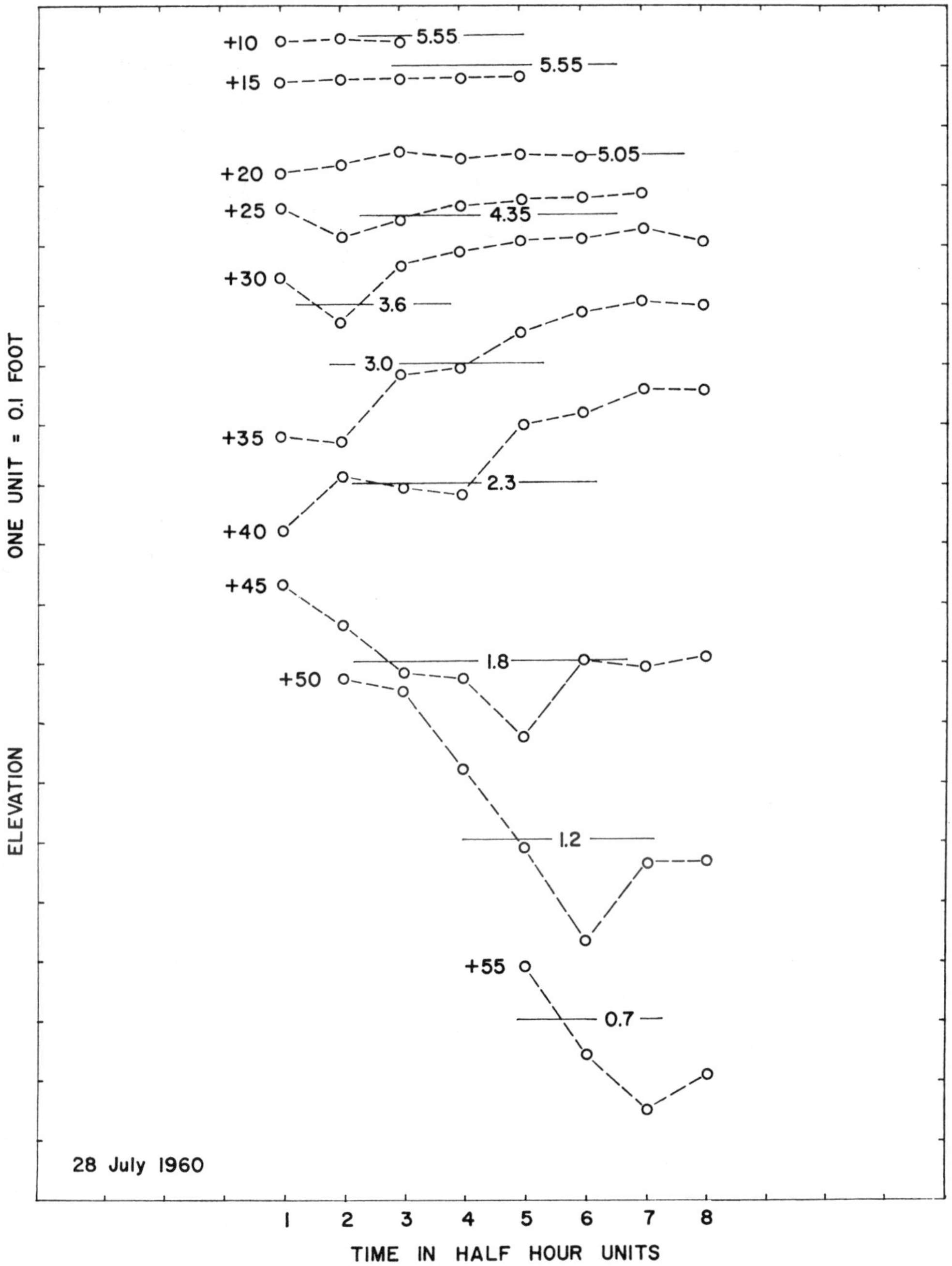

FIG. 6*B*.—Half-hourly changes in elevation, July 28, 1960, Kingmill Station

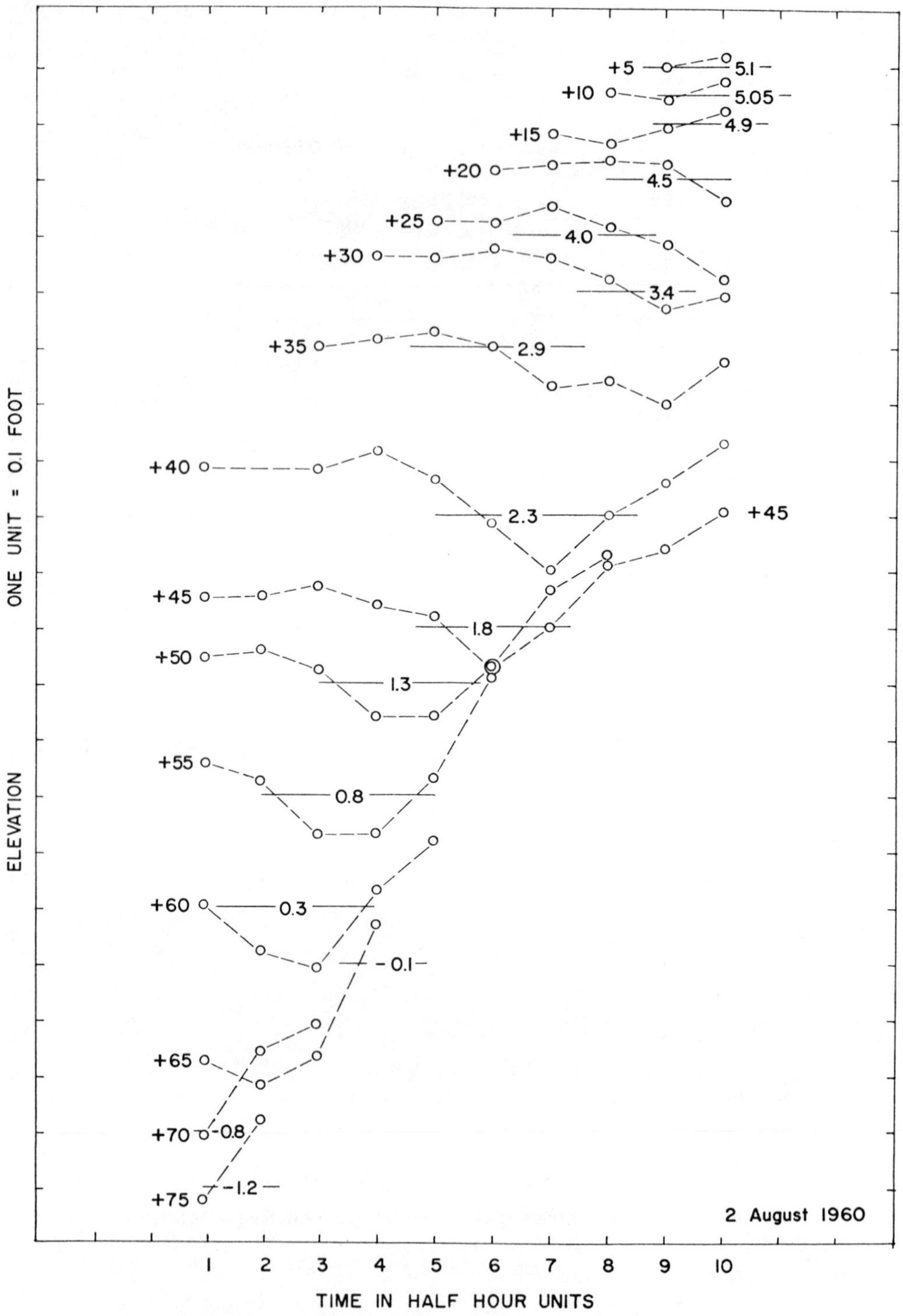

FIG. 6C.—Half-hourly changes in elevation, August 2, 1960, Kingmill Station

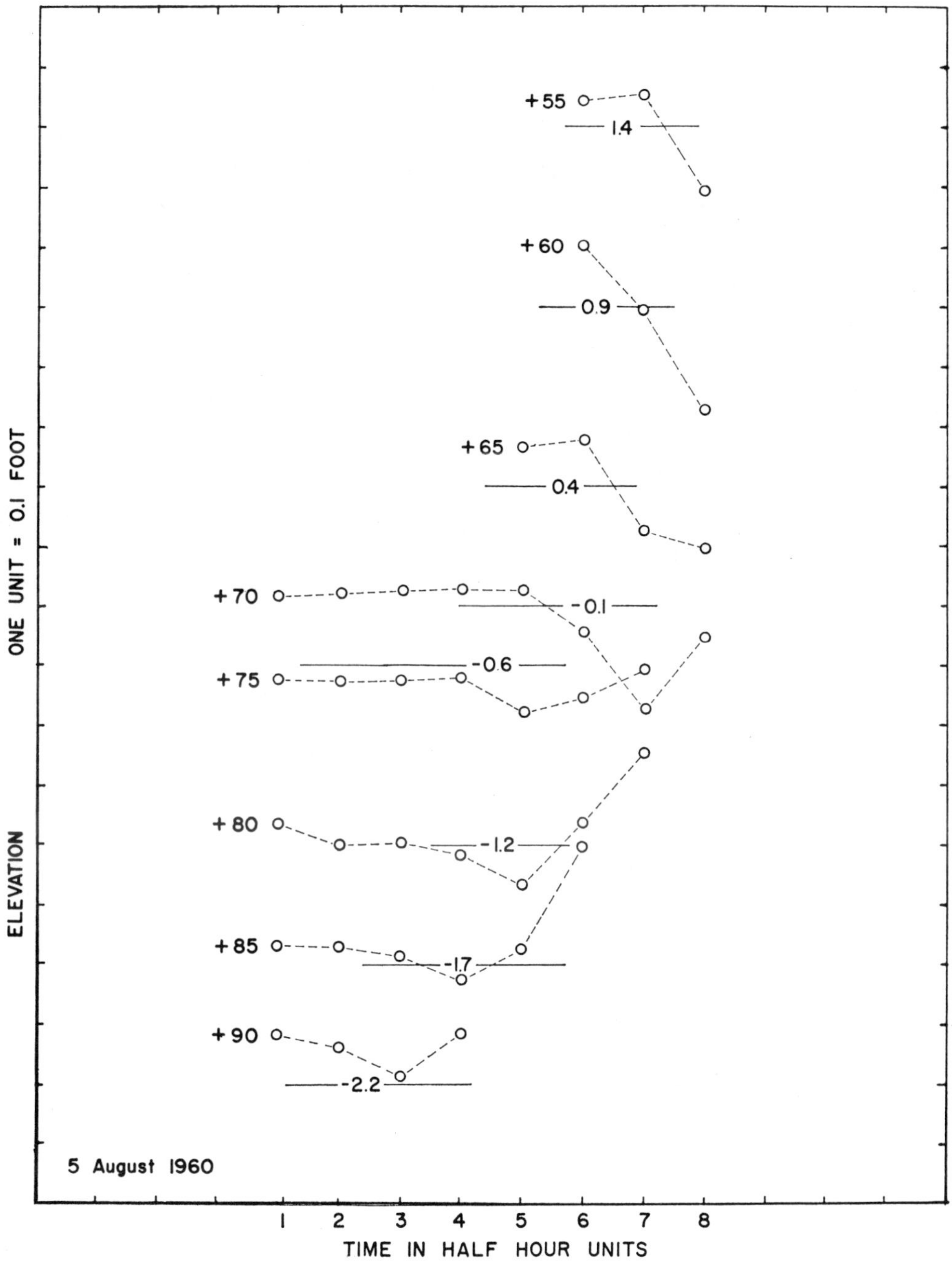

FIG. 6*D*.—Half-hourly changes in elevation, August 5, 1960, Kingmill Station

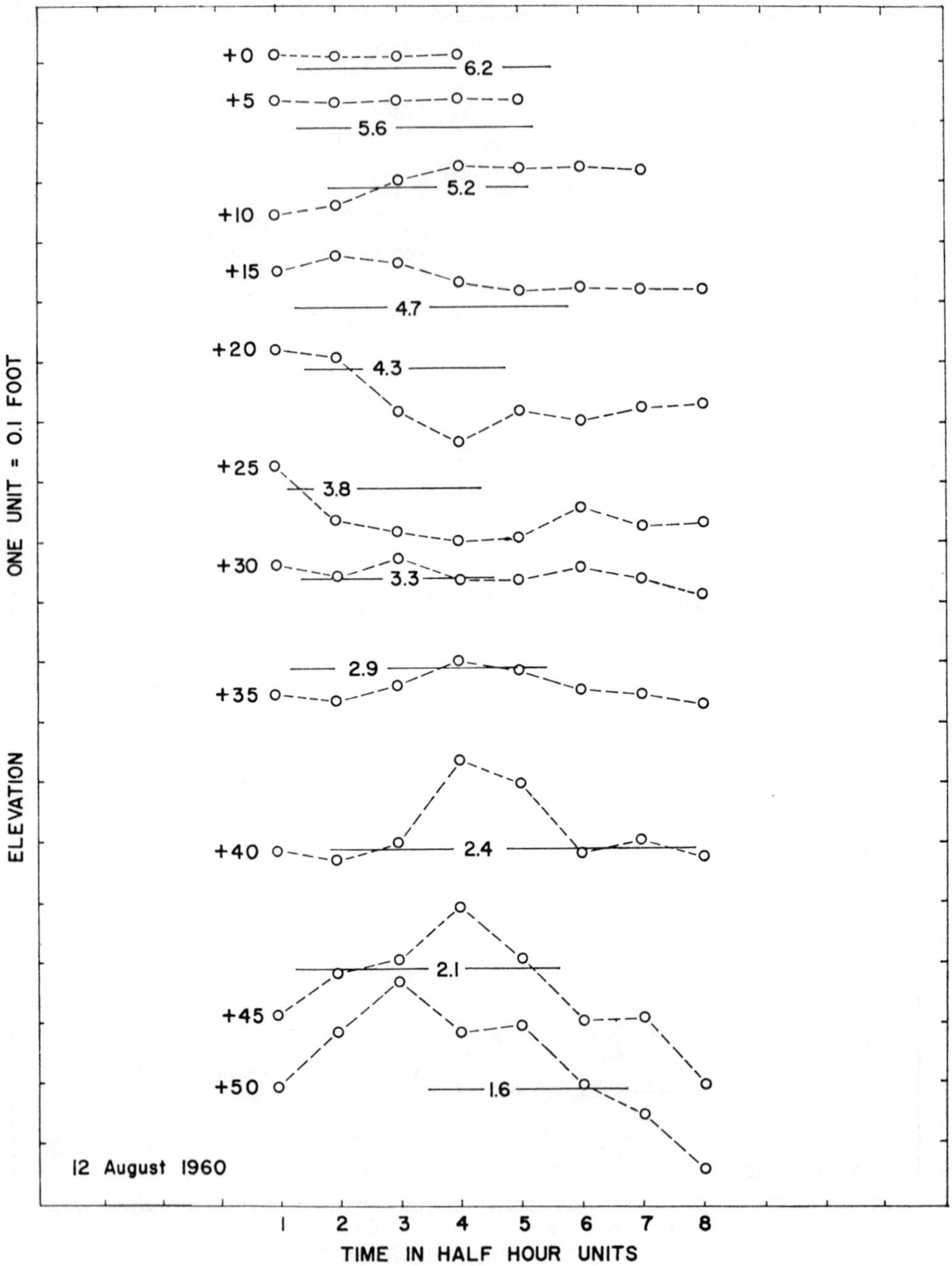

FIG. 6*E*.—Half-hourly changes in elevation, August 12, 1960, Kingmill Station

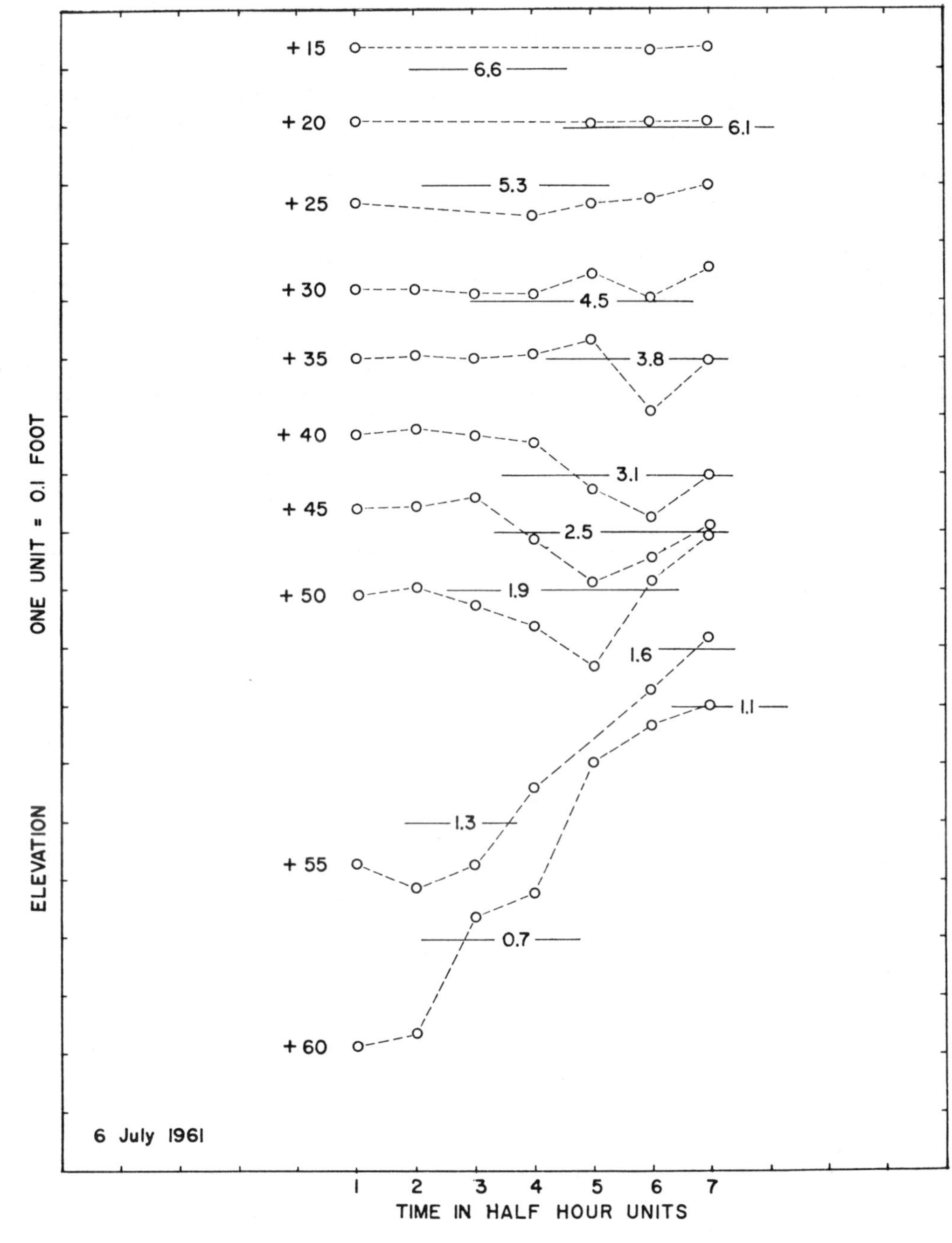

FIG. 6*F*.—Half-hourly changes in elevation, July 6, 1961, Kingmill Station

the changes seen during rising tide. Figure 6*A* also illustrates changes during a falling tide.

SYNTHESIS OF TIDAL CYCLE CHANGES

Using all observational data, an idealized system of changes during the semidaily tidal cycle has been synthesized as shown in figure 7. The solid lines running from left to right across the diagram represent elevations of survey points during the tidal cycle from one low water to the next. Relative horizontal distances separating the survey points are suggested by the distances +10, +15, +20, +30, etc., measured from the summer berm crest. Elevation in units of 0.1 foot is scaled on the ordinate. Elevation figures given for each distance are representative of the actual elevation that might be expected at that distance. The points marked *L* show the time of passage of inner limit of the swash. Obviously, points situated above the limit of swash show no change with time, hence are represented by straight horizontal portions of the lines.

Three phases of beach change are recognizable (fig. 8): (1) *Initial deposition*, accompanying the first quarter to half hour. During this phase, the swash of rising tide first acts upon the beach. Deposition is of a

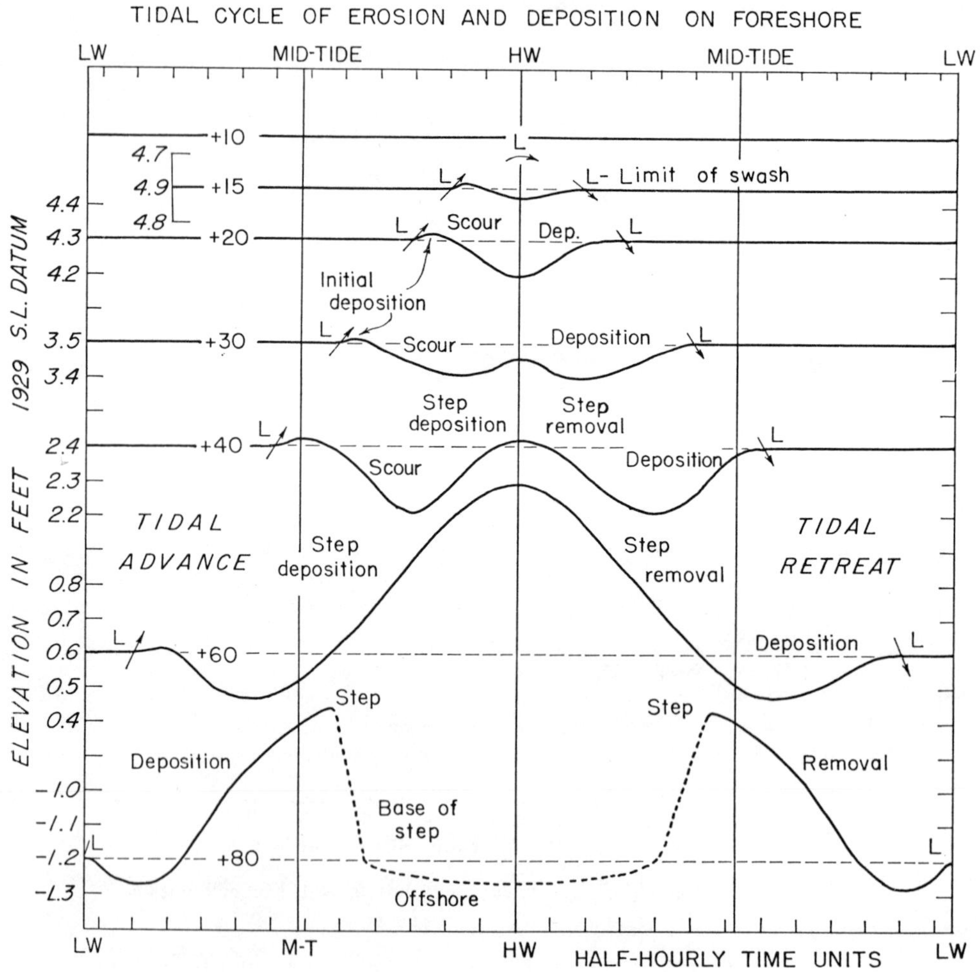

FIG. 7.—Synthesis of elevation changes in the tidal cycle

thickness on the order of 0.02 foot and is not always observed or recognized. (2) *Scour* in the upper part of zone of swash and backwash. In a period lasting about $1\frac{1}{2}$ hours, the beach is lowered until a sand layer about 0.2 foot is removed. (3) *Deposition* preceding the arrival of the step. As the step moves up the beach, it brings with it a wedge of deposition reaching to progressively higher levels of the beach. Thus, the scour phase is replaced by a depositional phase, and elevations increase rapidly.

the amount of rise of tide, hence is less at higher levels on the beach and is not felt at all higher than about 4.0 feet above mean sea level.

At a survey point sufficiently low on the beach that the step actually rises landward beyond the point, elevation changes show the type of curve illustrated by the line labeled +80 in figure 7. Because the seaward face of the step is very steep, passage of the step brings a abrupt decrease in elevation of about 0.5 foot. Most level sur-

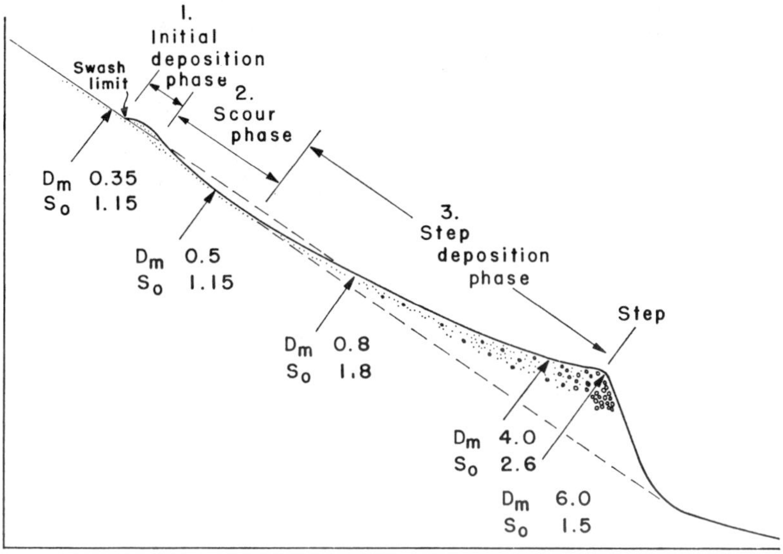

FIG. 8.—Phases of beach change in the tidal cycle

The material of the depositional wedge becomes progressively coarser through the admixture of pebble grades. Sorting becomes poorer as median diameter approaches the size of fine pebbles (4–8 mm.). Figure 9 shows changes in elevation, median diameter, and Trask sorting coefficient throughout a $2\frac{1}{2}$-hour period on August 2, 1960, at Kingmill Station at the +55-foot point (see fig. 6*C*). The total deposition of the advancing step is commonly on the order of 0.5 foot with respect to the initial beach elevation preceding the cycle. With respect to the preceding maximum scour level, deposition may be on the order of thickness of 0.7 foot. Of course, the step deposition is limited by

veys did not include the offshore region. It may be inferred that the elevation beyond the step will decline slightly to a minimum at high water.

As the tide falls, the zone of step deposition and the zone of scour retreat down the beach. Scour is restored by deposition of medium sand, so that the final swashes leave the beach surface in essentially the same condition of slope and particle size as existed at the start of the cycle. (This statement may not hold true if strong winds have acted upon the beach during the low-tide period of exposure. Evidence of selective removal of particles by deflation is commonly seen on these beaches.) During falling tide there seems to be no recognizable counterpart of the very small initial deposition phase seen during rising tide.

MECHANICS OF DEPOSITION AND SCOUR

Phases of deposition and scour can be explained descriptively in terms of relative high infiltration capacity of the beach and because the wetting of dry sand, bringing it to field capacity, withholds water from return flow. Thus a thin layer of sand, perhaps on the order of 0.01–0.02 foot, is deposited. Samples of surficial material taken before and after the action of the advancing swash

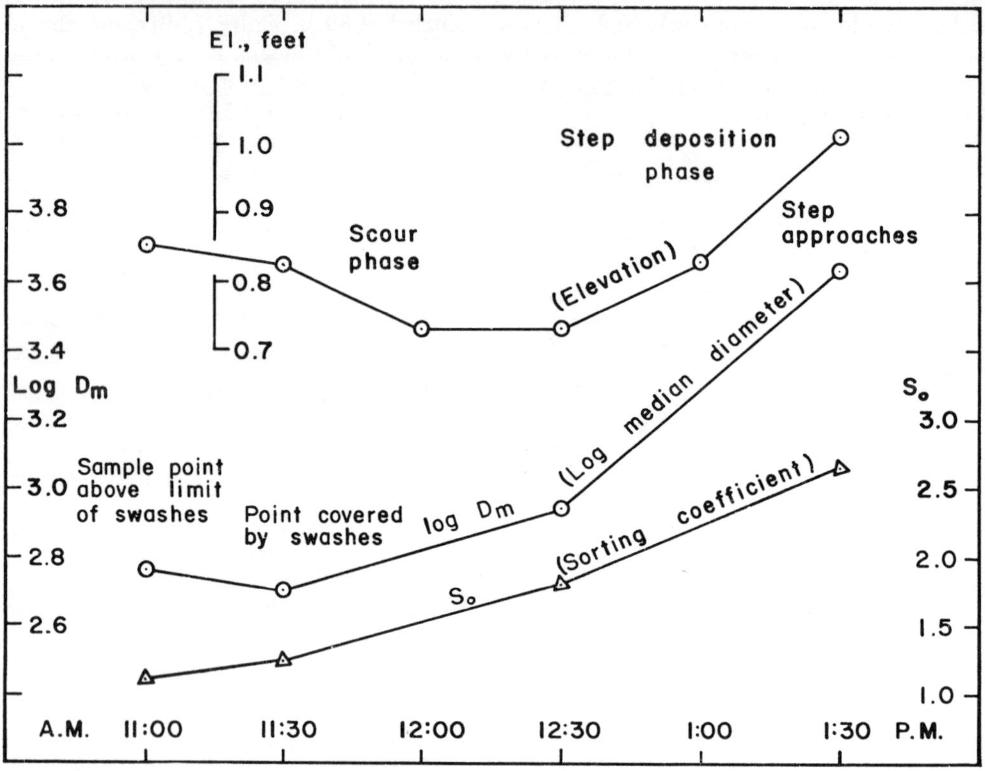

FIG. 9.—Changes in elevation, median diameter, and sorting coefficient, Kingmill Station, August 2, 1960.

capacity of the swash and backwash and by application of the principle of diffusion of suspended sediment in the breaker zone.

As the tide rises, those swashes which are somewhat deeper and faster moving than the average sweep over beach surfaces previously drained and dried in the low-tide period. Swashes of the first few minutes appear to have greater capacity to transport sand upbeach than the intervening backwashes have capacity to transport seaward. This disparity may result because of initially showed a slight reduction in median diameter (drop from 0.41 mm. to 0.34 mm.), but this change is of questionable significance.

As the swash zone moves farther up the beach, a reference point comes into the zone of scour, in which the capacity of the backwash to move sand seaward exceeds the capacity of the swash to carry it up the beach. Undermining of the base of the surveying rod and of the rod-man's feet is strongly felt with each backwash in this zone. Return capacity is the larger because little of the swash is lost to infiltration, and the initial beach slope is steeper than an equilibrium slope for the conditions of depth and velocity of the backwash. Restoration of sand by the succeeding swash is not quite equal to the quantity removed by the backwash, so that there is a net loss, on the average. Lowering at the rate of about 0.1 foot per hour may be expected, judging from the measured rates. Samples of sand taken from a point experiencing scour showed a strong rise in median diameter (from 0.51 mm. to 0.87 mm.) and a decrease in goodness of sorting (sorting index increased from 1.23 to 1.80) (fig. 9). The observed change is readily explained through the introduction of coarser fractions brought from the step zone and mized with sand being reworked by scour.

Deposition associated with the gravel step may be thought of as resulting from the upbeach translation of the entire step following the progressive landward movement of the breaker zone during rising tide. The step crest itself represents a ridge of gravel from which the finer constituents of the beach have been removed by diffusion to less turbulent regions on either side. The gravel remains concentrated in the step because it is readily dragged to the line of maximum turbulence across beds of finer material on either side. On the offshore bed, landward pulses accompanying orbital movement of shoaling waves drag and roll the pebbles toward the step. On the foreshore side, backwash readily rolls and drags the pebbles down the beach slope, returning them to the step. Within the step, the pebbles are repeatedly lifted and dropped by breaker turbulence. Many grains escape into the swash and are carried up the beach to become mixed with the sand before being dragged back to the step. We thus expect that the step will be bounded on the foreshore side by a wedge of mixed sand and pebbles and that the proportion of pebbles will decrease landward within the wedge. As noted above, analysis of sediment in the zone of step deposition confirms this expectation.

SUMMARY

Observations of the Atlantic beaches of Sandy Hook, New Jersey, throughout the summers of 1959, 1960, and 1961 showed that the beach regularly experienced accretion (progradation) throughout the early and middle summer and that a summer berm was gradually developed. From repeated profiling, it appeared that by late July the progradation process became very slow, and at some localities the profile became so closely fixed in position and slope that successive resurveys show negligible change in position of the relatively straight profile portion in the zone of swash and backwash at about mid-tide level. By the end of July, the beach profiles were largely stabilized, and the change throughout August and early September was close to zero, if changes in four adjacent profiles are averaged. The same beach profiles showed, however, that the height of the berm crest may be increased in response to occurrences of greater high waters of spring tides, while at the same time the profile can remain essentially fixed in position in the mid-tide zone of swash and backwash.

While beach equilibrium prevails under the summer regime of low swells and small waves and a persistent angle of wave incidence from the southeast, the rise and fall of tide results in rhythmic patterns of deposition and scour. As tide rises, a short phase of initial deposition near the inner limit of swash is followed by a scour phase in the zone of swash and backwash. As the wedgelike step deposit moves upbeach, the scour phase gives way to a deposition phase involving coarser, more poorly sorted material. Maximum deposition is reached at the step where well-sorted pebbles and gravel exist under the breaker point. Surveys at half-hour intervals, using level instrument, level rod, and tape, yielded measurements of the depths of deposition and scour in each phase of activity and permitted the entire cycle to be reconstructed with approximate specification of quantities involved. In most of the cycles observed, a layer of sand about 0.2 foot thick was completely removed and redeposited with each semidaily tidal cycle at the mid-tide level of the beach. Falling water brings the same succession of phases, but in reverse order, ending with restoration of the beach to its previous level and composition.

ACKNOWLEDGMENTS.—The studies described in this report formed one phase of a broader research project undertaken with the aim of applying quantitative methods of observation and analysis to problems of beach morphology and beach-forming processes. The project was supported by the Geography Branch of the Office of Naval Research (Project NR 388-057) under Contract Nonr 266 (68) with Columbia University.

Surveying operations described in this report were carried out by the writer, who served as project director. He was assisted by students of Columbia University. The courtesy of the Commanding Officer, U.S. Army, Fort Hancock Military Reservation, in permitting field parties to operate on Sandy Hook is greatly appreciated. Laboratory and storage facilities were provided at the Sandy Hook Marine Laboratory, Bureau of Sport Fisheries and Wildlife, U.S. Department of the Interior, through the courtesy of its director, Dr. Lionel Walford.

The writer is indebted to Dr. Warren E. Yasso for various forms of assistance in the field and laboratory and for critical reading of the manuscript. Mr. Gustavo A. Antonini prepared many of the illustrations contained in the report. Mrs. Joan K. Lubowe carried out the grade-size analysis of sand samples in the Geomorphology Laboratory of Columbia University.

AUTHOR CITATION INDEX

SUBJECT INDEX

About the Editors

JOHN S. FISHER is Assistant Professor of Environmental Sciences at the University of Virginia. He received the B.S. in mechanical engineering from the University of Florida in 1966, the M.S. in mechanical and ocean engineering from the University of Miami in 1968, and the Ph.D. in civil engineering from the Massachusetts Institute of Technology in 1972. Professor Fisher is interested in the impact of storms on dunes and beaches and is involved in field studies to quantify the sediment transport mechanics of these storms.

ROBERT DOLAN, Professor of Environmental Sciences at the University of Virginia, received the B.S. and M.S. in general science from Southern Oregon College in 1956, and Oregon State University in 1958, and the Ph.D. in 1965 in geomorphology and geology from Louisiana State University. He was on the staff of the Coastal Studies Institute for several years before joining the faculty at Virginia. His research interests are focused on beach processes, coastal geomorphology, and man's impact on barrier islands of the Atlantic Coast of the United States.